THE LEUKEMIA-LYMPHOMA CELL LINE *FactsBook*

Other books in the FactsBook Series:

Robin Callard and Andy Gearing
The Cytokine FactsBook

Steve Watson and Steve Arkinstall
The G-Protein Linked Receptor FactsBook

Rod Pigott and Christine Power
The Adhesion Molecule FactsBook

Shirley Ayad, Ray Boot-Handford, Martin J. Humphries, Karl E. Kadler and C. Adrian Shuttleworth
The Extracellular Matrix FactsBook, 2nd edn

Grahame Hardie and Steven Hanks
The Protein Kinase FactsBook
The Protein Kinase FactsBook CD-Rom

Edward C. Conley
The Ion Channel FactsBook
I: Extracellular Ligand-Gated Channels

Edward C. Conley
The Ion Channel FactsBook
II: Intracellular Ligand-Gated Channels

Edward C. Conley and William J. Brammar
The Ion Channel FactsBook
IV: Voltage-gated Channels

Kris Vaddi, Margaret Keller and Robert Newton
The Chemokine FactsBook

Marion E. Reid and Christine Lomas-Francis
The Blood Group Antigen FactsBook

A. Neil Barclay, Marion H. Brown, S.K. Alex Law, Andrew J. McKnight, Michael G. Tomlinson and P. Anton van der Merwe
The Leucocyte Antigen FactsBook, 2nd edn

Robin Hesketh
The Oncogene and Tumour Suppressor Gene FactsBook, 2nd edn

Jeffrey K. Griffith and Clare E. Sansom
The Transporter FactsBook

Tak W. Mak, Josef Penninger, John Rader, Janet Rossant and Mary Saunders
The Gene Knockout FactsBook

Bernard J. Morley and Mark J. Walport
The Complement FactsBook

Steven G.E. Marsh, Peter Parham and Linda D. Barber
The HLA FactsBook

Clare M. Isacke and Michael A. Horton
The Adhesion Molecule FactsBook, 2nd edn

THE LEUKEMIA-LYMPHOMA CELL LINE *FactsBook*

Hans G. Drexler
German Collection of Microorganisms and Cell Cultures
Braunschweig, Germany

ACADEMIC PRESS
A Harcourt Science and Technology Company
San Diego San Francisco New York Boston
London Sydney Tokyo

This book is printed on acid-free paper.

Academic Press
A Harcourt Science and Technology Company
Harcourt Place, 32 Jamestown Road, London NW1 7BY, UK
http://www.academicpress.com

Academic Press
A Harcourt Science and Technology Company
525 B Street, Suite 1900, San Diego, California 92101-4495, USA
http://www.academicpress.com

ISBN 0-12-221970-8

Library of Congress Catalog Card Number: 00-104886

A catalogue for this book is available from the British Library

Typeset by Mackreth Media Services, Hemel Hempstead, UK

Printed and bound in the United Kingdom
Transfered to Digital Printing, 2011

Contents

Section III MYELOMA AND PLASMA CELL LEUKEMIA CELL LINES

Section IV T-CELL LEUKEMIA AND T-CELL LYMPHOMA CELL LINES

Section V NATURAL KILLER CELL LEUKEMIA-LYMPHOMA CELL LINES

Section VI HODGKIN'S DISEASE CELL LINES

Section VII ANAPLASTIC LARGE CELL LYMPHOMA CELL LINES

Section VIII MYELOID LEUKEMIA CELL LINES

Section IX TABLES

Preface

A few years ago shortly after Christmas, I first conceived the idea for this FactsBook in a bookshop in the Italian city of Florence. My wife and I had sought refuge in the bookshop from an extremely icy wind that rendered walking through the streets virtually impossible. While warming up in the medical section of the bookshop, I came across one of the now very popular 'practical handbooks'. I thought that such a compendium of data should also be written for human leukemia-lymphoma cell lines. While I can trace back the initial idea for this volume precisely to a certain day, my interest for and knowledge in these cell lines developed over a long period of time, going back as far as 1983. During these years, I enjoyed the support and friendship of four scientists who significantly guided and influenced my career: Gerhard Gaedicke (Ulm, Germany), Jun Minowada (Chicago, USA and Okayama, Japan), Victor Hoffbrand (London, UK) and Nicole Muller-Bérat (Copenhagen, Denmark and Paris, France).

Over the years, literally hundreds of scientists had provided me with aliquots of their leukemia-lymphoma cell lines. I would like to take this opportunity and express my sincere appreciation of their support and generosity. I have also met many cell culturists and other scientists who have helped with advice and information. In particular, I would like to mention Zhen-Bo 'Jimbo' Hu (Toronto, Canada and Chicago, USA) and Yoshi Matsuo (Okayama, Japan), who also became close friends.

I am grateful to my colleagues at our institute for their hard work in setting up and running an excellent cell line bank and in sharing my vision on the uniqueness and usefulness of human leukemia-lymphoma cell lines, in particular Hilmar Quentmeier, Rod MacLeod, Cord Uphoff and Corinna Meyer. I thank Lilian Leung, Sutapas Bhattacharya and Tessa Picknett at Academic Press for their enthusiastic support and help at all stages in the preparation of this FactsBook.

Finally, I am especially grateful to Suzanne Gignac for her unwavering, patient and continuous support of my work over so many years.

I hope that there are a minimum of omissions and inaccuracies and that these can be rectified in later editions. I am happy to receive any constructive comments on this FactsBook. Please write either to The Editor, *The Human Leukemia-Lymphoma Cell Lines FactsBook*, Academic Press, Harcourt Place, 32 Jamestown Road, London NW1 7BY, UK or to Hans G. Drexler, Department of Human and Animal Cell Cultures, DSMZ-German Collection of Microorganisms and Cell Cultures, Mascheroder Weg 1B, D-38124 Braunschweig, Germany, Fax: +49-531-2616.150, E-mail: <hdr@dsmz.de>.

Abbreviations

ACP	Acid phosphatase
ADCC	Antigen-dependent cell-mediated cytotoxicity
AIDS	Acquired immunodeficiency syndrome
ALCL	Anaplastic large cell lymphoma
ALL	Acute lymphoblastic leukemia
ALP	Alkaline phosphatase
AML	Acute myeloid leukemia
ANAE	α-Naphthylacetate esterase
ANBE	α-Naphthylbutyrate esterase
AraC	Cytosine arabinoside
ATCC	American Type Culture Collection
ATLL	Adult T-cell leukemia-lymphoma
ATRA	All-trans retinoic acid
AUL	Acute undifferentiated leukemia
β-TG	β-Thromboglobulin
baso	Basophil
BC	Blast crisis (of CML)
BCGF	B-cell growth factor
BCP	B-cell precursor
bFGF	Basic fibroblast growth factor
BM	Bone marrow
BMT	Bone marrow transplantation
BNX	Beige/Nude/Xid (mice)
CAE	Naphthol AS-D chloroacetate esterase
cALL	Common acute lymphoblastic leukemia
CLL	Chronic lymphocytic leukemia
CM	Conditioned medium
CML	Chronic myeloid leukemia
CMML	Chronic myelomonocytic leukemia
CMV	Cytomegalovirus
CNTF	Ciliary neurotropic factor
CP	Chronic phase (of CML)
CSF	Colony-stimulating factor or cerebrospinal fluid
cyCD	cytoplasmatic expression of CD marker
cyIg	cytoplasmatic immunoglobulin
δ-ALA	δ-Aminolevulinic acid
DLCL	Diffuse large cell lymphoma
DMEM	Dulbecco's modified essential medium
DMSO	Dimethylsulfoxide
DSMZ	Deutsche Sammlung von Mikroorganismen und Zellkulturen
EBNA	Epstein-Barr virus nuclear antigen
EBV	Epstein-Barr virus
ECM	Extracellular matrix
EDF	Erythroid differentiation factor

EGF	Epidermal growth factor
ELISA	Enzyme-linked immunoassay
EMA	Epithelial membrane antigen
eosino	Eosinophil
EPO	Erythropoietin
ery	Erythroid
FAB	French–American–British morphological classifications
FBS	Fetal bovine serum
FCL	Follicular cell lymphoma
FGFR	Fibroblast growth factor receptor
FISH	Fluorescence *in situ* hybridization
FL	FLT3 ligand
FN	Fibronectin
G-CSF	Granulocyte CSF
GLC	β-Glucuronidase
GlyA	Glycophorin A
GM-CSF	Granulocyte-macrophage CSF
granulo	Granulocytic
HBV	Hepatitis B virus
HCL	Hairy cell leukemia
HCV	Hepatitis C virus
HGF	Hepatocyte growth factor
HHV	Human herpesvirus
HIV	Human immunodeficiency virus
HSV	Herpes simplex virus
HTLV	Human T-cell leukemia (or lymphotropic) virus
IFN	Interferon
IFO	Institute for Fermentation Osaka
Ig	Immunoglobulin
IGF	Insulin-like growth factor
IGH/K/L	Immunoglobulin heavy/κ light/λ light chain
IL	Interleukin
IMDM	Iscove's modified Dulbecco's medium
jCML	Juvenile chronic myelogenous leukemia
JCRB	Japanese Cancer Research Resources Bank
LCL	Lymphoblastoid cell line
LGL	Large granular lymphocytes
LIF	Leukemia inhibitory factor
lym	Lymphoid
M1	(Immature) myeloblastic AML
M2	Myeloblastic AML
M3	Promyelocytic AML
M4	Myelomonocytic AML
M5	Monocytic AML
M6	Erythroid AML
M7	Megakaryocytic AML
macro	Macrophage
MBR	Major breakpoint region
MCL	Mantle cell lymphoma

MCP-1	Monocyte chemotactic protein-1
M-CSF	Macrophage CSF
MDR	Multiple drug resistance
MDS	Myelodysplastic syndromes
meg	Megakaryocytic
MEM	Minimum essential medium
MGP	Methyl Green Pyronin
MIP	Macrophage inflammatory protein
MLR	Mixed lymphocyte reaction
mono	Monocytic
MPD	Myeloproliferative disorder
MPO	Myeloperoxidase
MSE	Monocyte-specific esterase
my	Myeloid
NBT	Nitroblue tetrazolium
NCS	Newborn calf serum
neutro	Neutrophil
NGF	Nerve growth factor
NHL	Non-Hodgkin's lymphoma
NK	Natural killer
NOD	Nonobese diabetic (mice)
OSM	Oncostatin M
PAS	Periodic acid Schiff
PDGF	Platelet-derived growth factor
PEL	Primary effusion lymphoma
PF-4	Platelet factor-4
Ph	Philadelphia chromosome
PHA-LCM	Phytohemagglutinin induced-leukocyte condition medium
PLL	Prolymphocytic leukemia
PPO	Platelet peroxidase
PWM	Pokeweed mitogen
RAEB	Refractory anemia with excess of blasts
RAEBT	Refractory anemia with excess of blasts in transformation
RARS	Refractory anemia with ring sideroblasts
RB1	Retinoblastoma (gene)
RCB	Riken Cell Bank
RPMI	RPMI 1640 medium
RT-PCR	Reverse transcriptase-polymerase chain reaction
SBB	Sudan black B
SCF	Stem cell factor
SCID	Severe combined immunodeficiency
sIg	Surface immunoglobulin
SKY	Spectral karyotyping
t-AML	Therapy-related AML
TCRA/B/D/G	T-cell receptor α, ß, δ, γ-chain
TGF	Transforming growth factor
TNF	Tumor necrosis factor
TPA	Phorbol ester 12-*O*-tetradecanoyl phorbol-13 acetate
TPO	Thrombopoietin

TRAP	Tartrate-resistant acid phosphatase
Vit. D3	Vitamin D3
vWF	von Willebrand factor

Section I

THE INTRODUCTORY CHAPTERS

1 Introduction

AIM OF THE BOOK

The main aim of this book is to provide a comprehensive source of information on all continuous human leukemia-lymphoma cell lines. The individual entries are designed to be succinct and to highlight the essential cell culture, clinical, immunological, genetic and functional features of the cell lines. Each entry contains a comprehensive listing of primary and relevant references for the cell line. Prior to the entries there are seven chapters which provide background and perspective for culturing malignant hematopoietic cell lines.

SCOPE OF THE BOOK

The number of human leukemia-lymphoma cell lines has increased dramatically in recent years to more than 1000. This book includes entries for 431 well-characterized leukemia-lymphoma cell lines. A significant number of additional cell lines is not sufficiently characterized and described in the literature. This book serves as an essential reference manual for all of the well-characterized cell lines as well as for the partly characterized leukemia-lymphoma cell lines. The spectrum of cell lines encompasses all major hematopoietic cell lineages and different stages of arrested differentiation.

In detail, entries for the well-characterized cell lines provide information on:

- sister cell lines and subclones
- cell culture aspects: establishment, medium, doubling time
- viral status
- authentication of derivation
- availability
- clinical data concerning age/sex of patient, diagnosis, treatment status, specimen source
- immunophenotype
- cytogenetic karyotype
- translocations and fusion genes
- receptor gene rearrangements and genetic alterations
- cytochemical profile
- functional features: differentiation induction, heterotransplantability, colony formation
- cytokine production and response to cytokines
- proto-oncogene and transcription factor expression or alteration
- special unique features
- key references.

STRUCTURE OF THE BOOK

The book is divided into three main parts:

- Section I: Introductory Chapters
- Sections II–VIII: Entries on Leukemia-Lymphoma Cell Lines
- Section IX: Tables.

Section I: Introductory Chapters

These introductory chapters deal with the history and classification, characterization and description, culture and expansion, quality and identity control, mycoplasma detection and elimination, and availability of cell lines. Although this is not a completely comprehensive summary of the cell culture of human leukemia-lymphoma cell lines, it is intended to provide information on the successful culture and application of such cell lines, the possibilities and necessities for characterizing and describing the cell lines, the various types and subtypes of cell lines, the potential problems inherent to the culture of the cell lines and possible solutions for these problems, and, finally, where and how to obtain such cell lines for research purposes.

Sections II–VIII: Entries on Leukemia-Lymphoma Cell Lines

The leukemia-lymphoma cell lines reviewed here are divided into two categories (Table 1): (1) well-characterized cell lines with a complete or nearly complete description, and (2) partly characterized cell lines with an incomplete description. The distinction between well- and partly characterized cell lines is based on the information reported in the literature regarding the clinical, cell culture, immunophenotypic, genetic and functional characterization of the cell lines. Well-characterized cell lines are those for which a comprehensive set of immunophenotypic and cytogenetic data has been published. Partly characterized cell lines are insufficiently described in that regard.

Data on the well-characterized cell lines are presented in Sections II–VIII. The sections are arranged according to the various cellular lineages and cell types (sections) and subtypes (parts). The criteria used for assigning individual cell lines to sections and parts are discussed in Section I, *Chapter 2*: History and Classification

Table 1 *Which human hematopoietic cell lines are included in this book?*

Cell lines included:
- Well-characterized leukemia-lymphoma cell lines (Sections II–VIII)
- Partly characterized leukemia-lymphoma cell lines (Tables I–XIII)
- Sister cell lines and relevant subclones of these cell lines

Cell lines not included:
- Uncharacterized leukemia-lymphoma cell lines
- Primary hematopoietic cell cultures (not immortalized)
- Normal EBV$^+$ or HTLV-I$^+$ hematopoietic cell lines (except for sister cell lines of included cell lines)
- Burkitt's lymphoma cell lines (except for a list of selected cell lines)
- ATLL cell lines (except for a list of selected cell lines)

of Hematopoietic Cell Lines. The partly characterized cell lines are reviewed in Section IX: Tables. Sections II–VIII comprise individual entries, each with the same layout (as described in detail below under *Organization of the Data*).

Primary short-term or long-term hematopoietic cell cultures (non-immortalized cells) are not included in this book (see also below *Cell Culture Definitions*). 'Normal' cells immortalized by transformation with EBV and HTLV are not described. Burkitt's lymphoma-derived and ATLL-derived cell lines are not presented, except for two panels of selected Burkitt's lymphoma and ATLL cell lines in Tables III and VII, respectively (see also Section I, *Chapter 3*: EBV- and HTLV-Positive Cell Lines).

Section IX: Tables

Tables I–XIII detail the partly characterized leukemia-lymphoma cell lines which are not included in the main sections of the book. Following the categorization of the cell lines into cell types and subtypes as in Sections II–VIII, each table comprises entries with the same layout: name of cell line; disease of the patient and status of treatment; age/sex of patient and specimen source; availability of core data on immunoprofile, karyotype and functional aspects; comments (special features, controversies, loss of cell line, relevant sister cell lines and subclones, etc.); and key references.

Table XIV summarizes all constitutively growth factor-dependent leukemia-lymphoma cell lines. Besides the cytokine(s) on which these cell lines are absolutely dependent, the table also presents the growth stimulatory or inhibitory effects of other cytokines on these cell lines. The absolute growth factor-dependency of these cell lines is defined as follows: in the absence of externally added cytokines, the cells stop proliferating and undergo apoptosis. However, under certain conditions, growth factor-independent subclones may arise in some cell lines.

Table XV lists those non-random chromosomal balanced translocations, inversions and insertions which have been described in leukemia-lymphoma cell lines. The genes which are altered by these chromosomal aberrations and, if applicable and/or known, further data on breakpoint regions, specific breakpoints, reverse fusion genes and gene overexpression are indicated as well. Involvement of the genes indicated has not been shown formally for each cell line, but can be assumed to have occurred in most cases given the specific chromosomal breakpoints.

Table XVI provides a list of false or misinterpreted cell lines: both the imposter cell line and its real identity are shown. See also Section I, *Chapter 6*: Authentication of Cell Lines.

Table XVII suggests a panel of reference cell lines arranged according to cell type and subtype categories. With few exceptions, only cell lines which can be obtained from public cell line banks are indicated. The most unique feature for use as an *in vitro* model system is described.

ORGANIZATION OF THE DATA

Each cell line entry includes information under the following headings (in general only positive data are listed – unless otherwise indicated).

Name of cell line

The name of the cell line is reproduced as written in the original publication. Particulary for often used cell lines, minor alterations of the cell line designation may be found in the literature (e.g. spelling or hyphenation, e.g. HL60 instead of the correct name HL-60). Even from a cursory glance at the names of cell lines, it should become clear that cell line nomenclature owes little to any systematic approach at naming biological entities. In other words, there are neither rules nor systems. Often the initials of the patient from whom the cell line was established and/or the initials of the individual who cultured the primary cells are incorporated in the cell line name. Very popular are also acronyms of the institute, university or the city (prominent examples are: CCRF-CEM for Children's Cancer Research Foundation + patient initials; OCI- for Ontario Cancer Institute; EU- for Emory University; SU- for Stanford University; K-562 for Knoxville, Tennessee; U-937 for Uppsala, Sweden). Fortunately, it is rare that independent cell lines are given the same names in duplicate by independent investigators; duplications are: A1 (one precursor B-cell line and one B-cell line), B1 (one precursor B-cell line and one B-cell line), HBL-1 (both mature B-cell lines), HBL-2 (both mature B-cell lines), HBL-3 (one precursor and one mature B-cell line), K-T1/KT-1 (one immature T-cell line and one myelocytic cell line), ME-1 (one monocytic and one plasma cell line), ME-2 (sublines of ATLL cell line ME and monocytic cell line ME-1), ME-F1/MEF-1 (one monocytic and one myeloma cell line), PLL 1/PLL1 (both B-cell lines), PLL 2/PLL2 (both B-cell lines), and TALL-1 (both immature T-cell lines).

Culture characterization

At the beginning of each entry is a brief description of salient features of the culture conditions and of several general, but important characteristics of the cell line.

Other name of cell line

Occasionally, cell lines are known by several other names or by certain abbreviations; for example the cell line Karpas 422 is also known as K422 (Section II, Part 2: Mature B-Cell Lines).

Sister cell line

The names of any simultaneous or serial sister cell lines are listed, eventually their non-neoplastic nature (EBV$^+$ B-LCLs), the cellular source from which they were established, and any major differences with the principal cell line; for example simultaneous sister cell line Mono Mac 1 with cytogenetics features different from those of its sister cell line Mono Mac 6 (Section VIII, Part 2: Monocytic Cell Lines).

Subclone

A large number of subclones have been established from the major cell lines, commonly cells which were selected for their resistance to a variety of biomodulating substances including chemotherapeutic agents; as such resistant subclones are usually not further characterized, and, as it is not known whether the expression of the divergent features is stable, such cell lines are not listed here.

However, there are some subclones from cell lines with interesting new characteristics which appeared to be worth mentioning, for example subclone KG-1a derived from the principal cell line KG-1 with a more immature immunophenotype (Section VIII, Part 1: Myelocytic Cell Lines).

Establishment

Commonly, cell lines were established by isolation and purification of the malignant cells from the residual normal cells in the original clinical specimen and direct incubation in appropriate culture media (in some instances with the addition initially of human peripheral blood or umbilical cord serum) and culture vessels (plastic flasks, 24- or 96-well plates, Petri dishes) under standard culture conditions (at 37°C, 5% CO_2 in air), but commonly without growth factors or feeder layers. Occasionally, this standard cell culture procedure, which is not mentioned in this rubric, was modified in order to give the original inoculum a certain growth advantage in a more suitable environment, e.g. initial culture *in vivo* in nude or SCID mice, culture on feeder layers, etc. Example: cell line CHRF-288-11 was first serially passaged by heterotransplantation into nude mice and then later adapted to growth *in vitro* (Section VIII, Part 3: Erythrocytic-Megakaryocytic Cell Lines).

Culture medium

Volume percentages and type of media and FBS plus obligatory additives such as conditioned medium or cytokines are indicated. Not listed are some non-specific non-obligatory additives such as bovine serum albumin, L-glutamine, hydrocortisone, insulin, 2-mercaptoethanol, pyruvate, α-thioglycerol, transferrin, antibiotics, essential/non-essential amino acids, etc. which, according to my experience, generally do not enhance further strong cell growth and do not improve or save badly growing cultures. Commonly, cell lines are incubated at 37°C in an atmosphere of 5% CO_2 in air; thus, these incubator conditions are not mentioned further in the main sections. Only the cell lines of the SUP- and UoC-series, established by Dr S.D. Smith's group at Stanford University Pediatrics (SUP) or University of Chicago (UoC), require an hypoxic atmosphere of 5% CO_2, 6% O_2 and 89% N_2 (but might be adaptable to standard incubator conditions).

Doubling time

Doubling times are indicated in hours or days. Obviously there is a range of doubling times even for the same cell line and cell culture depending on the growth phase and other external conditions (e.g. percentages of FBS, starting cell density, etc.).

Viral status

EBV is certainly the most frequently examined and most relevant virus in the field of hematological cell lines. The EBV status is important to discern 'normal B-lymphoblastoid cell lines' from neoplastic cell lines; on the other hand, EBV can also lead to immortalization of certain types of neoplastic cell lines. HTLV has been used to immortalize normal and malignant cells. Details on the biology of EBV^+ and $HTLV^+$ cell lines are presented below (see Section I, *Chapter 3*: EBV- and HTLV-Positive Cell Lines). The viral status of cell lines serves as the basis for their assignment to biological safety risk classes (biosafety levels): EBV^+ and $HHV\text{-}8^+$

cell lines must be handled at biosafety level 2 containment; HTLV-I/II$^+$ cell lines are considered to be in the biosafety level 3. While the wild-type hematological cell lines have never been found to be positive for HBV and HCV or HIV, positive cell lines (e.g. cells deliberately manipulated and infected with these organisms) must be handled under biosafety level 3 precautions. Thus, negativity for these viruses is also shown.

Authentication

Authentication of the proper derivation of the new cell line from the assumed patient is of utmost importance. This authentication process also allows for the exclusion of any cross-contamination. Most cell lines are, however, not properly authenticated. For details on authentication, see Section I, *Chapter 6*: Authentication of Cell Lines. Whether or not a cell line is authenticated is indicated.

Primary reference

Either the overall first or the first complete description of a new cell line is cited.

Availability

The availability of cell lines to investigators outside of the original laboratory is an important aspect. Four different possibilities regarding the availability of cell lines are discerned, the latter three based on my experience requesting these cell lines: (1) public non-profit cell line banks: cell lines are available to every requesting scientist, commonly for a certain fee (the accession number is indicated); (2) original authors: cell lines are available from the original laboratory upon request; (3) restricted: cell lines are available only to selected scientists and/or under specific conditions; (4) not known: either addressee did not respond to the request or cell line was not requested. See also Section I, *Chapter 8*: Availability of Cell Lines and Cell Line Banks.

Clinical characterization

The following headings provide clinical information on the patient from whom the cell line was established.

Patient

Sex and age of the patient at the time when the sample from which the cell line was established was obtained.

Disease diagnosis

The diagnosis of the disease is given as indicated in the original publication; as far as available and relevant further subclassifications or subtypes of the particular leukemia or lymphoma are listed as well. It should be appreciated that over the last decades some disease designations have been replaced by more accurate or simply more modern terms.

Treatment status

The treatment/disease status of the patient at the time when the cell material for cell culture was obtained is indicated.

Specimen site
The biological material from which the cell line was established is listed.

Year of establishment
The actual calendar year in which the cell line was established.

Immunophenotypic characterization

Immunophenotyping is the most useful analysis for assigning a given cell line to one of the major hematopoietic cell lineages. The cell lineage-specific or -associated immunodata are presented under the headings **T-/NK cell marker, B-cell marker, Myelomonocytic marker** and **Erythroid-megakaryocytic marker**. Other non-lineage-restricted immunomarkers are included under the headings **Progenitor/activation/other marker, Adhesion marker** and **Cytokine receptor**. As far as possible, antigen designations are listed following the CD nomenclature. A composite immunoprofile includes not only the expression of certain markers but also any lack of expression of some antigens. Thus, both positive and negative data are described under the immunophenotyping headings. Obviously, some of the CD markers may be listed under various cell lineage headings; for example CD56 qualifies both as an NK cell marker and as an adhesion marker.

Genetic characterization

Here classical cytogenetic and molecular genetic information is summarized.

Cytogenetic karyotype
There are few (if any) malignant hematopoetic cell lines with a normal karyotype. The cell lines are characterized by non-random numerical and structural chromosomal abnormalities. The karyotypes are presented as reported in the original publications with only minor modifications.

Unique translocation/fusion gene
Defined balanced translocations are associated with specific hematopoietic malignancies and often lead to the formation of new chimeric fusion genes. While it was not shown in all cell lines that the fusion genes were indeed generated, the presence of tell-tale translocations suggests alteration of the genes commonly involved in these chromosomal aberrations. All cell lines carrying such a specific translocation are summarized in Section IX, Table XV: Cell Lines with Unique Translocations and Fusion Genes.

Receptor gene rearrangement
In analogy to events occurring during physiological maturation processes in their normal counterpart cells, malignant T- and B-cells (but occasionally also malignant non-T non-B cells) rearrange their lymphoid antigen receptor genes, namely T-cell receptor α, β, δ, and γ and/or immunoglobulin heavy chain, κ, and λ genes. Here the existing data on the rearrangement status of these genes are described: R for rearrangement; D for deletion; G for germline.

Unique gene alteration
Molecular alterations of genes presumed to be involved in hematopoietic malignancy such as deletions, mutations and methylations are described.

Functional characterization

Besides the immunophenotypic and genetic characterization, the functional characterization provides further information regarding the uniqueness and usefulness of any given cell line. While the first two fields concern rather clearly and narrowly defined fields of research, the functional characterization combines a vast and rather diverse spectrum of biological parameters. The most often studied functional features of cell lines are described under the following headings.

Colony formation
The colony formation assays in semi-solid media are a traditional method thought to indicate the neoplastic nature of the cells.

Cytochemistry
Cytochemical staining represents a classical analysis providing valuable information on the cell lineage commitment and differentiation status of a cell. Both positive staining and lack thereof are indicated.

Cytokine production
Production of cytokines at either the protein or RNA level is detailed.

Cytokine response
The cytokines on which the cell line is constitutively dependent are reviewed. Growth stimulation and growth inhibition by cytokines are described. All cell lines which are constitutively dependent on externally added cytokines are also summarized in Section IX, Table XIV: Growth Factor-Dependent Cell Lines.

Heterotransplantation
Heterotransplantation and growth of cell lines in immunodeficient mice or hamsters is considered an indicator of malignancy.

Inducibility of differentiation
A characteristic feature of malignant hematopoietic cells is the arrest of differentiation. A variety of physiological and pharmacological agents can overcome this block and induce differentiation. The phenotypical effects and the direction of the induced differentiation are both inducer- and target cell context-dependent.

Proto-oncogene
Activation or overexpression of proto-oncogenes at the protein or mRNA level which are reported under this heading appear to be critical steps in malignant transformation.

Transcription factor
Overexpression or otherwise altered expression of transcription factors which are believed to function as critical regulators of the normal and neoplastic development of the hematopoietic system are described under this heading.

Special feature

Some cell lines display unique functional features which are not recorded under the other functional categories.

Comments

Cell line properties of particular interest are annotated under this heading. The type of cell line is summarized succinctly; the most useful scientific characteristics of the cell line are highlighted; availability from public cell line banks is indicated; other important points are commented upon.

References

The primary reference describing establishment and characterization of the new cell line is indicated. Further references which provide informative data on the cell line are listed as well. For some of the well-studied cell lines, it is not feasible to provide a comprehensive list of references (e.g. there are more than 5000 entries in Medline for the cell line HL-60). Therefore, a selection of only the most important references is listed. For some cell lines, reviews are given which should allow access to the rest of the literature.

CELL CULTURE DEFINITIONS

What is a cell line

The term 'cell line' is often further specified as 'continuous cell line' (synonyms: immortalized, permanent)[1]. The addition of the descriptive adjectives appears to be necessary to differentiate cultured cells with an unlimited number of cell doublings from cultured cells with a limited lifespan. A less ambiguous description for the latter type of cultured cells is short-term or long-term primary cell culture (Table 2). Normal T-lymphocytes can be grown for extended periods of time in the presence of IL-2; while the length of survival is variable, ultimately these cultures of normal T-cells possess only a limited *in vitro* lifespan. The cytogenetic literature in particular often uses the term cell line instead of cell culture or cell clone. Furthermore, occasionally the term cell line is misleadingly employed in lieu of cell lineage (for example: the monocytic cell line instead of the monocytic cell lineage).

The criterion for defining what constitutes an immortalized cell remains somewhat empirical. Primary cell cultures invariably senesce at less than 50–60 doublings. A culture capable of greater than 150–200 doublings may with confidence be considered to be immortal (another parameter may be continuous growth for at least one year). Primary hematopoietic cell cultures containing normal or malignant cells are not described in this book.

It should be recognized that an immortalized cell is not necessarily one that is neoplastically or malignantly transformed. Extending the notion to the field of leukemia-lymphoma cell lines, it must be pointed out that a cell line derived from a patient with a tumor is not necessarily a tumor cell line.

Table 2 *Definitions of terms 'cell line', 'subclone' and 'sister cell line'*

Term	Definition
Primary culture:	
• short-term culture	lifespan: only weeks; <10 doublings
• long-term culture	lifespan: only months; maximal 50–60 doublings
Secondary culture:	
• cell line	lifespan: immortalized; >150–200 doublings (>1 year continuous growth)
Cell line:	independent of the original organism, continuously growing, individual cells
Subclone:	derived from an original (parental) cell line and harboring divergent and unique features
Sister cell line:	
• simultaneous	established from the same patient at the same time, but possibly from different sites or the primary sample was split into several aliquots prior to culture
• serial/longitudial	established from the same patient, but at different time points, e.g. at diagnosis and at relapse

Subclones and sister cell lines

Subclones of any given cell line and sister cell lines must be discerned and properly presented (Table 2). A subclone has properties which are quantitatively or qualitatively different from those of the parental culture. Sister cell lines can be divided into simultaneous or serial sister cell lines depending on the time at which the specimen(s) was/were taken from the patient or how the primary cells were processed. Occasionally, simultaneous sister cell lines may display divergent properties. Serial sister cell lines (e.g. established at diagnosis and relapse from the same patient) may provide unique opportunities to study the molecular mechanisms involved in disease progression and transformation. In any event, subclones are by definition derived from the same clone and must be recognizable as such by clonal markers (e.g. by DNA fingerprinting, cytogenetic marker chromosomes). However, sister cell lines must not necessarily be derived from the same clone, but stem from the same patient.

Reference

[1] Schaeffer, W.I. (1990) In Vitro Cell. Dev. Biol. 26, 97–101.

2 History and Classification of Hematopoietic Cell Lines

HISTORY OF HEMATOPOIETIC CELL LINES

Burkitt's lymphoma cell lines were the first human hematopoietic cell lines

Cell culture has existed for nearly a century[1-3]. Although it represented a relatively primitive methodology in biomedical research, cell culture nevertheless generated some very impressive results. In 1951 at Johns Hopkins University in Baltimore (Maryland, USA), Gey and his colleagues established, for the first time, a human cell line (termed HeLa) from a uterine cervix carcinoma[4,5]. The subsequently established human cell lines derived from various solid tumors were all of the monolayer culture type, which is characterized by the proliferation of tumor cells adhering to the surface of the culture vessel. In 1963 at the University of Ibadan, Nigeria, Pulvertaft established the first continuous human hematopoietic cell lines, namely a series of cell lines derived from Nigerian patients with Burkitt's lymphoma in a suspension type cell culture: cell line Raji is the best known culture of this series[6]. Suspension cultures, which were new at the time, contain cells that are free-floating, singly or in clusters, in the nutrient medium.

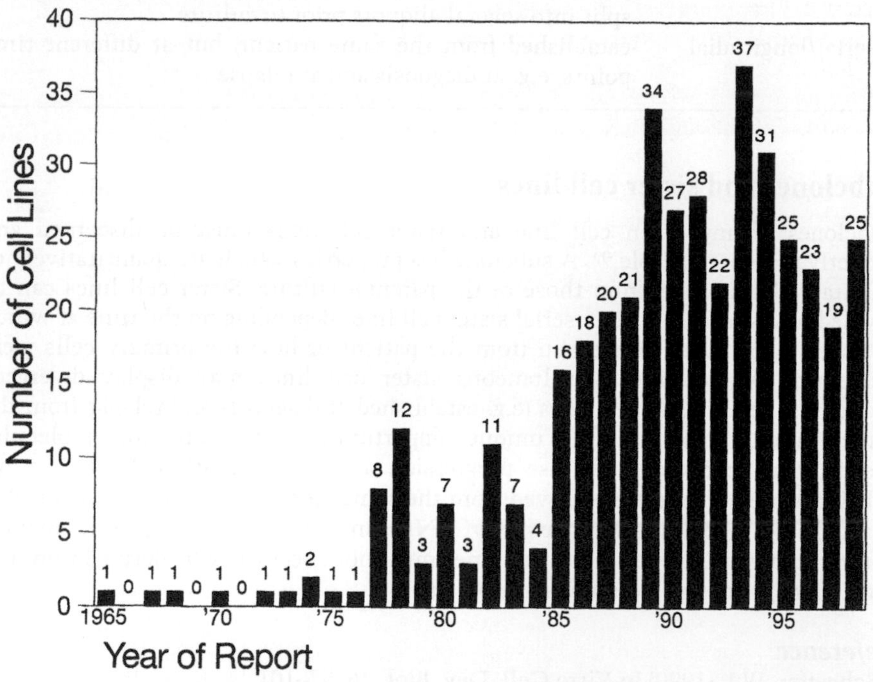

Figure 1 *Increase in the number of newly published human leukemia-lymphoma cell lines from 1965 to 1998. Shown are only the number of well-characterized cell lines published during that time period. Sister cell lines and subclones, Burkitt's lymphoma cell lines, HTLV-I⁺ ATLL cell lines and EBV⁺ B-LCLs are not included.*

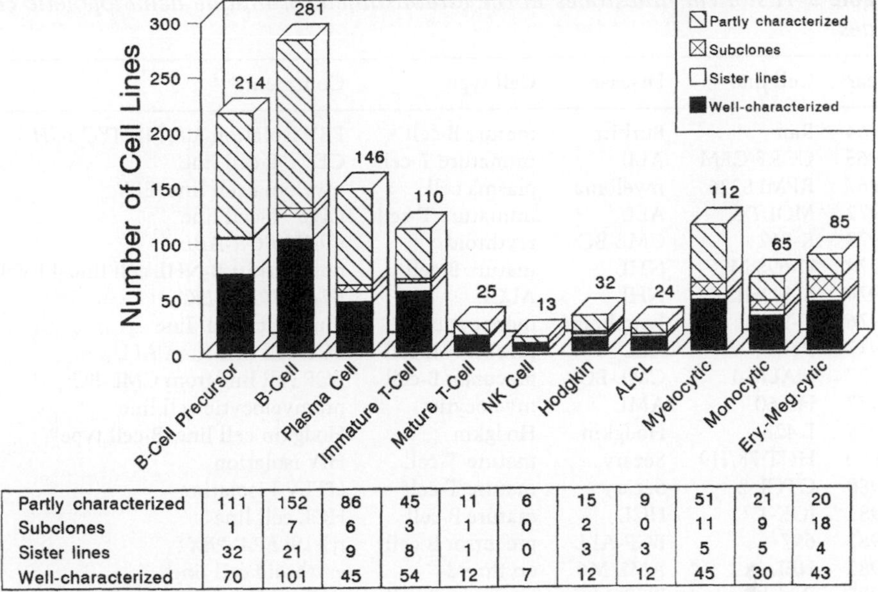

	B-Cell Precursor	B-Cell	Plasma Cell	Immature T-Cell	Mature T-Cell	NK Cell	Hodgkin	ALCL	Myelocytic	Monocytic	Ery.-Meg.cytic
Partly characterized	111	152	86	45	11	6	15	9	51	21	20
Subclones	1	7	6	3	1	0	2	0	11	9	18
Sister lines	32	21	9	8	1	0	3	3	5	5	4
Well-characterized	70	101	45	54	12	7	12	12	45	30	43

Figure 2 *Total numbers of leukemia-lymphoma cell lines according to cell line types. The cell line types are explained in Table 5. Shown are the numbers of well-characterized cell lines, their sister cell lines and relevant subclones, and of partly characterized cell lines. Burkitt's lymphoma cell lines, HTLV-I$^+$ ATLL cell lines and EBV$^+$ B-LCLs are not included.*

Erroneously, Osgood and Burke have been credited with establishing the first human hematopoietic cell line in 1955[7]. However, this cell line J-111 which was allegedly derived from a patient with AML was, in reality, found to be the cell line HeLa (and thus represents one of the first cross-contaminations) (Table XVI)[8].

Since the publication of the first Burkitt's lymphoma-derived cell lines, there has been a steady increase in the number of new leukemia-lymphoma cell lines published each year, even when excluding Burkitt's lymphoma cell lines, HTLV-I$^+$ ATLL cell lines and EBV$^+$ B-LCLs (Figure 1). These panels of cell lines now span almost the whole spectrum of hematopoietic cell lineages (except for dendritric cells) and the various stages of differentiation along the respective cell axes (Figure 2).

Milestones in the establishment of human hematopoietic cell lines

Table 3 lists some of the chronological milestones in the establishment of the first cell lines each representing the respective subtypes of human leukemias and lymphomas and several other cell lines that were instrumental in the detection of significant new scientific information (e.g. the isolation of viruses such as EBV, HTLV and HIV, the cloning of chromosomal translocation breakpoints and their new fusion genes, etc.)[9,10]. In recent years, a number of reviews have summarized specific groups or types/subtypes of human leukemia-lymphoma-derived cell lines[11–49].

Table 3 *Historical milestones in the establishment of unique hematopoietic cell lines*

Year[a]	Cell line	Disease	Cell type	Comments[b]
1964	Raji	Burkitt	mature B-cell	EBV isolation; t(8;14) *MYC-IGH*
1965	CCRF-CEM	ALL	immature T-cell	$CD2^-$ T-cell line
1967	RPMI 8226	myeloma	plasma cell	myeloma cell line (EBV^-)
1972	MOLT 3	ALL	immature T-cell	$CD2^+$ T-cell line
1973	K-562	CML-BC	erythroid	t(9;22) *BCR-ABL*
1974	U-698 M	NHL	mature B-cell	non-Burkitt B-NHL cell line (EBV^-)
1974	SU-DHL-1	NHL	ALCL	t(2;5) *NPM-ALK*
1976	U-937	lymphoma	monocytic	monocytic cell line
1977	Reh	ALL	precursor B-cell	BCP cell line from ALL
1977	NALM-1	CML-BC	precursor B-cell	BCP cell line from CML-BC
1977	HL-60	AML	myelocytic	promyelocytic cell line
1979	L 428	Hodgkin	Hodgkin	Hodgkin cell line (B-cell type)
1980	HUT 78/H9	Sézary	mature T-cell	HIV isolation
1980	CTCL-2	Sézary	mature T-cell	HTLV-I isolation
1981	JOK-1	HCL	mature B-cell	HCL cell line
1982	697	BCP-ALL	precursor B-cell	t(1;19) *E2A-PBX1*
1982	HEL	AML M6	erythroid	erythroid cell line
1985	RS4;11	BCP-ALL	precursor B-cell	t(4;11) *MLL-AF4*
1985	KU812	CML-BC	myelocytic	basophilic cell line
1985	MEG-01	CML-BC	megakaryocytic	megakaryocytic cell line
1985	YT	T-ALL	NK cell	NK cell line
1985	EoL-1	leukemia	myelocytic	eosinophilic cell line
1986	HDLM-2	Hodgkin	Hodgkin	Hodgkin cell line (T-cell type)
1988	HMC-1	leukemia	myelocytic	mast cell line
1988	M-07e	AML M7	megakaryocytic	cytokine-dependent
1989	TF-1	AML M6	erythroid	multi cytokine-dependent/responsive
1991	ME-1	AML M4eo	monocytic	inv(16) *CBFB-MYH11*
1991	Kasumi-1	AML M2	myelocytic	t(8;21) *AML1–ETO*
1991	NB4	AML M3	myelocytic	t(15;17) *PML-RARA*
1991	DoHH2	B-NHL	mature B-cell	t(14;18) *IGH-BCL2*
1994	MDS92	MDS	myelocytic	MDS cell line
1996	BC-1	PEL	mature B-cell	$HHV\text{-}8^+$ PEL cell line

Modified after refs. 9,10. A more extensive historical overview is given in these references.

[a] Year of publication.

[b] The particular scientific significance of the cell line is highlighted.

CARDINAL FEATURES OF LEUKEMIA-LYMPHOMA CELL LINES

Malignant hematopoietic cell lines

The advent of cell culturing technology and the application of continuous cell lines has provided entirely new approaches for dissecting the hematopoietic compartment. Clearly, leukemia-lymphoma cell lines have now become indispensable tools in research[31,41,50]. These cell lines have proven to be particularly informative in hematological, immunological, molecular biological, cytogenetic, pharmacological and virological studies, but also in several other areas of

Table 4 *Common features of leukemia-lymphoma cell lines*

Major advantages:

- Unlimited supply of cell material
- Infinite storability and recoverability

Common cellular characteristics:

- Monoclonal origin
- Differentiation arrest at a discrete maturation stage
- Sustained proliferation in culture
- Genetic alterations

biomedicine and biotechnology. The major advantage is the unlimited supply of cellular material (Table 4). Furthermore, cell lines can be stored in liquid nitrogen and recovered without any detrimental loss of cellular features or cell viability.

Common characteristics of leukemia-lymphoma cell lines

A detailed analysis of their general features demonstrates the following four common characteristics of human leukemia-lymphoma cell lines (Table 4): (1) they are presumed to be of monoclonal origin and are proven to originate from one cell (however, subclones may emerge during long-term culture); (2) their differentiation is arrested at a discrete stage during maturation in each cell lineage; (3) there is sustained, autonomous and external growth factor-independent proliferation of the cultured cells; this does, of course, not apply to new types of cell lines that were first developed in the late 1980s and which were deliberately established as constitutively growth factor-dependent cell lines[32,43,46,51]; and (4) they contain genetic alterations: a survey of the cell line entries in this book shows that among 429 well-characterized cell lines (excluding sister cell lines, subclones, EBV^+ B-LCLs, Burkitt's lymphoma- and ATLL-derived cell lines) for which karyotypes have been published, only two (0.5%) showed a normal karyotype without any structural or numerical aberrations (the immature T-cell line TALL-107 and the precursor B-cell line 207). Besides these gross alterations at the cytogenetic level, many cell lines also carry alterations which are detectable only at the molecular level, e.g. point mutations or deletions of the *P15INK4B*, *P16INK4A*, *P53*, *RAS*, *RB1* and other genes[52]. These genetic changes presumably provide the affected cell with either proliferative or survival advantages and are thought to play an important role in both the *in vivo* tumorigenesis and the *in vitro* establishment of the cell line[53].

Intrinsic differences between primary cells and resulting cell line

In general and under optimal cell culture conditions, leukemia-lymphoma cell lines very stably retain the major features of their original cells. It has occasionally been suggested that some cellular characteristics are acquired during establishment of the cell lines or during extended *in vitro* culture, thus representing *in vitro* artefacts. However, there are few if any unequivocal data forthcoming to support such contentions as a general property of cells in culture. An alternative view entertains the possibility that a given abnormality of the malignant *in vivo* cell may favor its immortalization (or may even represent a prerequisite for immortalization)[52,53].

However, these possibilities (whether these aberrations are acquired during growth *in vitro* or preexist in the primary material) cannot be discussed by comparing apples with oranges, i.e. comparing cohorts of primary cases with cohorts of cell lines, as these two study populations are not related to each other. This question can only be adequately addressed in unique experimental systems consisting of matched samples of fresh malignant cells and their respective derivative cell lines[53].

At first sight, this issue may appear to be a semantic dispute. However, because cell lines are increasingly frequently employed in experimental studies as prototype malignant cells that grow *in vitro*, it is clearly important to know whether, for example, abnormalities in critical tumor-associated genes such as p53 truly reflect the molecular lesion responsible for disease initiation or progression, development of chemotherapy resistance and many other pathophysiological mechanisms *in vivo*. If this were not the case, then it could be questioned whether or not cell lines represent bona fide model systems for such studies[53].

Finally, intentional or accidental suboptimal culture conditions with the ensuing stress on cells can lead to selection pressure. Deliberate experimental manipulations can also cause phenotypic and genotypic shifts.

CLASSIFICATION OF HEMATOPOIETIC CELL LINES

Distinction of malignant from 'normal' hematopoietic cell lines

Normal hematopoietic cells only survive *in vitro* for days or weeks, even in the presence of physiological or pharmacological stimulators. However, normal hematopoietic cells can be immortalized by certain viruses, the most prominent being EBV and HTLV-I. A more detailed discussion of this topic and the distinction between normal and malignant hematopoietic cell lines is given in Section I, *Chapter 3*: EBV- and HTLV-Positive Cell Lines.

Classification of leukemia-lymphoma cell lines

There are a number of possibilities for classifying leukemia-lymphoma cell lines: according to the diagnosis of the patient or according to immunophenotypes, specific cytogenetic abberrations, functional characteristics or other salient features of the cultured cells. The most often used (and clearly most practical) classification is based on the physiological spectrum of the normal hematopoietic cell lineages. The primary distinction separates lymphoid from myeloid cells (Table 5). Within the lymphoid and myeloid categories, B-cell, T-cell and NK cells on the one hand and myelocytic, monocytic, erythrocytic and megakaryocytic cells on the other hand are discerned. The basis for the even finer subclassification is described below. While it is still occasionally debated, there appears to be a consensus now that NK cells represent a third lymphoid lineage and do not belong to the T-cell lineage[54].

Table 5 *Classification of leukemia-lymphoma cell lines and scheme adopted for presentation in this book*

Main cell type	Physiological cell lineage	Type and subtype of of cell line	Section/Part in this book
Lymphoid	B-cell	Precursor B-cell line	Section II, Part 1
		Mature B-cell line	Section II, Part 2
		Plasma cell line	Section III
	T-cell	Immature T-cell line	Section IV, Part 1
		Mature T-cell line	Section IV, Part 2
	NK cell	NK cell line	Section V
Myeloid	Myelocytic	Myelocytic cell line	Section VIII, Part 1
	Monocytic	Monocytic cell line	Section VIII, Part 2
	Erythrocytic	Erythrocytic cell line [a]	Section VIII, Part 3
	Megakaryocytic	Megakaryocytic cell line [a]	Section VIII, Part 3

[a] It is often difficult, if not impossible, to assign a given cell line to either the erythrocytic or megakaryocytic cell lineage as most of these cell lines express features of both lineages, e.g. (hemo)globin, specific transcription factors, surface antigens, differentiation potential, etc; thus, it is preferable to use the term 'erythrocytic-megakaryocytic cell line' and to combine such cell lines under this heading.

Hodgkin's disease, ALCL and dendritic cell lines

Commonly the assignment of any given hematopoietic cell line to a cell lineage and stage of arrested differentiation, based on its immunological and other phenotypes, does not present any problems. Exceptions to this rule are the Hodgkin's disease- and ALCL-derived cell lines. Although the lymphoid nature of Hodgkin–Reed–Sternberg cells (which are the presumed neoplastic cells in this disease) appears to be firmly established[13,17,55], and thus *bona fide* Hodgkin's disease cell lines may be assigned to lymphoid T- or B-cell categories, their uniqueness and the fact that such cell lines display very unusual and often asynchronous marker profiles that are not found in normal physiological cell types justifies a separate category for Hodgkin's cell lines and the equally unique ALCL cell lines (listed in Sections VI and VII, respectively). Historically, the research area of Hodgkin's disease and the resulting cell lines have been very controversial. Several clearly non-lymphoid cell Hodgkin's disease cell lines had been described as well and are listed in the section on these cell lines. Continuous and confirmed human dendritic cell lines have not yet been described.

Classification based on immunophenotype and other features

The most useful technique for assigning a cell line to one of the major cell lineages is undoubtedly immunophenotyping. The more extensive and complete the immunoprofile, the more precise is the classification of the cell lineage derivation and status of arrested differentiation along this cell axis. Other techniques may add highly valuable information in cases of uncertain cell lineage assignment. The original diagnosis of the patient is also of importance for the cell lineage assignment.

Precursor B-cell lines

Precursor B-cell lines are distinguished from mature B-cell and plasma cell lines by their lack of surface immunoglobulin (sIg) expression. However, these cell lines display other B-cell-associated immunomarkers and their immunoglobulin genes are generally already rearranged[42]. All precursor B-cell lines fulfill the so-called EGIL criteria[56] for B-lineage-derivation, i.e. being positive for HLA-DR, CD19, CD22 (both cytoplasmic and surface) and CD79a/CD79b (cytoplasmic). Generally, the immunoprofiles of the cell lines parallel those of their respective primary samples. As suggested in the EGIL classification, the B-I pro-B cell line is CD10-negative, while B-II common-B and B-III pre-B cell lines are CD10+. The latter subgroup also shows cytoplasmic IgM (cyIgM). All three categories are sIg-negative. Precursor B-cell lines are commonly derived from ALL and CML in lymphoid blast crisis.

B-cell lines

B-cell lines are characterized by the membrane expression of sIg and other specific or associated B-cell markers such as CD19, CD20, CD21 and others. The Ig receptor genes are rearranged and the gene products are expressed at the mRNA and/or protein level. A large percentage of the B-cell lines carries chromosomal aberrations involving the loci of the *IGH, IGK* or *IGL* genes (on 14q32, 2p12 or 22q11, respectively) and secretes Igs. B-cell lines are derived from a wide spectrum of different but related B-cell malignancies including B-ALL, B-CLL, B-PLL, HCL and various B-NHLs (DLCL, FCL, MCL, PEL).

Plasma cell lines

Plasma cell lines are defined by their derivation from patients with multiple myeloma, plasma cell leukemia or plasmacytoma and their lack of sIg expression in the context of a B-cell immunoprofile. Many plasma cell lines are IL-6-dependent. In more than 80% of the cell lines analyzed, chromosome 14 band q32 (*IGH* locus) is affected[23]. The *IGH* genes are rearranged in all cell lines. Plasma cell lines can be subdivided according to their isotype (IgA, D, E, G or M and κ or λ).

Immature T-cell lines

Immature T-cell lines are defined by membrane or cytoplasmic CD3 expression. Commonly these cell lines also react with other T-cell-specific or -associated immunomarkers such as CD2, CD4, CD5, CD7 or CD8. Subdivision of the immature T-cell lines follows the EGIL classification which discerns four subgroups, all being CD3$^+$ or cyCD3$^+$ [56]: (1) T-I pro-T cells are CD7$^+$; (2) T-II pre-T cells are CD2$^+$ and/or CD5$^+$ and/or CD8$^+$ (CD4 was not considered in this classification as it is also found on monocytes); (3) T-III cortical T-cells are CD1$^+$; and (4) T-IV T-cells are CD1$^-$, CD3$^+$. Furthermore, two subgroups are identified according to the membrane expression of T-cell receptor (TCR) αβ chains and TCRγδ chains in association with CD3. Another classification scheme which discerns five subtypes of T-cell lines is based primarily on the expression of the markers CD10, HLA-DR and TdT[57]. With few exceptions one or more of the various *TCR* genes are rearranged and mRNA or protein may be expressed. Most immature T-cell lines are derived from patients with T-ALL or T-lymphoblastic lymphoma.

Mature T-cell lines

Mature T-cell lines express an immunophenotype distinct from that of immature T-cell lines ('post-thymic T-cells')[58]. These cell lines are derived from mature T-cell malignancies such as T-CLL, CTCL (mycosis fungoides, Sézary syndrome), and other T-NHLs (e.g. DLCL).

NK cell lines

NK cell lines display typical immunoprofiles and functional features. NK cell leukemia-lymphomas are also summarized under the term large granular lymphocyte (LGL) leukemia-lymphoma. NK LGL neoplasms must be discerned from so-called T-LGL leukemias[59]. Both T-LGL and NK LGL neoplastic cells share the features of morphologically detectable azurophilic granules and functionally detectable cytotoxic activity. The differences between T-LGL and NK LGL malignancies lie in their immunoprofiles, EBV status and *TCR* gene status. T-LGL cells (also termed cytotoxic T-cells – cell lines derived from such cells are listed here under immature T-cell lines) are commonly $CD3^+$, $CD4^-$, $CD8^+$, $CD16^+$, $CD56^-$, $CD57^+$, $TCR\alpha\beta^+$ or $TCR\gamma\delta^+$, EBV^-, *TCRs* rearranged. In contrast, NK cell diseases are generally $CD3^-$, $CD4^-$, $CD8^+$, $CD16^+$, $CD56^+$, $CD57^{+/-}$, $TCR\alpha\beta^-$, $TCR\gamma\delta^-$, $EBV^{+/-}$, *TCRs* germline. The phenotypes of the respective cell lines do not, however, correspond entirely to these descriptions.

Hodgkin's disease cell lines

Hodgkin's disease cell lines are assigned to this category solely on the basis of derivation from patients with Hodgkin's disease. Similar to the research field of primary Hodgkin–Reed–Sternberg cells which for decades was dominated by controversies[55], some of the 'Hodgkin cell lines' are also rather controversial[13,17,29]. Following the findings on primary Hodgkin–Reed–Sternberg cells, Hodgkin cell lines should express the CD15 and CD30 antigens and possess lymphoid characteristics. However, this profile is not specific for Hodgkin's disease, but is also found on various types of NHL cell lines.

ALCL cell lines

ALCL cell lines are established from patients with ALCL which was traditionally diagnosed as malignant histiocytosis, (Ki-1) large cell lymphoma or other now obsolete terms. Cells are by definition $CD30^+$. The subtypes T-cell, B-cell, and null-cell ALCL cell line can be discerned on the basis of surface marker expression[29]. Many, but not all, ALCL cell lines carry the specific t(2;5) chromosomal abnormality (or variants thereof).

Myelocytic cell lines

Myelocytic cell lines are identified by their expression of myeloid surface markers and their functional features. These cell lines are positive for one or several of the CD13, CD15, CD33 and CD65 antigens, but negative for all lymphoid (except CD4 and CD7), monocytic (CD14, CD68), erythroid (GlyA) and megakaryocytic (CD36, CD41, CD42b, CD61, PPO) immunomarkers. The cells show typical cytochemical staining profiles (positive for ANAE, ANBE, CAE, Lysozyme, MPO, SBB). Many myelocytic cell lines are cytokine-dependent or -responsive[28,51]. The *TCR* and *IGH* genes are commonly in germline configuration. Myelocytic cell lines carry a number of specific chromosomal aberrations, e.g. inv(3), t(6;9), t(8;21), t(15;17), t(9;22) and others[60]. The following subtypes can be discerned among the myelocytic cell lines: (1) promyelocytic cell lines carrying the specific t(15;17) (cell lines Ei501, HT93, NB4, UF-1) and/or promyelocytic features (HL-60)[19]; (2) eosinophilic cell lines with constitutively expressed or inducible eosinophilic features (AML14, EoL-1, HL-60, YJ); (3) basophilic cell lines with constitutively expressed or inducible basophilic features (HL-60, KU812, LAMA-84); and (4) a mast cell line (HMC-1). Most myelocytic cell lines are derived from patients with AML or CML in blast crisis.

Monocytic cell lines

Monocytic cell lines display one or several of the myeloid surface antigens, in particular the prototypical CD14 marker (and CD68). Typical functional parameters are phagocytosis and expression of the monocyte-specific esterase (MSE)[61]. Other features are similar to those of myelocytic cell lines. Monocytic cell lines are established from patients with AML M4 or AML5, rarely from patients with CML in blast crisis.

Erythrocytic-megakaryocytic cell lines

Erythrocytic-megakaryocytic cell lines are stained by the erythroid and megakaryocytic immunomarkers CD36, CD41, CD42b, CD61, GlyA, PPO, or vWF[27,30]. The specific functional features include benzidine staining, production of platelet-like particles, and expression of (hemo)globin at the mRNA or protein level, α-granules and demarcation membranes. Often cell lines express both erythrocytic and megakaryocytic markers equally strongly. In other instances there may be a predominance of one cell lineage at the expense of the other, but commonly the cells are still positive for one or two markers of the 'second' cell lineage. Erythrocytic-megakaryocytic cell lines are derived from cases with AML M6, AML 7 and CML in blast crisis.

References

1 Harrison, R.G. (1907) Proc. Soc. Exp. Biol. Med. 4, 140–143.
2 Carrel, A. (1912) J. Exp. Med. 15, 516–528.
3 Sanford, K.K. et al. (1948) J. Natl. Cancer Inst. 9, 229–246.
4 Gey, G.O. et al. (1952) Cancer Res. 12, 264–265.
5 Gold, M. (1986) A Conspiracy of Cells. State University of New York Press, Albany, NY.
6 Pulvertaft, R.J.V. (1964) Lancet i, 238–240.
7 Osgood, E.E. and Brooke, J.H. (1955) Blood 10, 1010–1022.
8 Nelson-Rees, W.A. and Flandermeyer, R.R. (1976) Science 191, 96–97.
9 Drexler, H.G. and Minowada, J. (1998) Leukemia Lymphoma 31, 305–316.
10 Drexler, H.G. and Matsuo, Y. (1999) Leukemia 13, 835–842.
11 Collins, S.J. (1987) Blood 70, 1233–1244.
12 Diehl, V. et al. (1982) Cancer Treat. Rep. 66, 615–632.
13 Diehl, V. et al. (1990) Semin. Oncol. 17, 660–672.
14 Drexler, H.G. et al. (1987) Leukemia 1, 629–637.
15 Drexler, H.G. et al. (1989) Hematol. Oncol. 7, 95–113.
16 Drexler, H.G. and Minowada, J. (1992) Human Cell 5, 42–53.
17 Drexler, H.G. (1993) Leukemia Lymphoma 9, 1–25.
18 Drexler, H.G. (1994) Leukemia Res. 18, 919–927.

[19] Drexler, H.G. et al. (1995) Leukemia Res. 19, 681–691.
[20] Drexler, H.G. et al. (1998) Leukemia 12, 1507–1517.
[21] Drexler, H.G. et al. (1999) Leukemia Res. 23, 207–215.
[22] Drexler, H.G. (2000) Leukemia Res. 24, 109–115.
[23] Drexler, H.G. and Matsuo, Y. (2000) Leukemia Res. (in press).
[24] Ferrero, D. and Rovera, G. (1984) Clin. Haematol. 13, 461–487.
[25] Gogusev, J. and Nezelof, C. (1998) Hematol. Oncol. Clin. North Am. 12, 445–463.
[26] Haluska, F.G. et al. (1994) Blood 84, 1005–1019.
[27] Hassan, H.T. and Freund, M. (1995) Leukemia Res. 19, 589–594.
[28] Hassan, H.T. and Drexler, H.G. (1995) Leukemia Lymphoma 20, 1–15.
[29] Herbst, H. et al. (1993) Crit. Rev. Oncogenesis 4, 191–239.
[30] Hoffman, R. (1989) Blood 74, 1196–1212.
[31] Hozumi, M. (1993) CRC Crit. Rev. Oncol. Hematol. 3, 235–277.
[32] Ihle, J.N. and Askew, D. (1989) Int. J. Cell Cloning 7, 68–91.
[33] Kalle, C. von and Diehl, V. (1992) Int. Rev. Exp. Pathol. 33, 185–203.
[34] Keating, A. (1987) Baillière's Clin. Haematol. 1, 1021–1029.
[35] Koeffler, H.P and Golde, D.W. (1980) Blood 56, 344–350.
[36] Koeffler, H.P. (1983) Blood 62, 709–721.
[37] Koeffler, H.P. (1986) Semin. Hematol. 23, 223–236.
[38] Lübbert, M. and Koeffler, H.P. (1988) Blood Rev. 2, 121–133.
[39] Lübbert, M. and Koeffler, H.P. (1988) Cancer Rev. 10, 33–62.
[40] Lübbert, M. et al. (1991) Blood 77, 909–924.
[41] Matsuo, Y. and Minowada, J. (1988) Human Cell 1, 263–274.
[42] Matsuo, Y. and Drexler, H.G. (1998) Leukemia Res. 22, 567–579.
[43] Mire-Sluis, A.R. et al. (1995) J. Immunol. Methods 187, 191–199.
[44] Nezelof, C. et al. (1992) Semin. Diagn. Pathol. 9, 75–89.
[45] Nilsson, K. (1992) Human Cell 5, 25–41.
[46] Oval, J. and Taetle, R. (1990) Blood Rev. 4, 270–279.
[47] Schaadt, M. et al. (1985) Int. Rev. Exp. Pathol. 27, 185–202.
[48] Schaadt, M. et al. (1988) Cancer Rev. 10, 108–122.
[49] Slivnick, D.J. et al. (1989) Hematol. Oncol. Clin. North Am. 3, 205–220.
[50] Drexler, H.G. et al. (1994) In: Atlas of Human Tumor Cell Lines (Hay, R.J. et al., eds). Academic Press, Orlando, FL, pp. 213–250.
[51] Drexler, H.G. et al. (1997) Leukemia 11, 701–708.
[52] Drexler, H.G. (1998) Leukemia 12, 845–859.
[53] Drexler, H.G. et al. (2000) Leukemia 14, 198–206.
[54] Spits, H. et al. (1995) Blood 85, 2654–2670.
[55] Drexler, H.G. (1992) Leukemia Lymphoma 8, 283–313.
[56] Bene, M.C. et al. (1995) Leukemia 9, 1783–1786.
[57] Minowada, J. (1988) Cancer Rev. 10, 1–18.
[58] Matutes, E. (1999) T-Cell Lymphoproliferative Disorders: Classification, Clinical and Laboratory Aspects. Harwood Academic Publishers, Chur, Switzerland.
[59] Lamy, T. and Loughran, T.P. Jr. (1998) Educational Program Book, ISH-EHA, Amsterdam, pp. 38–43.
[60] Drexler, H.G. et al. (1995) Leukemia 9, 480–500.
[61] Uphoff, C.C. et al. (1994) Leukemia 8, 1510–1526.

3 EBV- and HTLV-Positive Cell Lines

EPSTEIN-BARR VIRUS TRANSFORMED CELL LINES

EBV-transformed B-lymphoblastoid cell lines

A cell line established from a patient with leukemia is not necessarily a 'leukemia cell line'[1,2]. Early studies from the 1960s documented that EBV infection is capable of immortalizing certain B-cell subsets in normal human leukocyte cultures but also in samples from patients with malignant hematopoietic cells. The term 'lymphoblastoid' was introduced by Benyesh-Melnick *et al.* in 1963 to describe such cell cultures[3]. Unfortunately, some authors use this description indiscriminately for any type of lymphoid cell line (including proven malignant cell lines). Thus, the extended designation 'EBV$^+$ B-lymphoblastoid cell line (B-LCL)' represents an unequivocal term which was generally adopted to define this type of non-malignant lymphoid cell[4].

As long as 25 years ago, Nilsson and Pontén complained in their landmark paper about the state of affairs, remarking that 'the literature abounds with lines from leukemia, lymphoma and other neoplasms which are obviously lymphoblastoid but nevertheless treated as tumor cell lines'[5]. Innumerable publications still use EBV$^+$ B-LCLs in lieu of bona fide leukemia-lymphoma cell lines.

The biological activity of EBV lies in its capacity to transform resting B-cells into immortalized lymphoblastoid cells that proliferate indefinitely and autonomously, harboring the virus in a latent state. Various EBV genes are expressed as proteins in EBV$^+$ B-LCLs which divide continuously *in vitro*, possibly as a result of one or of both of the following two mechanisms. The first proposes that viral gene expression leads to activation of cytokine-cytokine receptor pathways that are similar to the normal physiological pathways; thus, the cell is responding in a similar way to similar effectors. The other suggests that the virus affects a very small number of key molecules (perhaps c-myc, RB, and p53) in the resting B-lymphocytes, and the panoply of effects seen reflects activation processes preprogrammed in resting B-cells[6].

It appears to be 10- to 100-fold easier to establish an EBV$^+$ B-LCL from a patient with leukemia than a neoplastic leukemia cell line[5]. EBV$^+$ B-LCLs can be obtained from non-neoplastic (bystander) cells from healthy as well as from diseased individuals. Obviously, leukemia-lymphoma cell lines are only obtained from the respective patients with this disease. Non-neoplastic cells are inevitably present as 'contaminants' in an explant from patients with leukemia or lymphoma (peripheral blood, bone marrow, lymph node, etc.). Since EBV$^+$ B-LCLs may grow out from normal tissue, it does not automatically follow that a cell line from a patient with a malignancy is a malignant cell line[5]. Culture in liquid media favors outgrowth by the most rapidly dividing cell. Thus, EBV$^+$ B-LCLs always have a selective advantage over the slow growing neoplastic cells as, at least initially, freshly explanted leukemia-lymphoma cells have rather long doubling times. A potential malignant cell line will therefore be lost once the outgrowth of the EBV-carrying cells begins[7]. EBV$^+$ B-LCLs derived from patients with hematopoietic malignancies were indistinguishable from B-LCLs established from non-neoplastic lymphoid tissues[5].

The clear distinction between a malignant hematopoietic cell line and an EBV$^+$ B-LCL is particularly relevant for cell lines established from patients with mature B-cell malignancies such as CLL, PLL, HCL, NHL, myeloma and other malignancies. The issue is further complicated by the fact that all EBV$^+$ B-LCLs will eventually become monoclonal and aneuploid, grow as colonies, and form tumors in immunodeficient rodents[5,8]. Still, in most cases 'true malignant hematopoietic cell lines' can be discerned from EBV$^+$ B-LCL by various parameters (Table 6).

Table 6 *Comparison of characteristics of leukemia-lymphoma cell lines and EBV$^+$ B-LCLs*

Feature	Leukemia-lymphoma cell lines	EBV$^+$ B-LCLs initially → long-term[a]
Cell line establishment:		
From healthy person	–	+
From patient with leukemia-lymphoma	+	+
Success rate	very low	high
Cell culture:		
Colony formation	mostly +	– → +
Cloning ability	mostly +	– → +
Doubling time	initially long	30–40 h
Morphology:		
Culture	in suspension, loose single cells or small clumps (2–5 cells), rarely loosely adherent	large dense clumps
Cellular	cell morphology typical for specific lineage (monomorph)	striking pleomorphology, hand-mirror cells
Functional/other features:		
Ig secretion	–[b]	+
EBV genome	–[c]	+
Clonality	monoclonal	polyclonal → monoclonal
Karyotype	aneuploid	diploid → aneuploid
Tumorigenicity in mice	mostly +	– → +
Lineage-specific functional features	mostly +	–
Immunoprofile:		
Lineage-specific immuno-markers	+	only B-cell markers (e.g. CD19$^+$, CD20$^+$, sIg$^+$)

Modified after refs. 5,9,10.

[a] There is a significant difference between EBV$^+$ B-LCLs shortly after establishment and after extended culture as one subclone with a growth advantage will overgrow other clones (polyclonality → monoclonality) and may even develop cytogenetic alterations (normal → abnormal karyotype).

[b] Except for cell lines derived from B-cell or plasma cell malignancies (e.g. B-CLL, B-NHL, multiple myeloma).

[c] Except for most African Burkitt's lymphoma cell lines, rare American Burkitt's lymphoma cell lines, some mature B-leukemia cell lines (B-CLL, B-PLL, HCL), some (oriental) NK leukemia-lymphoma cell lines, and rare T-leukemia cell lines (see also Table 7).

In summary, the interpretation of non-malignant EBV+ B-LCLs as representatives of the malignant *in vivo* clone is clearly misguided and raises some concerns regarding the conclusions drawn in studies using such cell lines. A combination of markers commonly allows for an accurate determination of the nature of the cell lines: cellular morphology, EBV status, karyotype, functional features and immunoprofile.

EBV-positive leukemia-lymphoma cell lines

The majority of African-type (endemic) Burkitt's lymphoma-derived cell lines are EBV+, while the American-type Burkitt's lymphoma cell lines are generally EBV−. The strong facilitating effect of EBV on cell line establishment *in vitro* may explain why Burkitt's lymphoma cells are much easier to grow *in vitro* than other malignant human cells[8]. Consequently, a huge number of Burkitt's lymphoma cell lines have been reported. It is beyond the scope of this book to describe or list all these cell lines. A panel of selected Burkitt's lymphoma cell lines is presented in Table III.

It appears that B-CLL, B-PLL and HCL cells cannot grow long-term *in vitro* unless they are infected with EBV[11]. As pointed out above, it must then be ascertained, using unequivocal parameters, that one is dealing with a bona fide leukemia-derived cell line and not with a B-LCL. Interestingly, plasma cells obtained from patients with myeloma or plasma cell leukemia can be grown as continuous cell lines without EBV transformation, though the success rate of cell line establishment is quite low. In fact, myeloma cells are usually refractory to EBV infection[10].

While it was thought for decades that EBV can only infect B-cells, recent evidence from the last decade shows that (malignant) T-cells and NK cells can also be infected by EBV[12–14]. Several NK leukemia-lymphoma cell lines are EBV+ (Section V). We found EBV sequences in the genome of one immature T-cell line (RPMI 8402)[15].

HUMAN T-CELL LEUKEMIA VIRUS TRANSFORMED CELL LINES

T-cells of normal peripheral blood do not grow spontaneously *in vitro* or become cell lines. However, normal T-cells can be induced to grow in liquid as well as in semi-solid culture medium after transfection with HTLV-I[16,17]. HTLV-I- or II-transformed normal T-cells produce a wide spectrum of cytokines (for example GM-CSF, IFN-γ, IL-1, IL-2, IL-3, IL-5, IL-6, IL-9, TNF-α). The cells actively transcribe HTLV RNA and thus produce virions. This contrasts with the lack of HTLV gene expression in ATLL tumor cells, suggesting that the process of *in vitro* transformation by HTLV does not directly parallel formation of a leukemic clone *in vivo*[18]. HTLV-I, although first isolated from a mature T-cell line derived from a patient with a cutaneous T-cell lymphoma (mycosis fungoides)[19,20], is almost exclusively associated *in vivo* with adult T-cell leukemia-lymphoma (ATLL)[21–23]. While not all specimens from ATLL cases can be immortalized, continuous cell lines are established with relative facility from the involved tissue of patients with ATLL. Thus, the large number of reported ATLL cell lines are not described here in detail. A panel of selected cell lines is presented in Table VII.

OTHER VIRUSES

Human herpesvirus-8

HHV-8 (also termed Kaposi's sarcoma-associated herpesvirus, KSHV) was first discovered in 1994[24]. Herpesviruses have been divided into three major subgroups, termed alpha, beta and gamma, on the basis of their general biological properties *in vivo* and *in vitro*. HHV-8 appears to be the first human member of the gamma-2 herpesviruses (genus *Rhadinovirus*). Detection of HHV-8 in lymph nodes and peripheral blood B-cells from patients suggested that it is a lymphotropic herpesvirus. Sequence analysis showed partial homology to EBV[25].

So far, in panels of several hundred neoplasms of T- and B-cell origin, HHV-8 has only been detected *in vivo* in a unique type of B-cell lymphoma which was designated primary effusion lymphoma (PEL) or body cavity-based lymphoma[26,27]. A spectrum of lymphoma cell lines has been established from patients with PEL (Section II, Part 2). Only these PEL cell lines are HHV-8$^+$ (often together with EBV) while more than 100 human leukemia-lymphoma cell lines other than PEL cell lines were found to be HHV-8$^-$ [28]. Thus, also *in vitro*, HHV-8 is selectively associated with PEL-derived cells.

Hepatitis B virus, hepatitis C virus and human immunodeficiency virus

These viruses are not found in 'wild-type' (unmanipulated) human hematopoietic cell lines. Using DNA PCR and RT-PCR, we have examined 118 leukemia-lymphoma cell lines and 12 B-LCLs for HBV, HCV and HIV. All 130 cell lines were unequivocally negative for these viruses[15,29]. Some T- and monocytic cell lines were reported to be permissive for HIV replication under certain experimental conditions (e.g. immature T-cell line JURKAT, mature T-cell line H9 which is a subclone of line HUT 78, and monocytic cell line U1 which is a subclone of line U-937)[30–34].

The constitutive occurrence of viral infection in leukemia-lymphoma cell lines and in normal hematopoietic cell lines is summarized in Table 7.

Table 7 *Viral infection of leukemia-lymphoma cell lines and normal hematopoietic cell lines*

Virus	Cell type/derivation of infected cell lines	Typical example
EBV	all normal B-cell lines (B-LCLs)	
	most African Burkitt's lymphoma cell lines	Raji
	rare American Burkitt's lymphoma cell lines	SU-Amb-2
	some mature B-leukemia cell lines (B-CLL, B-PLL, HCL)	MEC1
	some (oriental) NK leukemia-lymphoma cell lines	NK-YS
	rare T-leukemia cell lines	RPMI 8402
HBV	all cell lines constitutively negative	
HCV	all cell lines constitutively negative	

Continued overleaf

Table 7 *Continued*

Virus	Cell type/derivation of infected cell lines	Typical example
HHV-8	only unique B-NHL (PEL) cell lines	BC-1
HIV	all cell lines constitutively negative	
	some immature T-ALL cell lines are permissive	JURKAT
	some monocytic leukemia cell lines are permissive	U-937
HTLV-I	normal T-cell lines	
	rare mature T-lymphoma (CTCL) cell lines	HUT 102
	most ATLL cell lines	MT-1
HTLV-II	rare mature T-leukemia cell lines	Mo-T

References

1 Drexler, H.G. and Matsuo, Y. (1999) Leukemia 13, 835–842.
2 Drexler, H.G. et al. (1999) Leukemia 13, 1601–1607.
3 Benyesh-Melnick, M. et al. (1963) J. Natl. Cancer. Inst. 31, 1311–1325.
4 Moore, G.E. and Minowada, J. (1969) In Vitro 4, 100–114.
5 Nilsson, K. and Pontén, J. (1975) Int. J. Cancer 15, 321–341.
6 Kieff, E. (1996) In: Fields Virology, 3rd edn (Fields, B.N. et al. eds). Lippincott-Raven, Philadelphia, pp. 2343–2396.
7 Jernberg, H. et al. (1987) Blood 69, 1605–1612.
8 Nilsson, K. et al. (1977) Int. J. Cancer 19, 337–344.
9 Ben-Bassat, H. et al. (1977) Int. J. Cancer 19, 27–33.
10 Drexler, H.G. and Matsuo, Y. (2000) Leukemia Res. (in press).
11 Melo, J.V. et al. (1988) Clin. Exp. Immunol. 73, 23–28.
12 Jaffe, E.S. (1996) Blood 87, 1207–1210.
13 Oshimi, K. (1996) Int. J. Hematol. 63, 279–290.
14 Kanegane, H. et al. (1998) Leukemia Lymphoma 29, 491–498.
15 Drexler, H.G. et al. (1999) DSMZ Catalogue of Cell Lines, 7th edn, Braunschweig, Germany.
16 Aboud, M. et al. (1987) Blood 70, 432–436.
17 Miyoshi, I. et al. (1981) Nature 294, 770–771.
18 Cann, A.J. and Chen, I.S.Y. (1996) In: Fields Virology, 3rd edn (Fields, B.N. et al. eds). Lippincott-Raven, Philadelphia, pp. 1849–1880.
19 Poiesz, B.J. et al. (1980) Proc. Natl. Acad. Sci. USA 77, 6815–6819.
20 Poiesz, B.J. et al. (1980) Proc. Natl. Acad. Sci. USA 77, 7415–7419.
21 Uchiyama, T. et al. (1977) Blood 50, 481–492.
22 Hinuma, Y. et al. (1981) Proc. Natl. Acad. Sci. USA 78, 6476–6480.
23 Yoshida, M. et al. (1982) Proc. Natl. Acad. Sci. USA 79, 2031–2035.
24 Chang, Y. et al. (1994) Science 266, 1865–1869.
25 Schulz, T. et al. (1998) In: Human Tumor Viruses (McCane, D.J. ed). ASM Press, Washington, pp. 87–134.
26 Gaidano, G. et al. (1997) Leukemia Lymphoma 24, 257–266.
27 Drexler, H.G. et al. (1998) Leukemia 12, 1507–1517.

28 Uphoff, C.C. et al. (1998) Leukemia 12, 1806–1809.
29 Uphoff, C.C. and Drexler, H.G. (1999) unpublished data.
30 Banerjee, R. et al. (1992) Proc. Natl. Acad. Sci. USA 89, 9996–10000.
31 Popovic, M. et al. (1984) Science 224, 497–500.
32 Popovic, M. et al. (1984) Lancet ii, 1472–1473.
33 Mann, D.L. et al. (1989) AIDS Res. Hum. Retroviruses 5, 253–255.
34 Folks, T.M. et al. (1987) Science 238, 800–802.

4 Guidelines for Characterization of Cell Lines

DESCRIPTION OF NEW CELL LINES

We have discerned six cardinal requirements for the description and publication of new leukemia-lymphoma cell lines (Table 8)[1,2].

Table 8 *Cardinal requirements for new leukemia-lymphoma cell lines*

- Immortality of cells
- Verification of neoplasticity
- Authentication of derivation
- Scientific significance
- Characterization of cells
- Availability to other scientists

Modified after refs. 1,2.

Immortality of cells

A 'continuous cell line' should be grown in permanent uninterrupted culture for at least 6 months, even better for more than a year. This is an important aspect as upon addition of growth factors, primary neoplastic cells or normal cells can sometimes be kept in culture for several months before proliferation ceases. Such cultures cannot be regarded as 'continuous cell lines'. Furthermore, cells may live but not grow *in vitro* (for example B-cell lymphoma and myeloma cells may survive with virtually no proliferation for up to 6 months). Thus, it is essential that the time period of continuous culture be indicated for permanently proliferating cells. Continuous cell lines have been defined as cultures that are apparently capable of an unlimited number of population doublings (immortalization) (Table 2). It should further be recognized that an immortalized cell is not necessarily one that is neoplastically or malignantly transformed[3]. As pointed out above, a cell line established from a patient with a tumor is not necessarily a tumor cell line (Chapter 3).

Verification of neoplasticity

The neoplastic nature of the cell line should be demonstrated by functional assays or by the detection of clonal cytogenetic abnormalities. Given the overwhelming preponderance of karyotypically abnormal leukemia-lymphoma cell lines (>99%), the detection of an abnormal karyotype in a cell line can be regarded as a necessary (though not sufficient) condition of neoplasticity. Colony formation in methylcellulose or agar (clonogenic assays) is also considered an operational test of neoplasticity. Finally, xenotransplantability of cell lines into immunodeficient mice is often taken as a sign of the malignant nature and furthermore indicates the immortality of the transplanted cells. However, in long-term culture most EBV$^+$ B-LCLs will eventually become monoclonal and aneuploid, grow as colonies in semi-solid media, and form tumors in immunodeficient mice (Chapter 3). Therefore, the criteria listed for 'neoplasticity' are valid only when an EBV^+ B-LCL has been excluded.

Authentication of derivation

The cellular origin of a new cell line must be proven by authentication, i.e. it must be shown that the cultured cells are indeed derived from the presumed patient´s tumor and are not the result of a cross-contamination with an older cell line. We estimate that about 10–20% of human leukemia-lymphoma cell lines are misidentified or cross-contaminated by the original investigator (see also Chapter 6)[4,5]. These cell lines are clearly 'false cell lines'. A list of known and proven 'false cell lines' is presented in Table XVI. However, it is likely that this may only be the tip of the iceberg, as commonly cell lines are passed from laboratory to laboratory without any proper identity control. The method of choice for identity control is forensic-type DNA fingerprinting[6]. Microsatellite analysis does not appear to be sufficient as the loci seem to be prone to instability in certain tumor types. Immunophenotyping will not suffice either as cell lines of the same category will often have similar if not identical immunoprofiles. The presence of unique cytogenetic marker chromosomes or molecular biological data (e.g. identical clonal gene rearrangement patterns on Southern blots) might also provide unequivocal evidence for the derivation of the cell line from the corresponding patient.

Scientific significance and characterization of cells

With regard to novelty and scientific significance, the new cell line should carry features not yet detected in previously established cell lines. Thus, the scientific significance of a new cell line depends on the degree of its characterization. A thorough multi-parameter analysis of the cells (Table 9) will often unveil unique characteristics of cell lines attesting to their scientific importance.

Availability to other scientists

The sharing of cell lines with other scientists is of the utmost importance. Some scientific journals have adopted the policy that any readily renewable resources, including cell lines published in that journal, must be made available to all qualified investigators in the field, if not already obtainable from commercial sources. The policy stems from the long-standing scientific principle that authenticity requires reproducibility. While cell lines are proprietary and unique, suitable material transfer agreements can be drawn up between the provider and requester[8]. By providing authenticated and unique biological material, cell line banks play a major role in this regard (Chapter 8)[9,10]. Thus, authors are encouraged to deposit their cell lines in non-profit reference cell line collections.

CHARACTERIZATION OF CELL LINES

Detailed characterization of cellular features

Since leukemia-lymphoma cell lines commonly grow as single or clustered cells in suspension or only loosely adherent to the flask, single cell populations can be

Table 9 *Analytical characterization of leukemia-lymphoma cell lines*

Parameter	Details and examples
Most important data:	
Clinical data:	patient's data (see Table 10)
In vitro culture:	growth kinetics, proliferative characteristics (see Table 10)
Immunophenotyping:	surface marker antigens (fluorescence microscopy, flow cytometry) intracytoplasmic and nuclear antigens (immunoenzymatic staining)
Cytogenetics:	structural and numerical abnormalities specific chromosomal rearrangements
Further characterization:	
Morphology:	in-situ (flask, plate) under inverted microscope light microscopy (May-Grünwald-Giemsa-staining) electron microscopy (transmission and scanning)
Cytochemistry:	acid phosphatase, α-naphthyl acetate esterase, others
Genotyping:	Southern blot analysis of T-cell receptor (TCR) and immunoglobulin (Ig) heavy and light chain gene rearrangements Northern analysis of expression of TCR and Ig transcripts
Cytokines:	production of cytokines expression of cytokine receptors response to cytokines, dependency on cytokines
Functional aspects/specific features:	phagocytosis antigen presentation immunoglobulin production/secretion (hemo)globin synthesis capacity for (spontaneous or induced) differentiation positivity for EBV or HTLV-I or other viruses heterotransplantability into mice or other animals colony formation in agar/methylcellulose – clonogenicity production/secretion of specific proteins NK cytotoxic activity – ADCC oncogene expression transcription factor expression unique point mutations
Date of analysis:	age of cell line at time of analysis possible changes in the specific marker profile during prolonged culture

Adapted from refs. 1,2,7.

easily prepared and the cells can thus be characterized. Table 9 lists a variety of parameters useful for the description of the cells and a panel of possible tests applicable for the phenotypic and functional characterization of most cell lines. This necessary multiparameter examination of the cellular phenotype provides important information on the likely cell of origin, the variable stringency of maturation arrest, and any discrepancies in the pattern of normal gene expression. The list is not intended to cover comprehensively all possible informative parameters; with new techniques becoming available and research areas extending to new avenues, other or entirely new features might be of interest to scientists. Thus, only some of the features of the phenotypic profiles of cell lines that are most often studied are highlighted. It is also important to indicate when in the life of a cell line individual data were generated and also whether alterations in the phenotypic features of the cells might occur during prolonged culture.

Core data

Immunophenotypic analysis and cytogenetic karyotyping currently appear to be the most important and informative examinations (Table 9). While the scope and extent of the analytical characterization of leukemia-lymphoma cell lines is certainly variable, a core data set is obligatory and essential for the identification, description and culture of a cell line. These data include the clinical and cell culture description of the cell line (an example for the presumably most often used human leukemia cell line HL-60 is given in Table 10). Clearly, the origin of an established cell line must be sufficiently documented. However, a cardinal principle exists requiring that the informed consent of the donor of the malignant cell line be obtained and that the anonymity of the donor be upheld. It is important to check for mycoplasma contamination and to indicate whether the cells are EBV$^+$.

Sister cell lines and subclones

Subclones of any given cell line and sister cell lines must be distinguished and properly presented (Table 2). Serial or longitudinal sister cell lines (e.g. established at diagnosis and relapse of the same patient) may provide unique opportunities to study the molecular mechanisms involved in disease progression and transformation[11]. Of particular interest also are pairs of cell lines consisting of one cell line with diseased cells and one cell line with normal, albeit EBV-transformed cells from the same patient (which, as a bonus, provide reliable data allowing for subsequent authentication).

SYNOPSIS OF PRINCIPAL CELL LINE FEATURES

A summary of the salient characteristics of a new cell line are of great benefit. The entries in Sections II–VIII provide illustrative examples of such summary tables.

Table 10 *Clinical and cell culture data for leukemia-lymphoma cell lines*

Parameter	Example
Cell line:	
Name of cell line:	HL-60
Cell phenotype:	myelocytic cell
Clinical data:	
Original disease of patient:	initially AML M3, later corrected to AML M2
Disease status:	at diagnosis
Patient data (age, race, sex):	35-year-old Caucasian woman
Source of material:	peripheral blood
Year of establishment:	1976
Cell culture data:	
Culture medium:	90% RPMI 1640 + 10% FBS
Culture environment:	at 37°C with 5% CO_2 in air
Subcultivation routine:	maintain at 0.1–0.5 × 10^6 cells/ml; split ratio 1:2 to 1:5 every 1–2 days
Initial seeding:	at c. 1.0 × 10^6 cells/ml
Minimum cell density:	0.5–1.0 × 10^5 cells/ml
Maximum cell density:	1.5–2.0 × 10^6 cells/ml
Doubling time:	25–30 hours
Cell storage conditions:	70% RPMI 1640 + 20% FBS + 10% DMSO
In situ morphology:	round, single cells in suspension
Mycoplasma contamination:	negative (checked with PCR)
Viral status:	EBV−

Adapted from refs. 2,10.

References

1 Drexler, H.G. et al. (1998) Human Cell 11, 51–60.

2 Drexler, H.G. and Matsuo, Y. (1999) Leukemia 13, 835–842.

3 Schaeffer, W.I. (1990) In Vitro Cell. Dev. Biol. 26, 97–101.

4 Drexler, H.G. et al. (1999) Leukemia 13, 1601–1607.

5 MacLeod, R.A.F. et al. (1999) Int. J. Cancer 83, 555–563.

6 Dirks, W.G. et al. (1999) Cell. Mol. Biol. 45, 841–853.

7 Drexler, H.G. et al. (1994) In: Atlas of Human Tumor Cell Lines (Hay, R.J. et al., eds). Academic Press, Orlando, FL, pp. 213–250.

8 Kaushansky, K. (1998) Blood 91, 1–2.

9 Hay, R.J. et al. (1996) J. Cell. Biochem. 24, 107–130.

10 Drexler, H.G. et al. (1999) DSMZ Catalogue of Cell Lines, 7th edn, Braunschweig, Germany.

11 Zhang, L.Q. et al. (1993) Leukemia 7, 1865–1874.

5 Establishment and Culture of Cell Lines

ESTABLISHMENT OF CELL LINES

Introduction

It is still extremely difficult to establish new leukemia-lymphoma cell lines, and the majority of attempts fail. Seeding of neoplastic cells directly into suspension cultures is a common procedure in attempts to establish cell lines. The success rate for the establishment of continuous cell lines is low, and success is basically unpredictable. Table 11 gives an overview of the success rates for establishing cell lines from various types of hematopoietic tumors as reported in the literature. The success rates vary strongly between the various studies. However, the malignant nature of the derived cell lines and the immortalization of the alleged cell lines have not been demonstrated in all instances.

Table 11 *Success rates for establishing leukemia-lymphoma cell lines in various studies*

Disease	Success rate	Comments	Ref.
Precursor B-cell lines:			
BCP-ALL	3/30	pediatric peripheral blood/bone marrow samples	1
BCP-ALL	20/123	pediatric cases; unique method	2
BCP-ALL	21/150	pediatric cases	3
BCP-ALL	16/125	pediatric cases; unique method	4,5
BCP-ALL	9/42	adult cases	5
Mature B-cell lines:			
B-CLL	1/23	deliberate EBV infection	6
B-PLL	2/6	deliberate EBV infection	7
B-CLL	1/10		8
B-CLL/B-PLL	4/72	deliberate EBV infection; 10/95 EBV+ B-LCLs	9
HCL	2/7	deliberate EBV infection	10
B-CLL	0/12	deliberate EBV infection	11
B-PLL	6/6	deliberate EBV infection	11
HCL	2/2	deliberate EBV infection	11
B-CLL	1/5	deliberate EBV infection	12
B-CLL	3/37	deliberate EBV-MYC-RAS in-/transfection	13
B-ALL	5/16	adult cases	5
Plasma cell lines:			
Myeloma	2/80	peripheral blood samples	14
Myeloma	6/71	in liquid suspensions; 14/71 colony growth in semi-solid media	15
Myeloma	0/100	bone marrow samples; in 70% overgrowth by EBV+ B-LCLs	16

Continued overleaf

Table 11 *Continued*

Disease	Success rate	Comments	Ref.
Myeloma	2/73		17
Myeloma	4/40		18
Myeloma	2/95	5/95 EBV+ B-LCLs	19
Myeloma	10/10	extramedullary samples; with IL-6 + GM-CSF	20
Myeloma	0/11	bone marrow samples	20
Immature T-cell lines:			
T-ALL	2/7	pediatric peripheral blood/bone marrow samples	1
T-ALL	15/18	pediatric cases; unique method	21
T-ALL	5/9	5/6 with chromosomal alterations	22
T-ALL	3/25	pediatric cases; unique method	4,5
Mature T-cell lines:			
CTCL	0/32	6/32 EBV+ B-LCLs	23
ATLL	10/11	HTLV-I+	24
T-CLL	1/8		25
Myelocytic-monocytic cell lines:			
AML	3/13	pediatric peripheral blood/bone marrow samples	1
Erythrocytic-megakaryocytic cell lines:			
AML M7/ MDS	1/21		26
Various cell lines:			
ALL/NHL	3/64	2 BCP cell lines, 1 immature T-cell line, 5 EBV+ B-LCLs	27
NHL	16/69	12 mature B-cell lines, 2 mature T-cell lines, 2 EBV+ B-LCLs	28
ALL	12/26	5 BCP cell lines, 4 T-cell lines, 3 undetermined cell lines	29,30
Normal cell lines:			
CML	4/22	Ph+ EBV+ B-LCLs	31

Reports are arranged in chronological order in each cell type category. Some of the cell lines (in particular the B-cell lines) are not always proven to be derived from malignant cells and may thus represent EBV+ B-LCLs; other 'cell lines' may not be continuous cell lines.

It has been purported that precursor B-cells derived from patients at relapse or from cases with poor prognostic features have an enhanced growth potential *in vitro* in comparison to samples obtained at presentation or from children with good prognostic parameters[2,4,27]. However, this notion is not supported by the results of other studies[1,5,30]. A recent review reported that on aggregate the success rate for BCP-cell lines established from patients at diagnosis was 6%, whereas the success rate was 29% for relapse samples[5]. There seems to be a higher success rate in cases which have *a priori* certain chromosomal aberrations or gene mutations.

This notion has been confirmed in T-cell lines carrying an (8;14)(q24;q11) translocation and in a variety of cell lines carrying alterations of the *P53*, *P15INK4B* or *P16INK4A* genes[22,32–34].

One particular cell culture method promoted by Smith *et al.* deserves further comment: when leukemic T-cells were cultured in wells with a feeder layer consisting of complete media, human serum and agar in an hypoxic environment with supplemental insulin-like growth factor I, 10 out of 12 frozen samples and five out of six fresh samples grew and were established as cell lines[21,35]. However, when BCP-ALL cells were cultured under these conditions, rapid cell death occurred concomitant with the outgrowth of normal mononuclear cells. Upon applying the monocyte toxin l-leucine methyl ester and insulin instead of IGF-I, the success rate increased from 11% to 40%[2]. It should be noted that some of these BCP-cell lines have extremely long doubling times (10–14 days), which clearly limits their usefulness. The reproducibility of this method in other hands and the long-term growth (immortalization) of these cultures is not known.

The reasons for the frequent failure to establish cell lines remain unclear. According to general experience, the major causes appear to be culture deterioration with cessation of multiplication of the neoplastic cells and overgrowth by fibroblasts, macrophages or lymphoblastoid cells. While the lymphoblastoid cells may give rise to a continuous cell line (an EBV^+ B-LCL), human fibroblast and macrophage cultures are commonly not immortalized.

Despite the fact that the proliferation of malignant hematopoietic cells *in vivo* seems to be independent of the normal regulatory mechanisms, these cells usually fail to proliferate autonomously *in vitro* even for short periods of time. *In vivo*, at least initially, these cells seem to require one or probably several hematopoietic growth factors for proliferation. The addition of regulatory proteins, e.g. so-called hematopoietic growth factors such as erythropoietin (EPO), granulocyte colony-stimulating factor (G-CSF), granulocyte-macrophage CSF (GM-CSF), interleukin-2 (IL-2), IL-3 or IL-6, or stem cell factor (SCF), mitogens such as phytohemagglutinin (PHA), or conditioned medium (CM) secreted by certain tumor cell lines (which often contains various factors), is a culturing technique that appears to increase the frequency of success by overcoming the 'crisis' period in which the neoplastic cells cease proliferating. These molecules enable the leukemia cells from the majority of patients to multiply for about 2–4 weeks. Out of these short-term cultures a few continuous cell lines which are derived from the malignant cells can be established.

Taken together, the efficiency of cell line establishment is still rather low and the deliberate establishment of new leukemia-lymphoma cell lines remains by and large an unpredictable random process. Clearly, difficulties in establishing continuous cell lines may be due to the inappropriate selection of nutrients and growth factors for these cells. Thus, a suitable microenvironment for hematopoietic cells, either malignant or normal, cannot yet be created *in vitro*. Further work is required to achieve significant improvements in the success rates of leukemia-lymphoma cell line immortalization.

In the following, some of the more promising techniques for establishing new leukemia-lymphoma cell lines are described. Table 12 provides a statistical insight into three clinical and cell culture parameters of the well-characterized cell lines listed in Sections II–VIII that might have some influence on the success rate: choice of culture medium, specimen site of the primary cells, and status of the patient at the time of sample collection.

Table 12 *Statistics on leukemia-lymphoma cell line establishment: media, specimen sites, treatment status*

Culture medium:		
RPMI 1640	315/432	73%
IMDM	52/432	12%
McCoy's 5A	29/432	7%
α-MEM	21/432	5%
Others [a]	15/432	3%
Specimen site:		
Peripheral blood	200/417	48%
Bone marrow	101/417	24%
Pleural effusion	56/417	13%
Ascites	23/417	5%
Tumor	13/417	3%
Lymph node	12/417	3%
Cerebrospinal fluid	4/417	1%
Pericardial effusion	3/417	1%
Spleen	2/417	1%
Liver	1/417	
Meninges	1/417	
Tonsil	1/417	
Treatment status:		
At relapse/refractory/terminal	210/395	53%
At diagnosis/presentation	139/395	35%
At blast crisis	39/395	10%
During therapy	7/395	2%

Literature data were not available in all categories for all cell lines. Counted were only data on well-characterized cell lines (without sister cell lines or subclones).

[a] Other media used: Cosmedium, DMEM, Eagle's MEM, Fischer's, Ham's F10, Ham's F12, L-15.

Methods

The most commonly used specimens are peripheral blood or bone marrow samples as these are relatively easily obtained from patients. Other liquid specimens (such as pleural effusion, ascites, cerebrospinal fluid, etc.) and solid tissue samples (lymph node, tonsil, spleen, etc.) are less often used but can be processed in a similar way to peripheral blood or bone marrow samples. Solid tissues require a prior step whereby the tissue is dissociated mechanically with scissors, scalpel, mesh or similar means. In most instances, the cells are immediately processed upon receipt. However, cryo-preserved specimens can also be used. It appears to be of advantage to isolate the mononuclear cells prior to cryo-preservation. All solutions and utensils must be sterile. Work in a laminar flow cabinet (class II) under sterile conditions is recommended.

The heparinized or otherwise anti-coagulated peripheral blood, bone marrow or other samples containing cells in suspension are diluted with culture medium (the most commonly used media are listed in Table 12) at a ratio of 1:2. Isolation of cells from a leukapheresis collection requires dilution of the sample with culture medium at 1:4. The mixture of medium and sample is layered slowly over a Ficoll-Hypaque density gradient solution (density 1.077 g/L). Equal volumina of sample mixture and Ficoll-Hypaque solution should be used. Following centrifugation for 20–30 min at 450 × **g** at room temperature (with the centrifuge brakes switched off), a layer containing mononuclear cells is visible on top of the Ficoll-Hypaque phase as these cells have a lower density than the Ficoll-Hypaque solution. The anucleated erythrocytes and the polynucleated granulocytes are concentrated as pellet below the Ficoll-Hypaque layer. The interface layer is harvested and the cells are washed twice with culture medium plus at least 2% FBS (centrifugation for 5–10 min at 200 × **g** at room temperature). The FBS should be inactivated prior to use in a 56°C waterbath for 30–45 min.

The washed mononuclear cells are resuspended in culture medium with 20% FBS plus any supplements (e.g. 10% CM from cytokine-producing cell lines, recombinant growth factors, etc.). The cells are counted, and the viability is determined by Trypan blue vital staining. The final cell concentration is adjusted to 2–5 × 10^6 cells/ml. Cell yields depend on the number of malignant cells in the original specimen and are highly variable from patient to patient, in the case of peripheral blood being correlated obviously with the white blood cell count.

There are various sizes of plastic culture vessels: flasks (for example 25 cm^2, 80 cm^2, 175 cm^2) or plates (with 12-, 24-, 96-wells). The number of flasks or wells used depends on the number of leukemia-lymphoma cells available. As many of the malignant cells as possible should be used in attempts to establish a cell line. In theory, a cell line starts from one single cell. Thus, the more attempts, the higher the chances. It is absolutely mandatory to freeze aliquots of the original cells and to store them in appropriate locations for later documentation, authentication and comparisons. Most cell lines were established using this 'direct method' of placing cells into liquid culture and incubating in a humidified incubator at 37°C and 5% CO_2 in air. Alternative approaches are inoculation in semi-solid media (methylcellulose, soft agar), initial heterotransplantation and serial passage in immunodeficient mice with subsequent adaptation to *in vitro* culture, or culture on temporary feeder layer (e.g. on fibroblasts).

In suspension cultures, the cells may be expanded by exchanging half of the spent culture volume with culture medium plus 20% FBS plus 10% CM (or with appropriate concentrations of growth factors) once a week. After 4 h, some cells become adherent. These adherent cells are the source of colony-stimulating factors for both normal and leukemia cells. During the first 2 weeks, it is not necessary to remove the adherent cells from the culture unless there is a specific reason to do so, for example because of the addition of a purified growth factor to the medium in order to obtain a unique type of cell line. After 2 weeks, if the suspension cells grow very rapidly, the adherent cells can be removed simply by transferring the suspension cells into new culture vessels in order to reduce the potential for overgrowth of fibroblasts and normal lymphoblastoid cells.

During the first weeks, the neoplastic cells may appear to proliferate actively. If the medium becomes acidic quickly (yellow in the case of RPMI 1640 medium), it is necessary to change half of the volume of medium at 2- to 3-day intervals. If the number of the cells increases rapidly, the cells are readjusted weekly to a concentration of at least 1×10^6/ml in fresh complete medium by dilution or subdivision into new flasks. The neoplastic cells from the majority of the patients with leukemia-lymphoma undergo as many as four doublings in 2 weeks, but after 2–3 weeks most malignant cells cease proliferating. Following a lag time of 2–4 weeks (crisis period), a small percentage of cells from the total population may still proliferate actively and may continue to grow forming a cell line.

If the malignant cells continue to proliferate for more than 2 months, there is a high possibility of generating a leukemia-lymphoma cell line. In these cases, the work of characterizing the proliferating cells should be begun as soon as possible. Prior to the characterization of the cells, it is necessary to freeze ampoules of the proliferating cells containing a minimum of 3×10^6 cells/ampoule in liquid nitrogen in order to avoid loss of the cells due to occasional contaminations or other accidents.

Limiting dilution leads to the generation of monoclonal cell lines. After prolonged culture *in vitro*, the cell line will become oligoclonal or monoclonal due to the outgrowth of selected cell clones. In most cases, it is not absolutely necessary to subclone the cell line by limiting dilution. In some types of leukemia cell lines, e.g. immature T- and precursor B-cell lines, it might be very difficult or virtually impossible to 'clone' the cells.

For various reasons, a newly established leukemia-lymphoma cell line needs to be characterized: (1) to determine that the cell line is derived from the original primary cells; (2) to investigate whether the established cell line was transformed by viruses such as EBV or HTLV-1/-2; and (3) to identify the basic biological features of the cultured cells.

CHARACTERIZATION OF NEW CELL LINES

Newly established cell lines need to be categorized and characterized. The cardinal requirements for the description and publication of new leukemia-lymphoma cell lines and examples of the detailed characterization of the cells are discussed in detail in Chapter 4.

MAINTENANCE AND EXPANSION OF CELL LINES

Introduction

The procedure outlined below refers to cell lines growing in standard culture media supplemented with FBS. Other cell cultures that require special media and unique growth supplements can be maintained similarly, applying the necessary changes to the culture system.

Materials and methods

All solutions and utensils that come into contact with cells must be sterilized prior to use and sterile techniques must be applied throughout the procedure. Commonly leukemia-lymphoma cell lines are stored frozen in liquid nitrogen (see below). Cell lines may also be kept as active long-term culture. Frozen cells must be thawed carefully in order to minimize cell loss.

After removal of the frozen ampoule from liquid nitrogen, the cells should be thawed rapidly in a 37°C waterbath by gently shaking the ampoule in the water. It is important that the frozen cell solution be thawed in about 1 min. Rapid warming is necessary so that the frozen cells pass quickly through the temperature zone between −50°C to 0°C where most cell damage is believed to occur. Slow thawing will harm the cells by formation of ice crystals in the cells, causing hypertonicity and destruction of cellular organelles.

The vial content is diluted by the slow addition of about 10 ml culture medium plus 10% FBS. Cells frozen with dimethylsulfoxide (DMSO) are usually dehydrated. During the washing steps with medium, water will diffuse into the cells. Diluting the suspension slowly is thought to reduce the loss of electrolytes, to counteract extreme pH changes, and to prevent denaturation of cellular proteins.

Cells are washed twice by centrifugation and resuspended in complete medium at a general cell concentration of *c.* 0.2–1.0 × 10^6 cells/ml. It appears that most leukemia-lymphoma cell lines grow better at higher cell concentrations than at lower ones. Some cell lines (e.g. some precursor B-cell lines) prefer a concentration higher than 1.0 × 10^6/ml. Usually, the optimal concentration of a cell line for expansion must be explored empirically. If after 2–3 days of culture the cells do not grow well (perhaps due to the presence of many dead cells), it might be useful to concentrate the cells and to culture them at a higher cell density. It is recommended that the cells be resuspended first in medium containing 20% FBS; should the cells start to multiply and resume their expected growth activity, the percentage of FBS can be decreased stepwise.

Suspension cell lines may be cultured in flasks or 24-well plates. General recommendations are 5–10 ml suspension into a 50 ml flask, 20–40 ml suspension into a 260 ml flask, or 1–2 ml suspension into each well of a 24-well plate. When dealing with 'difficult' cell lines, it may be advantageous to suspend some cells in a flask and another aliquot in a 24-well plate (or even a 96-well microplate). There are distinct differences between flask and plate regarding exposure to CO_2, accessibility to microscopic observation, and possibilities of manipulation. Human cell lines are incubated in a humidified 37°C incubator with 5% CO_2 in air. The top of the flask is slightly loosened to allow for free gaseous exchange into and out of the flask.

The cells are fed by exchanging half of the culture volume with culture medium plus FBS at 2- to 3-day intervals. If the cells proliferate actively, the culture medium will soon change color due to a pH change caused by cellular metabolism. In this case, it is necessary to change the medium more frequently. Should the cells have doubled, the culture may be subdivided from the original flask into a second flask by diluting the suspension 1:2 with new medium. When changing the medium, it is important to calculate the total cell number by determining the density as well as the viability of the cells using Trypan blue dye exclusion. A careful documentation of all manipulations, macroscopic and microscopic observations, intentional and accidental changes in the cellular conditions, and data on cell density, viability and total cell number at different time points is mandatory.

The cell lines can be maintained as long as required. The cells can be harvested at any time for different uses. If the cells proliferate more quickly than needed, the cell growth can be kept at a slower pace by decreasing the FBS to a lower percentage in the medium, changing the medium at longer intervals, or discarding a certain amount of the cells (up to 75%) during the exchange of medium. There is a fundamental difference between 'expansion' and 'maintenance' of a leukemia-lymphoma cell line. Some cell lines will deteriorate over long-time culture under maintenance conditions. In such cases, it might be better to freeze and rethaw the cells when needed. As after longer usage culture flasks will often contain a certain amount of unused ingredients of the medium, metabolized molecules and cell detritus, it is recommendable that the plastic flask be changed once every 1–2 months.

STORAGE OF CELL LINES

Introduction

It is generally assumed that leukemia-lymphoma cell lines can be stored at −196°C in liquid nitrogen for more than 10 years without any significant changes in their biological features. The viable cell lines can be recovered at any time when needed. Prior to freezing in liquid nitrogen, the cells are suspended traditionally in the appropriate medium containing 20% FBS and 10% DMSO, which can lower the freezing point in order to protect the frozen cells from damage caused by ice crystals. Glycerol has been used as an alternative to DMSO. It appears that no single suspending medium and procedure will be ideal for processing and cryogenic storage of all cell cultures. However, the procedures described here are suitable for most of the human leukemia-lymphoma cell lines. When the procedure is carried out properly, it seems to be compatible with prolonged preservation of viability and other characteristics of the cell lines.

Methods

It is important that the cells be harvested in the logarithmic growth phase. It must be remembered that freezing and storing the cells will not improve the status and quality of the cell culture prior to freezing; at best, the status quo will be preserved, but most often the condition of the cells will be diminished to various degrees.

The freezing medium consists of RPMI 1640 medium containing 20% FBS and 10% DMSO. The total cell number is determined using Trypan blue dye solution, the cells are harvested by centrifugation, and the supernatant is discarded. The freezing medium should be added quickly to the cells. Long-time exposure to DMSO at room temperature can trigger significant cellular changes such as activation and so-called induced differentiation. Keeping cells in DMSO-containing media on ice could minimize the effect of DMSO on the cells. It is not necessary to sterilize the DMSO solution as pure DMSO is lethal to bacteria.

The cells are distributed into freezing ampoules (plastic cryo vials) with 1 ml per ampoule, containing at least 5×10^6 cells. The ampoules must be properly labeled with the name of the cell line and the date of freezing. More cells per ampoule can be frozen if needed: depending on the cell type up to 50×10^6/ml/ampoule.

The freezing process can be performed in a computer-controlled cooling apparatus (cryo freezing system). An initial cooling rate of 1°C per minute appears to be optimal. When the temperature reaches −25°C, the cooling rate can be increased to 5–10°C per minute. When the temperature of the specimen reaches −100°C, the ampoules can be transferred quickly to a liquid nitrogen container for storage. Permanent storage should be in the liquid phase of the liquid nitrogen. Alternatively, if only a few ampoules are frozen, the sealed ampoules can be placed in a plastic box with an inset for ampoules; the box is half filled with isopropanol and stored in a −70°C freezer for at least 4 h. The ampoules must later be transferred quickly to liquid nitrogen. With this method, a 1°C per minute cooling rate can be achieved as well. Long-term storage of cell lines at −70°C (beyond 1 week) cannot be recommended as the cells will die under these conditions.

FURTHER CONSIDERATIONS

General information

Although *in vivo* the malignant cells enjoy a selective growth advantage over normal hematopoietic cells, *in vitro* leukemia-lymphoma cells are so difficult to grow and to maintain that attempts to establish cell lines meet much more often with failure than with success. Although currently there is no one single cell culture system that assures consistent establishment of malignant hematopoietic cell lines, several methods for immortalizing neoplastic cells have been developed. The technique of seeding cells in suspension cultures as described in this chapter is certainly the most often used. Other methods recommended by several researchers have their advantages and might meet with success in some attempts.

Growth of leukemia-lymphoma cells in soft agar or methylcellulose offers the advantage that the colonies formed are well fixed and can be easily removed from the supporting medium for further culture in other environments. Thus, a cell line might be established by passaging single colonies. Some lymphocytic leukemia cells can be immortalized using transforming viruses. EBV can promote growth of malignant B-cell lines from some patients with B-CLL or other mature B-cell malignancies. But the EBV can also transform normal B-lymphocytes. HTLV-I and HTLV-II allow the growth of malignant T-cells by inducing the IL-2 receptor. Considering the fact that EBV and HTLV can also transform normal cells, it is necessary to ascertain the malignant origin of the established cell lines by means of karyotype and molecular genetic analysis and, possibly, the demonstration of the absence of the virus genome.

The growth of malignant and normal hematopoietic cells *in vitro* and *in vivo* is the result of complex interactions between growth factors and their respective receptors. The addition of some factors into the culture medium can support the proliferation of the neoplastic cells and induce the formation of cell lines. The most often used molecules are EPO, G-CSF, GM-CSF, SCF, TPO, and several interleukins (IL-2, IL-3, IL-5, IL-6). Another protocol used insulin-like growth factor 1 (IGF-1). The IGF-1-induced cell proliferation appeared to be restricted to a low oxygen environment and was blocked at high oxygen concentrations[2,35].

As purified or recombinant factors are expensive, the CM of some malignant human or murine cell lines can be used alternatively. Such cell lines are, for instance, 5637 (an adherent cell line from a patient with bladder carcinoma), Mo-T

(HTLV-II transformed T-leukemia cell line), HDLM-3 (Hodgkin's disease-derived cell line), and WEHI-3 (mouse monocytic cell line)[1,36–38]. These cell lines generate several growth factors (e.g. G-CSF, GM-CSF, IL-1, IL-3, IL-6, IL-9, SCF and others) that are secreted into the culture supernatant. The supernatant from cultures of these tumor cell lines can be stored at −20°C for several months prior to use. These CM should be used at a final concentration of 10–20% (vol). CM from PHA-stimulated lymphocytes is also an ideal and inexpensive source of these biomodulators.

A number of completely synthetic media such as RPMI 1640, Minimum Essential Medium (Eagle's MEM or α-MEM), Dulbecco's Modified Eagle's Medium (DMEM), Iscove's Modified Dulbecco's Medium (IMDM), Ham's F-10 and F-12, L-15, McCoy's 5A, and others, including several media designed specifically for unique types of leukemia cells, have been used by researchers for establishment and maintenance of cell lines in suspension cultures (Table 12). It appears that no single medium is well suited for the growth of all types of leukemia-lymphoma cells. Although the most widely used of these media is RPMI 1640 medium, which is usually employed together with 10–20% FBS, IMDM, McCoy's 5A and α-MEM are also often employed. Should one medium fail to support cell growth, it might become necessary to try another kind of medium.

FBS is the standard supplement in the suspension culture system of leukemia cell lines. It is commonly used in concentrations of 5–20%. Leukemia-lymphoma cell lines that do not require FBS in the culture medium have been described; however, they appear to be rather rare. Prior to usage, it is recommended that batches of serum be pretested for their ability to support vigorous cell growth and for viral (in particular for bovine viral diarrhea virus), mycoplasmal and other bacterial contamination. If possible, a large supply of the FBS from a pretested batch that supports cell growth well and has no contamination should be purchased and stored at −20°C for future use. Alternatives are newborn calf serum (NCS, usually at only 25% of the price of the expensive FBS) or serum-free media. However, not all cell lines will grow in NCS as well as in FBS. Although serum-free media do not provide financial advantages, they do allow for certain experimental manipulations that are not possible with FBS as the investigator is able to control all the substances to which the cells are exposed. FBS is known to contain many unidentified ingredients at variable concentrations.

Culture environment

A minimal amount of oxygen is essential for the growth of most types of leukemia-lymphoma cells in suspension cultures. The majority of cell lines grow well when they are incubated in a humidified 37°C incubator with 5% CO_2 in air. However, the partial pressure of oxygen (pO_2) in the normal body fluids is significantly less than that of air. Studies have shown that the pO_2 in human bone marrow is 2–5%. This is considerably lower than the pO_2 (15–20%) existing in the typical cell culture incubator maintained at 5% CO_2 in air. It has been reported that growth of cultured cells could be improved by reducing the percentage of oxygen in the gaseous phase to between 1 and 10%. Growing leukemia cells under low oxygen conditions of 6% CO_2, 5% O_2 and 89% N_2 may be a useful method for the establishment of leukemia-lymphoma cell lines[2,21,35].

Biological safety

When working with human blood or other tissues including fresh leukemia-lymphoma cells, established leukemia-lymphoma cell lines, and pathogenic and infectious agents, the biosafety practices must be followed rigorously. Some cell lines are virally infected: for example, EBV and HHV-8 are assigned to biological safety risk category 2; HTLV-I/-II and HIV fall into risk category 3. Fresh primary material may contain HBV or HCV (both in risk category 3).

Common cell culture problems

In the attempts to establish leukemia-lymphoma cell lines, the overgrowth of fibroblasts and normal EBV$^+$ lymphoblastoid cells is the most common problem (see also *Chapter* 3: EBV- and HTLV-Positive Cell Lines). Should the nutrients in the medium become exhausted too quickly, the adherent cells should be removed by passaging the suspension cells into new flasks containing fresh medium. The overgrowth of EBV$^+$ B-lymphocytes can become visible as early as 2 weeks after seeding of the new culture. The EBV$^+$ B-cells look small and have irregular contours with some short villi (Table 6). They proliferate preferentially in big floating clusters or colonies. The colonies can be picked out with a Pasteur pipette and the EBV genome should be detected as soon as possible.

It is always necessary to freeze aliquots (at least two ampoules) of the fresh primary cells in liquid nitrogen before culture. If a cell line should subsequently become established, the original cells can be used as a control for characterization of the established cell line. In case of failure to immortalize a cell line, the fresh cells can be used for another attempt.

Maintenance of cell lines requires careful attention. Every cell line appears to have its optimal growth environment; factors such as culture medium, cell density, nutrition supplements and pH all play a major role. If cell growth becomes suboptimal or cells inexplicably die during culture, some of the following problems should be considered: suitability of the culture medium or the growth supplements for this particular cell line; proper functioning of the incubator at the appropriate temperature, humidity and CO_2 levels; and selection of the adequate cell density. Some cell lines clearly grow better in 24-well plates than in culture flasks.

Contamination with mycoplasma, other bacteria, fungus, viruses and other 'foreign' cells is the most common problem encountered in the maintenance of leukemia-lymphoma cell lines. Therefore, analyses at regular intervals must be undertaken to ensure a contamination-free environment for cell growth. All solutions and utensils coming into contact with cells must be sterilized prior to use; sterile techniques and good laboratory practices must be followed strictly. Although antibiotics can be added to the culture medium to prevent bacterial infection, they do not usually inhibit virus, fungus or mycoplasma infection. Therefore, antibiotics such as penicillin and streptomycin are not necessary if care is taken regarding cell culture techniques. Because contamination can cause the loss of valuable cell lines, it is important to cryopreserve a sufficient amount of cell material from each cell line for future use. In order to prevent cellular contamination and misidentification, it is mandatory to use a separate bottle of medium for each cell line. Furthermore, cell culturists should not deal with and feed more than one cell line at the same time.

Further details on the two most common problems in cell culture, namely cross-contamination with other cell lines and mycoplasma contamination, are addressed in *Chapter 6*: Authentication of Cell Lines and *Chapter 7*: Mycoplasma Detection and Elimination, respectively.

Time frame

Generation of a leukemia-lymphoma cell line may require 2–6 months. However, a cell culture should not be considered a continuous cell line until the cells have been passaged and expanded for at least half a year, better one year. Expansion of the cell line takes 2–3 weeks. Depending on the parameters analyzed, 2–4 months might be needed for a thorough characterization of the established cell line.

References

[1] Lange, B. et al. (1987) Blood 70, 192–199.
[2] Zhang, L.Q. et al. (1993) Leukemia 7, 1865–1874.
[3] Saito, M. et al. (1995) Leukemia 9, 1508–1516.
[4] Zhou, M. (1998) personal communication.
[5] Matsuo, Y. and Drexler, H.G. (1998) Leukemia Res. 22, 567–579.
[6] Rickinson, A.B. et al. (1982) Clin. Exp. Immunol. 50, 347–354.
[7] Finerty, S. et al. (1982) Int. J. Cancer 30, 1–7.
[8] Caligaris-Cappio, F. et al. (1987) Leukemia Res. 11, 579–588.
[9] Melo, J.V. et al. (1988) Clin. Exp. Immunol. 73, 23–28.
[10] Faguet, G.B. et al. (1988) Blood 71, 422–429.
[11] Walls, E.V. et al. (1989) Int. J. Cancer 44, 846–853.
[12] Saltman, D. et al. (1990) Leukemia Res. 14, 381–387.
[13] Zheng, C.Y. et al. (1996) Brit. J. Haematol. 93, 681–683.
[14] Katagiri, S. et al. (1985) Int. J. Cancer 36, 241–246.
[15] Takahashi, T. et al. (1985) J. Clin. Oncol. 3, 1613–1623.
[16] Jernberg, H. et al. (1987) Blood 69, 1605–1612.
[17] Eton, O. et al. (1989) Leukemia 3, 729–735.
[18] Okuno, Y. et al. (1991) Leukemia 5, 585–591.
[19] Scibienski, R.J. et al. (1992) Leukemia 6, 940–947.
[20] Zhang, X.G. et al. (1994) Blood 83, 3654–3663.
[21] Smith, S.D. et al. (1989) Blood 73, 2182–2187.
[22] O'Connor, R. et al. (1991) Blood 77, 1534–1545.
[23] Gazdar, A.F. et al. (1980) Blood 55, 409–417.
[24] Hoshino, H. et al. (1983) Proc. Natl. Acad. Sci. USA 80, 6061–6065.
[25] Hori, T. et al. (1987) Blood 70, 1069–1072.
[26] Kitamura, T. et al. (1989) J. Cell. Physiol. 140, 323–334.
[27] Schneider, U. et al. (1977) Int. J. Cancer 19, 621–626.
[28] Tweeddale, M.E. et al. (1987) Blood 69, 1307–1314.
[29] Kees, U.R. et al. (1987) Leukemia Res. 11, 489–498.
[30] Kees, U.R. et al. (1996) Oncogene 12, 2235–2239.
[31] Nitta, M. et al. (1985) Blood 66, 1053–1061.
[32] Lange, B.J. et al. (1992) Leukemia 6, 613– 618.
[33] Drexler, H.G. (1998) Leukemia 12, 845–859.
[34] Drexler, H.G. et al. (2000) Leukemia 14, 198–206.
[35] Smith, S.D. et al. (1984) Cancer Res. 44, 5657–5660.

[36] Quentmeier, H. et al. (1997) Leukemia Res. 21, 343–350.
[37] Drexler, H.G. (1993) Leukemia Lymphoma 9, 1–25.
[38] Gruss, H.J. et al. (1994) Crit. Rev. Oncogenesis 5, 473–538.

Further Reading

Baserga, R. (ed). (1989) Cell Growth and Division: A Practical Approach. IRL Press, Oxford.
Drexler, H.G. et al. (1994) In: Atlas of Human Tumor Cell Lines (Hay, R.J. et al. eds). Academic Press, San Diego, pp. 213–250.
Drexler, H.G. and Minowada, J. (1998) Leukemia Lymphoma 31, 305–316.
Drexler, H.G. et al. (1998) Human Cell 11, 51–60.
Drexler, H.G. and Matsuo, Y. (1999) Leukemia 13, 835–842.
Drexler, H.G. and Uphoff, C.C. (2000) In: The Encyclopedia of Cell Technology (Spier, E. ed). Wiley, New York, pp. 609–627.
Freshney, R.I. ed. (1993) Culture of Animal Cells. Wiley-Liss, New York.
Freshney, R.I. et al. eds. (1994) Culture of Hematopoietic Cells. Wiley-Liss, New York.
Hu, Z.B. and Drexler, H.G. (2000) In: Cell and Tissue Culture for Medical Research (Doyle, A. and Griffiths, J.B. eds). John Wiley, New York (in press).
Jones, G.E. ed. (1996) Human Cell Culture. Humana Press, Totowa, New Jersey.
Klaus, G.G.B. ed. (1987) Lymphocytes: A Practical Approach. IRL Press, Oxford.
Pollard, J.W. and Walker, J.M. (eds). (1990) Animal Cell Culture. Humana Press, Clifton, New Jersey.

6 Authentication of Cell Lines

THE TWO BIGGEST PROBLEMS IN CELL LINE CULTURE

The two biggest problems in cell culture of continuous cell lines involve contaminations: (1) contamination with microorganisms, in particular with mycoplasmas (some 20–30% of the cell lines are believed to be mycoplasma-positive); and (2) cross-contamination with other cell lines. *Chapter 7*: Mycoplasma Detection and Elimination, deals with the problem of mycoplasma contamination of cell lines and describes methods for the elimination and strategies for the avoidance of this contamination.

Even more disturbing than the high prevalence of mycoplasma-positive cell lines is the high percentage of cell line cross-contamination, a problem that does not appear to be sufficiently recognized in the scientific community. Even if the contaminating cells have only a slight growth advantage, the intruding cells may overgrow and completely replace the original cell line. The most notorious culprit of such cross-contaminations is, of course, the solid tumor cell line HeLa[1]. The history of cell cross-contamination in cell cultures is long and goes back as far as 1957[2–7]. The advent of forensic-type DNA fingerprinting and its application to cell line identity control now allows for the detection and identification of cell line cross-contamination with levels of sensitivity and accuracy greatly surpassing those achievable in classical surveys.

It has been estimated that more than one third of cell cultures in use are cross-contaminated either with cells from other species (interspecies contamination) or with unrelated cells from the same species (intraspecies contamination)[7]. Such cell cultures may be grown, maintained, and used for years, and experimental results may be published without documented authenticity of the cells. The potential problems and even dangers in using cross-contaminated cell cultures for the quality of research and production in virtually any scientific or biomedical area cannot be overemphasized (for instance, invalid research results, financial loss, bodily harm, etc.).

A recent study reporting on cell lines obtained between 1990 and 1998 documented that a large percentage of allegedly new human cell lines (cell lines derived from solid tumors and hematopoietic malignancies) which were received from the original investigators (in contrast to cell cultures obtained from secondary sources) were cross-contaminated with other cell lines: (1) 45/252 (18%) of the human cell lines overall were false; and (2) 27/93 (29%) of the original donors supplied false cell lines[8].

INCIDENCE OF CROSS-CONTAMINATION

Data on cross-contaminated cell lines may represent only the tip of the iceberg. Commonly cell lines are passed from laboratory to laboratory and here the percentages of false cell lines are thought to be higher. While undoubtedly the majority of cases go unnoticed or are occasionally detected and then silently corrected, the frequency of intra- and interspecies contamination events can be extrapolated from several publications that offer data on cultures submitted specifically for monitoring and excluding cross-contamination.

Nelson-Rees *et al.* found that 41/253 (16%) cell cultures from 45 laboratories were not as they were purported to be. Cultures were sent for routine monitoring or because the cultures were suspected of being contaminated[2–4,9]. In another series, Nelson-Rees and co-workers reported that 16% of cell lines obtained from 466 laboratories were incorrectly identified[10]. Stuhlberg and colleagues examined 246 cell lines for evidence of cross-contamination or mislabeling[11]. They reported a frequency of 14% for interspecies contamination and 25% for intraspecies contamination; overall nearly 30% of the cell lines were incorrectly designated. Hukku *et al.* summarized data from 275 cultures sent to their laboratory for analysis of cross-contamination[12]. They found that overall 35% of the cultures received were contaminated; 36% of the human cell lines were cross-contaminated (25% by cells of another species and 11% by another human cell line).

The preferential inclusion of suspect cell cultures may have caused a certain bias leading to a slight overestimation of the problem in these studies. Nevertheless, the prevalence of cross-contaminated human cell lines seems to be in the range of 10–30% and is clearly a major concern.

METHODS FOR CELL LINE AUTHENTICATION

Isoenzyme analysis

A traditional method is isoenzyme analysis, taking advantage of the different banding patterns and relative migration distances for the individual isoforms of intracellular enzymes with similar substrate specificity, but different molecular structures (such as lactate dehydrogenase, purine nucleoside phosphorylase, glucose-6-phosphate dehydrogenase, malate dehydrogenase, and others) in agarose gel electrophoresis[13,14]. This technique can only be used for interspecies contamination that should be detectable if the contaminating cells represent at least 10% of the total cell population[15].

DNA fingerprinting

The technique of DNA fingerprinting has great potential for the authentication of cell cultures and identification of cross-contamination. Each individual's DNA gives specific banding patterns with exclusion rates of 99% or higher. Various single-locus and multi-locus probes can be used (for reviews see refs 16–18). A polymerase chain reaction-based minisatellite typing assay in a multiplex format has recently been designed[19,20]. DNA fingerprinting was introduced in 1985[21] and three years later was first applied for identification of cell line cross-contamination[22]. Since then, DNA fingerprinting has become the most often used methodology for determination of cell line authenticity[23–26]. The detection level varies depending on the technique and probes used and is in the range of 5–30% contaminating cells. Of utmost importance for cell line banks is the development of a library of fingerprints stored in modern searchable database systems[20].

Cytogenetic analysis (karyotyping)

Chromosome analysis is currently considered the most sensitive method for identification of intraspecies contamination. Although routinely lower sensitivity can be achieved, an experienced cytogeneticist may detect a contamination of 1% [7,8,27]. However, chromosome analysis is a labor-intensive, time-consuming and rather expensive procedure. Interpretation of the cytogenetic data requires a high degree of skill and experience. An additional problem is the selection of which cells to analyze based on the methods applied. Still, cytogenetics is without question the method currently offering the highest versatility in the characterization of a cell line (not only specifically for the purpose of identification/authentication)[27].

DNA fingerprinting and cytogenetic analysis have their advantages and disadvantages, but they complement rather than exclude each other. The DNA profiling technology appears to be destined to become the method of choice for routine authentication.

CROSS-CONTAMINATED HEMATOPOIETIC CELL LINES

The problem of false cell lines does, of course, also afflict hematological research where leukemia-lymphoma-derived cell lines are extremely important research tools. The first case of a false hematopoietic cell line is J-111 (initially termed 'Oregon J-111') published in 1955 as an 'acute monocytic leukemia cell line'[28]. However, later it was found that J-111 is in reality the cervix carcinoma cell line HeLa[3]. Nevertheless, there are still a number of publications using J-111 as a cell line model for 'monocytes, histiocytes or monocytic leukemia' in the 1990s. The morphology of the J-111 cells in culture is that of a (strongly adherent, epithelial-like) solid tumor cell line and totally different from commonly used monocytic reference cell lines, e.g. THP-1 and U-937, or any other hematopoietic cell line[29].

Another illustrious case concerned Hodgkin's disease-derived cell lines. The initially inadvertent cross-contamination of human Hodgkin's disease-derived cell cultures with monkey cells (cell lines FQ, RB and SpR) and the subsequent deliberate fraud using these cell lines caused quite a scandal in 1980. This case was reported in the journals *Science* and *Nature* [5,30] and in the *New York Times*, led to a hearing before a subcommittee of the U.S. Congress, and also found its entry into the book *Betrayers of the Truth: Fraud and Deceit in Science*[31].

Besides further cases reports and one smaller study[32–34], there is only one systematic study on the prevalence of false human leukemia-lymphoma cell lines[35]. In that work covering about 10 years of analysis, the investigators received 189 cell cultures representing 170 human hematopoietic cell lines (in 19 instances, the same cell lines were obtained a second or third time from independent sources): 117 cell lines came from the original investigators and 72 from secondary sources. Using a combination of DNA fingerprinting and cytogenetic analysis, it was found that 17/117 (14.5%) cell lines from the original source and 11/72 (15.3%) cell lines from a secondary source were cross-contaminated with another hematopoietic cell line (total: 28/189 = 14.8%). Figure 3 shows these results in their updated version. Thus, overall 313 leukemia-lymphoma cell lines have been examined: 18.3% of these cell lines were found to be cross-contaminated (19.9% of

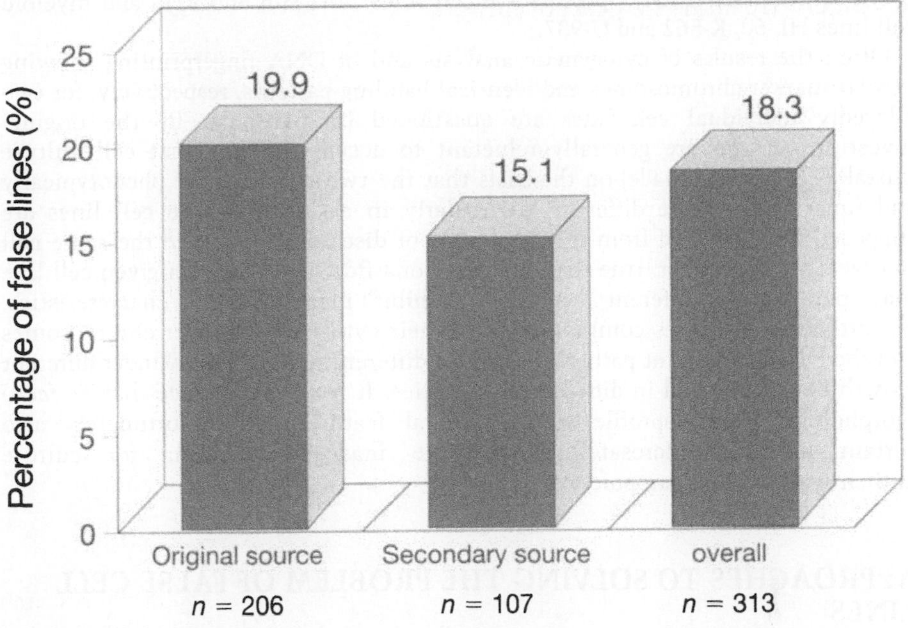

Figure 3 *Percentages of cross-contaminated human leukemia-lymphoma cell lines received at the DSMZ. Shown are the incidences of false leukemia-lymphoma cell lines which were obtained from the original investigators (*n *= 206) or from a secondary source (*n *= 107) at the German Collection of Microorganisms and Cell Cultures (DSMZ), Braunschweig, Germany. Overall, 313 cell lines were examined by DNA fingerprinting, cytogenetic or immunophenotyping analysis. Data were taken from ref. 35 and updated (December 1999).*

the cell lines received from the original source and 15.1% of the cell lines received from a secondary source).

Table XVI lists the hematopoietic cell lines that were cross-contaminated at the original source and for which it is known that no correct culture under that name exists. In most instances, the karyotype in the original publication corresponds already to the karyotype of the cross-contaminating cell line. There are a number of cell lines that were found to be cross-contaminated and which masquerade under the pseudonym of another existing cell line, but for which there is no clearcut proof of 'early cross-contamination'. In these cases, the bona fide correct cell lines with their proper designations do indeed still exist or may still exist. The detection of such a cross-contaminated culture for which the correct parental cells may still be grown or stored elsewhere is annotated in the rubric 'Comments' in the respective cell line entry in this book. The study cited above is certainly not all-inclusive as only a fraction of the more than 1000 human malignant hematopoietic cell lines described in the literature has been tested.

The most commonly detected contaminating intruders are several well-known 'old' cell lines that are widely used in many laboratories: immature T-cell lines

CCRF-CEM and JURKAT; precursor B-cell lines NALM-6 and Reh; and myeloid cell lines HL-60, K-562 and U-937.

Often the results of cytogenetic analysis and of DNA fingerprinting showing similar marker chromosomes and identical banding patterns, respectively, for two allegedly individual cell lines are questioned (in particular by the original investigators who are generally reluctant to accept the fact that cell culture mistakes have been made) on the basis that the two cell lines are phenotypically and functionally quite different (particularly in cases where the cell lines are supposed to be derived from the same type of disease or belong to the same cell lineage). It is, no doubt, true that different bona fide subclones of a given cell line may perform in different ways and exhibit many distinct characteristics. Nevertheless, cell lines commonly retain their cytogenetic marker chromosomes and the DNA fingerprint pattern, in spite of different passage levels under different growth conditions and in different laboratories. It is our experience that *de facto* morphology, immunoprofile and functional features (and unfortunately also certain unstable microsatellite loci) are inadequate criteria for culture authentication of hematopoietic cell lines.

APPROACHES TO SOLVING THE PROBLEM OF FALSE CELL LINES

Authentication of new cell lines

In order to avoid problems before they arise, namely during the establishment, propagation and dissemination of new cell lines, a rigorous authentication of any new cell line is absolutely mandatory. In other words, it must be unequivocally shown that the cultured cells are indeed derived from the presumed patient's tumor and are not the result of a cross-contamination by an older cell line. The method of choice for this identity control is DNA fingerprinting. Microsatellite analysis using only certain loci does not appear to be sufficient as some loci seem to be prone to instability in certain tumor types. Immunophenotyping will not suffice either as cell lines of the same category will often have similar if not identical immunoprofiles. The presence of unique cytogenetic marker chromosomes or molecular biological analyses (e.g. identical clonal gene rearrangement patterns on Southern blots) might also provide unequivocal evidence for the derivation of the cell line from its patient[36].

Cell line banks

It is advisable to obtain established cell lines from well-documented and quality-controlled sources such as established, non-profit culture collections rather than those passed from one laboratory to another (a practice which also raised the mycoplasma problem to epidemic proportions). Naturally, it is important that cell lines be deposited with such culture collections by the original investigator. See also *Chapter 8*: Availability of Cell Lines and Cell Line Banks. The 'standard' practice of carrying actively growing cell lines across country borders in jacket pockets should be avoided in order to protect the integrity of the cell lines and to avoid legal recrimination, as some countries require import permits for cell lines.

Publications

It appears to be obvious that contributors to, as well as editors, referees and readers of scientific journals should attempt to avoid disseminating data based on inaccurately specified cell lines. In any event, the source of cell lines used should be clearly indicated in the Materials and Methods section. See also *Chapter 4*: Guidelines for Characterization of Cell Lines.

Good cell culture practice

The need for scientists and technical staff using cell lines to adhere to the strictest culture procedures and monitor, if possible, regularly for purity and identity cannot be overemphasized. General preventive measures to lower the risk of cross-contamination go hand-in-hand with prevention of mycoplasma contamination. See also *Chapter 7*: Mycoplasma Detection and Elimination.

Awareness and vigilance

Awareness of the seriousness of the problem and continuous critical vigilance in this area might represent the most important steps to decrease the frequency of false leukemia-lymphoma cell lines which are in use all over the world and to improve the current situation. The appearance of cells with phenotypic or other features different from those first noted in the primary cultures or in earlier passages is a characteristic occurrence in the course of cross-contamination of cultures and should ring an alarm bell. On the other hand, cross-contamination with a similar type of cell line may not necessarily lead to a phenotypic or genotypic shift in the cultured cells.

SUMMARY

Cells cultured *in vitro* are a widely used and important resource in biomedical science. Their scientific significance and overall value are definitely diminished when they are not properly and frequently monitored for identity and expression of their unique characteristics. While working with cell lines, it is important to realize that inadvertent mixing of cultures can and does occur. There are endless opportunities for cross-contamination of cell lines. However, it is important to know that there are measures that may be taken (1) to prevent or decrease the risk of cross-contamination and (2) to establish the identity of cells grown in culture. The problem may be solved or at least mitigated by increasing the awareness of its seriousness and by introducing adequate cell line identity controls. Continuous vigilance is needed to avoid these cell culture mistakes, which result in confusion and a waste of both time and money.

References

1 Gold, M.A. (ed). (1986) Conspiracy of Cells: One Woman's Legacy and the Medical Scandal. State University of New York Press, Albany, New York.
2 Nelson-Rees, W.A. et al. (1974) Science 184, 1093–1096.
3 Nelson-Rees, W.A. and Flandermeyer, R.R. (1976) Science 191, 96–98.

[4] Nelson-Rees, W.A. and Flandermeyer, R.R. (1977) Science 195, 1343–1344.
[5] Harris, N.L. et al. (1981) Nature 289, 228–230.
[6] Reid, Y. et al. (1995) J. Leuk. Biol. 57, 804.
[7] Markovic, O. and Markovic, N. (1998) In Vitro Cell. Dev. Biol. 34, 1–8.
[8] MacLeod, R.A.F. et al. (1999) Int. J. Cancer 83, 555–563.
[9] Nelson-Rees, W.A. et al. (1981) Science 212, 446–452.
[10] Nelson-Rees, W.A. et al. (1978) Prog. Clin. Biol. Res. 26, 25.
[11] Stuhlberg, C.S. et al. (1976) Am. J. Hematol. 1, 237–242.
[12] Hukku, B. et al. (1984) Adv. Exp. Med. Biol. 172, 13–31.
[13] Drexler, H.G. et al. (1999) DSMZ Catalogue of Cell Lines, 7th edn, Braunschweig, Germany.
[14] Steube, K.G. et al. (1995) In Vitro Cell. Dev. Biol. 31, 115–119.
[15] Nims, R.W. et al. (1998) In Vitro Cell. Dev. Biol. 34, 35–39.
[16] Bär, W. et al. (eds). (1994) Advances in Forensic Haematogenetics 5. Springer-Verlag, Berlin, Germany.
[17] Möller, A. and Brinkmann, B. (1995) Cell. Mol. Biol. 41, 715–724.
[18] Yan, R. et al. (1996) In Vitro Cell. Dev. Biol. 32A, 656–662.
[19] Dirks, W.G. et al. (2000) In Vitro Cell. Dev. Biol. 35, 558–559.
[20] Dirks, W. et al. (1999) Cell. Mol. Biol. 45, 841–853.
[21] Jeffreys, A.J. et al. (1985) Nature 214, 67–73.
[22] Masters, J.R. et al. (1988) Brit. J. Cancer 57, 284–286.
[23] Thacker, J. et al. (1988) Somat. Cell. Mol. Genet. 14, 519–525.
[24] Hampe, J. et al. (1992) Hum. Antibodies Hybridomas 3, 186–190.
[25] Stacey, G.N. et al. (1992) Cytotechnology 9, 211–216.
[26] Stacey, G.N. et al. (1992) Nature 357, 261–262.
[27] Hay, R.J. et al. (1996) J. Cell. Biochem. 24 (Suppl), 107–130.
[28] Osgood, E.E. and Brooke, J.H. (1955) Blood 10, 1010–1022.
[29] Drexler, H.G. et al. (1993) Leukemia 7, 2077–2079.
[30] Wade, N. (1981) Science 211, 1022–1025.
[31] Broad, W. and Wade, N. (eds). (1985) Betrayers of the Truth: Fraud and Deceit in Science. Oxford University Press, Oxford, pp. 89–96.
[32] Häne, B. et al. (1992) Leukemia 6, 1129–1133.
[33] Gignac, S.M. et al. (1993) Leukemia Lymphoma 10, 359–368.
[34] MacLeod, R.A.F. et al. (1997) Leukemia 11, 2032–2038.
[35] Drexler, H.G. et al. (1999) Leukemia 13, 1601–1607.
[36] Drexler, H.G. and Matsuo, Y. (1999) Leukemia 13, 835–842.

7 Mycoplasma Detection and Elimination

INTRODUCTION

The contamination of cell cultures by mycoplasmas is certainly one of the major problems occurring in biological research and biotechnology using cultured cells. Mycoplasmas can produce extensive changes and growth arrest in the cultures they infect. The possible effects of mycoplasma contamination are legion. These organisms are resistant to many of the antibiotics that are commonly used in cell cultures. This problem has become more widely appreciated since the advent of accurate, sensitive and reliable detection methods. The application of robust and successful elimination methods provides the possibility to cleanse mycoplasma-positive cell cultures.

BIOLOGY OF MYCOPLASMAS

The mycoplasmas represent a large group of microorganisms which are all characterized by their lack of a rigid cell wall, a standard attribute in all other types of bacteria. Therefore, a distinct class within the prokaryotes, appropriately named *Mollicutes,* was created. For historical and practical reasons, the trivial terms mycoplasmas and mollicutes are often used as synonyms. Initially mycoplasmas were termed 'pleuropneumonia-like organisms'. For specific details regarding the biology and taxonomy of mycoplasmas, the reader is referred to specialist textbooks[1–5].

It has long been assumed that mycoplasmas exist only on the outside of the eukaryocytic cell membrane (cytadherence). However, recent studies have demonstrated the intracellular location of certain mollicutes (*M. fermentans, M. genitalium* and *M. pneumoniae*), not only after phagocytosis by granulocytes and monocytes, but also in non-phagocytic epithelial cells. A new mycoplasma species capable of entering a variety of human cells *in vivo* and *in vitro* was discovered and was named *M. penetrans*. Mycoplasma cytadherence and invasion appear to be active, but separable processes. Extensive invasion of cells by *M. penetrans* eventually leads to cell disruption. The percentage of the mycoplasma population able to invade the cells seems to depend on the mycoplasma species and may be influenced by the eukaryotic cell type and by culture conditions. While the great majority of the infecting mycoplasma population is definitely located extracellularly, the intracellular location, even for only a short period, sequesters mycoplasmas and may protect them effectively from mycoplasmacidal therapies. This phenomenon may also explain the difficulty of eradicating mycoplasmas from all infected cell cultures[6].

MYCOPLASMA CONTAMINATION OF CELL CULTURES

Incidence of mycoplasma contamination

One mycoplasma cell can grow to 10^6 so-called colony-forming units (CFU)/ml within 3–5 days in an infected cell culture. Eukaryotic cell cultures contaminated with mycoplasma have titers in the range of 10^6 to 10^8 CFU/ml.

Frequently, there are 100 to 1000 mycoplasmas attached to each infected cell. Contamination may initially go undetected because mycoplasma infections do not produce overt turbid growth, as commonly seen with other bacterial or fungal contamination. Often mycoplasmas can achieve very high densities in cell cultures without causing a change in the pH (and therefore without a color change of most culture media). Some mycoplasmas produce very little overt cytopathology, and apparent contamination may remain undetected for months.

In general, primary cell cultures and cultures in early passage are less frequently contaminated than continuous cell lines: primary cultures and early passage cultures are contaminated on the order of 1% and 5%, respectively; continuously cultured cell lines in the range of 15–35%. Several large series on several thousands of cell cultures analyzed over several decades (1960s–1980s) found an incidence of *c.* 15%. However, more recent studies on smaller series documented significantly higher infection rates of cell cultures, commonly in the range of 15–35%, but also as high as 65–80%[7-9].

The ever expanding application of cell lines in research and biotechnology (with the resulting exchange of non-authenticated and mycoplasma-positive lines between scientists) and the increasing use of certain antibiotics (mostly penicillin plus streptomycin, which merely serve to mask but do not remove mycoplasmas) in routine culture have presumably led to this increase in mycoplasma contamination of cell cultures, which has now reached epidemic proportions. It appears that contamination rates are higher for cultures continuously grown with such 'cell culture antibiotics'. Using a panel of first three and later four different mycoplasma detection assays (see below), 407 human hematopoietic cell lines were examined at the DSMZ from 1989 to 1999. While infections with other bacteria and fungus or yeast accounted for about 2% of all cell line contaminations, a staggering 31% of the cell lines examined were found to be mycoplasma-positive (Figure 4)[10].

Most common contaminating mycoplasma species

Mycoplasmas can be found nearly ubiquitously, as long as the environment offers sufficient amounts of nutrients and the essential compounds. Within the last two decades, mycoplasmas have been recognized as one of the major cell culture problems. Consequently their biology and physiology have become topics of intense research.

While at least 20 distinct species have been isolated from contaminated cell lines, by far the largest portion of infections is caused by a relatively small number of *Mycoplasma* and *Acholeplasma* species: 90–95% of the contaminants were identified as either *M. orale* (frequency 20–40%; natural host: human), *M. hyorhinis* (10–40%; swine), *M. arginini* (20–30%; bovine), *M. fermentans* (10–20%; human), *M. hominis* (10–20%; human) or *A. laidlawii* (5–20%; bovine). Depending on the study, the individual percentages of these six species may vary. Although many mycoplasmas are host-specific *in vivo*, the various species are not restricted to cell cultures derived from their natural hosts. In other words, 'bovine' and 'swine' mycoplasmas also grow in human cell cultures.

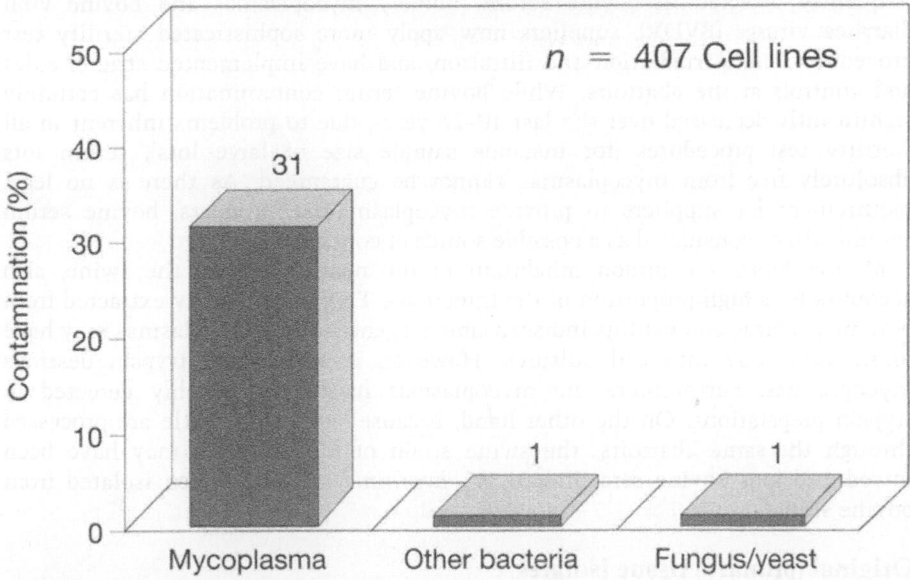

Figure 4 *Incidence of mycoplasma contamination of human leukemia-lymphoma cell lines at the DSMZ. Shown are the percentages of the 407 cell lines which upon arrival at the institute were found to be infected with mycoplasma, other bacteria, fungus or yeast (study period from 1989–1999)* [10]. *Nine of these 407 cell lines had a double infection.*

Sources of mycoplasma contamination

Laboratory personnel

Mycoplasmas from human sources are the most prevalent group and account for approximately one third to half of all strains isolated. Generally *M. orale*, which is the most common mycoplasma species in the oral cavity of clinically healthy humans, also represents the single most common isolate. Other non-pathogenic mycoplasma species from the normal human microbial flora of the oropharynx that are seen in cell cultures are *M. fermentans* and *M. hominis*. As the largest percentage of mycoplasma found in cell cultures are of human origin, it appears to be apparent that the laboratory personnel is directly involved and represents the major source of contamination.

Culture reagents

The bovine group of mycoplasmas accounts for about another third of all strains isolated from cell cultures. Here the most frequent infectants are *M. arginini* and *A. laidlawii*. These cell culture contaminants are believed to derive from bovine sources, as in the early days of cell culture (1950s–1970s) the bovine sera were not routinely and as strictly screened for mycoplasma contamination as they are today. Studies in the 1960s and 1970s showed that 25–40% of the serum lots provided by commercial suppliers were contaminated[9]. In order to prevent contaminations and to eliminate those adventitious agents that are most frequently detected in bovine serum, namely mycoplasmas and bovine viral diarrhea viruses (BVDV), suppliers now apply more sophisticated sterility test procedures, use γ-irradiation and filtration, and have implemented stricter rules and controls at the abattoirs. While bovine serum contamination has certainly significantly decreased over the last 10–20 years, due to problems inherent in all sterility test procedures (for instance sample size of large lots), serum lots absolutely free from mycoplasmas cannot be guaranteed. As there is no legal requirement for suppliers to provide mycoplasma-free products, bovine serum should still be considered as a possible source of contamination.

M. hyorhinis, a common inhabitant of the nasal cavity of the swine, also accounts for a high proportion of the infections. Trypsin is usually extracted from porcine pancreas and via this indispensable reagent, swine mycoplasmas may have found their way into cell cultures. However, it seems that trypsin destroys mycoplasmas. Furthermore, no mycoplasmas have been reliably detected in trypsin preparations. On the other hand, because swine and cattle are processed through the same abattoirs, the swine strain of *M. hyorhinis* may have been introduced into bovine sera. Indeed, *M. hyorhinis* has often been isolated from bovine sera.

Original (primary) tissue isolates

The frequency of infection in primary cell cultures is low, on the order of 1% [7,9]. Furthermore, murine, avian and canine species of mycoplasma only account for 0.5–1% within the panel of mycoplasmas isolated from contaminated cultures – despite the wide use of murine cell lines.

Cross-contamination from infected cultures

In laboratories with contaminated cells, most or all cultures are positive containing the same mycoplasma species[7,10]. Mycoplasma-infected cell lines are themselves the single most important source for further spreading of the contamination. This is due to the ease of droplet generation during handling of cell cultures, the high concentration of mycoplasmas in infected cultures (10^6 to 10^8 CFU/ml of supernatant), and the prolonged survival of dried mycoplasmas. Operator-induced contamination is a complex problem. Mycoplasmas are spread by using laboratory equipment, media, or reagents that have been contaminated by previous use in processing mycoplasma-infected cells.

Effects of mycoplasma contamination

Mycoplasma infections can have a myriad of different effects on the contaminated cell cultures. However, this multitude of different effects does not affect the various cells in the same manner and to the same degree (Table 13). Many mycoplasma species produce severe cytopathic effects while others may cause very little overt cytopathology. There can be qualitative and quantitative differences in the same parameter, depending on the infecting mycoplasma species, the culture conditions, the type of the infected cell culture, the intensity and duration of the infection, an additional infection with viruses, and other parameters. Thus, contaminations can interfere with virtually every parameter measured in cell cultures during routine cultivation or in experimental investigations. Consequently, the mycoplasmas in these cultures cannot simply be ignored or regarded as harmless bystander organisms. Besides the loss of an important culture, in the worst case all experiments might be influenced by the infections and artefacts are produced. Because of the virtually unlimited number of reported mycoplasmal effects on cultured cells, only some of the most important among the diversity of possible effects are highlighted in Table 13. Clearly, the term mycoplasma infection is a rather general term similar to virus infection[11].

Table 13 *Effects of mycoplasma contamination on cell cultures*

General effects on eukaryotic cells:
Altered levels of protein, RNA and DNA synthesis
Alteration of cellular metabolism
Induction of chromosomal alterations (numerical and structural aberrations)
Change in cell membrane composition (surface antigen and receptor expression)
Alteration of cellular morphology
Induction or inhibition of lymphocyte activation
Induction or suppression of cytokine expression
Increase or decrease in virus propagation
Interference with various biochemical and biological assays
Influence on signal transduction
Promotion of cellular transformation
Alteration of proliferation characteristics (growth, viability)
Induction of malignant transformation
Total culture degeneration and loss

Specific effects on hybridomas:
Inhibition of cell fusion
Influence on selection of fusion products
Interference in screening of monoclonal antibody reactivity
Generation of monoclonal antibody against mycoplasma instead of target antigen
Reduced yield of monoclonal antibody
Impaired conservation of hybridoma

Modified from ref. 5.

DETECTION OF MYCOPLASMA CONTAMINATION

Various mycoplasma detection methods

Over the last four decades a variety of techniques have been developed to detect mycoplasma contamination of cell cultures (Table 14). Most of these methods are relatively lengthy, involve subjective assessments and use measurements that are often quite complex in nature. The two classical detection methods, DNA fluorochrome staining and microbiological colony assay, together with the recently introduced and highly specific and sensitive methods, RNA hybridization and polymerase chain reaction (PCR), appear to be the most reliable and practical and will be described below.

Table 14 *Mycoplasma detection methods*

Histological staining:
Histochemical staining and light microscopy
Electron microscopy:
Transmission electron microscopy
Scanning electron microscopy
Biochemical methods:
Enzyme assays
Gradient or electrophoresis separation of labeled RNA
Protein analysis
Immunological procedures:
Fluorescence or enzymatic staining with antibodies
ELISA
Autoradiography
DNA fluorochrome staining:
DAPI staining
Hoechst 33258 staining
Microbiological culture:
Colony formation on agar
RNA hybridization:
Filter hybridization
Liquid hybridization
Polymerase chain reaction:
Species- or genus-specific PCR primers
Universal PCR primers

Modified from refs. 5,12.

Traditionally, mycoplasmologists discerned direct and indirect detection methods. While the term 'direct' method referred to the classical microbiological colony growth of mycoplasmas on agar, 'indirect' detection techniques included procedures that measure a gene product that is associated with mycoplasmas rather than with the mammalian cells in culture. Furthermore, tests may be performed directly on the specimen taken from a given cell culture or indirectly using the so-called indicator cell culture procedure whereby the specimen is inoculated into another cell culture known to be free of mycoplasmas. Use of an indicator cell culture allows for standardization and includes positive and negative controls in each assay.

Most reliable and practical mycoplasma detection methods

Until the arrival of molecular methods, DNA fluorochrome staining and microbiological culture had been regarded as the 'gold standards' for detection of mycoplasma contamination[7,9]. While the classical methods are certainly still useful, the new approach to a highly sensitive, specific and rapid diagnosis of mycoplasmal infection is based on the development of gene or DNA probes, which were first introduced in the 1980s. The principle is based on the identification of genomic sequences that are specific for a single species or a particular group of mycoplasma or universally for nearly all mycoplasmas. These probes are used for DNA or RNA hybridization. The more recent development of PCR enables the amplification of the target DNA in the specimen using specific synthetic oligonucleotides complementary to conserved rRNA sequences and increases the sensitivity by several orders of magnitude[13,14].

DNA fluorochrome staining

Fluorescent dyes binding to DNA were first used to detect mycoplasmal infection in the 1970s. The two primary DNA-binding dyes are 4′-6-diamidino-2–phenylindole (DAPI) and bisbenzamide (Hoechst 33258). The fluorescent dyes bind to eukaryotic and prokaryotic DNA, forming strongly fluorescent complexes with high specificity. Under fluorescence microscopy, an uncontaminated cell culture shows only nuclear fluorescence against a dark background. Mitochondrial DNA does bind the fluorochrome, but at levels imperceptible by routine microscopy. Mycoplasmas, however, which have approximately ten times the DNA content of mitochondria, are readily detected as bright foci over the cytoplasm, lining the cell membrane or in intercellular spaces. This procedure is not diagnostic for mycoplasmas as other prokaryotic non-mycoplasma contaminants will also be detected.

The use of an indicator cell line such as murine cell line 3T6 or monkey cell line Vero is possible. Specimens are inoculated into the indicator cell culture which, after incubation, is examined under UV-light.

DNA fluorochrome staining will detect titers of 10^5 organisms or greater per milliliter. False-positive results can be caused by staining artefacts or cell detritus. This test is dependent on the intensity of the infection since only strongly contaminated cultures can be identified, thus leading easily to false-negatives. The majority of false-positive and false-negative readings are due to equivocal stainings with the difficulties in the subjective interpretation of the results and borderline decisions. Use of an indicator cell line significantly enhanced sensitivity, specificity, accuracy and predictive value of the DNA fluorochrome assays[12,15]. The main disadvantage appears to be the necessity to permanently carry an adherent cell line. Advantages of the fluorochrome tests are that they are inexpensive, simple and rapid and can be applied for regular screening and long-term monitoring.

Microbiological culture

Specimens are inoculated into mycoplasma broth and onto agar. Anaerobic incubation is recommended as aerobic incubation yields a lower detection rate. Broths are transferred to agar plates after 4–7 days of incubation. Most mycoplasmas produce microscopic colonies (100–400 μm in diameter) with a 'fried egg' appearance growing embedded beneath the surface of the agar. Because the contaminants grow embedded, they maintain their morphological shape. Consequently, they are easily distinguishable from bacterial colonies. Certain mycoplasmas may produce a more diffuse, granular type of colony[5].

This procedure has the advantage of ease of manipulation and visual recognition of colonies. Some artefacts are occasionally seen on agar after inoculation of cell culture specimens: colonies must be distinguished from cell clumps by their eventual increase in size; pseudocolonies (e.g. crystals, air bubbles) can be a problem for the inexperienced observer since they can increase in size and can actually be transferred. There are wide variations in size, morphology and speed of growth of the mycoplasma colonies isolated from different cell cultures. Colonies usually become detectable after an average of 3–6 days, but detection only after longer incubation times is also possible. *M. hyorhinis* is a 'non-cultivable' mycoplasma species that grows poorly or not at all in broth and agar media.

While the microbiological culture has the advantages of being inexpensive, highly sensitive with a high detection rate, and of representing an established and important reference method, the disadvantages are the long incubation time, the need for a subjective and experienced interpretation, and the fact that not all mycoplasmas can be successfully cultured.

Molecular RNA hybridization

The assays of this category of mycoplasma detection are based on the principle of nucleic acid hybridization: complementary nucleic acid strands come together to form stable double-stranded complexes. The highest sensitivity is achieved in DNA–RNA hybridization, which is commonly used, but also DNA–DNA hybridization would work satisfactorily. The availability of rRNA sequences in databases allows for the construction of species-specific, group-specific or universal oligonucleotide probes to cover all mycoplasmas commonly isolated from cell cultures. Two unique features make rRNA the most suitable target for probes: the general organization of the molecule with conserved, semi-conserved and variable regions; and the high copy number of rRNA present in each mycoplasma cell (about 1×10^4). The probes represent sequences of the 16S or 23S rRNA genes. Some bacterial species may also be detected by these tests. RNA from eukaryotic cells will not hybridize with the probes.

The liquid solution hybridization (best exemplified by the commercially available kit 'Gen-Probe', Mycoplasma T.C.) uses a [^{3}H]-labeled single-stranded DNA probe homologous to 16S and 23S rRNA sequences. After the rRNA is released from the organism, the [^{3}H]-DNA probe combines with the target rRNA to form a stable DNA–RNA hybrid which is then separated from the non-hybridized DNA probe; positive signals are measured in a scintillation counter. The second method uses filter hybridization, whereby cell culture samples are simply heat-fixed on the membranes, hybridized with the [^{32}P]dCTP-labeled probe derived from 16S or 23S rRNA sequences, and incubated for autoradiography. The detection limits are in the range of 10^3 to 10^4 organisms. The filter method is reportedly more sensitive than the liquid hybridization.

The advantages of these assays are broad specificity, high sensitivity, convenient sample preparation, processing of large sample numbers and rapid results. Validity and reproducibility of the 'Gen-Probe' test are very high[12,15]. Drawbacks are the use of radiolabeled probes and the relatively high costs.

Polymerase chain reaction

Nearly all mycoplasmic 16S rRNA sequences have been determined. Computer alignment studies of mollicute 16S rRNA sequences reveal regions with sequence variability or conservation at the species, genus or class level, allowing for the selection of appropriate oligonucleotides (primers) for detection and identification of mycoplasmas. The highly conserved regions of the genes enable the selection of primers of wide specificity ('universal primers') which will react with DNA of any mycoplasma or even with the DNA of other prokaryotes[14]. This is satisfactory for detection of mycoplasma cell culture infection where the goal is just to screen the cultures for contamination. Besides the conserved regions of mycoplasmal 16S rRNA genes, the 16S-23S intergenic regions are also quite useful for mycoplasma detection[13,16]. It is important that no cross-reactions with the eukaryotic DNA of the cell lines occur.

As PCR can be achieved with frozen and lyophilized material, this provides the possibility for retrospective analyses and facilitates the transport, collection and storage of samples. The amplification may be performed as a single-step PCR or as a two-step (nested) PCR[17]. The latter approach increases the sensitivity and specificity considerably, but also increases the risk of contamination by DNA carryover. Southern blotting of PCR products and hybridization with a specific radiolabeled internal probe is another possibility to improve sensitivity. However, a very high sensitivity level may not usually be required in routine diagnosis as acute and particularly chronic cell culture mycoplasma infections involve a large number of organisms (commonly 10^6 or higher). The very high sensitivity may be advantageous at a very early phase of infection or under conditions where mycoplasma growth is suppressed (e.g. in the testing and monitoring of cell cultures post-treatment with antibiotics for elimination of the contaminants).

The detection limit using a set of nested universal primers was determined to be 1 fg mycoplasmal DNA, which is equivalent to 1–2 genome copies of the 16S coding region (mollicute genomes carry only one or two rRNA gene sets). The ability to detect a single mycoplasma cell makes PCR the most sensitive detection method available. In theory, a positive cell culture may be derived from a single mycoplasma cell. In practice, however, due to several reasons (including mycoplasma cell aggregates, multinuclear filamentous forms, defective or non-viable cells), a 'successful' infection requires an inoculum equivalent to about 100 to 1000 cells[13].

The high sensitivity of PCR may cause problems in producing false-positive results due to contamination with target DNA. Another possible problem is false-negative data caused by the inhibition of the *Taq* polymerase by components in the samples[14]. However, once all PCR-related problems are properly addressed, single-step or double-step PCR are clearly superior to other mycoplasma detection methods in many respects as this method combines simplicity and speed with high specificity and extreme sensitivity, in addition to objectivity, accuracy and reproducibility[14,18]. Furthermore, PCR is not limited by the ability of an organism to grow in culture. In the realm of cell culture quality control, this molecular nucleic acid amplification may eventually replace biological amplification (i.e. growth in artificial media), a feature of paramount importance considering the fastidious nature of mycoplasmas[13]. Thus, PCR should prove to be the technique of the future for mycoplasma detection in cell cultures. Several PCR kits are commercially available.

ELIMINATION OF MYCOPLASMA CONTAMINATION

Various elimination methods

Ever since mycoplasma contamination of cell cultures was first reported, attempts have been made to develop methods for elimination of the mycoplasma. It has been suggested that efforts to eradicate mycoplasmas from contaminated cells should be considered as a last resort (in order to prevent spread of the contaminant) and that it would be often far better to eliminate the problem completely by autoclaving the infected cultures and replacing them with fresh stocks known to be mycoplasma-free[8]. However, often the cell line is not replaceable with a mycoplasma-free aliquot and purging such cultures of mycoplasmas is the only resort to save the cell line.

Four general types of procedures have been used to eliminate mycoplasmas from infected cell cultures: physical, chemical, immunological and chemotherapeutic treatment (Table 15). Many of the methods were shown to be unreliable. Some techniques may apply to some, but not all mycoplasma species; some of them are too laborious or simply impractical. Elimination is typically time-consuming, often unsuccessful and poses risks of secondary infection to other cell cultures. Methods of elimination should ideally be simple, rapid and efficient, reliable and inexpensive, have minimal effect on the eukaryotic cell and result in no loss of specialized characteristics; accidental cloning selection of treated cells also should not occur. There is clearly not a single method available that is both 100% effective and fulfills all the ideal requirements.

Table 15 *Mycoplasma elimination methods*

Physical methods:
Heat treatment
Filtration through microfilters
Induction of chromosomal or cell membrane damage through photosensitization

Chemical methods:
Exposure to detergents
Washings with ether-chloroform
Treatment with methyl glycine buffer
Incubation with sodium polyanethol sulfonate
Culture in 6–methylpurine deoxyriboside

Immunological methods:
Co-cultivation with macrophages
In vivo passage through nude mice
Culture with specific anti-mycoplasma antisera
Exposure to complement
Cell cloning

Chemotherapeutic treatments:
Antibiotic treatment in standard culture
Antibiotic treatment plus hyperimmune sera or co-cultivation with macrophages
Soft agar cultivation with antibiotics

Modified from refs. 5,19.

The effectiveness of some elimination methods has been investigated only in experimentally contaminated cell cultures, yet experimentally infected cultures may not realistically reflect the laboratory situation since chronic infections certainly result in complex interactions between mycoplasmas and cells. If a clean-up is attempted, it is imperative to monitor closely the effectiveness of treatment relative to mycoplasma elimination and eukaryotic cytotoxicity. A variety of procedures have been described and utilized. Administration of antibiotics is by far the most common and efficient approach and will be discussed in greater detail below.

Antibiotic treatment of mycoplasma contamination

Mycoplasmas, which lack a cell wall and are incapable of peptidoglycan synthesis, are theoretically not susceptible to antibiotics such as penicillin and its analogues, which are effective against most bacterial contaminants of cell cultures. However, it has been reported that several bacteriostatic antimicrobial agents inhibit the growth of mycoplasmas. Thus they may not eradicate the contaminants, but simply suppress the full-blown picture of an infection and tend to mask the presence of mycoplasmas. A number of different antibiotics has been used for mycoplasma control[5]. The contaminant strains, however, often developed resistance to certain antibiotics which were thus completely ineffective. Other antibiotics (for instance some aminoglycosides and lincosamides) were moderately to highly effective in eliminating mycoplasmas, but only at concentrations that had detrimental effects on the eukaryotic cells, such as marked cytotoxicity[5].

Ideally, a basic procedure should involve isolating, speciating, and determining the antibiotic susceptibility of the contaminants to the arsenal of possible reagents to maximize success; then the cultures should be exposed to the effective antibiotics. However, this approach is extremely time-consuming, labor-intensive and requires certain expertise.

Tetracyclines are generally effective anti-mycoplasmal agents. Also quinolones were found to be highly effective mycoplasmacidal antibiotics. Recently, these new antibiotics have been introduced for purging of mycoplasmas from cell cultures and are marketed commercially. Of particular note are the quinolones ciprofloxacin (distributed as Ciprobay by Bayer), enrofloxacin (Baytril; Bayer) and an unpublished quinolone reagent available as Mycoplasma Removal Agent (MRA; Flow Laboratories ICN); the product BM-Cyclin (Roche) combines the macrolide tiamulin (a pleuromutilin-derivative) and the tetracycline minocycline.

Tetracyclines inhibit protein synthesis by binding to subunits of ribosomes, thereby blocking peptide chain elongation. Tetracyclines inhibit both prokaryotic and eukaryotic ribosomal protein synthesis. The mode of action of the quinolones involves the binding to and inhibition of the bacterial DNA gyrase which is essential for DNA replication, transcription, repair and recombination. The quinolones may also exert an inhibitory effect on eukaryotic DNA polymerase α, topoisomerases and DNA deoxynucleotidyl transferases. The activity of these enzymes is especially high in dividing cells. High doses of quinolones induced double-strand DNA breaks in human cells. Selectivity of the quinolones for the bacterial cell is at least partly due to the far greater sensitivity of the bacterial enzymes compared to the mammalian enzymes.

Antibiotic treatment of mycoplasma contamination in cell cultures has a high degree of efficiency as reported in the literature: 65–74% of the mycoplasma-positive cultures were cured with MRA; 74–100% with Ciprobay; 76–91% with Baytril; and 82–100% with BM-Cyclin[20]. Our results using these four anti-mycoplasma antibiotics are presented in Figure 5; the recommended treatment periods and the concentrations of the antibiotics are listed in Table 16.

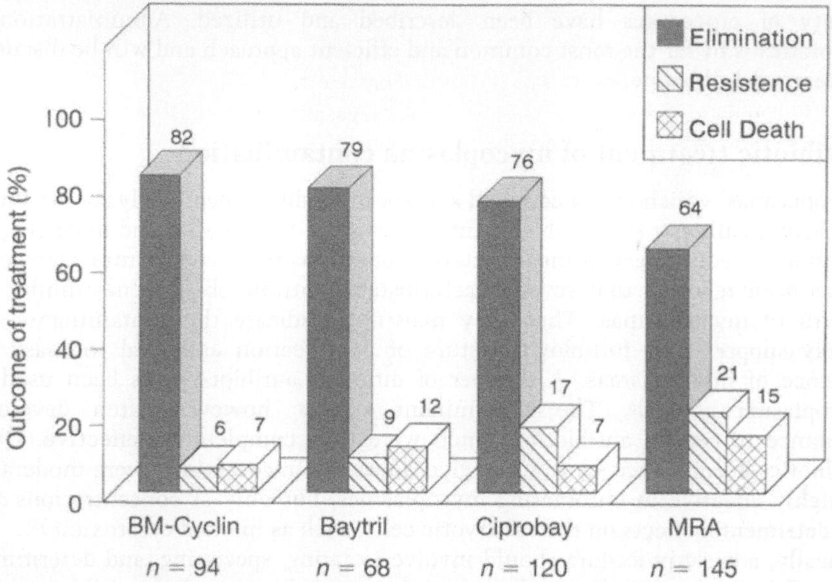

Figure 5 *Outcome of treatment of mycoplasma-positive human leukemia-lymphoma cell lines with antibiotics. Cells were treated with either BM-Cyclin, Baytril, Ciprobay or MRA as indicated in Table 16. All treated cultures were retested with highly sensitive and specific mycoplasma detection methods after an extensive antibiotic-free post-treatment period. The results are shown as percent of cultures that were either cured or that remained mycoplasma contaminated (due to resistance) or that were lost during the treatment period (due to cytotoxicity). The number of cultures treated in each category is indicated (n). Data were taken from refs 19–23 and updated (December 1999)* [10].

Table 16 *Recommended anti-mycoplasma antibiotic treatment*

Brand name (commercial source)	Generic name (antibiotic category)	Treatment period	Final concentration
Baytril (Bayer)	enrofloxacin (quinolone)	7 d	25 μg/ml
BM-Cyclin I (Roche)	tiamulin (macrolide)	3 d × 3[a]	10 μg/ml
BM-Cyclin II (Roche)	minocycline (tetracycline)	4 d × 3[a]	5 μg/ml
Ciprobay (Bayer)	ciprofloxacin (quinolone)	14 d	10 μg/ml
MRA (Flow ICN)	not known (quinolone)	7 d	0.5 μg/ml

For further details see refs. 19–23.

[a] Alternating treatment with 3 days of BM-Cyclin I and 4 days of BM-Cyclin II (3 cycles).

Two-thirds to three-quarters of all cultures treated can be cleansed by one of these anti-mycoplasma chemotherapeutic regimens. Besides cure, two other possible outcomes are loss of the culture and resistance. Culture death, occurring in 5–15% of the cases, is presumably caused by cytotoxic effects of the reagents. BM-Cyclin (containing a tetracycline) shows the greatest growth-inhibiting effect which may be either cytostatic or cytotoxic. Generally, 1 week after cessation of treatment, cell growth will return to normal. Cytostatic and cytotoxic effects of the antibiotics may be enhanced by the poor condition of cell cultures commonly found in chronically infected cells. This situation is clearly different from that of experimentally contaminated cell cultures. It was found that increasing the serum concentration and incubating the cells at higher densities (at or near clonal densities) was advantageous for the cell cultures[19].

With regard to resistance, the quinolones showed cross-resistance. Sequential administration of BM-Cyclin to the same cells that were first exposed to a quinolone can still result in eradication of the resistant infectant. Higher concentrations of the antibiotics may be more effective in purging mycoplasma-contaminated cultures, decreasing the rate of resistance, but this success would be counterbalanced by significantly higher cytotoxicity. It is not known whether the resistance of cell culture mycoplasmas to antibiotics is acquired during treatment or exists already prior to exposure to these reagents. Anti-mycoplasma antibiotics should be reserved for the specific situation of mycoplasma eradication in an infected cell culture and should not be routinely added as a standard supplement to the culture medium, as this would run a substantial risk of selecting resistant mycoplasma strains.

Anti-mycoplasma treatment conditions are certainly stressful to the eukaryotic cells. Thus, cells might no longer express the desired properties as a result of antibiotic administration. Outgrowth of a selected clone is another possibility. Some data suggest that cured cells generally preserve their characteristics. Still, any alterations to the cell lines induced by antibiotic treatment are obviously a matter of concern and require further detailed studies.

Taken together, the technically simplest option for mycoplasma decontamination with the most promising results is antibiotic treatment. The convenience of use and general availability of the reagents render it a reliable routine laboratory procedure. However, antibiotic mycoplasma elimination is laborious and time-consuming as the duration of the treatment plus the minimum antibiotic-free post-treatment period ranges from 3 to 5 weeks depending on the protocol used. Furthermore, special attention must be placed on possible cytotoxic effects or effects that alter the characteristics of the cell line.

PROTOCOLS FOR DETECTION, ELIMINATION AND PREVENTION OF MYCOPLASMA CONTAMINATION

There are three main aspects with regard to the problem of mycoplasma contamination of cell cultures: detection, elimination and above all prevention of mycoplasma infection. Mycoplasmal contamination is a potential threat to all cell cultures. Overall, detection methods have dramatically improved in recent years and the availability of testing kits has increased. Instead of discarding mycoplasma-infected cultures, elimination of the contaminants has become a realistic and practical possibility. Critical to maintaining cell cultures free of mycoplasma (and other adventitious organisms) is the establishment of routine operating and testing procedures by the cell culturist. The primary difficulty in controlling the epidemic mycoplasma contamination lies in convincing the culture operator of the necessity for validating the absence of these covert organisms and in creating a certain level of awareness for possible contamination.

Protocol for mycoplasma detection

The high frequency of mycoplasma-contaminated cell cultures and the rapid spread of the infection within the laboratory imposes the need for rigorous mycoplasma testing of any new incoming cell culture, and also for regular screening of every permanently cultured cell line. The best test will fail if the sample submitted for testing is not representative or is not treated properly. Antibiotics can mask mycoplasma infection, resulting in false-negative test results. While it is highly desirable that cell lines be cultured generally without any antibiotics, for mycoplasma testing purposes it is essential that cell cultures be maintained antibiotic-free for several passages. Furthermore, cultures should be assayed a sufficient time after the last passage or exchange of medium. This allows organisms, if present, to grow to high titers. While PCR is definitely the test of choice, this method can also produce false-positive or false-negative results. Analysis of the same sample with at least two different techniques is suggested in order to obtain a reliable diagnosis.

Newly arriving cell lines should be quarantined in a separate cell culture laboratory. The initial mycoplasma testing determines the mycoplasma contamination status of the cells. If the culture is contaminated and no clean replacement is available, a decontamination might be attempted followed by strict retesting post-treatment. It is absolutely mandatory that clean cultures also be regularly assayed for mycoplasma contamination, for instance at monthly intervals. PCR with its extreme sensitivity and DNA fluorochrome staining as an inexpensive, simple and rapid method can be applied for regular screening.

Protocol for mycoplasma elimination

The advised procedure is to replace a mycoplasma-positive cell line whenever possible. Frozen stocks can be examined or cell culture repositories can be contacted. An invaluable or irreplaceable cell line can be cured as decontamination with various methods has become a practical and highly successful endeavor. However, a number of factors should be weighed carefully. The treatment protocol plus the obligatory intensive monitoring thereafter is a laborious and time-consuming task. The possibility exists that upon exposure to potentially toxic conditions, the cell line will be subjected to selective pressure and the cured population might differ from the original. Also, the procedure includes extensive work with mycoplasma-positive cultures, thus carrying the risk of further spread of the contamination.

Chemotherapeutic purging of cultures applying various anti-mycoplasma antibiotics has been successful. As a general rule, if cell cultures are diagnosed as being positive for mycoplasma, as quickly as possible, they should be either discarded, frozen and stored in liquid nitrogen, or treated with anti-mycoplasma procedures in order to prevent spread of the contaminant. The whole decontamination process should be undertaken in a quarantined culture laboratory until sterility is attained. It is advisable to freeze some cells for further elimination attempts, in case the first treatment fails to achieve freedom from mycoplasma contamination. The basic principle of antibiotic treatment involves exposure of the positive culture to a recommended antibiotic concentration for a given number of days, followed by maintenance of the cells for at least 2 weeks in antibiotic-free medium, and then the close monitoring of the mycoplasma status by adequate tests to detect residual or new contaminants. Resistant cell lines may be rescued by subsequent incubation with a non-cross-resistant antibiotic (for instance quinolone-resistant cells can be cured with BM-Cyclin). An alternative would be a second attempt with an untreated, mycoplasma-positive aliquot of the same culture using different antibiotics or higher concentrations. The application of elimination methods other than antibiotics remains a further possibility.

The recommendation by commercial suppliers that mycoplasmacidal antibiotics should be used permanently as a preventive measure to avoid contamination must be strongly rejected. Antibiotics in general and anti-mycoplasma reagents in particular should not be used as a substitute for good cell culture technique.

Protocol for prevention of mycoplasma contamination

Of primary importance in preventing mycoplasma contamination of cell cultures is an awareness of sources of infection combined with a program of cell maintenance, regular quality and identity checks, and repeated characterization of the cells[24]. The protocol for prevention of mycoplasma contamination should be designed to minimize the risks of exposure of cells to mycoplasmas. As pointed out above, human sources and bovine serum constitute two of the main infectant reservoirs. Lateral spread from one cell line to another accounts for the majority of infections. A variety of specific steps can be adopted to prevent cell culture contamination with mycoplasmas (Table 17)[7,8].

Any sterile cell culture work should be performed in a vertical laminar-flow biohazard hood. It is critical to disinfect all work surfaces before and after culture manipulations, including the various devices entering the laminar flow hood. Mycoplasmas are extremely sensitive to most disinfectants, but can show prolonged survival in a dried state. The main recommendation is to keep the laboratory and its inventory as clean as possible. It is important that cell culture laboratories establish effective and scheduled mycoplasma testing procedures in the form of a routine screening program for all forms of microbial contamination, including mycoplasmas. Sera, media and supplements (and also cell lines whenever possible) should be purchased from reputable suppliers that adequately test for mycoplasmal contamination.

All incoming cell lines should be quarantined until the contamination status is determined. Mycoplasma-free cultures should be segregated from infected cultures by time and place of handling. Separate reagents for the two sets of cultures should be used. The general addition of antibiotics is not recommended except in special

Table 17 *Protocol for prevention of mycoplasma contamination*

Cell culture facility:

Facility should be designed and equipped for aseptic culture procedures.
Certified laminar flow biological safety cabinets should be used and their function should be regularly examined.
Work surfaces should be chemically disinfected prior to and following work and thoroughly cleaned at regular intervals (monthly).
Incubators should be regularly controlled and cleaned (monthly).
Discarded glass- and plastic-ware and spent media should be carefully disinfected.
Cell culture materials should be properly disposed of by central sterilization.
Effective housekeeping procedures should be followed to minimize contamination of the environment (e.g. floor, sinks, faucets, water baths).
Unauthorized persons should not be allowed entry.
Animals should not be kept in the cell culture room.
Laboratory should be kept clean.

Quality control:

A defined quality, identity and characterization control program for the cell lines should be established.
Mycoplasma testing should be performed at the time of arrival of the culture and at regular intervals (monthly).
Reliable mycoplasma detection methods should be established and performed.
Medium components (especially serum) should be tested for sterility before use.
Strict aseptic techniques and good laboratory procedures should be followed.

Cell cultures:

Cell cultures should be obtained from reputable cell repositories.
Incoming cell cultures should be held in quarantine until proof of sterility (or at least separated in time and space from sterile cultures).
Aliquots free of mycoplasma (and other adventitious agents) should be stored in a master and working stock system in liquid nitrogen.
Antibiotic-free media should be used whenever possible.
Periodic standardization checks of the cell culture (quality, identity and characterization control program) should be made.
Mycoplasma-positive cultures should be immediately discarded or treated with mycoplasmacidal measures.

Culture culturist:

Handwashing prior to and following work should be required.
Mouth pipetting should be prohibited.
There should be no unnecessary talking or traffic at the laminar flow cabinet or in the immediate work area.
Protective clothing should be used to protect both the culture and the culturist.
Jewelry (rings, bracelets, wristwatches) should be taken off.
Written laboratory records for every cell culture should be made.
The same medium aliquot should not be used for different cell lines.
Medium should not be poured from bottles or flasks, but pipetted.
Only one cell line should be handled at a given time.

Modified from ref. 5.

applications and then only for short durations. Use of antibiotics may lead to lapses in aseptic technique, to selection of drug-resistant organisms, and to delayed detection of low-level infection by either mycoplasmas or other bacteria. Master stocks of mycoplasma-free cell lines should be frozen and stored to provide a continuous supply of cells should working stocks become contaminated. Mycoplasma-infected cell lines should be discarded or treated with mycoplasmacidal measures as quickly as possible in order to prevent lateral spread.

Strict adherence of the cell culturist to aseptic culture techniques is another fundamental aspect in mycoplasma control. Cell culturists should continually be aware of the possibility of contaminating clean cultures with aerosols from mycoplasma-containing cultures manipulated in the same area. For example, the following procedures with liquid media generate droplets: pipetting, decanting, centrifuging, sonicating. These relatively large droplets do not remain airborne but settle within seconds in the immediate environment where they remain viable for days. Furthermore, mouth pipetting and unnecessary talking are not acceptable. The prohibition of eating, drinking, smoking, or application of make-up in the laboratory is obvious.

References

1 Razin, S. and Freundt, E.A. (1984) In: Bergey´s Manual of Systematic Bacteriology, Vol. 1 (Holt, J.G. ed). Williams & Wilkins, New York, pp. 740–794.

2 Razin, S. (1991) In: The Prokaryotes, 2nd edn (Balows, A. et al. eds). Springer-Verlag, New York, pp. 1937–1958.

3 Tully, J.G. (1992) In: Encyclopedia of Microbiology (Lederberg, J. ed). Academic Press, San Diego, pp. 181–215.

4 Razin, S. (1995) In: Molecular and Diagnostic Procedures in Mycoplasmology, Vol. 1 (Razin, S. and Tully, J.G. eds). Academic Press, San Diego, pp. 1–25.

5 Drexler, H.G. and Uphoff, C.C. (2000) In: The Encyclopedia of Cell Technology (Spier, E. ed). Wiley, New York, pp. 609–627.

6 Razin, S. et al. (1998) Microbiol. Mol. Biol. Reviews 62, 1094–1156.

7 McGarrity, G.J. et al. (1985) Am. Soc. Microbiol. News 51, 170–183.

8 Hay, R.J. et al. (1989) Nature 339, 487–488.

9 Barile, M.F. and Rottem, S. (1993) In: Rapid Diagnosis of Mycoplasmas (Kahane, I. and Adoni, A. eds). Plenum Press, New York, pp. 155–193.

10 Drexler, H.G. and Uphoff, C.C. (1999) unpublished data.

11 McGarrity, G.J. et al. (1992) In: Mycoplasmas – Molecular Biology and Pathogenesis (Maniloff, J. et al. eds). American Society for Microbiology, Washington, DC, pp. 445–454.

12 Uphoff, C.C. et al. (1992) J. Immunol. Methods 149, 43–53.

13 Razin, S. (1994) Mol. Cell. Probes 8, 497–511.

14 Uphoff, C.C. and Drexler, H.G. (2000) Human Cell 12, 229–236.

15 Uphoff, C.C. et al. (1992) Leukemia 6, 335–341.

16 Rawadi, G. and Dussurget, O. (1995) PCR Methods Applications 4, 199–208.

17 Hopert, A. et al. (1993) In Vitro Cell. Dev. Biol. 29A, 819–821.

18 Hopert, A. et al. (1993) J. Immunol. Methods 164, 91–100.

19 Uphoff, C.C. et al. (1992) J. Immunol. Methods 149, 55–62.

20 Drexler, H.G. et al. (1994) In Vitro Cell. Dev. Biol. 30A, 344–347.

21 Gignac, S.M. et al. (1991) Leukemia 5, 162–165.

[22] Gignac, S.M. et al. (1992) Leukemia Res. 16, 815–822.
[23] Fleckenstein, E. et al. (1994) Leukemia 8, 1424–1434.
[24] Drexler, H.G. et al. (1999) DSMZ Catalogue of Cell Lines, 7th edn, Braunschweig, Germany.

8 Availability of Cell Lines and Cell Line Banks

AVAILABILITY OF CELL LINES

Availability of cell lines from originators or secondary sources

The availability of cell lines to investigators outside the original laboratory where the cell line was established is of the utmost importance. There are three possibilities for obtaining a cell line: (1) directly from the original investigators; (2) indirectly from other scientists who have been provided with the cell line by the original investigators; and (3) indirectly from cell line banks. Thus, one must discern the original source from secondary sources. Obtaining a cell line indirectly from a third party appears to potentially cause several problems: the forwarding of cell lines might be done without the approval of the original investigator; the correct identity of the cell line might be even less guaranteed than in the original laboratory (although the data presented in Figure 3 in *Chapter 6*: Authentication of Cell Lines, do not seem to support this contention); the secondary laboratory may take less care with regard to the quality of the cell culture (viability, mycoplasma contamination, etc.); and the lack of optimal culture conditions may lead to the outgrowth of subclones with somewhat different biological features. The third option for obtaining cell lines, namely from cell line banks, is discussed below.

While it appears to be logical to go directly to the original source, unfortunately this is all too often not an easy and convenient task. Figure 6 shows that, according to my experience, only 54% of the cell lines that were requested were, indeed, provided by the original investigators. The second largest contingent (29%) did not respond at all to the requests, followed by an initial pledge to provide the cells (but the authors never did so) (7%), outright refusal to provide the cells (5%), admission that the cell line no longer existed (2%), and finally referral to other scientists or cell banks (3%).

There are a number of reasons that may intervene to prevent scientists from providing cell lines (at least 41%) to another independent scientist. These fall into two broad categories reflecting problems with the cells themselves (negligence and 'premature publication') or their originators (change of address, commercial interest, co-authorship, indifference):

1. the cell line may be unrecoverable, for example due to faulty cryopreservation or microbial contamination, or fail to maintain proliferation *in vitro*, perhaps having never been truly immortalized;
2. originators cannot be contacted due to change of address; they may also withhold release actively because of perceived loss of proprietorial interest (for example commercial, rights to co-authorship, uncontrolled release to competitors), or passively due to overriding constraints on time, material and personnel.

Some scientific journals (e.g. the premier hematological journal *Blood*) have adopted the policy that any readily renewable resources, including cell lines published in that journal, must be made available to all qualified investigators in the field, if not already obtainable from commercial sources (in the case of cell lines from cell line banks). The policy stems from the long-standing scientific principle

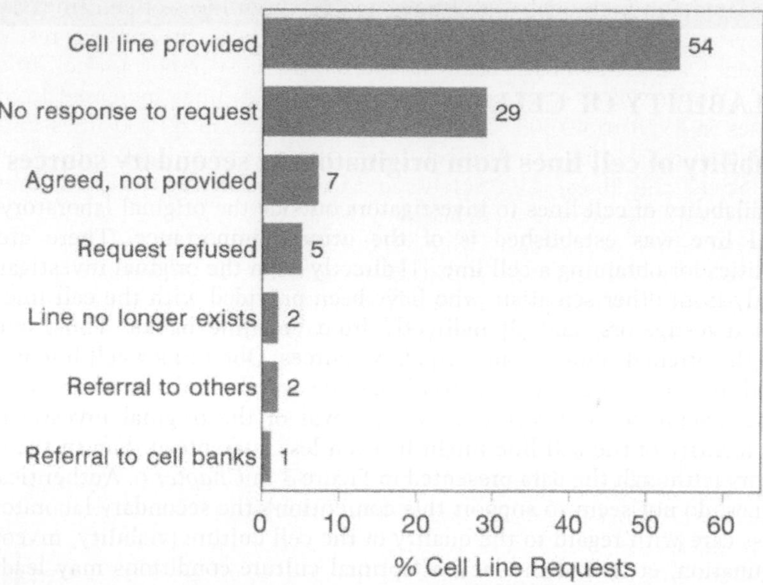

Figure 6 *Response to leukemia-lymphoma cell line requests. Shown are the various types of response by the original investigators to my request for provision of their cell lines. A total of 578 cell lines were requested: 54% of the cell lines were provided; in 29% of the cases, there was no response to the request; for 7% of the cell lines, the original investigators agreed to provide the cells, but never did so; in 5% of the cases, the scientists refused to provide the cell lines; 2% of the cell lines no longer existed (cells were lost or were not immortalized); for 3% of the cell lines, the original authors referred to other scientists or to cell banks (where the cells had not been deposited). Not included here are 63 cell lines that were obtained from the original investigators as they are close collaborators of mine (Drs J. Minowada and Y. Matsuo).*

that authenticity requires reproducibility (see also *Chapter 4*: Guidelines for Characterization of Cell Lines). While cell lines are proprietary and unique, suitable material transfer agreements can be drawn up between the provider and requester[1].

CELL LINE BANKS

The role of cell line banks

The principal role of centralized cell banks lies in providing reference cultures for cancer research. At least for the major cell banks, the procedures applied in development, maintenance, expansion, quality and identity controls, and characterization of these resources are standardized. The use of cell lines in biomedical research and biotechnology increased exponentially in the last two decades. One parameter for documenting this development is the distribution of cell lines by cell banks. The distribution profile clearly reflects both the need for authenticated and high quality cell lines and the usefulness of cell lines as either model systems in basic research or important tools in applied research or production. Data are available from the oldest and most important cell line bank, the ATCC in the USA: the distribution of ATCC cell lines increased from some 5000 lines in 1978 to 40 000 lines in 1988 to more than 60 000 lines in 1998[2,3]. These figures refer to all types of cell lines, human and animal, solid tumor and hematological cell lines; data solely on the distribution of human leukemia-lymphoma cell lines are not available from the ATCC.

The availability of stocks of characterized reference cell lines for basic and clinical research carries the following advantages[4]: (1) scientists can receive contaminant-free reference cell lines at the same passage level from time to time over the years for systematic experimentation; (2) cell lines whose origin, history and characteristic features have been described and documented are readily available; (3) some cell types that may normally be difficult to acquire can be obtained; (4) the original investigator is released from time-consuming and costly long-term serial subcultivation and supply of cell lines; and (5) both individual scientists and scientific institutions have insurance against the total loss of valuable cells through contamination, alterations, or accidents.

The advantages of working with properly identified cell lines in a system free from contaminating organisms are apparent enough in terms of conservation of time, effort, and research money. An added overall benefit to the entire scientific community is the provision of a common source of reference cell lines at similar passage levels being available for repeated use in perpetuity. It is known that the features and properties of cells sometimes change markedly during cultivation over extended periods *in vitro*. Thus, cell populations maintained in different laboratories may vary dramatically[4].

Table 18 lists particulars including acronyms, postal addresses, phone and fax numbers, e-mail and website addresses of the major public cell line banks and a

Table 18 *Reference cell line collections for human leukemia-lymphoma cell lines*

Major cell line banks:[a]

Germany	**DSMZ** – Deutsche Sammlung von Mikroorganismen und Zellkulturen Mascheroder Weg 1B, 38124 Braunschweig, Germany Tel +49-531-2616.161 Fax +49-531-2616.150 E-Mail <mutz@dsmz.de> Website <www.dsmz.de>
Japan	**JCRB/HSRRB** **JCRB** – Japanese Collection of Research Bioresources Cellbank at National Institute of Health Sciences 1-18-1 Kami-Yoga, Setagaya, Tokyo 158-8501, Japan Tel +81-3-3700.1141 Fax +81-3-3707.6950 E-Mail <cellbank@nihs.go.jp> Website <www.nihs.go.jp>

Continued overleaf

Table 18 *Continued*

	HSRRB – Health Science Research Resources Bank (previously at: **IFO** – Institute for Fermentation) 1-1-43 Hoenzaka, Chuo-Ku, Osaka 540-0006, Japan Tel +81-6-945.2869 Fax +81-6-945.2872 E-Mail <hsrrb@nihs.go.jp Website <www.nihs.go.jp>
Japan	**RIKEN** Cell Bank 3-1-1 Koyadai, Tsukuba Science City, Ibaraki 305, Japan Tel +81-298-36.3611 Fax +81-298-36.9130 E-Mail <cellbank@rtc.riken.go.jp> Website <www.rtc.riken.go.jp>
United States	**ATCC** – American Type Culture Collection 10801 University Boulevard, Manassas, Virginia 20110-2209, USA Tel +1-703-365.2700 Fax +1-703-365.2701 E-Mail <help@atcc.org> Website <www.atcc.org>
Other cell line banks:[b]	
Brazil	**BCRJ** – Banco de Células do Rio de Janeiro University Hospital Clementino Frago Filho Federal University of Rio de Janeiro, Cidade Universitaria, Ilha do Fundao CP 68021, 21941-970 Rio de Janeiro, Brazil Tel/Fax +55-21-590.8736 E-Mail <flabra@ccard.com.br>
China	**TCCCAS** – Type Culture Collection of Chinese Academy of Sciences Shanghai Institute of Cell Biology of Chinese Academy of Sciences 320 Yue Yang Road, Shanghai 200031, People's Republic of China Tel +86-21-6431.5030 ext. 2130/2052 Fax +86-21-6433.1090
China	**CBCCTCC** – The Cell Bank of China Center for Type Culture Collection School of Life Science, University of Wuhan Wuhan 430072, People's Republic of China Tel +86-27-722.157 Fax +86-27-713.833
England	**ECACC** – European Collection of Cell Cultures

Table 18 *Continued*

	Centre for Applied Microbiology and Research Porton Down, Salisbury, Wiltshire SP4 0JG, UK Tel +44-1980-612.512 Fax +44-1980-611.315 E-Mail <ecacc@camr.org.uk> Website <www.camr.org.uk>
India	**NFATCC** – National Facility for Animal Tissue and Cell Culture Department of Biotechnology, Government of India Jopasana, Paud Road, Kothrud, Pune 411 029, India Tel +91-212-335.928 Fax +91-212-369.501
Italy	**ICLC** – Interlab Cell Line Collection Advanced Biotechnology Center, Department of Biotechnology Largo Rosanna Benzi 10, 16132 Genova, Italy Tel +39-010-573.7474 Fax +39-010-573.7295 E-Mail <iclc@ist.unige.it> Website <www.biotech.ist.unige.it>
Russia	**RACCC** – Russian Animal Cell Culture Collection Institute of Cytology, Department of Cell Culture Tikoretsky av. 4, 194064 St. Petersburg, Russia Tel +7-812-247.4296 Fax +7-812-247.0341 E-Mail <root@cell.spb.su>
South Korea	**KCLRF** – Korean Cell Line Research Foundation Cancer Research Institute, Seoul National University College of Medicine 28 Yongun-dong, Chongno-ku, Seoul 110-744, South Korea Tel +82-2-760.3380 Fax +82-2-742.4727
United States	**CIMR** – Coriell Institute for Medical Research Coriell Cell Repositories 401 Haddon Avenue, Camden, New Jersey 08103, USA Tel +1-856-757.4848 Fax +1-856-757.9737 E-Mail <ccr@arginine.umdnj.edu> Website <locus.umdnj.edu>

[a] These major cell line banks have programs for quality and identity control, albeit performed at varying levels of stringency.

[b] These additional or semi-commercial (ECACC) cell line banks do not or cannot perform stringent quality and identity control programs.

panel of additional cell line banks all over the world. The stringency and extent of the programs for the quality and identity controls and characterization of the cell lines varies considerably between these banks.

Human leukemia-lymphoma cell lines in cell line banks

The development of leukemia-lymphoma cell line collections within several public cell line banks, with the ensuing availability of large panels of cell lines, has tremendously enhanced research in this area. A primary function of cell line banks should be to provide authenticated, clean and well-characterized cell material. The requirement for a stringent program of quality, identity and characterization analyses of cells is assured in only a few cell banks[5,6]. The most extensive non-profit collection of hematopoietic cell lines has been established at the DSMZ in Germany (109 leukemia-lymphoma cell line holdings). Other major cell line banks also hold a limited number of the most often used hematopoietic cell lines: ATCC (54 holdings), JCRB (38 holdings), IFO (12 holdings), and RIKEN (23 holdings) (Table 18). Table XVII recommends a list of well-characterized and authenticated reference leukemia-lymphoma cell lines which are, for the most part, available from these cell line banks.

While cell banks are certainly eager to acquire new cell lines, the various functions of expansion, authentication and distribution of cell lines are very costly processes requiring considerable support from the government and other sources. For this reason, evidence of utility and demand for each particular cell line must be considered before a given cell line can be accessioned.

References

1 Kaushansky, K. (1998) Blood 91, 1–2.
2 Hay, R.J. (1984) In: Uses and Standardization of Vertebrate Cell Cultures (Patterson, M.K.Jr. ed). In Vitro Monograph No. 5, Tissue Culture Association, Gaithersburg, MD, USA, pp. 215–224.
3 Hay, R.J. (1999) personal communication.
4 Hay, R.J. et al. (1996) J. Cell. Biochem., Suppl. 24, 107–130.
5 Drexler, H.G. et al. (1999) DSMZ Catalogue of Cell Lines, 7th edn, Braunschweig, Germany.
6 Hay, R.J. (1998) Methods Cell Biol. 57, 31–47.

Section II

B-CELL LEUKEMIA AND B-CELL LYMPHOMA CELL LINES

Part 1
Precursor B-Cell Lines

Culture characterization[1]

Other name of cell line	**EU-2**
Culture medium	90% RPMI 1640 or IMDM + 10% FBS
Doubling time	60–70 h
Viral status	EBV$^-$
Authentication	no
Primary reference	Findley et al.[1]
Availability of cell line	original authors

Clinical characterization[1]

Patient	10-year-old male
Disease diagnosis	cALL
Treatment status	at 2nd relapse
Specimen site	bone marrow
Year of establishment	1980

Immunophenotypic characterization[1,2]

T-/NK cell marker	CD1$^-$, CD2$^-$, CD3$^-$, CD4$^-$, CD5$^+$, CD7$^-$, CD8$^-$, CD57$^+$
B-cell marker	CD10$^+$, CD19$^+$, cyIgM$^+$, sIg$^-$
Myelomonocytic marker	CD15$^-$
Progenitor/activation/other marker	CD71$^+$, HLA-DR$^+$, TdT$^+$

Genetic characterization[1,3]

Cytogenetic karyotype	46, XY, no chromosomal abnormalities
Unique gene alteration	*P16INK4A* deletion, *P53* mutation

Functional characterization[1,4]

Colony formation	in methylcellulose
Cytochemistry	ACP$^+$, ANAE$^-$, CAE$^-$, MPO$^-$, PAS$^+$
Proto-oncogene	mRNA$^+$: BAX, BCLXL, MDM2

Comments

- Precursor B-cell line of type B-III (pre B-cell line).
- Three different 207 cultures which were received from the original investigators and from a secondary source were found to be cross-contaminated (real identities of these cultures: CCRF-CEM, Reh, SUP-B8)[5]. It is not known whether the original, correct 207 still exists.

References

[1] Findley, H.W. et al. (1982) Blood 60, 1305–1309.
[2] Zhou, M. et al. (1995) Blood 85, 1608–1614.
[3] Zhou, M. et al. (1995) Leukemia 9, 1159–1161.
[4] Findley, H.W. et al. (1997) Blood 89, 2986–2993.
[5] Drexler, H.G. et al. (1999) Leukemia 13, 1601–1607.

Culture characterization[1,2]

Culture medium	80% RPMI 1640 + 20% FBS
Doubling time	3 d
Viral status	EBV⁻, HBV⁻, HCV⁻, HHV-8⁻, HIV⁻, HTLV-I/II⁻
Authentication	no
Primary reference	Pegoraro et al.[1]
Availability of cell line	DSM ACC 39

Clinical characterization[1]

Patient	15-year-old male
Disease diagnosis	ALL (L2)
Treatment status	at relapse
Specimen site	peripheral blood
Year of establishment	1983

Immunophenotypic characterization[1,2]

T-/NK cell marker	$CD1^-$, $CD2^-$, $CD3^-$, $CD4^-$, $CD8^-$
B-cell marker	$CD10^+$, $CD19^+$, $CD37^+$, $cyIgM^-$, sIg^-
Myelomonocytic marker	$CD13^-$
Progenitor/activation/other marker	$CD71^+$, $HLA\text{-}DR^+$, TdT^+

Genetic characterization[1-3]

Cytogenetic karyotype	near-diploid, 1.8% polyploidy; 46(43–47)<2n XY, −14, t(8;14;18)(q24;q32;q21), +der(14)t(8;14;18)(q24;q32;q21)
Unique translocation/fusion gene	(1) t(8;14)(q24;q32) → *MYC-IGH* genes altered (2) t(14;18)(q32;q21) → *IGH-BCL2* genes altered
Receptor gene rearrangement	*IGH* RR, *IGK* DD, *IGL* RR
Unique gene alteration	*P53* mutation

Functional characterization[1]

Cytochemistry	ALP^-, $ANAE^-$, MPO^-, PAS^-

Comments

- Precursor B-cell line of type B-II (common B-cell line).
- Carries t(8;14)(q24;q32) affecting the *MYC-IGH* genes (typically seen in B-ALL or Burkitt lymphoma).
- Carries t(14;18)(q32;q21) affecting the *BCL2-IGH* genes (typically seen in follicular B-cell lymphoma).
- Cells are very difficult to culture.
- Available from cell line bank.

References

[1] Pegoraro, L. et al. (1984) Proc. Natl. Acad. Sci. USA 81, 7166–7170.

[2] Drexler, H.G. et al. (1999) DSMZ Catalogue of Cell lines, 7th edn, Braunschweig, Germany.

[3] van Ooteghem, R.B.C. et al. (1994) Cancer Genet. Cytogenet. 74, 87–94.

697

Culture characterization[1,2]

Other names of cell line	**EU-3**
Culture medium	80% RPMI 1640 or IMDM + 20% FBS
Doubling time	30–40 h
Viral status	EBV⁻, HBV⁻, HCV⁻, HHV-8⁻, HIV⁻, HTLV-I/II⁻
Authentication	yes (by cytogenetics)
Primary reference	Findley et al.[1]
Availability of cell line	DSM ACC 42

Clinical characterization[1]

Patient	12-year-old male
Disease diagnosis	null ALL (non-T non-B ALL)
Treatment status	at relapse
Specimen site	bone marrow
Year of establishment	1979

Immunophenotypic characterization[1-4]

T-/NK cell marker	$CD1^-$, $CD2^-$, $CD3^-$, $CD4^-$, $CD5^-$, $CD7^-$, $CD8^-$, $CD57^+$
B-cell marker	$CD10^+$, $CD19^+$, $CD20^-$, $CD24^+$, $CD37^{(+)}$, $cyIgM^+$, sIg^-
Myelomonocytic marker	$CD13^-$, $CD15^+$
Progenitor/activation/other marker	$CD71^+$, HLA-DR^+, TdT^+

Genetic characterization[1,2,5]

Cytogenetic karyotype	46(45–48)<2n>XY, t(1;19)(q23;p13), del(6)(q21)
Unique translocation/fusion gene	t(1;19)(q23;p13) → *E2A-PBX1* fusion gene
Unique gene alteration	*P53* mutation

Functional characterization[1,6,7]

Colony formation	in methylcellulose
Cytochemistry	ACP^+, $ANAE^-$, CAE^-, MPO^-, PAS^+
Cytokine response	growth inhibition: TNF-α
Proto-oncogene	$mRNA^+$: BAX, BCL2, BCLXL, MDM2

Comments

- Precursor B-cell line of type B-III (pre B-cell line).
- Carries t(1;19)(q23;p13) leading to *E2A-PBX1* fusion gene.
- Available from cell line bank.

References

[1] Findley, H.W. et al. (1982) Blood 60, 1305–1309.

[2] Drexler, H.G. et al. (1999) DSMZ Catalogue of Cell Lines, 7th edn, Braunschweig, Germany.

[3] Kim, D.H. et al. (1996) Blood 88, 785–794.

[4] Zhou, M. et al. (1995) Blood 85, 1608–1614.
[5] Barker, P.E. et al. (1987) Cancer Genet. Cytogenet. 25, 379–380.
[6] Zhou, M. et al. (1991) Blood 77, 2002–2007.
[7] Findley, H.W. et al. (1997) Blood 89, 2986–2993.

Culture characterization[1,2]

Other name of cell line	**ALL-A**
Culture medium	90% α-MEM + 10% FBS
Doubling time	48–61 h
Viral status	EBV⁻
Authentication	no
Primary reference	Kamel-Reid et al.[1]
Availability of cell line	restricted

Clinical characterization[1,3]

Patient	male child
Disease diagnosis	pre B-ALL
Treatment status	at relapse
Specimen site	peripheral blood

Immunophenotypic characterization[1,3]

B-cell marker	$CD10^-$, $CD19^+$, $CD20^-$, $cyIgM^-$, sIg^-
Progenitor/activation/other marker	$HLA\text{-}DR^+$
Adhesion marker	$CD44^+$

Genetic characterization[1,3–5]

Cytogenetic karyotype	45, XY, multiple numerical + structural abnormalities
Receptor gene rearrangement	*IGH* RG, *IGK* G, *IGL* G
Unique gene alteration	*P53* mutation

Functional characterization[1,3,4]

Cytokine response	growth stimulation: GM-CSF
Heterotransplantation	into SCID mice (not nude mice)

Comments

- Precursor B-cell line of type B-I (pro B-cell line).

References

[1] Kamel-Reid, S. et al. (1989) Science 246, 1597–1600.
[2] Brown, G.A. et al. (1995) Cancer Res. 55, 78–82.
[3] Freedman, M.H. et al. (1993) Blood 81, 3068–3075.
[4] Kamel-Reid, S. et al. (1992) Leukemia 6, 8–17.
[5] Lam, V. et al. (1999) Leukemia Res. 23, 871–880.

ALL-1

Culture characterization[1]

Culture medium	90% IMDM + 10% FBS
Viral status	EBNA$^-$, HTLV-II$^-$
Authentication	yes (partly by cytogenetics)
Primary reference	Lange et al.[1]
Availability of cell line	original authors

Clinical characterization[1]

Patient	6-year-old female
Disease diagnosis	pre B-ALL (L1)
Treatment status	at diagnosis
Specimen site	bone marrow

Immunophenotypic characterization[1–3]

T-/NK cell marker	CD2$^-$, CD3$^-$, CD5$^-$, CD7$^-$
B-cell marker	CD10$^+$, CD19$^+$
Myelomonocytic marker	CD13$^-$, CD14$^-$, CD15$^-$, CD33$^-$
Progenitor/activation/other marker	HLA-DR$^+$, TdT$^+$
Adhesion marker	CD18$^-$, CD44$^+$, CD54$^-$

Genetic characterization[1–4]

Cytogenetic karyotype	47, XX, t(9;22)(q34;q11), +Ph
Unique translocation/fusion gene	Ph$^+$ t(9;22)(q34;q11) → *BCR-ABL* b3-a2 fusion gene
Receptor gene rearrangement	*IGH* RD, *IGK* RR, *IGL* G, *TCRB* G

Functional characterization[1,2]

Cytochemistry	ANAE$^-$, CAE$^-$, MPO$^-$, PAS$^+$
Heterotransplantation	into SCID mice

Comments

- Precursor B-cell line of undetermined type.
- Carries Philadelphia chromosome leading to *BCR-ABL* b3-a2 fusion gene.

References

[1] Lange, B. et al. (1987) Blood 70, 192–199.
[2] Cesano, A. et al. (1991) Blood 77, 2463–2474.
[3] Erikson, J. et al. (1986) Proc. Natl. Acad. Sci. USA 83, 1807–1811.
[4] Drexler, H.G. et al. (1999) Leukemia Res. 23, 207–215.

Culture characterization[1,2]

Subclone	**ALL-2B** (nearly identical)
Culture medium	90% IMDM + 10% FBS
Viral status	EBNA⁻, HTLV-II⁻
Authentication	yes (by cytogenetics)
Primary reference	Lange et al.[1]
Availability of cell line	not known

Clinical characterization[1,2]

Patient	10-year-old female
Disease diagnosis	pre B-ALL (L1)
Treatment status	at relapse

Immunophenotypic characterization[1]

T-/NK cell marker	$CD2^-$, $CD3^-$, $CD5^-$, $CD7^-$
B-cell marker	$CD10^+$, $CD19^+$
Myelomonocytic marker	$CD13^-$, $CD14^-$, $CD15^-$, $CD33^-$
Progenitor/activation/other marker	HLA-DR^+
Adhesion marker	$CD18^+$, $CD44^+$, $CD54^-$

Genetic characterization[1,2]

Cytogenetic karyotype	ALL-2B: 45, XX, −9, t(3;?)(q29;?), t(4;11)(q21;q23), del(6)(q23), del(17)(p11), der(19)t(1;19)(q23;p13)
Unique translocation/fusion gene	(1) t(1;19)(q23;p13) → *E2A-PBX1* fusion gene (2) t(4;11)(q21;q23) → *MLL-AF4* fusion gene
Receptor gene rearrangement	*IGH* RR, *IGK* DD, *IGL* G, *TCRB* G

Functional characterization[1,2]

Cytochemistry	$ANAE^-$, CAE^-, MPO^-, PAS^-
Heterotransplantation	into SCID mice

Comments

- Precursor B-cell line carrying two cytogenetic fusion genes.
- Carries t(1;19)(q23;p13) leading to *E2A-PBX1* fusion gene.
- Carries t(4;11)(q21;q23) leading to *MLL-AF4* fusion gene.

References

[1] Lange, B. et al. (1987) Blood 70, 192–199.
[2] Cesano, A. et al. (1991) Blood 77, 2463–2474.

Culture characterization[1]

Culture medium	90% RPMI 1640 + 10% FBS
Doubling time	36–48 h
Viral status	EBNA⁻
Authentication	yes (by *IGH* rearrangement)
Primary reference	Higa et al.[1]
Availability of cell line	original authors

Clinical characterization[1]

Patient	74-year-old female
Disease diagnosis	ALL
Treatment status	during therapy
Specimen site	peripheral blood
Year of establishment	1986

Immunophenotypic characterization[1,2]

T-/NK cell marker	$CD1^-$, $CD2^-$, $CD3^-$, $CD4^-$, $CD7^-$, $CD8^-$, $CD56^-$
B-cell marker	$CD10^+$, $CD19^+$, $CD20^+$, $CD21^-$, $CD22^+$, $cyIgM^-$, sIg^-
Myelomonocytic marker	$CD13^-$, $CD14^-$, $CD15^-$, $CD33^-$
Erythroid-megakaryocytic marker	$CD36^-$, $CD41^-$, $GlyA^-$
Progenitor/activation/other marker	$CD34^+$, $CD38^+$, $HLA\text{-}DR^+$
Adhesion marker	$CD11b^-$
Cytokine receptor	$CD25^-$, $CD122^-$, $CD124^+$

Genetic characterization[1,3]

Cytogenetic karyotype	46, XX, t(9;22)(q34;q11)
Unique translocation/fusion gene	Ph^+ t(9;22)(q34;q11) → *BCR-ABL* e1-a2 fusion gene
Receptor gene rearrangement	*IGH* RR, *IGK* DR

Functional characterization[1,4]

Cytochemistry	$ANBE^-$, CAE^-, MPO^-, PAS^+, SBB^-
Cytokine response	growth stimulation: IL-7; growth inhibition: IFN-α, IFN-β, IFN-γ, IL-4, TNF-α
Inducibility of differentiation	TPA → mono differentiation

Comments

- Precursor B-cell line of type B-II (common B-cell line).
- Carries Ph chromosome leading to *BCR-ABL* e1-a2 fusion gene.
- Described to be responsive to monocytic differentiation induction.

References

[1] Higa, T. et al. (1994) Leukemia Lymphoma 12, 287–296.
[2] Toba, K. et al. (1996) Exp. Hematol. 24, 894–901.
[3] Drexler, H.G. et al. (1999) Leukemia Res. 23, 207–215.
[4] Okabe, M. et al. (1992) Leukemia Lymphoma 8, 57–63.

B1

Culture characterization[1,2]

Other name of cell line	**ALL-B**
Culture medium	90% α-MEM + 10% FBS
Doubling time	40–50 h
Viral status	EBV⁻
Authentication	yes (by cytogenetics)
Primary reference	Cohen et al.[1]
Availability of cell line	restricted

Clinical characterization[1]

Patient	14-year-old male
Disease diagnosis	ALL
Treatment status	at relapse
Specimen site	bone marrow

Immunophenotypic characterization[1]

T-/NK cell marker	$CD2^-$, $CD3^-$
B-cell marker	$CD10^-$, $CD19^+$, $CD20^-$, $cyIgM^-$, sIg^-
Myelomonocytic marker	$CD13^-$, $CD14^-$, $CD33^+$
Progenitor/activation/other marker	$HLA\text{-}DR^+$
Adhesion marker	$CD11b^-$

Genetic characterization[1]

Cytogenetic karyotype	45, XY, −4, +6, −9, der(11)t(4;11)(q21;q23), der(1)t(1;8)(p36;q13), der(10)t(1;10)(q11;p15)
Unique translocation/fusion gene	t(4;11)(q21;q23) → *MLL-AF4* fusion gene
Receptor gene rearrangement	*IGH* R

Functional characterization[1,3,4]

Colony formation	in methylcellulose
Cytochemistry	ACP^+, $ANAE^+$, PAS^+, SBB^-
Cytokine production	secretion: IL-1β
Cytokine receptor	$RT\text{-}PCR^+$: IFN-γR, IL-1R, IL-6Rα, IL-7Rα, TNF-αR
Cytokine response	growth stimulation: IL-1α, IL-1b; growth inhibition: IFN-γ, IL-6, IL-7, TNF-α
Inducibility of differentiation	IL-6 → myeloid differentiation

Comments

- Precursor B-cell line of type B-I (pro B-cell line).
- Carries t(4;11)(q21;q23) leading to *MLL-AF4* fusion gene.
- Responsive to various cytokines.

References

[1] Cohen, A. et al. (1991) Blood 78, 94–102.

[2] Brown, G.A. et al. (1995) Cancer Res. 55, 78–82.

[3] Cohen, A. et al. (1992) Leukemia Res. 16, 751–760.

[4] Freedman, M.H. et al. (1993) Blood 81, 3068–3075.

BLIN-1

Culture characterization[1]

Culture medium	75% RPMI 1640 + 25% FBS (or serum-free)
Doubling time	4–5 d
Viral status	EBNA⁻
Authentication	yes (by cytogenetics)
Primary reference	Wörmann et al.[1]
Availability of cell line	original authors

Clinical characterization[1]

Patient	11-year-old male
Disease diagnosis	ALL
Treatment status	at diagnosis
Specimen site	bone marrow
Year of establishment	1986

Immunophenotypic characterization[1]

T-/NK cell marker	$CD2^-$, $CD5^+$
B-cell marker	$CD10^+$, $CD19^+$, $CD20^{(+)}$, $CD21^-$, $CD22^+$, $CD24^+$, $cyIgM^+$ $sIgM^{(+)}$
Progenitor/activation/other marker	$HLA\text{-}DR^+$, TdT^+

Genetic characterization[1,2]

Cytogenetic karyotype	46, XY, −9, der(9)t(8;9)(q?21.2;p2?2)
Receptor gene rearrangement	*IGH* R, *IGK* R, *IGL* R, *TCRB* G, *TCRG* G
Unique gene alteration	*P15INK4B* deletion, *P16INK4A* deletion, *P19ARF* deletion

Functional characterization[1]

Colony formation	in methylcellulose
Cytochemistry	$ANAE^-$, CAE^-, PAS^+, SBB^-
Cytokine response	growth stimulation: BCGF, IL-3, IL-7
Inducibility of differentiation	spontaneous B-cell differentiation (TPA no effect)

Comments

- Precursor B-cell line of type B-III (pre B-cell line).

References

[1] Wörmann, B. et al. (1989) J. Immunol. 142, 110–117.

[2] Shah, N. et al. (1998) Blood 92, 3817–3828.

BLIN-2

Culture characterization[1]

Culture medium	RPMI 1640 + bone marrow stromal cells (feeder layer)
Doubling time	2 d
Authentication	yes (by *IGH* rearrangement)
Primary reference	Shah et al.[1]
Availability of cell line	original authors

Clinical characterization[1]

Patient	3-year-old female
Disease diagnosis	BCP-ALL
Treatment status	at diagnosis
Specimen site	bone marrow
Year of establishment	1993

Immunophenotypic characterization[1]

T-/NK cell marker	$CD7^-$
B-cell marker	$CD10^+$, $CD19^+$, $CD20^+$, $CD21^-$, $CD22^+$, $CD40^-$, $sIgM^{(+)}$
Progenitor/activation/other marker	$CD34^-$, $CD38^+$, HLA-DR^+
Adhesion marker	$CD49d^+$, $CD49e^+$

Genetic characterization[1]

Cytogenetic karyotype	46, XX, +8, dic(9;20)(p11;q11.1)
Receptor gene rearrangement	*IGH* R
Unique gene alteration	*P16INK4A* deletion, *P19ARF* deletion

Functional characterization[1]

Cytokine response	stromal cell feeder layer-dependent

Comments

- Precursor B-cell line of type B-III (pre B-cell line).
- Constitutively dependent on some unknown growth factor.

Reference

[1] Shah, N. et al. (1998) Blood 92, 3817–3828.

Culture characterization[1]

Culture medium	85% IMDM + 15% FBS
Doubling time	24–48 h
Viral status	EBNA⁻
Authentication	yes (by cytogenetics)
Primary reference	Katz et al.[1]
Availability of cell line	not known

Clinical characterization[1]

Patient	12-day-old male
Disease diagnosis	BCP-ALL (L1)
Treatment status	at diagnosis
Specimen site	bone marrow
Year of establishment	1989

Immunophenotypic characterization[1]

T-/NK cell marker	$CD2^-$, $CD7^-$
B-cell marker	$CD10^-$, $CD19^+$, $CD20^-$, $CD21^-$, $CD24^+$, $cyIgM^-$
Myelomonocytic marker	$CD13^+$, $CD14^-$, $CD15^-$, $CD33^-$
Progenitor/activation/other marker	$CD34^-$, $HLA\text{-}DR^+$, TdT^+
Adhesion marker	$CD11c^-$

Genetic characterization[1]

Cytogenetic karyotype	46, XY, t(11;19)(q23;p13), t(11;19)(q13;q13)
Unique translocation/fusion gene	t(11;19)(q23;p13) → *MLL-ENL* fusion gene?
Receptor gene rearrangement	*IGH* R, *IGK* G, *TCRB* G, *TCRG* G, *TCRD* G

Functional characterization[1]

Cytochemistry	ACP^-, $ANAE^{(+)}$, CAE^-, PAS^+, SBB^-
Inducibility of differentiation	TPA → myelomono differentiation

Comments

- Precursor B-cell line of type B-I (pro B-cell line) established from an infant.
- Carries t(11;19)(q23;p13) possibly leading to the *MLL-ENL* fusion gene.
- Described to be responsive to myelomonocytic differentiation induction.

Reference

[1] Katz, F. et al. (1991) Leukemia Lymphoma 4, 397–404.

BV173

Culture characterization[1,2]

Culture medium	90% RPMI 1640 + 10% FBS (or serum-free)
Doubling time	30–48 h
Viral status	EBV⁺, HBV⁻, HCV⁻, HHV-8⁻, HIV⁻, HTLV-I/II⁻
Authentication	no
Primary reference	Pegoraro et al.[1]
Availability of cell line	DSM ACC 20

Clinical characterization[1]

Patient	45-year-old male
Disease diagnosis	CML
Treatment status	at lymphoid blast crisis
Specimen site	peripheral blood
Year of establishment	1980

Immunophenotypic characterization[1–6]

T-/NK cell marker	$CD1b^-$, $CD2^-$, $CD3^-$, $cyCD3^-$, $CD4^-$, $CD5^-$, $CD7^-$, $CD8^-$, $CD28^-$, $CD56^-$, $CD57^-$, $TCR\alpha\beta^-$, $TCR\gamma\delta^-$
B-cell marker	$CD9^+$, $CD10^+$, $CD19^+$, $CD20^-$, $CD21^-$, $CD22^+$, $CD23^-$, $CD24^-$, $CD37^{(+)}$, $cyCD79a^+$, $cyCD79b^+$, $CD85^+$, $cyIgM^-$, sIg
Myelomonocytic marker	$CD13^+$, $CD14^-$, $CD15^{(+)}$, $CD33^+$, $CD34^-$
Erythroid-megakaryocytic marker	$CD31^-$, $CD36^-$, $CD41a^-$, $CD61^-$, $GlyA^+$
Progenitor/activation/other marker	$CD34^{(+)}$, $CD38^+$, $CD71^+$, $HLA\text{-}DR^+$, TdT^+
Adhesion marker	$CD11b^-$, $CD11c^-$, $CD29^+$, $CD44^+$, $CD49a^-$, $CD49b^+$, $CD49c^+$, $CD49d^+$, $CD49e^+$, $CD49f^+$, $CD51^-$, $CD54^+$, $CD58^+$
Cytokine receptor	$CD25^-$, $CD105^-$, $CD115^-$, $CD117^-$, $CD119^+$, $CD124^-$, $CD126^-$, $CD127^-$, $CD135^+$

Genetic characterization[1,2,6–8]

Cytogenetic karyotype	47(46–48)<2n>X/XY, −9, +22, +mar, add(1)(q42), add(8)(p23), t(9;22)(q34;q11), der(22)t(9;22)(q34;q11), der(?)t(9;?)(?p11;?)
Unique translocation/fusion gene	Ph⁺ t(9;22)(q34;q11) → *BCR-ABL* b2-a2 fusion gene
Receptor gene rearrangement	*IGH* RR, *TCRB* GG, *TCRG* GG
Unique gene alteration	*P15INK4B* deletion, *P16INK4A* deletion

Functional characterization[1,6,9]

Cytochemistry ANAE$^-$, CAE$^-$, MPO$^-$, PAS$^-$
Heterotransplantation into SCID mice
Inducibility of differentiation resistant to differentiation induction with natrium butyrate, retinoic acid, TPA
Special features MPO mRNA$^+$

Comments

- Well-characterized precursor B-cell line of type B-II (common B-cell line) (reference cell line).
- Carries Ph chromosome with *BCR-ABL* b2-a2 fusion gene.
- Co-expression of myeloid surface antigens.
- Available from cell line bank.

References

[1] Pegoraro, L. et al. (1983) J. Natl. Cancer Inst. 70, 447–453.
[2] Drexler, H.G. et al. (1999) DSMZ Catalogue of Cell Lines, 7th edn, Braunschweig, Germany.
[3] Minowada, J. and Matsuo, Y. (1999) unpublished data.
[4] Inoue, K. et al. (1997) Blood 89, 1405–1412.
[5] Segat, D. et al. (1994) Blood 83, 1586–1594.
[6] Matsuo, Y. and Drexler, H.G. (1998) Leukemia Res. 22, 567–579.
[7] Drexler, H.G. et al. (1999) Leukemia Res. 23, 207–215.
[8] Aguiar, R.C.T. et al. (1997) Leukemia 11, 233–238.
[9] Uckun, F.M. (1996) Blood 88, 1135–1146.

CEMO-1

Culture characterization[1]

Culture medium	80–90% RPMI 1640 + 10–20% FBS
Doubling time	48 h
Viral status	EBNA$^-$
Authentication	yes (by cytogenetics)
Primary reference	Silva et al.[1]
Availability of cell line	restricted

Clinical characterization[1]

Patient	30-year-old male
Disease diagnosis	pre B-ALL (L1)
Treatment status	at diagnosis
Specimen site	peripheral blood
Year of establishment	1993

Immunophenotypic characterization[1]

T-/NK cell marker	CD2$^-$, CD7$^-$
B-cell marker	CD10$^+$, CD19$^+$, CD20$^{(+)}$, CD22$^+$, cyIgM$^+$, sIg$^-$
Myelomonocytic marker	CD13$^-$, CD15$^-$
Progenitor/activation/other marker	CD34$^-$, HLA-DR$^+$

Genetic characterization[1,2]

Cytogenetic karyotype	46, t(1;14)(q21;q32), t(9;9)(p24;q32)
Unique translocation/fusion gene	t(1;14)(q21;q32) → *BCL9-IGH* genes altered
Receptor gene rearrangement	*IGH* RR

Comments

- Weakly characterized precursor B-cell line of type B-III (pre B-cell line).
- Carries t(1;14)(q21;q32) affecting the *BCL9-IGH* genes.

References

[1] Silva, M.L.M. et al. (1996) Leukemia 10, 575–578.
[2] Willis, T.G. et al. (1998) Blood 91, 1873–1881.

DUNATIS

Culture characterization[1]

Culture medium	90% RPMI 1640 + 10% FBS
Doubling time	5–7 d
Authentication	yes (by cytogenetics)
Primary reference	Renard et al.[1]
Availability of cell line	original authors

Clinical characterization[1]

Patient	67-year-old male
Disease diagnosis	BCP-ALL (L1)
Treatment status	refractory
Specimen site	peripheral blood

Immunophenotypic characterization[1,2]

B-cell marker	CD10$^+$, CD19$^+$, CD20$^+$, CD24$^+$, CD40$^-$, cyIgM$^-$, sIg$^-$
Progenitor/activation/other marker	CD34$^+$, CD38$^+$, HLA-DR$^+$, TdT$^+$
Cytokine receptor	CD124$^+$

Genetic characterization[1]

Cytogenetic karyotype	43, XY, −7, −9, −21, t(9;22)(q34;q11), der(13)t(9;13)(q21;q34), del(16)(q23)
Unique translocation/fusion gene	Ph$^+$ t(9;22)(q34;q11) → *BCR-ABL* fusion gene?

Functional characterization[1,2]

Cytokine response	growth inhibition: IL-4

Comments

- Incompletely characterized precursor B-cell line of type B-II (common B-cell).
- Carries Ph chromosome (type of fusion gene is not known).
- Extremely slow growth.

References

[1] Renard, N. et al. (1997) Leukemia Res. 21, 1037–1046.
[2] Renard, N. et al. (1994) Blood 84, 2253–2260.

EU-1

Culture characterization[1,2]

Establishment	initially as colonies in methylcellulose
Culture medium	90% RPMI 1640 + 10% FBS
Viral status	EBV⁻
Authentication	yes (by cytogenetics)
Primary reference	Zhou et al.[1]
Availability of cell line	original authors

Clinical characterization[1]

Patient	16-year-old male
Disease diagnosis	ALL
Treatment status	at 2nd relapse
Specimen site	bone marrow

Immunophenotypic characterization[1-3]

T-/NK cell marker	$CD2^-$, $CD3^-$, $CD7^-$
B-cell marker	$CD10^+$, $CD19^+$, $cyIgM^-$, sIg^-
Myelomonocytic marker	$CD13^{(+)}$, $CD14^{(+)}$, $CD15^-$, $CD33^{(+)}$
Progenitor/activation/other marker	$HLA\text{-}DR^+$, TdT^+

Genetic characterization[1,2,4]

Cytogenetic karyotype	45, X, −Y, del(3)(p21p24), del(4)(q?31.3;q?35), del(5)(q13q35), der(12)inv(12)(p11.2q21), add(12)(q12), add(12)(p13), add(16) (q24)×2, t(18;21)(q12.2;p11.2)
Unique gene alteration	*P16INK4A* deletion

Functional characterization[1,2,5]

Cytochemistry	ANAE⁻, CAE⁻, MPO⁻, SBB⁻
Cytokine production	TNF-α
Inducibility of differentiation	DMSO → myeloid differentiation
Proto-oncogene	mRNA⁺: BAX, BCL2, BCLXL, MDM2
Special features	MPO mRNA⁺

Comments

- Precursor B-cell line of type B-II (common B-cell line).
- Spontaneous and induced expression of myeloid features.

References

1 Zhou, M. et al. (1994) Leukemia 8, 659–663.
2 Zhou, M. et al. (1991) Blood 77, 2002–2007.
3 Zhou, M. et al. (1995) Blood 85, 1608–1614.
4 Zhou, M. et al. (1995) Leukemia 9, 1159–1161.
5 Findley, H.M. et al. (1997) Blood 89, 2986–2993.

Culture characterization[1–3]

Culture medium	90% RPMI 1640 + 10% FBS
Doubling time	75–85 h
Viral status	EBNA⁻
Authentication	no
Primary reference	Gu et al.[1]
Availability of cell line	restricted

Clinical characterization[1–3]

Patient	1-year-old child
Disease diagnosis	pre B-ALL
Treatment status	at 1st relapse
Specimen site	bone marrow

Immunophenotypic characterization[1,2]

T-/NK cell marker	$CD2^-$, $CD7^-$
B-cell marker	$CD9^+$, $CD10^-$, $CD19^+$, $cyIgM^-$, sIg^-
Myelomonocytic marker	$CD13^+$, $CD15^+$, $CD33^+$
Progenitor/activation/other marker	$HLA\text{-}DR^+$
Cytokine receptor	$CD126^+$

Genetic characterization[1–3]

Cytogenetic karyotype	+8, t(4;11)(q21;q23), i(7q)
Unique translocation/fusion gene	t(4;11)(q21;q23) → *MLL-AF4* fusion gene?
Unique gene alteration	*P16INK4A* deletion

Functional characterization[1–4]

Colony formation	in methylcellulose
Cytokine response	growth stimulation: IL-6
Heterotransplantation	into SCID mice
Proto-oncogene	$mRNA^+$: BAX, BCLXL

Comments

- Precursor B-cell line of type B-I (pro B-cell line).
- Carries t(4;11)(q21;q23) leading presumably to *MLL-AF4* fusion gene.
- Co-expression of myeloid surface markers.

References

[1] Gu, L. et al. (1997) Leukemia 11, 1779–1786.
[2] Zhou, M. et al. (1995) Blood 85, 1608–1614.
[3] Zhou, M. et al. (1995) Leukemia 9, 1159–1161.
[4] Findley, M.H. et al. (1997) Blood 89, 2986–2993.

Culture characterization[1,2]

Other name of cell line ALL-G
Culture medium 90% α-MEM + 10% FBS (or serum-free)
Doubling time 72–74 h
Viral status EBV⁻
Authentication no
Primary reference Kamel-Reid et al.[1]
Availability of cell line restricted

Clinical characterization[1]

Patient child
Disease diagnosis pre B-ALL
Treatment status at 2nd relapse
Specimen site peripheral blood

Immunophenotypic characterization[1,3]

T-/NK cell marker $CD2^-$, $CD3^-$, $CD4^-$, $CD8^-$
B-cell marker $CD10^-$, $CD19^+$, $CD20^-$
Progenitor/activation/other marker $HLA\text{-}DR^+$

Genetic characterization[1]

Cytogenetic karyotype 46, XX, −11, +mar, der(5)t(5;?)(q15;?), der(9)t(9;?)(p13;?), der(14)t(14;?)(q22;?)

Functional characterization[1,3]

Colony formation in methylcellulose
Cytokine production GM-CSF; $mRNA^+$: IL-1β, TNF-α
Cytokine receptor $mRNA^+$: GM-CSFRα, TNF-αR
Cytokine response growth stimulation: GM-CSF, IL-3, IL-6, IL-7; growth inhibition: IFN-γ, IL-4, TNF-α
Heterotransplantation into SCID mice

Comments

- Precursor B-cell line of undetermined subtype.
- Cytokine-responsive cell line.

References

[1] Kamel-Reid, S. et al. (1992) Leukemia 6, 8–17.
[2] Brown, G.A. et al. (1995) Cancer Res. 55, 78–82.
[3] Freedman, M.H. et al. (1993) Blood 81, 3068–3075.

HAL-01

Culture characterization[1]

Culture medium	75% RPMI 1640 + 25% FBS
Doubling time	32 h
Viral status	EBNA⁻
Authentication	yes (by cytogenetics)
Primary reference	Ohyashiki et al.[1]
Availability of cell line	RCB 0540

Clinical characterization[1]

Patient	17-year-old female
Disease diagnosis	ALL (L2)
Specimen site	peripheral blood
Year of establishment	1990

Immunophenotypic characterization[1,2]

T-/NK cell marker	$CD1^-$, $CD2^-$, $CD3^-$, $CD4^-$, $CD5^-$, $CD7^-$, $CD8^-$, $CD56^-$, $CD57^-$
B-cell marker	$CD10^+$, $CD19^+$, $CD20^-$, $CD22^+$, $cyCD79a^+$, $cyCD79b^+$, $CD85^+$, $cyIgM^-$, sIg^-
Myelomonocytic marker	$CD13^+$, $CD14^-$, $CD33^-$
Erythroid-megakaryocytic marker	$CD36^-$, $CD41^-$, $CD42b^-$, $GlyA^-$
Progenitor/activation/other marker	$CD34^-$, $HLA\text{-}DR^+$, TdT^+
Adhesion marker	$CD11b^-$
Cytokine receptor	$CD25^-$, $CD105^-$, $CD115^-$, $CD117^-$, $CD119^+$, $CD124^-$, $CD126^-$, $CD127^-$, $CD135^+$

Genetic characterization[1,3]

Cytogenetic karyotype	46, XX, t(1;7)(p34;q21), t(17;19)(q21;p13)
Unique translocation/fusion gene	t(17;19)(q21;p13) → *E2A-HLF* fusion gene
Receptor gene rearrangement	*IGH* RD, *IGK* DD, *TCRB* GG
Unique gene alteration	*P15INK4B* deletion, *P16INK4A* deletion

Functional characterization[1,2]

Colony formation	in methylcellulose
Cytochemistry	$ANBE^-$, CAE^-, MPO^-, PAS^-, SBB^-
Cytokine response	growth stimulation: G-CSF; growth inhibition: IL-3
Heterotransplantation	into nude mice
Inducibility of differentiation	no effect of TPA
Special features	MPO $mRNA^+$

Comments

- Well-characterized precursor B-cell line of type B-II (common B-cell line).
- Carries t(17;19)(q21;p13) leading to *E2A-HLF* fusion gene.

- Co-expression of myeloid surface antigens.
- Available from cell line bank.

References

[1] Ohyashiki, K. et al. (1991) Leukemia 5, 322–331.
[2] Matsuo, Y. and Drexler, H.G. (1998) Leukemia Res. 22, 567–579.
[3] Maloney, K.W. et al. (1997) Blood 90, 218b.

HBL-3

Culture characterization[1]

Culture medium	80% RPMI 1640 + 20% FBS
Doubling time	48–72 h
Viral status	EBNA⁻
Authentication	no
Primary reference	Abe et al.[1]
Availability of cell line	not known

Clinical characterization[1]

Patient	9-year-old female
Disease diagnosis	pre B-ALL
Treatment status	at diagnosis
Specimen site	bone marrow
Year of establishment	1985

Immunophenotypic characterization[1]

T-/NK cell marker	$CD1^-$, $CD2^-$, $CD3^-$, $CD4^-$, $CD5^-$, $CD8^-$, $CD57^-$
B-cell marker	$CD10^-$, $CD19^+$, $CD20^-$, $CD21^-$, $CD24^+$, $cyIgM^-$, sIg^-
Myelomonocytic marker	$CD13^-$, $CD14^-$, $CD15^-$, $CD33^-$
Progenitor/activation/other marker	$HLA\text{-}DR^+$, TdT^+
Cytokine receptor	$CD25^-$

Genetic characterization[1]

Cytogenetic karyotype	46, XX, −3, −9, −9, +der(3)t(3;?), +der(9)t(9;?), +der(9)t(1;9)
Receptor gene rearrangement	*IGH* R, *IGK* G, *TCRB* G, *TCRG* G

Functional characterization[1]

Cytochemistry	ACP^-, ALP^-, $ANBE^-$, CAE^-, MPO^-

Comments

- Precursor B-cell line of type B-I (pro B-cell line).
- Note that there is a mature B-cell line also termed **HBL-3** which was derived independently from an unrelated patient in New York, USA [reference: Gaidano, G. et al. (1993) Leukemia 7, 1621–1629].

Reference

[1] Abe, M. et al. (1990) Virchows Archiv B Cell. Pathol. 59, 107–113.

JKB-1

Culture characterization[1,2]

Sister cell line/Subclone	**JKB-2** (not specified)
Culture medium	90% RPMI 1640 + 10% FBS (or serum-free)
Doubling time	24 h
Viral status	EBV⁻
Authentication	yes (by cytogenetics)
Primary reference	Urashima et al.[1]
Availability of cell line	restricted

Clinical characterization[1]

Patient	16-year-old female
Disease diagnosis	pre B-ALL (L2)
Treatment status	at relapse
Specimen site	bone marrow
Year of establishment	1992

Immunophenotypic characterization[1,2]

T-/NK cell marker	$CD1^-$, $CD2^-$, $CD3^-$, $CD4^-$, $CD5^-$, $CD7^-$, $CD8^-$, $CD56^-$
B-cell marker	$CD10^+$, $CD19^+$, $CD20^-$, $CD21^-$, $CD22^-$, $CD23^-$, $CD24^+$, $CD40^-$, $CD80^-$, $CD86^+$, $cyIgM^-$, sIg^-
Myelomonocytic marker	$CD13^-$, $CD14^-$, $CD33^-$
Progenitor/activation/other marker	$CD34^-$, $CD38^+$, $HLA\text{-}DR^+$, TdT^+
Cytokine receptor	$CD25^-$, $CD127^+$

Genetic characterization[1,2]

Cytogenetic karyotype	46, XX, t(9;14)(p21;q32)
Receptor gene rearrangement	*IGH* R, *IGK* G
Unique gene alteration	*P15INK4B* deletion, *P16INK4A* deletion

Functional characterization[1,2]

Colony formation	in methylcellulose
Cytochemistry	$ANBE^-$, CAE^-, MPO^-, PAS^-
Cytokine response	growth stimulation: IL-3, SCF; growth inhibition: IL-7, TGF-β1
Inducibility of differentiation	bone marrow stromal layer → B-cell differentiation

Comments

- Precursor B-cell line of type B-II (common B-cell line).

References

[1] Urashima, M. et al. (1994) Am. J. Hematol. 46, 112–119.
[2] Urashima, M. et al. (1996) Leukemia 10, 1576–1583.

KH-3A

Culture characterization[1]

Sister cell line **KH-3B** (serial sister cell line – at 2nd relapse – similar features)
Establishment initially on autologous bone marrow stroma cells
Culture medium 80% RPMI 1640 + 20% FBS
Doubling time 48 h
Viral status EBNA$^-$
Authentication no
Primary reference Nagasaka-Yabe et al.[1]
Availability of cell line not known

Clinical characterization[1]

Patient 18-year-old female
Disease diagnosis cALL
Treatment status at 1st relapse
Specimen site bone marrow
Year of establishment 1983

Immunophenotypic characterization[1]

T-/NK cell marker CD2$^-$, CD5$^-$
B-cell marker CD10$^+$, CD20$^+$, cyIgM$^-$, sIg$^-$
Progenitor/activation/other marker CD38$^+$, CD71$^+$, HLA-DR$^+$, TdT$^-$
Adhesion marker CD11b$^-$

Genetic characterization[1]

Cytogenetic karyotype 46(44–47)XX, 2q+, +13
Receptor gene rearrangement *IGH* R, *TCRB* R

Functional characterization[1]

Colony formation in methylcellulose
Cytochemistry ACP$^{(+)}$, ANAE$^-$, MPO$^-$, PAS$^{(+)}$, SBB$^-$
Inducibility of differentiation TPA → B-cell differentiation

Comments

- Precursor B-cell line of type B-II (common B-cell line).
- Serial sister cell lines from 1st and 2nd relapse.

Reference

[1] Nagasaka-Yabe, M. et al. (1988) Jpn. J. Cancer Res. 79, 59–68.

KLM-2

Culture characterization[1]

Culture medium 90% RPMI 1640 + 10% FBS
Authentication no
Primary reference Matsuo and Drexler[1]
Availability of cell line original authors

Clinical characterization[1]

Patient male
Disease diagnosis AML?
Specimen site peripheral blood

Immunophenotypic characterization[1,2]

T-/NK cell marker CD1⁻, CD2⁻, CD3⁻, cyCD3⁻, CD4⁻, CD5⁻, CD7⁻, CD8⁻, CD28⁺, CD57⁻, TCRαβ⁻, TCRγδ⁻
B-cell marker CD9⁻, CD10⁺, CD19⁺, CD20⁺, CD21⁺, CD22⁺, CD24⁻, cyCD79a⁺, cyCD79b⁺, CD85⁺, cyIgM⁺, sIg⁻
Myelomonocytic marker CD13⁻, CD14⁻, CD15⁻, CD33⁻
Erythroid-megakaryocytic marker CD41a⁻, CD61
Progenitor/activation/other marker CD34⁻, CD71⁺, HLA-DR⁺, TdT⁻
Adhesion marker CD11b⁽⁺⁾
Cytokine receptor CD25⁻, CD105⁻, CD115⁻, CD117⁻, CD119⁺, CD124⁻, CD126⁻, CD127⁻, CD135⁻

Genetic characterization[1,2]

Cytogenetic karyotype 47, XY, −4, −6, +7, −8, −9, −9, −14, +16, −17, −18, +mar, +der(4)t(4;?)(pter → q3?1::?), +der(6)t(6;18)(6qter → 6p21::18q1?1 →18qter), +der(9)t(9;17)(9qter → 9p2?1::17q21 → 17qter), +der(9)t (9;11?)(9pter→9q34::11?q1?3→11?qter), +der(14or8)t(14;8) (14qter → 14p1?1::8q1?1 → 8q24), +der(14)t(14;8)(14pter → 14q32: :8q24 → 8qter)
Receptor gene rearrangement *IGH* R, *TCRG* GG, *TCRD* GG

Comments

- Precursor B-cell line of type B-III (pre B-cell line).
- Authentication of cell line required.

References

[1] Matsuo, Y. and Drexler, H.G. (1998) Leukemia Res. 22, 567–579.
[2] Minowada, J. and Matsuo, Y. (1999) unpublished data.

KMO-90

Culture characterization[1]

Culture medium	90% RPMI 1640 + 10% FBS
Doubling time	72 h
Viral status	EBNA$^-$
Authentication	yes (by cytogenetics)
Primary reference	Sotomatsu et al.[1]
Availability of cell line	restricted

Clinical characterization[1]

Patient	12-year-old female
Disease diagnosis	ALL (L1)
Treatment status	at diagnosis
Specimen site	bone marrow
Year of establishment	1990

Immunophenotypic characterization[1]

T-/NK cell marker	CD1a$^-$, CD2$^-$, CD3$^-$, CD4$^-$, CD5$^-$, CD7$^-$, CD8$^-$
B-cell marker	CD10$^+$, CD19$^+$, CD20$^-$, CD22$^+$, cyIgM$^+$, sIg$^-$
Myelomonocytic marker	CD13$^-$, CD14$^-$, CD33$^-$
Erythroid-megakaryocytic marker	CD36$^-$
Progenitor/activation/other marker	HLA-DR$^+$
Adhesion marker	CD11b$^-$

Genetic characterization[1,2]

Cytogenetic karyotype	48, XX, +8, +19, t(1;19)(q23;p13)
Unique translocation/fusion gene	t(1;19)(q23;p13) → *E2A-PBX1* fusion gene
Unique gene alteration	*P53* mutation

Functional characterization[1]

Cytochemistry	ACP$^+$, ANBE$^-$, CAE$^-$, MPO$^-$, PAS$^-$

Comments

- Precursor B-cell line of type B-III (pre B-cell line).
- Carries t(1;19)(q23;p13) leading to *E2A-PBX1* fusion gene.

References

[1] Sotomatsu, M. et al. (1993) Leukemia 7, 1615–1620.
[2] Kawamura, M. et al. (1995) Blood 85, 2546–2552.

KOPN-8

Culture characterization[1]

Culture medium	90% RPMI 1640 + 10% FBS
Viral status	EBNA$^-$
Authentication	no
Primary reference	Matsuo and Drexler[1]
Availability of cell line	original authors

Clinical characterization[1,2]

Patient	3-month-old female
Disease diagnosis	ALL
Specimen site	peripheral blood
Year of establishment	1977

Immunophenotypic characterization[1,3]

T-/NK cell marker	CD5$^-$, CD7$^-$, TCR$\alpha\beta^-$, TCR$\gamma\delta^-$
B-cell marker	CD9$^+$, CD10$^+$, CD19$^+$, CD20$^+$, CD21$^-$, CD22$^+$, CD24$^+$, cyCD79a$^+$, cyCD79b$^+$, CD85$^+$, cyIgM$^+$, sIg$^-$
Myelomonocytic marker	CD13$^-$, CD33$^-$
Progenitor/activation/other marker	CD34$^-$, CD38$^+$, HLA-DR$^+$, TdT$^-$
Cytokine receptor	CD25$^-$, CD105$^-$, CD115$^-$, CD117$^-$, CD119$^+$, CD124$^-$, CD126$^-$, CD127$^+$, CD135$^+$

Genetic characterization[3]

Cytogenetic karyotype	45, XX, −1, −13, −14, t(11;19)(q23;p13), t(8;13)(q24;q22), +der(1)t(1;?)(?;?), +der(13)t(13;14)(p11;q11)
Unique translocation/fusion gene	t(11;19)(q23;p13) → *MLL-ENL* fusion gene

Comments

- Precursor B-cell line of type B-III (pre B-cell line).
- Carries t(11;19)(q23;p13) leading to *MLL-ENL* fusion gene.

References

[1] Matsuo, Y. and Drexler, H.G. (1998) Leukemia Res. 22, 567–579.
[2] Nakazawa, S. et al. (1978) Clin. Hematol. 20, 189.
[3] Minowada, J. and Matsuo, Y. (1999) unpublished data.

KOPN-30bi

Culture characterization[1]

Culture medium	90% RPMI 1640 + 10% FBS
Authentication	no
Primary reference	Miyashita et al.[1]
Availability of cell line	restricted

Clinical characterization[1]

Patient	8-year-old male
Disease diagnosis	cALL (L2)
Specimen site	bone marrow

Immunophenotypic characterization[1-3]

T-/NK cell marker	CD1⁻, CD2⁻, CD3⁻, CD4⁻, CD5⁻, CD7⁻, CD8⁻, CD56⁻
B-cell marker	CD10⁺, CD19⁺, CD20⁻, CD22⁺, CD79a⁺
Myelomonocytic marker	CD13⁺, CD14⁻, CD33⁻
Erythroid-megakaryocytic marker	CD36⁻, CD41⁻, CD42⁻
Progenitor/activation/other marker	CD34⁺, HLA-DR⁺
Adhesion marker	CD11b⁻

Genetic characterization[1-3]

Cytogenetic karyotype	46, XY, −2, +mar, t(9;22)(q34;q11)
Unique translocation/fusion gene	Ph⁺ t(9;22)(q34;q11) → *BCR-ABL* m-bcr fusion gene
Receptor gene rearrangement	*IGH* R, *TCRB* R, *TCRG* R, *TCRD* R

Functional characterization[1,3]

Cytochemistry	ANBE⁻, CAE⁻, MPO⁻

Comments

- Precursor B-cell line of unknown B-type (type B-II or B-III).
- Carries Ph chromosome leading to *BCR-ABL* m-bcr fusion gene.

References

[1] Miyashita, T. et al. (1993) Leukemia 7, 586–592.
[2] Saito, M. et al. (1995) Leukemia 9, 1508–1516.
[3] Kojika, S. et al. (1996) Leukemia 10, 994–999.

LAZ 221

Culture characterization[1]

Establishment	initially in hypoxic environment
Culture medium	80% RPMI 1640 + 20% FBS
Doubling time	4 d
Viral status	EBNA$^-$
Authentication	yes (by cytogenetics)
Primary reference	Lazarus et al.[1]
Availability of cell line	not known

Clinical characterization[1]

Patient	24-year-old female
Disease diagnosis	null-ALL
Treatment status	at diagnosis
Specimen site	peripheral blood
Year of establishment	1977

Immunophenotypic characterization[1-3]

T-/NK cell marker	CD1b$^-$, CD2$^-$, CD3$^-$, CD4$^-$, CD5$^+$, CD7$^-$, CD8$^-$, TCR$\alpha\beta^-$, TCR$\gamma\delta^-$
B-cell marker	CD9$^+$, CD10$^+$, CD19$^+$, CD20$^+$, CD21$^-$, CD22$^+$, CD24$^+$, cyCD79a$^+$, cyCD79b$^+$, CD85$^+$, cyIgM$^+$, sIg$^-$
Myelomonocytic marker	CD13$^-$, CD14$^-$, CD15$^-$, CD33$^-$
Progenitor/activation/other marker	CD34$^{(+)}$, CD38$^+$, HLA-DR$^+$, TdT$^-$
Adhesion marker	CD11b$^{(+)}$
Cytokine receptor	CD25$^-$, CD105$^-$, CD115$^-$, CD117$^-$, CD119$^+$, CD124$^-$, CD126$^-$, CD127$^+$, CD135$^+$

Genetic characterization[1]

Cytogenetic karyotype	45, XX, −9, −12, +t(9q12q)
Receptor gene rearrangement	*IGH* RR, *TCRG* GG, *TCRD* GG

Comments

- Weakly characterized precursor B-cell line of type B-III (pre B-cell line).

References

[1] Lazarus, H. et al. (1978) Cancer Res. 38, 1362–1367.
[2] Minowada, J. and Matsuo, Y. (1999) unpublished data.
[3] Matsuo, Y. and Drexler, H.G. (1998) Leukemia Res. 22, 567–579.

Culture characterization[1]

Establishment	initially on feeder layer (normal bone marrow stroma cells or rat BRL cell line)
Culture medium	90% RPMI 1640 + 10% FBS
Viral status	EBV⁻
Authentication	yes (by DNA fingerprinting)
Primary reference	Grausz et al.[1]
Availability of cell line	restricted

Clinical characterization[1]

Patient	37-year-old female
Disease diagnosis	ALL (L1)
Treatment status	at diagnosis
Specimen site	peripheral blood
Year of establishment	1988

Immunophenotypic characterization[1]

T-/NK cell marker	$CD1^-$, $CD2^-$, $CD3^-$, $CD4^-$, $CD5^-$, $CD7^-$, $CD8^-$
B-cell marker	$CD9^+$, $CD10^-$, $CD19^+$, $CD20^+$, $CD22^+$, $CD23^+$, $CD24^+$, $cyIgG\kappa^+$, sIg^-
Myelomonocytic marker	$CD13^-$, $CD15^+$, $CD33^-$
Erythroid-megakaryocytic marker	$CD36^-$, $CD41^-$, $CD42^-$
Progenitor/activation/other marker	$CD38^+$, $CD71^+$, $HLA\text{-}DR^+$
Adhesion marker	$CD11b^-$
Cytokine receptor	$CD25^-$

Genetic characterization[1]

Cytogenetic karyotype	46, XX, t(9;22)(q34;q11), Ph
Unique translocation/fusion gene	Ph^+ t(9;22)(q34;q11) → *BCR-ABL* m-bcr fusion gene

Comments

- Precursor B-cell line of type B-III (pre B-cell line).
- Carries Ph chromosome leading to *BCR-ABL* m-bcr fusion gene.

Reference

[1] Grausz, D. et al. (1990) Leukemia 4, 359–364.

MIELIKI

Culture characterization[1]

Culture medium	90% RPMI 1640 + 10% FBS
Doubling time	5–6 d
Authentication	yes (by cytogenetics)
Primary reference	Renard et al.[1]
Availability of cell line	original authors

Clinical characterization[1]

Patient	13-month-old female
Disease diagnosis	pre B-ALL (L1)
Treatment status	at diagnosis
Specimen site	bone marrow

Immunophenotypic characterization[1,2]

T-/NK cell marker	CD1⁻, CD2⁻, CD3⁻, CD4⁻, CD5⁻, CD7⁻
B-cell marker	CD10⁺, CD19⁺, CD20⁺, CD21⁻, CD22⁺, CD23⁺, CD24⁺, CD40⁻, cyIgM⁻, sIg⁻
Myelomonocytic marker	CD13⁻, CD15⁺, CD33⁻
Progenitor/activation/other marker	CD34⁻, CD38⁺, HLA-DR⁺, TdT⁺
Cytokine receptor	CD124⁺, CD127⁺, CD132⁺

Genetic characterization[1]

Cytogenetic karyotype	t(7;9)(q?;q?)
Receptor gene rearrangement	*IGH* RR, *IGK* RR, *IGL* G

Functional characterization[1]

Cytokine response	growth inhibition: IL-4, IL-6, IL-7

Comments

- Precursor B-cell line of type B-II (common B-cell line).
- Cells are very difficult to culture (extremely long doubling time).
- Further characterization required.

References

[1] Renard, N. et al. (1995) Leukemia 9, 1219–1226.
[2] Renard, N. et al. (1994) Blood 84, 2253–2260.

MR-87

Culture characterization[1]

Sister cell line	**MR-B** (= EBV$^+$ B-LCL – Ph chromosome negative)
Culture medium	90% RPMI 1640 + 10% FBS
Doubling time	120–144 h
Viral status	EBV$^-$
Authentication	yes (by *IGH* rearrangement)
Primary reference	Okamura et al.[1]
Availability of cell line	original authors

Clinical characterization[1]

Patient	4-year-old male
Disease diagnosis	hybrid acute leukemia
Treatment status	at diagnosis
Specimen site	bone marrow
Year of establishment	1986

Immunophenotypic characterization[1-3]

T-/NK cell marker	CD1$^-$, CD2$^-$, CD3$^-$, CD4$^-$, CD5$^-$, CD7$^-$, CD8$^-$, CD28$^-$, CD56$^+$, CD57$^-$, TCR$\alpha\beta^-$
B-cell marker	CD9$^+$, CD10$^+$, CD19$^+$, CD20$^-$, CD21$^-$, CD22$^-$, cIg$^-$, sIg$^-$
Myelomonocytic marker	CD13$^+$, CD14$^-$, CD15$^-$, CD33$^{(+)}$
Erythroid-megakaryocytic marker	CD36$^-$, CD41a$^-$, CD61$^-$, GlyA$^-$
Progenitor/activation/other marker	CD34$^+$, CD38$^-$, CD71$^+$, HLA-DR$^+$, TdT$^+$
Cytokine receptor	CD25$^-$, CD122$^-$, CD124$^-$, CD127$^-$

Genetic characterization[1,4]

Cytogenetic karyotype	46,XY, 9p–, 17p?, t(9p?q+;22q–)
Unique translocation/fusion gene	Ph$^+$ t(9;22)(q34;q11) → *BCR-ABL* e1-a2 fusion gene
Receptor gene rearrangement	*IGH* R, *TCRB* G

Functional characterization[1]

Cytochemistry	ANAE$^-$, MPO$^+$, MSE$^-$, PAS$^-$, SBB$^+$

Comments

- Immature cell line co-expressing lymphoid and myeloid immuno- and other markers.
- *IGH* rearrangement suggests a precursor B-cell line (type B-II, common B-cell line).
- Carries Ph chromosome leading to the *BCR-ABL* e1-a2 fusion gene.

References

[1] Okamura, J. et al. (1988) Blood 72, 1261–1268.
[2] Minowada, J. and Matsuo, Y. (1999) unpublished data.
[3] Toba, K. et al. (1996) Exp. Hematol. 24, 894–901.
[4] Drexler, H.G. et al. (1999) Leukemia Res. 23, 207–215.

MUTZ-1

Culture characterization[1,2]

Culture medium	90% RPMI 1640 + 10% FBS
Doubling time	48–55 h
Viral status	EBV+, HBV−, HCV−, HHV-8−, HIV−, HTLV-I/II−
Authentication	yes (by DNA fingerprinting)
Primary reference	Steube et al.[1]
Availability of cell line	DSM ACC 13

Clinical characterization[1]

Patient	5-year-old female
Disease diagnosis	MDS (Fanconi → RAEB) → AML M2
Treatment status	at terminal stage
Specimen site	peripheral blood
Year of establishment	1994

Immunophenotypic characterization[1,2]

T-/NK cell marker	CD1a−, CD2−, CD3−, CD4−, CD5−, CD7+, CD8−, CD56−, TCRαβ−, TCRγδ−
B-cell marker	CD10+, CD19+, CD20−, CD21+, CD37+, cyIgM+, sIg−
Myelomonocytic marker	CD13−, CD14−, CD15−, CD33−, CD65−, CD68−
Erythroid-megakaryocytic marker	CD41a−, CD42b−, GlyA+
Progenitor/activation/other marker	CD30−, CD34−, CD38+, CD71+, HLA-DR+
Adhesion marker	CD11b−
Cytokine receptor	CD25−

Genetic characterization[1,2]

Cytogenetic karyotype	near-triploid, 12% polyploidy; 68(68–73)<3n>X, der(X), −X, add(1)(p11), der(2)t(2;?11)(q22;q23), der(3)t(3;5)(q27;q14)t(5;22) (q35;q12)×2, del(3)(p11), del(3)(q11), del(5)(q13/q33)×1–2, der(6)t (3;6)(p23;p21.2)×2, der(7)t(7;11)(q35;q13), der(8)t(1;8)(q11;q24), add(9)(q34), del(9)(p13), der(14)t(14;?)(q23;?), der(14)t(14;?) (p11;?)t(1;?)(q11;?), add(14)(p11), der(15)t(5;15)(p13;p11)×2, der(16)t(7;16)(q21.1;q24), add(18)(p11), add(21)(p11)
Receptor gene rearrangement	*IGH* RR, *IGK* GR, *IGL* GR, *TCRB* GR, *TCRG* GG, *TCRD* GG
Unique gene alteration	*P53* mutation

Functional characterization[1]

Cytochemistry	ACP$^+$, ANAE$^-$, MPO$^-$, MSE$^-$, PAS$^-$, TRAP$^-$
Inducibility of differentiation	growth arrest: DMSO, Dolastatin 10, TPA
Proto-oncogene	mRNA$^+$: BCL2, MYC

Comments

- Precursor B-cell line of type B-III (pre B-cell line).
- Available from cell line bank.

References

[1] Steube, K.G. et al. (1997) Leukemia Lymphoma 25, 345–363.
[2] Drexler, H.G. et al. (1999) DSMZ Catalogue of Cell Lines, 7th edn, Braunschweig, Germany.

Culture characterization[1]

Culture medium	90% α-MEM + 10% FBS
Doubling time	48 h
Authentication	yes (by cytogenetics)
Primary reference	Inokuchi et al.[1]
Availability of cell line	original authors

Clinical characterization[1]

Patient	52-year-old female
Disease diagnosis	ALL
Treatment status	at relapse (refractory)
Specimen site	bone marrow

Immunophenotypic characterization[1]

T-/NK cell marker	$CD2^-$, $CD3^-$, $CD5^-$, $CD7^-$
B-cell marker	$CD10^+$, $CD19^+$, $CD20^-$
Myelomonocytic marker	$CD13^+$, $CD14^-$, $CD33^-$
Erythroid-megakaryocytic marker	$CD41^-$, $GlyA^-$
Progenitor/activation/other marker	$CD34^+$, $HLA\text{-}DR^+$

Genetic characterization[1]

Cytogenetic karyotype	46, X, t(X;12)(p11.2;p13), t(2;3)(p16;q29), der(5)(5pter → 5q11.2: :22q11→22qter), der(9)(9pter → 9q34::5q11.2→5q15::10q23→10qter), der(10)(10pter→10q23::5q15→5qter), der(22)(22pter→22q11::9q34→9qter), del(22)(q11)
Unique translocation/fusion gene	Ph^+ t(9;22)(q34;q11) → *BCR-ABL* e1-a3 fusion gene

Functional characterization[1]

Heterotransplantation	into nude mice

Comments

- Precursor B-cell line of undetermined type.
- Carries Ph chromosome leading to *BCR-ABL* e1-a3 fusion gene.

Reference

[1] Inokuchi, K. et al. (1998) Genes Chromosomes Cancer 23, 227–238.

NALM-1

Culture characterization[1,2]

Culture medium	80–90% RPMI 1640 + 10–20% FBS
Doubling time	48–72 h
Viral status	EBV⁻, HBV⁻, HCV⁻, HHV-8⁻, HIV⁻, HTLV-I/II⁻
Authentication	no
Primary reference	Minowada et al.[1]
Availability of cell line	DSM ACC 131, JCRB 0064

Clinical characterization[1]

Patient	3-year-old female
Disease diagnosis	CML
Treatment status	at lymphoid blast crisis
Specimen site	peripheral blood
Year of establishment	1975

Immunophenotypic characterization[1–4]

T-/NK cell marker	CD1b⁻, CD2⁻, CD3⁻, CD4⁻, CD5⁻, CD7⁻, CD8⁻, CD28⁻, CD57⁻, TCRαβ⁻, TCRγδ⁻
B-cell marker	CD9⁺, CD10⁺, CD19⁺, CD20⁺, CD21⁻, CD22⁺, CD24⁺, CD37⁺, cyCD79a⁺, cyCD79b⁺, CD85⁺, cyIgM⁺, sIg⁻
Myelomonocytic marker	CD13⁻, CD14⁻, CD15⁻, CD33⁻
Erythroid-megakaryocytic marker	CD41a⁻, CD61⁻
Progenitor/activation/other marker	CD34⁺, CD38⁺, HLA-DR⁺, TdT⁺
Adhesion marker	CD11b⁻
Cytokine receptor	CD25⁻, CD105⁺, CD115⁻, CD117⁻, CD119⁺, CD124⁻, CD126⁻, CD127⁺, CD135⁺

Genetic characterization[1,2,4–7]

Cytogenetic karyotype	hypodiploid, 5% polyploidy; 45(42–47)<2n>X, −X, der(9)t(9;22)(q24;q12) ×1–2, dup(13)(q21), der(22)t(9;22)(q34;q12)
Unique translocation/fusion gene	Ph⁺ t(9;22)(q34;q11) → *BCR-ABL* b3-a2 fusion gene
Receptor gene rearrangement	*IGH* RR, *TCRA* G, *TCRB* R, *TCRD* DD

Functional characterization[1,4,8]

Cytochemistry	ANAE⁻, MPO⁻, PAS⁻, SBB⁻
Heterotransplantation	into nude mice
Special features	MPO mRNA⁺

Comments

- Widely distributed precursor B-cell line of type B-III (pre B-cell line).
- First cell line derived from lymphoid blast crisis of CML (reference cell line).
- Carries Ph chromosome leading to *BCR-ABL* b3-a2 fusion gene.
- Available from cell line banks.
- Cells grow well.

References

[1] Minowada, J. et al. (1977) J. Natl. Cancer Inst. 59, 83–87.
[2] Drexler, H.G. et al. (1999) DSMZ Catalogue of Cell Lines, 7th edn, Braunschweig, Germany.
[3] Minowada, J. and Matsuo, Y. (1999) unpublished data.
[4] Matsuo, Y. and Drexler, H.G. (1998) Leukemia Res. 22, 567–579.
[5] Minowada, J. et al. (1979) Leukemia Res. 3, 261–266.
[6] Drexler, H.G. et al. (1999) Leukemia Res. 23, 207–215.
[7] Sangster, R.N. et al. (1986) J. Exp. Med. 163, 1491–1508.
[8] Kubonishi, I. et al. (1980) Cancer 45, 2324–2329.

NALM-6

Culture characterization[1,2]

Sister cell lines	(1) **NALM-7, -8, -9, -10, -11, -12, -13, -14** (simultaneous sister cell lines – similar features) (2) **B85, B86** (= EBV$^+$ B-LCLs)
Culture medium	90% RPMI 1640 + 10% FBS
Doubling time	36–40 h
Viral status	EBV$^-$, HBV$^-$, HCV$^-$, HHV-8$^-$, HIV$^-$, HTLV-I/II$^-$
Authentication	no
Primary reference	Hurwitz et al.[1]
Availability of cell line	DSM ACC 128

Clinical characterization[1]

Patient	19-year-old male
Disease diagnosis	non-T non-B ALL
Treatment status	at relapse
Specimen site	peripheral blood
Year of establishment	1976

Immunophenotypic characterization[1–4]

T-/NK cell marker	CD1$^-$, CD2$^-$, CD3$^-$, cyCD3$^-$, CD4$^-$, CD5$^-$, CD7$^-$, CD8$^-$, CD28$^-$, CD57$^-$, TCR$\alpha\beta^-$, TCR$\gamma\delta^-$
B-cell marker	CD9$^+$, CD10$^+$, CD19$^+$, CD20$^-$, CD21$^-$, CD22$^{(+)}$, CD23$^-$, CD24$^+$, CD37$^-$, cyCD79a$^+$, cyCD79b$^+$, CD85$^+$, PC-1$^-$, PCA-1$^-$, cyIgM$^+$, sIg$^-$
Myelomonocytic marker	CD13$^-$, CD14$^-$, CD15$^-$, CD33$^-$
Erythroid-megakaryocytic marker	CD41a$^-$, CD61$^-$
Progenitor/activation/other marker	CD34$^-$, CD38$^+$, CD45$^-$, CD95$^-$, HLA-DR$^+$, TdT$^+$
Adhesion marker	CD11b$^-$
Cytokine receptor	CD25$^-$, CD105$^+$, CD115$^-$, CD117$^{(+)}$, CD119$^+$, CD124$^-$, CD126$^-$, CD127$^+$, CD135$^+$

Genetic characterization[1,2,4–6]

Cytogenetic karyotype	46(43–47) <2n> XY, t(5;12)(q33.2;p13.2)
Receptor gene rearrangement	*IGH* RG, *TCRB* GG, *TCRG* GG

Functional characterization[1,7]

Cytochemistry	ACP$^-$, ANAE$^-$, CAE$^-$, MPO$^-$, Oil Red O$^+$, PAS$^-$, SBB$^+$
Heterotransplantation	into SCID mice

Comments

- One of the oldest and most wide-spread precursor B-cell lines (reference cell line).
- Precursor B-cell line of type B-III (pre B-cell line).
- Available from cell line bank.
- Cells grow well.

References

[1] Hurwitz, R. et al. (1979) Int. J. Cancer 23, 174–180.
[2] Drexler, H.G. et al. (1999) DSMZ Catalogue of Cell Lines, 7th edn, Braunschweig, Germany.
[3] Minowada, J. and Matsuo, Y. (1999) unpublished data.
[4] Tani, A. et al. (1996) Leukemia 10, 1592–1603.
[5] Matsuo, Y and Drexler, H.G. (1998) Leukemia Res. 22, 567–579.
[6] Wlodarska, I. et al. (1997) Blood 89, 1716–1722.
[7] Uckun, F.M. (1996) Blood 88, 1135–1146.

NALM-16

Culture characterization[1]

Culture medium	90% RPMI 1640 + 10% FBS
Doubling time	36 h
Viral status	EBNA$^-$
Authentication	yes (by cytogenetics)
Primary reference	Kohno et al.[1]
Availability of cell line	original authors

Clinical characterization[1]

Patient	12-year-old female
Disease diagnosis	ALL
Treatment status	at relapse
Specimen site	peripheral blood
Year of establishment	1977

Immunophenotypic characterization[1,2]

T-/NK cell marker	CD1$^-$, CD2$^-$, CD3$^-$, cyCD3$^-$, CD4$^-$, CD5$^-$, CD7$^-$, CD8$^-$, CD28$^-$, CD57$^-$, TCR$\alpha\beta^-$, TCR$\gamma\delta^-$
B-cell marker	CD9$^+$, CD10$^+$, CD19$^+$, CD20$^+$, CD21$^-$, CD22$^+$, CD23$^-$, CD24$^+$, cyCD79a$^+$, cyCD79b$^+$, CD85$^-$, cyIgM$^-$, sIg$^-$
Myelomonocytic marker	CD13$^-$, CD14$^-$, CD15$^-$, CD33$^-$
Erythroid-megakaryocytic marker	CD41a$^-$, CD61$^-$
Progenitor/activation/other marker	CD34$^+$, CD38$^+$, CD71$^+$, CD95$^-$, HLA-DR$^+$, TdT$^+$
Adhesion marker	CD11b$^-$
Cytokine receptor	CD25$^-$, CD105$^-$, CD115$^-$, CD117$^-$, CD119$^+$, CD124$^-$, CD126$^-$, CD127$^+$, CD135$^+$

Genetic characterization[1,3,4]

Cytogenetic karyotype	27, X, +10, +14, +18, +21, 7p+
Receptor gene rearrangement	*IGH* RR, *TCRB* RG, *TCRG* RG

Comments

- Precursor B-cell line of type B-II (common B-cell line).
- Unusual near-haploid karyotype.

References

[1] Kohno, S. et al. (1980) J. Natl. Cancer Inst. 64, 485–493.
[2] Minowada, J. and Matsuo, Y. (1999) unpublished data.
[3] Tani, A. et al. (1996) Leukemia 10, 1592–1603.
[4] Matsuo, Y. and Drexler, H.G. (1998) Leukemia Res. 22, 567–579.

NALM-19

Culture characterization[1]

Sister cell lines	**B239, B240** (= EBV$^+$ B-LCLs)
Culture medium	90% RPMI 1640 + 10% FBS
Doubling time	29 h
Authentication	no
Primary reference	Matsuo et al.[1]
Availability of cell line	original authors

Clinical characterization[1]

Patient	26-year-old male
Disease diagnosis	AUL
Treatment status	at diagnosis
Specimen site	peripheral blood
Year of establishment	1988

Immunophenotypic characterization[1-3]

T-/NK cell marker	CD1$^-$, CD2$^-$, CD3$^-$, cyCD3$^-$, CD4$^-$, CD5$^-$, CD7$^-$, CD8$^-$, CD28$^-$, CD57$^-$, TCRαβ$^-$, TCRγδ$^-$
B-cell marker	CD9$^+$, CD10$^-$, CD19$^+$, CD20$^-$, CD21$^-$, CD22$^+$, CD24$^+$, cyCD79a$^+$, cyCD79b$^+$, CD85$^+$, cyIgM$^-$, sIg$^-$
Myelomonocytic marker	CD13$^+$, CD14$^-$, CD15$^+$, CD16$^-$, CD33$^-$
Erythroid-megakaryocytic marker	CD41a$^-$, CD61$^-$
Progenitor/activation/other marker	CD34$^-$, CD71$^+$, HLA-DR$^+$, TdT$^+$
Adhesion marker	CD11b$^-$
Cytokine receptor	CD25$^-$, CD105$^-$, CD115$^-$, CD117$^-$, CD119$^+$, CD124$^-$, CD126$^-$, CD127$^+$, CD135$^+$

Genetic characterization[1-3]

Cytogenetic karyotype	46, XY, −11, +der(11)t(11;?)(11pter-11q23::?)
Receptor gene rearrangement	*IGH* RR, *TCRB* GG, *TCRG* GG, *TCRD* DD

Functional characterization[1]

Cytochemistry	MPO$^-$, NBT$^-$

Comments

- Precursor B-cell line of type B-I (pro B-cell line).

References

[1] Matsuo, Y. et al. (1991) Human Cell 4, 257–260.
[2] Minowada, J. and Matsuo, Y. (1999) unpublished data.
[3] Matsuo, Y. and Drexler, H.G. (1998) Leukemia Res. 22, 567–579.

NALM-20

Culture characterization[1]

Sister cell lines (1) **B250** (simultaneous sister cell line = EBV+ B-LCL)
(2) **NALM-21, NALM-22, NALM-23** (serial sister cell lines – at relapse – similar features)
Culture medium 90% RPMI 1640 + 10% FBS
Doubling time 72 h
Authentication no
Primary reference Matsuo et al.[1]
Availability of cell line original authors

Clinical characterization[1]

Patient 62-year-old male
Disease diagnosis AUL
Treatment status at diagnosis
Specimen site peripheral blood
Year of establishment 1989

Immunophenotypic characterization[1,2]

T-/NK cell marker CD1−, CD3−, CD4−, CD5−, CD7−, CD8−, TCRαβ−, TCRγδ−
B-cell marker CD9+, CD10+, CD19+, CD20−, CD22+, CD24+, cyCD79a+, cyCD79b+, CD85+, cyIgM−, sIg−
Myelomonocytic marker CD13+, CD14−, CD15−, CD33(+)
Progenitor/activation/other marker CD34−, HLA-DR+, TdT+
Cytokine receptor CD25−, CD105−, CD115−, CD117(+), CD119+, CD124−, CD126−, CD127+, CD135+

Genetic characterization[1,3]

Cytogenetic karyotype 46, XY, +2, −8, t(9;22)(q34;q11)
Unique translocation/fusion gene Ph+ t(9;22)(q34;q11) → *BCR-ABL* e1-a2 fusion gene
Receptor gene rearrangement *IGH* RR, *TCRB* RR, *TCRG* RR, *TCRD* DD

Functional characterization[1,2,4]

Cytochemistry MPO−, NBT−
Cytokine response growth stimulation: IL-7
Special features MPO mRNA+

Comments

- Precursor B-cell line of type B-II (common B-cell line).
- Simultaneous EBV+ B-LCL and sequential sister cell lines established at relapse are available.

- Carries Ph chromosome leading to *BCR-ABL* e1-a2 fusion gene.
- Co-expression of myeloid surface antigens.

References

[1] Matsuo, Y. et al. (1991) Human Cell 4, 335–338.
[2] Matsuo, Y. and Drexler, H.G. (1998) Leukemia Res. 22, 567–579.
[3] Drexler, H.G. et al. (1999) Leukemia Res. 23, 207–215.
[4] Ohyashiki, K. et al. (1993) Leukemia 7, 1034–1040.

NALM-24

Culture characterization[1]

Sister cell lines	(1) **NALM-25** (simultaneous sister cell line – similar features) (2) **B262** (= EBV$^+$ B-LCL)
Culture medium	90% RPMI 1640 + 10% FBS
Doubling time	72 h
Authentication	no
Primary reference	Matsuo et al.[1]
Availability of cell line	original authors

Clinical characterization[1]

Patient	42-year-old female
Disease diagnosis	ALL
Treatment status	at diagnosis
Specimen site	peripheral blood
Year of establishment	1990

Immunophenotypic characterization[1,2]

T-/NK cell marker	CD1$^-$, CD3$^-$, CD4$^+$, CD5$^-$, CD7$^-$, CD8$^-$, TCR$\alpha\beta^-$, TCR$\gamma\delta^-$
B-cell marker	CD9$^+$, CD10$^+$, CD19$^+$, CD20$^-$, CD22$^+$, CD24$^+$, cyCD79a$^+$, cyCD79b$^+$, CD85$^+$, cyIgM$^-$, sIg$^-$
Myelomonocytic marker	CD13$^+$, CD14$^-$, CD15$^+$, CD33$^{(+)}$
Progenitor/activation/other marker	CD34$^+$, HLA-DR$^+$, TdT$^+$
Cytokine receptor	CD25$^-$, CD105$^-$, CD115$^-$, CD117$^-$, CD119$^+$, CD124$^-$, CD126$^-$, CD127$^+$, CD135$^+$

Genetic characterization[1,2]

Cytogenetic karyotype	45, XX, −15, −20, +mar, t(9;22)(q34;q11), del(9)(p21)
Unique translocation/fusion gene	Ph$^+$ t(9;22)(q34;q11) → *BCR-ABL* fusion gene
Receptor gene rearrangement	*IGH* DD, *TCRB* RR, *TCRG* RR, *TCRD* RD
Unique gene alteration	*P15INK4B* deletion, *P16INK4A* deletion

Functional characterization[1-3]

Cytochemistry	MPO$^-$, NBT$^-$
Cytokine response	growth stimulation: IL-7
Special features	MPO mRNA$^+$

Comments

- Precursor B-cell line of type B-II (common B-cell line).

- Carries Ph chromosome leading to *BCR-ABL* fusion gene (with unknown breakpoint region).
- Co-expression of various myeloid markers.

References

[1] Matsuo, Y. et al. (1991) Human Cell 4, 339–341.
[2] Matsuo, Y. and Drexler, H.G. (1998) Leukemia Res. 22, 567–579.
[3] Ohyashiki, K. et al. (1993) Leukemia 7, 1034–1040.

NALM-26

Culture characterization[1]

Culture medium	90% RPMI 1640 + 10% FBS
Doubling time	5 d
Authentication	no
Primary reference	Matsuo et al.[1]
Availability of cell line	original authors

Clinical characterization[1]

Patient	24-year-old male
Disease diagnosis	pre B-ALL
Treatment status	at diagnosis
Specimen site	peripheral blood
Year of establishment	1992

Immunophenotypic characterization[1,2]

T-/NK cell marker	CD1⁻, CD3⁻, CD4⁻, CD5⁽⁺⁾, CD7⁻, CD8⁻, CD28⁻, TCRαβ⁻, TCRγδ⁻
B-cell marker	CD9⁺, CD10⁺, CD19⁺, CD20⁻, CD22⁽⁺⁾, cyCD22⁺, cyCD79a⁺, cyCD79b⁺, CD85⁽⁺⁾, cyIgM⁺, sIg⁻
Myelomonocytic marker	CD13⁻, CD14⁻, CD15⁻, CD33⁻
Progenitor/activation/other marker	CD34⁻, HLA-DR⁺, TdT⁺
Cytokine receptor	CD25⁻, CD105⁺, CD115⁻, CD117⁻, CD119⁺, CD124⁻, CD126⁻, CD127⁺, CD135⁺

Genetic characterization[1]

Cytogenetic karyotype	45, X, −9, −19, +mar, dic(1;11)(p13;q25), +del(1)(p13), +del(1) (q21), i(9)(q10), add(14)(q24)

Functional characterization[1]

Cytochemistry	MPO⁻
Inducibility of differentiation	TPA → B-cell differentiation

Comments

- Precursor B-cell line of type B-III (pre B-cell line).

References

[1] Matsuo, Y. et al. (1994) Human Cell 7, 221–226.

[2] Matsuo, Y. and Drexler, H.G. (1998) Leukemia Res. 22, 567–579.

NALM-27

Culture characterization[1-3]

Sister cell lines (1) **NALM-28** (simultaneous sister cell line – similar features)
(2) **NALM-30, NALM-31, NALM-32** (serial sister cell lines – at relapse – different immunological and cytogenetic features)
Culture medium 90% RPMI 1640 + 10% FBS
Doubling time 4–5 d
Authentication no
Primary reference Ariyasu et al.[1]
Availability of cell line original authors

Clinical characterization[1]

Patient 38-year-old male
Disease diagnosis biphenotypic ALL
Treatment status at diagnosis
Specimen site peripheral blood
Year of establishment 1995

Immunophenotypic characterization[1-3]

T-/NK cell marker CD3⁻, CD4⁻, CD5⁻, CD7⁻, TCRαβ⁻, TCRγδ⁻
B-cell marker CD10⁺, CD19⁺, CD20⁻, CD21⁻, CD22⁺, CD23⁻, CD24⁺, cyCD79a⁺, cyCD79b⁺, CD85⁺, cyIgM⁻, sIg⁻
Myelomonocytic marker CD13⁺, CD14⁻, CD33⁽⁺⁾
Progenitor/activation/other marker CD34⁺, HLA-DR⁺, TdT⁺
Cytokine receptor CD105⁻, CD115⁻, CD117⁻, CD119⁻, CD124⁻, CD126⁻, CD127⁻, CD135⁺

Genetic characterization[1-4]

Cytogenetic karyotype 46, XY, t(9;22;10)(q34;q11;q22)
Unique translocation/fusion gene Ph⁺ t(9;22)(q34;q11) → *BCR-ABL* b3-a2 fusion gene
Unique gene alteration *P53* mutation

Functional characterization[1-3]

Cytochemistry MPO⁻
Special features MPO mRNA⁺

Comments

- Precursor B-cell line of type B-II (common B-cell).
- Carries unusual three-way Ph chromosome abnormality leading to *BCR-ABL* b3-a2 fusion gene.
- Co-expression of myeloid-associated immunomarkers.
- Serial sister cell lines established at diagnosis and at relapse available.

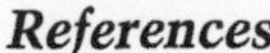

References

[1] Ariyasu, T. et al. (1998) Human Cell 11, 43–50.
[2] Matsuo, Y. et al. (1999) Human Cell 11, 221–230.
[3] Matsuo, Y. and Drexler, H.G. (1998) Leukemia Res. 22, 567–579.
[4] Drexler, H.G. et al. (1999) Leukemia Res. 23, 207–215.

NALM-29

Culture characterization[1]

Establishment	initially on bone marrow stroma cells as feeder layer
Culture medium	90% RPMI 1640 + 10% FBS
Doubling time	46 h
Viral status	EBV⁻
Authentication	yes (by DNA fingerprinting)
Primary reference	Matsuo et al.[1]
Availability of cell line	original authors

Clinical characterization[1]

Patient	46-year-old male
Disease diagnosis	ALL
Treatment status	at diagnosis
Specimen site	peripheral blood
Year of establishment	1994

Immunophenotypic characterization[1,2]

T-/NK cell marker	CD1⁻, CD2⁻, CD3⁻, CD4⁻, CD5⁻, CD7⁻, CD8⁻, CD28⁻, CD56⁻, CD57⁻
B-cell marker	CD9⁺, CD10⁺, CD19⁺, CD20⁻, CD21⁻, CD22⁺, CD23⁻, CD24⁺, CD37⁻, CD40⁻, cyCD79a⁺, cyCD79b⁺, CD80⁻, CD85⁺, CD86⁺, cyIgM⁻, sIg⁻
Myelomonocytic marker	CD13⁺, CD14⁻, CD15⁻, CD16⁻, CD32⁺, CD33⁻, CD64⁻, CD65⁻
Erythroid-megakaryocytic marker	CD31⁻, CD36⁻, CD41⁻, CD42a⁻, CD42b⁻, CD61⁻, CD62P⁻
Progenitor/activation/other marker	CD30⁻, CD34⁺, CD38⁺, CD44⁺, CD45⁺, CD45RA⁺, CD45RO⁺, CD71⁺, CD90⁻, CD95⁽⁺⁾, HLA-DR⁺, TdT⁺
Adhesion marker	CD11a⁺, CD18⁽⁺⁾, CD49d⁺, CD54⁺, CD58⁺, CD62L⁻
Cytokine receptor	CD25⁻, CD105⁻, CD115⁻, CD116⁻, CD117⁻, CD119⁺, CD120A⁻, CD120B⁻, CD122⁻, CD124⁻, CD126⁻, CD127⁺, CD130⁽⁺⁾, CD135⁺

Genetic characterization[1,2]

Cytogenetic karyotype	46, XY, del(6)(q15q21), t(9;22)(q34;q11)
Unique translocation/fusion gene	Ph⁺ t(9;22)(q34;q11) → *BCR-ABL* e1-a2 fusion gene
Receptor gene rearrangement	*IGH* R
Unique gene alteration	*P15INK4B* deletion

Functional characterization[1,2]

Cytochemistry	MPO⁻
Special features	MPO mRNA⁺

Comments

- Precursor B-cell line of type B-II (common B-cell line).
- Carries Ph chromosome leading to *BCR-ABL* e1-a2 fusion gene.
- Co-expression of myeloid immunomarkers.

References

[1] Matsuo, Y. et al. (1999) Leukemia Res. 23, 731–740.
[2] Matsuo, Y. and Drexler, H.G. (1998) Leukemia Res. 22, 567–579.

NALM-33

Culture characterization[1]

Sister cell line	**NALM-34** (simultaneous sister cell line – identical features)
Culture medium	90% RPMI 1640 + 10% FBS
Doubling time	72 h
Viral status	EBV⁻
Authentication	yes (by DNA fingerprinting)
Primary reference	Matsuo et al.[1]
Availability of cell line	original authors

Clinical characterization[1]

Patient	72-year-old male
Disease diagnosis	BCP-ALL
Treatment status	at relapse
Specimen site	peripheral blood
Year of establishment	1996

Immunophenotypic characterization[1,2]

T-/NK cell marker	$CD2^-$, $CD3^-$, $CD4^-$, $CD5^-$, $CD7^-$, $CD8^-$
B-cell marker	$CD9^+$, $CD10^+$, $CD19^+$, $CD20^-$, $CD21^-$, $CD22^+$, $CD23^-$, $CD39^+$, $CD40^+$, $cyCD79a^+$, $cyCD79b^+$, $CD85^-$, $cyIgM^-$, sIg^-
Myelomonocytic marker	$CD13^-$, $CD14^-$, $CD15^-$, $CD33^-$
Progenitor/activation/other marker	$CD34^-$, $CD38^+$, $CD71^+$, $HLA\text{-}DR^+$, TdT^+
Cytokine receptor	$CD116^-$, $CD117^-$, $CD119^+$, $CD120A^-$, $CD120B^-$, $CD123^-$, $CD124^-$, $CD126^-$, $CD127^-$, $CD130^-$, $CD132^-$, $CD135^{(+)}$

Genetic characterization[1,2]

Cytogenetic karyotype	46–48, XY, +1, +21, add(1)(q32), t(8;14)(q24.1;q32), del(13) (q22q32), add(15)(p11.2), add(19)(q?13.3)
Unique translocation/fusion gene	t(8;14)(q24;q32) → *MYC-IGH* genes altered?
Receptor gene rearrangement	*IGH* G, *IGK* R, *IGL* R

Functional characterization[1,2]

Cytochemistry	MPO⁻

Comments

- Precursor B-cell line of type B-II (common B-cell line).
- Carries t(8;14)(q24;q32) affecting the *MYC-IGH* genes (typically seen in B-ALL or Burkitt lymphoma).

References

[1] Matsuo, Y. et al. (1999) Leukemia Lymphoma 35, 513–526.
[2] Matsuo, Y. and Drexler, H.G. (1998) Leukemia Res. 22, 567–579.

OM9;22

Culture characterization[1]

Culture medium	80% RPMI 1640 + 20% FBS
Doubling time	80 h
Authentication	yes (*BCR* rearrangement)
Primary reference	Ohyashiki et al.[1]
Availability of cell line	original authors

Clinical characterization[1]

Patient	19-year-old female
Disease diagnosis	ALL (L2)
Treatment status	at relapse (post-BMT)
Specimen site	bone marrow
Year of establishment	1987

Immunophenotypic characterization[1,2]

T-/NK cell marker	$CD5^-$, $CD7^-$, $TCR\alpha\beta^-$, $TCR\gamma\delta^-$
B-cell marker	$CD10^+$, $CD19^+$, $CD20^-$, $CD22^+$, $cyCD79a^+$, $cyCD79b^+$, $CD85^+$, $cyIgM^-$, sIg^-
Myelomonocytic marker	$CD13^+$, $CD14^-$, $CD33^+$
Erythroid-megakaryocytic marker	$CD36^-$
Progenitor/activation/other marker	$CD34^+$, $HLA\text{-}DR^+$, TdT^+
Adhesion marker	$CD11b^-$
Cytokine receptor	$CD105^-$, $CD115^-$, $CD117^-$, $CD119^{(+)}$, $CD124^+$, $CD127^+$, $CD135^+$

Genetic characterization[1-3]

Cytogenetic karyotype	45, X, −X, −4, −8, −15, −16, −17, +4mar, +der(1)t(1;?)(q25;?), del(3)(q26), t(9;22)(q34;q11)
Unique translocation/fusion gene	Ph^+ t(9;22)(q34;q11) → *BCR-ABL* e1-a2 fusion gene
Receptor gene rearrangement	*IGH* RD, *IGK* G, *IGL* G, *TCRD* D

Functional characterization[1,2]

Colony formation	in methylcellulose
Cytochemistry	$ANBE^-$, CAE^-, MPO^-, PAS^-
Cytokine response	growth stimulation: IL-7; growth inhibition: IL-4
Special features	MPO $mRNA^+$

Comments

- Precursor B-cell line of type B-II (common B-cell line).
- Carries Ph chromosome leading to *BCR-ABL* e1-a2 fusion gene.
- Co-expression of myeloid surface antigens.

References

[1] Ohyashiki, K. et al. (1993) Leukemia 7, 1034–1040.

[2] Matsuo, Y. and Drexler, H.G. (1998) Leukemia Res. 22, 567–579.

[3] Drexler, H.G. et al. (1999) Leukemia Res. 23, 207–215.

Culture characterization[1]

Culture medium	90% RPMI 1640 + 10% FBS
Doubling time	36 h
Authentication	yes (by cytogenetics)
Primary reference	Nagai et al.[1]
Availability of cell line	original authors

Clinical characterization[1]

Patient	32-year-old male
Disease diagnosis	B-NHL (diffuse large cell)
Treatment status	at diagnosis
Specimen site	bone marrow
Year of establishment	1993

Immunophenotypic characterization[1]

T-/NK cell marker	CD2$^-$, CD3$^-$, CD4$^-$, CD5$^-$, CD8$^-$
B-cell marker	CD10$^+$, CD19$^+$, CD20$^-$, CD24$^+$, sIg$^-$
Myelomonocytic marker	CD13$^-$, CD33$^-$
Progenitor/activation/other marker	CD38$^+$, HLA-DR$^+$

Genetic characterization[1,2]

Cytogenetic karyotype	49, XY, del(2)(q35), der(8)t(8;12)(p21;q13.3), add(10)(p11.12), +add(12)(q24), t(14;18)(q32;q21) add(17)(p11)
Unique translocation/fusion gene	t(14;18)(q32;q21) → *IGH-BCL2* genes altered
Unique gene alteration	*P15INK4B* deletion, *P16INK4A* deletion, *P53* alteration

Comments

- Immunologically incompletely characterized precursor B-cell line (apparently type B-II or B-III due to lack of surface immunoglobulin expression).
- Carries t(14;18)(q32;q21) affecting the *IGH-BCL2* genes which is typically seen in mature B-cell lines.

References

[1] Nagai, M. et al. (1998) Hematol. Oncol. 15, 109–119.
[2] Gombart, A.F. et al. (1995) Blood 86, 1534–1539.

P30/Ohkubo

Culture characterization[1]

Culture medium	90% RPMI 1640 + 10% FBS
Doubling time	30 h
Viral status	EBNA$^-$
Authentication	no
Primary reference	Hirose et al.[1]
Availability of cell line	JCRB 0094

Clinical characterization[1]

Patient	11-year-old female
Disease diagnosis	ALL (L2)
Treatment status	at 3rd relapse (post-BMT)
Specimen site	bone marrow
Year of establishment	1980

Immunophenotypic characterization[1–3]

T-/NK cell marker	CD1$^-$, CD2$^-$, CD3$^-$, cyCD3$^-$, CD4$^-$, CD5$^-$, CD7$^-$, CD8$^-$, CD28$^+$, CD57$^-$, TCR$\alpha\beta^-$, TCR$\gamma\delta^-$
B-cell marker	CD10$^+$, CD19$^+$, CD20$^-$, CD21$^-$, CD22$^-$, CD24$^+$, cyCD79a$^+$, cyCD79b$^+$, CD85$^+$, cyIgM$^+$, sIg$^-$
Myelomonocytic marker	CD13$^-$, CD14$^-$, CD15$^-$, CD33$^-$
Erythroid-megakaryocytic marker	CD41a$^-$, CD61$^-$
Progenitor/activation/other marker	CD34$^-$, CD38$^+$, CD71$^+$, HLA-DR$^+$, TdT$^+$
Adhesion marker	CD11b$^-$
Cytokine receptor	CD25$^-$, CD105$^-$, CD115$^-$, CD117$^-$, CD119$^+$, CD124$^-$, CD126$^-$, CD127$^+$, CD135$^+$

Genetic characterization[1,3–5]

Cytogenetic karyotype	46(45), X(X), del(2)(p23), del(9)(p12-q31), t(11;12)(q25;q13), inv(12)(q13q24)
Receptor gene rearrangement	*TCRA* GG, *TCRB* GG, *TCRG* GG, *TCRD* GG
Unique gene alteration	*P16INK4A* deletion

Functional characterization[1]

Cytochemistry	ACP$^+$, ALP$^-$, ANAE$^-$, CAE$^-$, MPO$^-$, PAS$^+$
Special features	MPO mRNA$^+$

Comments

- Precursor B-cell line of type B-III (pre B-cell line).
- Originally described as immature T-cell line, but cell line does not carry any T-cell features.
- Available from cell line bank.

References

[1] Hirose, M. et al. (1982) Gann 73, 600–605.
[2] Minowada, J. and Matsuo, Y. (1999) unpublished data.
[3] Matsuo, Y. and Drexler, H.G. (1998) Leukemia Res. 22, 567–579.
[4] Ogawa, S. et al. (1994) Blood 84, 2431–2435.
[5] Sangster, R.N. et al. (1986) J. Exp. Med. 163, 1491–1508.

PALL-2

Culture characterization[1]

Culture medium	90% RPMI 1640 + 10% FBS
Viral status	EBNA⁻
Authentication	no
Primary reference	Miyagi et al.[1]
Availability of cell line	restricted

Clinical characterization[1]

Patient	45-year-old male
Disease diagnosis	pre B-ALL (L2)
Treatment status	at relapse
Specimen site	peripheral blood
Year of establishment	1987

Immunophenotypic characterization[1]

T-/NK cell marker	$CD1^-$, $CD2^-$, $CD3^-$, $CD4^-$, $CD5^-$, $CD7^-$, $CD8^-$
B-cell marker	$CD10^+$, $CD19^+$, $CD20^-$, $cyIgM^-$, sIg^-
Myelomonocytic marker	$CD13^+$, $CD14^-$, $CD15^-$, $CD33^-$
Progenitor/activation/other marker	$CD34^+$, $CD38^-$, $HLA\text{-}DR^+$, TdT^-
Adhesion marker	$CD11b^-$

Genetic characterization[1,2]

Cytogenetic karyotype	47, XY, +8, 7p−, 10p−, 14q-, 17q+, 18p+, t(9;22)(q34;q11)
Unique translocation/fusion gene	Ph^+ t(9;22)(q34;q11) → *BCR-ABL* e1-a2 fusion gene
Receptor gene rearrangement	*IGH* RR, *TCRB* RR

Functional characterization[1]

Cytochemistry	$ANBE^-$, MPO^-, PAS^-, SBB^-
Heterotransplantation	into nude mice

Comments

- Precursor B-cell line of type B-II (common B-cell line).
- Carries Ph chromosome leading to *BCR-ABL* e1-a2 fusion gene.

References

[1] Miyagi, T. et al. (1993) Int. J. Cancer 53, 457–462.
[2] Drexler, H.G. et al. (1999) Leukemia Res. 23, 207–215.

PC-53

Culture characterization[1]

Sister cell line PC-53A (serial sister cell line – at final, refractory stage – identical features)
Culture medium 90% IMDM + 10% FBS
Doubling time 70–80 h
Viral status EBV⁻
Authentication no
Primary reference Tarella et al.[1]
Availability of cell line original authors

Clinical characterization[1]

Patient 33-year-old female
Disease diagnosis cALL (L2)
Treatment status at 3rd relapse
Specimen site bone marrow
Year of establishment 1985

Immunophenotypic characterization[1]

T-/NK cell marker $CD3^-$, $CD5^-$
B-cell marker $CD9^+$, $CD10^+$, $CD19^+$, $CD20^-$, $CD21^-$, $CD24^+$, PCA-1^-, $cyIgM^-$, sIg^-
Progenitor/activation/other marker $CD38^+$, $CD71^+$, HLA-DR^+, TdT^+
Adhesion marker $CD11b^-$

Genetic characterization[1]

Cytogenetic karyotype 45, XX, −1, −1, −3, −14, −17, +3mar, +der(1)(pter→q32 or 44: :1p31→pter)
Receptor gene rearrangement *IGH* RR, *TCRB* G

Functional characterization[1]

Cytochemistry $ANAE^-$, CAE^-, PAS^+, SBB^-

Comments

- Precursor B-cell line of type B-II (common B-cell line).

Reference

[1] Tarella, C. et al. (1990) Leukemia Res. 14, 177–184.

PER-278

Culture characterization[1]

Culture medium	90% RPMI 1640 + 10% FBS
Doubling time	62–68 h
Viral status	EBNA⁻
Authentication	yes (by *IGH* rearrangement)
Primary reference	Kees et al.[1]
Availability of cell line	not known

Clinical characterization[1]

Patient	10-year-old male
Disease diagnosis	pre B-ALL (L1)
Treatment status	at diagnosis
Specimen site	bone marrow
Year of establishment	1987

Immunophenotypic characterization[1]

T-/NK cell marker	$CD7^-$
B-cell marker	$CD9^+$, $CD10^+$, $CD19^+$, $CD20^+$, $CD22^{(+)}$, $cyIgM^+$
Progenitor/activation/other marker	$HLA\text{-}DR^+$, TdT^+
Adhesion marker	$CD11b^-$

Genetic characterization[1]

Cytogenetic karyotype	46, XY, −9, −19, +der(9)t(1;9)(q23;p13), +der(19)t(1;19) (q23;p13)
Unique translocation/fusion gene	t(1;19)(q23;p13) → *E2A-PBX1* fusion gene
Receptor gene rearrangement	*IGH* DR, *IGK* G, *IGL* G, *TCRB* G

Functional characterization[1]

Colony formation	not clonable
Cytochemistry	ACP^-, $ANAE^-$, MPO^-, PAS^+

Comments

- Precursor B-cell line of type B-III (pre B-cell line).
- Carries t(1;19)(q23;p13) leading to *E2A-PBX1* fusion gene.

Reference

[1] Kees, U.R. et al. (1990) Cancer Genet. Cytogenet. 46, 201–208.

Pre-Alp

Culture characterization[1]

Culture medium	90% RPMI 1640 + 10% FBS (or serum-free)
Doubling time	24 h
Viral status	EBNA$^-$
Authentication	yes (by cytogenetics)
Primary reference	Pandrau et al.[1]
Availability of cell line	original authors

Clinical characterization[1]

Patient	6-year-old female
Disease diagnosis	pre B-ALL (L1)
Treatment status	at diagnosis
Specimen site	bone marrow
Year of establishment	1990

Immunophenotypic characterization[1]

B-cell marker	CD10$^+$, CD19$^+$, CD20$^+$, CD24$^+$, cyIgM$^+$, sIg$^-$
Progenitor/activation/other marker	CD34$^-$, HLA-DR$^+$, TdT$^+$

Genetic characterization[1,2]

Cytogenetic karyotype	45, XX, −2, t(1;19)(q23;p13.3)
Unique translocation/fusion gene	t(1;19)(q23;p13) → *E2A-PBX1* fusion gene
Receptor gene rearrangement	*IGH* R, *IGL* G
Unique gene alteration	*P16INK4A* deletion

Functional characterization[1]

Cytokine response	growth stimulation: IL-7; growth inhibition: TGF-β
Special features	RT-PCR$^+$: RAG1, RAG2, VPREB, γ5

Comments

- Precursor B-cell line of type B-III (pre B-cell line).
- Carries t(1;19)(q23;p13) leading to *E2A-PBX1* fusion gene.
- Immunophenotype requires further analysis.

References

[1] Pandrau, D. et al. (1993) Leukemia 7, 635–642.

[2] Kim, D.H. et al. (1996) Blood 88, 785–794.

RCH-ACV

Culture characterization[1]

Culture medium	80% RPMI 1640 + 20% FBS
Doubling time	36 h
Viral status	EBNA$^-$
Authentication	yes (by cytogenetics)
Primary reference	Jack et al.[1]
Availability of cell line	restricted

Clinical characterization[1]

Patient	8-year-old female
Disease diagnosis	cALL
Treatment status	at relapse
Specimen site	bone marrow

Immunophenotypic characterization[1,2]

T-/NK cell marker	CD7$^-$
B-cell marker	CD9$^+$, CD10$^+$, CD19$^+$, CD20$^-$, CD24$^+$, cyIgM$^+$, sIg$^-$
Myelomonocytic marker	CD14$^-$
Progenitor/activation/other marker	CD71$^+$, HLA-DR$^+$

Genetic characterization[1,2]

Cytogenetic karyotype	47, XX, +8, t(1;19)(q23;p13)
Unique translocation/fusion gene	t(1;19)(q23;p13) → *E2A-PBX1* fusion gene
Receptor gene rearrangement	*IGH* RR, *IGK* DD, *IGL* G

Functional characterization[1]

Colony formation	clonable

Comments

- Precursor B-cell line of type B-III (pre B-cell line).
- Carries t(1;19)(q23;p13) leading to *E2A-PBX1* fusion gene.

References

[1] Jack, I. et al. (1986) Cancer Genet. Cytogenet. 19, 261–269.
[2] Kim, D.H. et al. (1996) Blood 88, 785–794.

Reh

Culture characterization[1,2]

Culture medium	90% RPMI 1640 + 10% FBS
Doubling time	35–70 h
Viral status	EBV$^-$, HBV$^-$, HCV$^-$, HHV-8$^-$, HIV$^-$, HTLV-I/II$^-$
Authentication	yes (by cytogenetics)
Primary reference	Rosenfeld et al.[1]
Availability of cell line	ATCC CRL 8286, DSM ACC 22

Clinical characterization[1]

Patient	15-year-old female
Disease diagnosis	ALL
Treatment status	at relapse
Specimen site	peripheral blood
Year of establishment	1974

Immunophenotypic characterization[1-5]

T-/NK cell marker	CD1$^-$, CD2$^-$, CD3$^-$, cyCD3$^-$, CD4$^-$, CD5$^-$, CD7$^-$, CD8$^-$, CD28$^-$, CD57$^-$, TCRαβ$^-$, TCRγδ$^-$
B-cell marker	CD9$^+$, CD10$^+$, CD19$^+$, CD20$^-$, CD21$^-$, CD22$^+$, CD23$^-$, CD24$^+$, CD37$^-$, cyCD79a$^+$, cyCD79b$^+$, CD85$^-$, CD138$^+$, cyIgM$^-$, sIg$^-$
Myelomonocytic marker	CD13$^-$, CD14$^-$, CD15$^-$, CD33$^-$
Erythroid-megakaryocytic marker	CD41a$^-$, CD61$^-$
Progenitor/activation/other marker	CD34$^-$, CD38$^+$, CD71$^+$, CD95$^-$, HLA-DR$^+$, TdT$^+$
Adhesion marker	CD11b$^-$
Cytokine receptor	CD25$^-$, CD105$^-$, CD115$^-$, CD117$^+$, CD119$^-$, CD124$^-$, CD126$^+$, CD127$^-$, CD135$^+$

Genetic characterization[1-2,6-9]

Cytogenetic karyotype	46(44–47)<2n>X, −X, +16, del(3)(p22), t(4;12;21;16)(q32;p13;q22;q24.3)inv(12)(p13q22), t(5;12)(q31-q32;p12), der(16)t(16;21) (q24.3;q22); sideline with inv(5)der(5)(p15q31), +18
Unique translocation/fusion gene	t(12;21)(p13;q22) → *TEL-AML1* fusion gene
Receptor gene rearrangement	*IGH* RR
Unique gene alteration	*P15INK4B* deletion, *P16INK4A* deletion

Functional characterization[1,4,10]

Proto-oncogene	mRNA$^+$: BAX, BCL2, BCLXL, MDM2
Special features	MPO mRNA$^+$

Comments

- First precursor B-cell line established (reference cell line).
- Precursor B-cell line of type B-II (common B-cell line).
- Carries t(12;21)(p13;q22) leading to *TEL-AML1* fusion gene (reference cell line).
- Available from cell line banks.
- A Reh subclone termed **Reh-6**, which was in widespread circulation, was found to be cross-contaminated with murine cells[11].

References

[1] Rosenfeld, C. et al. (1977) Nature 267, 841–843.
[2] Drexler, H.G. et al. (1999) DSMZ Catalogue of Cell Lines, 7th edn, Braunschweig, Germany.
[3] Minowada, J. and Matsuo, Y. (1999) unpublished data.
[4] Matsuo, Y. and Drexler, H.G. (1998) Leukemia Res. 22, 567–579.
[5] Zhou, M. et al. (1995) Blood 85, 1608–1614.
[6] Venuat, A.M. et al. (1981) Cancer Genet. Cytogenet. 3, 327–334.
[7] Uphoff, C.C. et al. (1997) Leukemia 11, 441–447.
[8] Kim, D.H. et al. (1996) Blood 88, 785–794.
[9] Zhou, M. et al. (1995) Leukemia 9, 1159–1161.
[10] Findley, H.W. et al. (1997) Blood 89, 2986–2993.
[11] Drexler, H.G. et al. (1999) Leukemia 13, 1601–1607.

RS4;11

Culture characterization[1]

Establishment	initially in hypoxic environment
Culture medium	90% α-MEM + 10% FBS
Doubling time	60–70 h
Viral status	EBNA⁻
Authentication	yes (by cytogenetics)
Primary reference	Stong et al.[1]
Availability of cell line	ATCC CRL 1873

Clinical characterization[1]

Patient	32-year-old female
Disease diagnosis	ALL (L2)
Treatment status	at relapse
Specimen site	bone marrow

Immunophenotypic characterization[1]

T-/NK cell marker	$CD1^-$, $CD2^-$, $CD3^-$, $cyCD3^-$, $CD4^-$, $CD5^-$, $CD7^-$, $CD8^-$
B-cell marker	$CD9^+$, $CD10^-$, $CD19^+$, $CD20^-$, $cyIgM^-$, sIg^-
Myelomonocytic marker	$CD13^-$, $CD14^-$, $CD15^+$, $CD33^-$
Erythroid-megakaryocytic marker	$CD41b^-$
Progenitor/activation/other marker	$CD38^+$, $CD71^+$, $HLA\text{-}DR^+$, TdT^+
Adhesion marker	$CD11b^-$

Genetic characterization[1–4]

Cytogenetic karyotype	46, XX, i(7q), t(4;11)(q21;q23)
Unique translocation/fusion gene	t(4;11)(q21;q23) → *MLL-AF4* fusion gene
Receptor gene rearrangement	*IGH* RG, *IGK* RD, *TCRB* G, *TCRG* G
Unique gene alteration	*P15INK4B* deletion, *P16INK4A* deletion

Functional characterization[1,2,5]

Colony formation	in methylcellulose
Cytochemistry	ACP^+, $ANAE^-$, CAE^-, MPO^-, SBB^-
Heterotransplantation	into SCID mice
Inducibility of differentiation	TPA → mono differentiation

Comments

- Precursor B-cell line of type B-I (pro B-cell line).
- First cell line known to carry t(4;11)(q21;q23) leading to *MLL-AF4* fusion gene.
- Available from cell line bank.

References

[1] Stong, R.C. et al. (1985) Blood 65, 21–31.
[2] Iida, S. et al. (1992) Leukemia Res. 16, 1155–1163.
[3] Yamamoto, K. et al. (1994) Blood 83, 2912–2921.
[4] Okuda, T. et al. (1995) Blood 85, 2321–2330.
[5] Uckun, F.M. (1996) Blood 88, 1135–1146.

SEM

Culture characterization[1]

Establishment	initially supplemented with CM of hepatoma SK-HEP-1 cell line
Culture medium	90% IMDM + 10% FBS (or serum-free)
Doubling time	48 h
Viral status	EBNA$^-$
Authentication	yes (by cytogenetics)
Primary reference	Greil et al.[1]
Availability of cell line	original authors

Clinical characterization[1]

Patient	5-year-old female
Disease diagnosis	ALL
Treatment status	at relapse
Specimen site	peripheral blood
Year of establishment	1990

Immunophenotypic characterization[1]

T-/NK cell marker	CD1$^-$, CD2$^-$, CD3$^-$, CD5$^-$, CD7$^-$
B-cell marker	CD9$^+$, CD10$^-$, CD19$^+$, CD20$^-$, CD22$^+$, CD23$^-$, CD24$^-$, CD75$^+$, cyIgM$^-$, sIg$^-$
Myelomonocytic marker	CD13$^+$, CD14$^-$, CD15$^+$, CD32$^+$, CD33$^+$, CD65$^+$
Progenitor/activation/other marker	CD34$^-$, HLA-DR$^+$, TdT$^+$

Genetic characterization[1,2]

Cytogenetic karyotype	46, XX, t(4;11)(q21;q23), del(7)(p15), del(13)(q12)
Unique translocation/fusion gene	t(4;11)(q21;q23)→*MLL-AF4* fusion gene
Receptor gene rearrangement	*IGH* R, *IGK* G, *TCRB* G, *TCRG* R, *TCRD* R

Functional characterization[1]

Cytochemistry	ANAE$^-$, MPO$^-$, PAS$^-$, SBB$^-$
Cytokine receptor	RT-PCR$^+$: IL-7R
Cytokine response	growth stimulation: IL-7; growth inhibition: IFN-α, IFN-γ, IL-4, TNF-α

Comments

- Precursor B-cell line of type B-I (pro B-cell line).
- Carries t(4;11)(q21;q23) leading to *MLL-AF4* fusion gene.
- Co-expression of several myeloid-associated surface antigens.

References

[1] Greil, J. et al. (1994) Brit. J. Haematol. 86, 275–283.
[2] Marschalek, R. et al. (1997) Leukemia Lymphoma 27, 417–428.

SUP-B2

Culture characterization[1]

Establishment	initially on feeder layer (complete media, human serum, agar) in hypoxic environment
Culture medium	90% McCoy's 5A + 10% FBS
Doubling time	30 h
Viral status	EBV⁻
Authentication	no
Primary reference	Zhang et al.[1]
Availability of cell line	restricted

Clinical characterization[1]

Patient	5-year-old male
Disease diagnosis	BCP-ALL
Treatment status	at relapse

Immunophenotypic characterization[1–3]

T-/NK cell marker	$CD1^-$, $CD2^-$, $CD3^-$, $CD4^-$, $CD5^-$, $CD7^-$, $CD8^-$
B-cell marker	$CD10^-$, $CD19^+$, $CD20^-$, $CD24^+$, $cyIgM^-$, sIg^-
Myelomonocytic marker	$CD15^-$
Progenitor/activation/other marker	$CD34^-$, $CD38^-$, $CD71^+$, HLA-DR^+, TdT^+
Cytokine receptor	insulin-R^+, IGF-IR^+, IGF-IIR^+

Genetic characterization[1,4]

Cytogenetic karyotype	44, XY, −14, −21, t(2;4;12)(q13;q25;p13), t(3;7)(q25;p15), inv(5) (q13q33), del(6)(q23q27), t(11;17)(p11.2;p11.2), del(17)(p11p13), dic(21;21)(q21;q21)
Unique translocation/fusion gene	t(6;12)(q23;p13) → *STL-TEL/ETV6* fusion gene

Functional characterization[1–3,5]

Cytokine response	growth stimulation: insulin
Proto-oncogene	$mRNA^+$: BAX, BCLXL, MDM2

Comments

- Precursor B-cell line of type B-II (common B-cell line).
- Carries t(6;12)(q23;p13) leading to *STL-TEL/ETV6* fusion gene.

References

[1] Zhang, L.Q. et al. (1993) Leukemia 7, 1865–1874.
[2] Neely, E.K. et al. (1992) Leukemia 6, 1134–1142.
[3] Zhou, M. et al. (1995) Blood 85, 1608–1614.
[4] Suto, Y. et al. (1997) Genes Chromosomes Cancer 18, 254–268.
[5] Findley, H.W. et al. (1997) Blood 89, 2986–2993.

SUP-B7

Culture characterization[1]

Establishment	initially on feeder layer (complete media, human serum, agar) in hypoxic environment
Culture medium	90% McCoy's 5A + 10% FBS
Doubling time	4–7 d
Viral status	EBV$^-$
Authentication	yes (by cytogenetics)
Primary reference	Smith et al.[1]
Availability of cell line	restricted

Clinical characterization[1]

Patient	2-year-old female
Disease diagnosis	cALL (L1)
Treatment status	at diagnosis
Specimen site	bone marrow
Year of establishment	1983

Immunophenotypic characterization[1-3]

T-/NK cell marker	CD1$^-$, CD2$^-$, CD3$^-$, CD4$^-$, CD5$^-$, CD7$^-$, CD8$^-$
B-cell marker	CD9$^+$, CD10$^+$, CD19$^+$, CD24$^+$, cyIgM$^-$, sIg$^-$
Myelomonocytic marker	CD14$^-$, CD15$^+$
Progenitor/activation/other marker	CD38$^+$, CD71$^+$, HLA-DR$^+$, TdT$^+$
Cytokine receptor	insulin-R$^+$, IGF-IR$^+$, IGF-IIR$^+$

Genetic characterization[1,4,5]

Cytogenetic karyotype	46, XX, del(3)(q26q28)
Receptor gene rearrangement	*IGH* RD, *IGK* G, *IGL* G
Unique gene alteration	*P16INK4A* deletion

Functional characterization[1,2,3,6]

Cytochemistry	ACP$^+$, ANAE$^-$, CAE$^-$, PAS$^+$, SBB$^-$
Cytokine response	growth stimulation: insulin
Proto-oncogene	mRNA$^+$: BAX, BCLXL, MDM2

Comments

- Precursor B-cell line of type B-II (common B-cell line).

References

[1] Smith, S.D. et al. (1987) Cancer Res. 47, 1652–1656.
[2] Neely, E.K. et al. (1992) Leukemia 6, 1134–1142.
[3] Zhou, M. et al. (1995) Blood 85, 1608–1614.
[4] Zhou, M. et al. (1995) Leukemia 9, 1159–1161.
[5] Kim, D.H. et al. (1996) Blood 88, 785–794.
[6] Findley, H.W. et al. (1997) Blood 89, 2986–2993.

SUP-B13

Culture characterization[1]

Establishment	initially on feeder layer (complete media, human serum, agar) in hypoxic environment
Sister cell line	**SUP-B15** (serial sister cell line – at 2nd relapse – similar features)
Culture medium	85% McCoy's 5A + 15% FBS
Doubling time	35 h
Viral status	EBV⁻
Authentication	yes (by *IGH* rearrangement)
Primary reference	Naumovski et al.[1]
Availability of cell line	SUP-B15: ATCC CRL 1929, DSM ACC 389

Clinical characterization[1]

Patient	9-year-old male
Disease diagnosis	pre B-ALL (L1)
Treatment status	at 1st relapse
Specimen site	bone marrow
Year of establishment	1984

Immunophenotypic characterization[1–3]

T-/NK cell marker	CD1⁻, CD2⁻, CD3⁻, CD4⁻, CD5⁻, CD7⁻, CD8⁻
B-cell marker	CD10⁺, CD19⁺, cyIgM⁺, sIg⁻
Myelomonocytic marker	CD13⁻, CD15⁺, CD33⁺
Progenitor/activation/other marker	CD38⁺, CD71⁺, HLA-DR⁺, TdT⁺
Cytokine receptor	insulin-R⁺, IGF-IR⁺, IGF-IIR⁺

Genetic characterization[1,4–7]

Cytogenetic karyotype	SUP-B13: 46, XY, t(1;4)(p22;q33), t(9;22)(q34;q11), der(14q), 16p+ SUP-B15: 46<2n>XY, der(1)t(1;1)(p11;q31), add(3)(q2?7), der(4)t(1;4)(p11;q35), t(9;22)(q34;q11), add(10)(q25), ?del(14) (q23q31), der(16)t(9;16)(q11;p13)
Unique translocation/fusion gene	Ph⁺t(9;22)(q34;11) → *BCR-ABL* e1-a2 fusion gene
Receptor gene rearrangement	*IGH* RD
Unique gene alteration	*P16INK4A* deletion

Functional characterization[1–3,8]

Cytochemistry	ACP⁺, ANAE⁻, CAE⁻, PAS⁻, SBB⁻
Cytokine response	growth stimulation: insulin
Proto-oncogene	mRNA⁺: BAX, BCL2, BCLXL, MDM2

Comments

- Precursor B-cell line of type B-III (pre B-cell line).
- Carries Ph chromosome leading to *BCR-ABL* e1-a2 fusion gene (reference cell line).
- Serial sister cell lines established at first and second relapse.
- Available from cell line banks.

References

[1] Naumovski, L. et al. (1988) Cancer Res. 48, 2876–2879.
[2] Neely, E.K. et al. (1992) Leukemia 6, 1134–1142.
[3] Zhou, M. et al. (1995) Blood 85, 1608–1614.
[4] Zhou, M. et al. (1995) Leukemia 9, 1159–1161.
[5] Kim, D.H. et al. (1996) Blood 88, 785–794.
[6] Drexler, H.G. et al. (1999) Leukemia Res. 23, 207–215.
[7] Drexler, H.G. et al. (1999) DSMZ Catalogue of Cell lines, 7th edn, Braunschweig, Germany.
[8] Findley, H.W. et al. (1997) Blood 89, 2986–2993.

SUP-B24

Culture characterization[1]

Establishment	initially on feeder layer (complete media, human serum, agar) in hypoxic environment
Sister cell line	**SUP-B31** (serial sister cell line – at relapse – similar features)
Culture medium	90% McCoy's 5A + 10% FBS
Doubling time	72 h
Viral status	EBV$^-$
Authentication	no
Primary reference	Zhang et al.[1]
Availability of cell line	restricted

Clinical characterization[1]

Patient	3-year-old female
Disease diagnosis	pre B-ALL
Treatment status	at diagnosis
Specimen site	bone marrow
Year of establishment	1987

Immunophenotypic characterization[1,2]

T-/NK cell marker	CD1$^-$, CD3$^-$, CD4$^-$, CD5$^-$, CD7$^-$, CD8$^-$
B-cell marker	CD10$^+$, CD19$^+$, CD20$^-$, CD24$^+$, cyIgM$^+$, sIg$^-$
Myelomonocytic marker	CD15$^-$
Progenitor/activation marker	CD34$^-$, CD38$^+$, CD71$^+$, HLA-DR$^+$, TdT$^+$
Cytokine receptor	insulin-R$^+$, IGF-IR$^+$, IGF-IIR$^+$

Genetic characterization[1,3,4]

Cytogenetic karyotype	47, XY, −2, −5, −9, −15, −17, +6mar, der(3), inv(7)(p13p22), del(12)(p11.2p13), del(15), del(16)(q12q24)
Unique gene alteration	*P15INK4B* deletion, *P16INK4A* deletion

Functional characterization[1,2]

Cytokine response	growth stimulation: insulin

Comments

- Precursor B-cell line of type B-III (pre B-cell line).
- Serial sister cell lines established at diagnosis and at relapse.

References

[1] Zhang, L.Q. et al. (1993) Leukemia 7, 1865–1874.
[2] Neely, E.K. et al. (1992) Leukemia 6, 1134–1142.
[3] Jagasia, A.A. et al. (1996) Leukemia 10, 624–628.
[4] Kim, D.H. et al. (1996) Blood 88, 785–794.

SUP-B26

Culture characterization[1]

Establishment	initially on feeder layer (complete media, human serum, agar) in hypoxic environment
Sister cell line	**SUP-B28** (serial sister cell line – at 2nd relapse from bone marrow – similar features)
Culture medium	85% McCoy's 5A + 15% FBS
Viral status	EBV⁻
Authentication	yes (by *IGH* rearrangement)
Primary reference	Zhang et al.[1]
Availability of cell line	restricted

Clinical characterization[1]

Patient	5-year-old male
Disease diagnosis	BCP-ALL
Treatment status	at 1st relapse
Specimen site	CSF
Year of establishment	1987

Immunophenotypic characterization[1]

T-/NK cell marker	$CD5^-$, $CD7^-$
B-cell marker	$CD10^+$, $CD19^+$, $CD20^-$, $CD24^+$, $cyIgM^-$, sIg^-
Myelomonocytic marker	$CD15^-$
Progenitor/activation marker	$CD34^-$, $CD38^-$, $HLA\text{-}DR^+$, TdT^+

Genetic characterization[1–3]

Cytogenetic karyotype	48, XY, +21, +21, der(5)t(5;?9)(q33;q12), del(8)(q22), der(12)t (?5;12)(q33;p13)
Unique translocation/fusion gene	*TEL-AML1* fusion gene
Receptor gene rearrangement	*IGH* R
Unique gene alteration	SUP-B28: *P15INK4B* deletion, *P16INK4A* deletion

Comments

- Precursor B-cell line of type B-II (common B-cell line).
- Carries *TEL-AML1* fusion gene.
- Serial sister cell lines established at first relapse (from CSF) and second relapse (from bone marrow).

References

[1] Zhang, L.Q. et al. (1993) Leukemia 7, 1865–1874.
[2] Kim, D.H. et al. (1996) Blood 88, 785–794.
[3] Jagasia, A.A. et al. (1996) Leukemia 10, 624–628.

SUP-B27

Culture characterization[1]

Establishment	initially on feeder layer (complete media, human serum, agar) in hypoxic environment
Culture medium	85% McCoy's 5A + 15% FBS
Viral status	EBV⁻
Authentication	yes (by *IGH* rearrangement)
Primary reference	Zhang et al.[1]
Availability of cell line	original authors

Clinical characterization[1]

Patient	15-year-old male
Disease diagnosis	pre B-ALL
Treatment status	at relapse

Immunophenotypic characterization[1]

T-/NK cell marker	$CD5^-$, $CD7^-$
B-cell marker	$CD10^+$, $CD19^-$, $CD20^-$, $cyIgM^+$, sIg^-
Myelomonocytic marker	$CD15^-$
Progenitor/activation/other marker	$CD34^-$, $CD38^+$, $HLA\text{-}DR^+$, TdT^+

Genetic characterization[1-3]

Cytogenetic karyotype	47, XY, +8, t(1;19)(q23;p13.3)
Unique translocation/fusion gene	t(1;19)(q23;p13) → *E2A-PBX1* fusion gene
Receptor gene rearrangement	*IGH* RR

Comments

- Precursor B-cell line of type B-III (pre B-cell line).
- Carries t(1;19)(q23;p13) leading to *E2A-PBX1* fusion gene.

References

[1] Zhang, L.Q. et al. (1993) Leukemia 7, 1865–1874.
[2] Nourse, J. et al. (1990) Cell 60, 535–545.
[3] Yoshinari, M. et al. (1998) Cancer Genet. Cytogenet. 101, 95–102.

TA-1

Culture characterization[1]

Culture medium	80% RPMI 1640 + 20% FBS
Doubling time	96–120 h
Viral status	EBNA⁻
Authentication	yes (by *IGH* rearrangement)
Primary reference	Tanabe et al.[1]
Availability of cell line	not known

Clinical characterization[1]

Patient	26-year-old male
Disease diagnosis	AML M1
Treatment status	at relapse
Specimen site	peripheral blood
Year of establishment	1987

Immunophenotypic characterization[1]

T-/NK cell marker	$CD1^-$, $CD2^-$, $CD3^-$, $CD4^-$, $CD5^-$, $CD7^-$, $CD8^-$
B-cell marker	$CD10^-$, $CD19^+$, $CD20^+$, $CD21^-$, $cyIgM^-$, sIg^-
Myelomonocytic marker	$CD13^+$, $CD14^-$, $CD15^-$, $CD33^+$
Erythroid-megakaryocytic marker	$CD41^-$, $GlyA^-$
Progenitor/activation/other marker	$CD30^+$, $CD34^+$, $HLA\text{-}DR^+$, TdT^+
Adhesion marker	$CD11b^-$
Cytokine receptor	$CD25^-$

Genetic characterization[1]

Cytogenetic karyotype	47, XY, −6, +8, −9, +der(6)t(1;6)(q11;p22), +der(9)t(6;9)(p22;p21)
Receptor gene rearrangement	*IGH* RR, *IGK* GG, *TCRB* GR/GD, *TCRG* DD/RR

Functional characterization[1]

Colony formation	not clonable
Cytochemistry	$ANAE^-$, MPO^+
Inducibility of differentiation	TPA → B-cell differentiation
Special features	cyIgM $mRNA^+$

Comments

- Predominantly precursor B-cell line of type B-I (pro B-cell line).
- Co-expression of several myeloid markers.
- Cytoplasmic IgM expression can be induced.

Reference

[1] Tanabe, S. et al. (1993) Int. J. Hematol. 57, 229–243.

TC78

Culture characterization[1]

Culture medium	90% RPMI 1640 + 10% FBS
Authentication	no
Primary reference	Jenski et al.[1]
Availability of cell line	not known

Clinical characterization[1]

Patient	7-year-old male
Disease diagnosis	ALL (L2)
Treatment status	at 1st relapse
Specimen site	bone marrow
Year of establishment	1978

Immunophenotypic characterization[1]

T-/NK cell marker	$CD1a^-$, $CD2^-$, $CD3^-$, $CD4^-$, $CD8^-$, $CD57^-$
B-cell marker	$CD9^+$, $CD10^+$, $CD20^+$, $CD21^-$, $CD24^+$, $cyIgM^-$, sIg^-
Myelomonocytic marker	$CD13^-$, $CD15^+$, $CD33^-$
Erythroid-megakaryocytic marker	$CD36^-$
Progenitor/activation/other marker	$CD38^-$, $HLA\text{-}DR^+$, TdT^+
Adhesion marker	$CD11b^-$

Genetic characterization[1]

Cytogenetic karyotype	48, XY, 2q+, 5q+, 6q−, 7q−, 8q+, 9p−, 12p−

Functional characterization[1]

Colony formation	in agarose
Cytochemistry	$ANAE^-$, CAE^-, MPO^-, $PAS^{(+)}$
Heterotransplantation	into nude mice

Comments

- Precursor B-cell line of type B-II (common B-cell line).

Reference

[1] Jenski, L.J. et al. (1985) Leukemia Res. 9, 1497–1506.

TMD5

Culture characterization[1]

Culture medium	90% α-MEM + 10% FBS
Doubling time	50 h
Viral status	EBNA$^-$
Authentication	yes (by *IGH* rearrangement)
Primary reference	Tohda et al.[1]
Availability of cell line	original authors

Clinical characterization[1]

Patient	19-year-old male
Disease diagnosis	ALL (L2)
Treatment status	at diagnosis
Specimen site	peripheral blood
Year of establishment	1995

Immunophenotypic characterization[1]

T-/NK cell marker	CD3$^-$, CD5$^-$, CD7$^-$
B-cell marker	CD10$^+$, CD19$^+$, CD20$^+$, CD22$^-$, cyIgM$^+$, sIg$^-$
Myelomonocytic marker	CD13$^+$, CD14$^-$, CD33$^-$
Erythroid-megakaryocytic marker	CD41a$^-$
Progenitor/activation/other marker	CD34$^-$, HLA-DR$^+$

Genetic characterization[1,2]

Cytogenetic karyotype	47, XY, t(9;22)(q34;q11), +der(22)t(9;22)(q34;q11)
Unique translocation/fusion gene	Ph$^+$ t(9;22)(q34;q11) → *BCR-ABL* m-bcr fusion gene
Receptor gene rearrangement	*IGH* R

Functional characterization[1]

Cytochemistry	ANAE$^-$, CAE$^-$, MPO$^-$, PAS$^{(+)}$
Cytokine receptor	RT-PCR$^+$: GM-CSFRα, IL-3Rα
Inducibility of differentiation	resistant to dibutyryl cAMP, retinoic acid, TPA

Comments

- Precursor B-cell line of type B-III (pre B-cell line).
- Carries rare double Ph chromosome leading to *BCR-ABL* m-bcr fusion gene.

References

[1] Tohda, S. et al. (1999) Leukemia Res. 23, 255–261.
[2] Drexler, H.G. et al. (1999) Leukemia Res. 23, 207–215.

TOM-1

Culture characterization[1,2]

Culture medium 80% RPMI 1640 + 20% FBS
Doubling time 50–60 h
Viral status EBV⁻, HBV⁻, HCV⁻, HHV-8⁻, HIV⁻, HTLV-I/II⁻
Authentication no
Primary reference Okabe et al.[1]
Availability of cell line restricted

Clinical characterization[1]

Patient 54-year-old female
Disease diagnosis ALL
Treatment status refractory
Specimen site bone marrow
Year of establishment 1983

Immunophenotypic characterization[1-3]

T-/NK cell marker $CD2^-$, $CD3^-$, $CD4^-$, $CD7^-$, $CD8^-$
B-cell marker $CD10^+$, $CD19^+$, $CD20^+$, $CD37^-$, $cyIgM^-$, sIg^-
Myelomonocytic marker $CD13^+$, $CD14^-$, $CD33^-$
Progenitor/activation/other marker $CD34^+$, HLA-DR^+, TdT^+
Adhesion marker $CD11b^-$
Cytokine receptor $CD124^+$

Genetic characterization[1-4]

Cytogenetic karyotype hyperdiploid, 7% polyploidy; 47(47–48)<2n>X, −X, +8, +16, del(7)(p15), del(9)(q11), der(22)t(9;22)(q34;q11)
Unique translocation/fusion gene Ph^+ t(9;22)(q34;q11) → *BCR-ABL* e1-a2 fusion gene
Receptor gene rearrangement *IGH* RR, *IGK* GG, *TCRB* GG

Functional characterization[1,2,5]

Cytochemistry ACP^-, $ANBE^-$, CAE^-, MPO^-, PAS^+, SBB^-
Cytokine response growth stimulation: IL-7; growth inhibition: IFN-α, IFN-β, IFN-γ, IL-4, TNF-α
Inducibility of differentiation TPA → mono differentiation

Comments

- Precursor B-cell line of type B-II (common B-cell line).
- First cell line to carry the Ph chromosome leading to *BCR-ABL* e1-a2 fusion gene.
- Cells are very difficult to culture.

References

[1] Okabe, M. et al. (1987) Blood 69, 990–998.
[2] Drexler, H.G. et al. (1999) DSMZ Catalogue of Cell Lines, 7th edn, Braunschweig, Germany.
[3] Higa, T. et al. (1994) Leukemia Lymphoma 12, 287–296.
[4] Drexler, H.G. et al. (1999) Leukemia Res. 23, 207–215.
[5] Ogawa, M. et al. (1992) Leukemia Lymphoma 8, 57–63.

TS-2

Culture characterization[1]

Culture medium	90% RPMI 1640 + 10% FBS
Authentication	no
Primary reference	Yoshinari et al.[1]
Availability of cell line	not known

Clinical characterization[1]

Patient	3-year-old female
Disease diagnosis	pre B-ALL (L1)
Treatment status	at relapse
Specimen site	peripheral blood
Year of establishment	1994

Immunophenotypic characterization[1]

T-/NK cell marker	$CD3^-$, $CD4^-$, $CD8^-$
B-cell marker	$CD10^-$, $CD19^+$, $CD20^-$, $CD22^+$, cyIgM^+, sIg^-
Myelomonocytic marker	$CD33^-$
Progenitor/activation/other marker	$CD34^+$, HLA-DR^+
Adhesion marker	$CD29^+$

Genetic characterization[1]

Cytogenetic karyotype	47, X, −X, +6, +mar, t(1;19)(q23;p13), add(3)(q22), del(5) (q13), del(6)(q12q23), del(9)(q13q22), del(11)(p11), add(15)(q22)
Unique translocation/fusion gene	t(1;19)(q23;p13), but no *E2A-PBX1* fusion mRNA
Receptor gene rearrangement	*IGH* R

Functional characterization[1]

Cytochemistry	$ANBE^-$, CAE^-, MPO^-, PAS^-

Comments

- Precursor B-cell line of type B-III (pre B-cell line).
- Cells carry the t(1;19)(q23;p13), but not the *E2A-PBX1* fusion gene.

Reference

[1] Yoshinari, M. et al. (1998) Cancer Genet. Cytogenet. 101, 95–102.

UoC-B1

Culture characterization[1]

Establishment	initially on feeder layer (complete media, human serum, agar) in hypoxic environment
Culture medium	85% McCoy's 5A + 15% FBS
Viral status	EBV⁻
Authentication	yes (by cytogenetics)
Primary reference	Zhang et al.[1]
Availability of cell line	restricted

Clinical characterization[1]

Patient	15-year-old female
Disease diagnosis	BCP-ALL
Treatment status	at 2nd relapse

Immunophenotypic characterization[1,2]

T-/NK cell marker	$CD2^-$, $CD5^-$, $CD7^-$
B-cell marker	$CD10^+$, $CD19^+$, $CD20^-$, $CD24^+$, $cyIgM^-$, sIg^-
Myelomonocytic marker	$CD15^+$
Progenitor/activation/other marker	$CD34^-$, $CD38^+$, HLA-DR^+, TdT^+

Genetic characterization[1,3–6]

Cytogenetic karyotype	46, X, −X, −1, −6, −17, +4mar, t(7;11)(q22;q13), add(12)(q24), der(12)t(1;12)(q21;q13), der(19)t(17;19)(q22;p13)
Unique translocation/fusion gene	t(17;19)(q22;p13) → *E2A-HLF* fusion gene
Receptor gene rearrangement	*IGH* RR
Unique gene alteration	*P15INK4B* deletion, *P16INK4A* deletion

Functional characterization[2,3,7,8]

Heterotransplantation	into SCID mice
Proto-oncogene	$mRNA^+$: BAX, BCL2, BCLXL, MDM2

Comments

- Precursor B-cell line of type B-II (common B-cell line).
- Carries t(17;19)(q22;p13) leading to *E2A-HLF* fusion gene.

References

[1] Zhang, L.Q. et al. (1993) Leukemia 7, 1865–1874.
[2] Zhou, M. et al. (1995) Blood 85, 1608–1614.
[3] Stranks, G. et al. (1995) Blood 85, 893–901.
[4] Zhou, M. et al. (1995) Leukemia 9, 1159–1161.
[5] Kim, D.H. et al. (1996) Blood 88, 785–794.
[6] Yuille, M.A.R. et al. (1996) Leukemia 10, 1492–1496.
[7] Zhou, M. et al. (1998) Leukemia 12, 1756–1763.
[8] Findley, H.W. et al. (1997) Blood 89, 2986–2993.

Culture characterization[1]

Establishment	initially on feeder layer (complete media, human serum, agar) in hypoxic environment
Sister cell line	**UoC-B5** (serial sister cell line – at relapse – similar features)
Culture medium	85% McCoy's 5A + 15% FBS
Viral status	EBV⁻
Authentication	yes (by *IGH* rearrangement)
Primary reference	Zhang et al.[1]
Availability of cell line	restricted

Clinical characterization[1]

Patient	14-year-old female
Disease diagnosis	pre B-ALL
Treatment status	at diagnosis

Immunophenotypic characterization[1,2]

T-/NK cell marker	$CD2^-$, $CD5^-$, $CD7^-$
B-cell marker	$CD10^+$, $CD19^+$, $CD20^-$, $CD24^+$, $cyIgM^+$, sIg^-
Myelomonocytic marker	$CD15^-$
Progenitor/activation/other marker	$CD34^-$, $CD38^+$, $HLA\text{-}DR^+$, TdT^+

Genetic characterization[1,3,4]

Cytogenetic karyotype	46, XX, der(9)t(9;?9)(p1?2;q22), der(19)t(1;19)(q23;p13.3)
Unique translocation/fusion gene	t(1;19)(q23;p13) → *E2A-PBX1* fusion gene
Receptor gene rearrangement	*IGH* RR
Unique gene alteration	UoC-B5: *P15INK4B* deletion

Functional characterization[2,5]

Proto-oncogene	$mRNA^+$: BAX, BCLXL, MDM2

Comments

- Precursor B-cell line of type B-III (pre B-cell line).
- Carries t(1;19)(q23;p13) leading to *E2A-PBX1* fusion gene.
- Serial sister cell lines established at diagnosis and at relapse.

References

[1] Zhang, L.Q. et al. (1993) Leukemia 7, 1865–1874.
[2] Zhou, M. et al. (1995) Blood 85, 1608–1614.
[3] Zhou, M. et al. (1995) Leukemia 9, 1159–1161.
[4] Jagasia, A.A. et al. (1996) Leukemia 10, 624–628.
[5] Findley, H.W. et al. (1997) Blood 89, 2986–2993.

Culture characterization[1]

Establishment	initially on feeder layer (complete media, human serum, agar) in hypoxic environment
Sister cell line	**UoC-B6** (serial sister cell line – at 4th relapse from bone marrow – similar features)
Culture medium	85% McCoy's 5A + 15% FBS
Viral status	EBV⁻
Authentication	yes (by *IGH* rearrangement)
Primary reference	Zhang et al.[1]
Availability of cell line	restricted

Clinical characterization[1]

Patient	7-year-old female
Disease diagnosis	BCP-ALL
Treatment status	at 3rd relapse (1st relapse post-BMT)
Specimen site	CSF
Year of establishment	1991

Immunophenotypic characterization[1,2]

T-/NK cell marker	$CD2^-$, $CD5^-$, $CD7^-$
B-cell marker	$CD10^+$, $CD19^+$, $CD20^-$, $CD24^+$, $cyIgM^-$, sIg^-
Myelomonocytic marker	$CD15^+$
Progenitor/activation/other marker	$CD34^-$, $CD38^+$, $HLA\text{-}DR^+$, TdT^+

Genetic characterization[1,3–6]

Cytogenetic karyotype	47, XX, +21, t(2;14)(p11;q32), del(6)(q15q22), t(7;15)(q32;q15 or q21), add(16)(p13), del(8)(p11p21 or p21p23)
Receptor gene rearrangement	*IGH* RR
Unique gene alteration	*P16INK4A* deletion

Functional characterization[3,7]

Heterotransplantation	into SCID mice
Proto-oncogene	$mRNA^+$: BAX, BCL2, BCLXL, MDM2

Comments

- Precursor B-cell line of type B-II (common B-cell line).
- Serial sister cell lines established at different time points and from different tissues.

References

[1] Zhang, L.Q. et al. (1993) Leukemia 7, 1865–1874.
[2] Zhou, M. et al. (1995) Blood 85, 1608–1614.
[3] Stranks, G. et al. (1995) Blood 85, 893–901.
[4] Zhou, M. et al. (1995) Leukemia 9, 1159–1161.
[5] Kim, D.H. et al. (1996) Blood 88, 785–794.
[6] Jagasia, A.A. et al. (1996) Leukemia 10, 624–628.
[7] Findley, M.H. et al. (1997) Blood 89, 2986–2993.

Culture characterization[1,2]

Other name of cell line	ALL-W
Culture medium	90% α-MEM + 10% FBS
Doubling time	29 h
Viral status	EBV
Authentication	yes (partly by cytogenetics)
Primary reference	Lau et al.[1]
Availability of cell line	restricted

Clinical characterization[1,3]

Patient	2-year-old male
Disease diagnosis	jCML
Treatment status	at lymphoid blast crisis
Specimen site	peripheral blood

Immunophenotypic characterization[1,3]

T-/NK cell marker	$CD2^-$, $CD3^-$, $CD4^-$, $CD8^-$
B-cell marker	$CD10^+$, $CD19^+$, $CD20^{(+)}$, $CD21^-$
Myelomonocytic marker	$CD13^-$, $CD14^-$
Progenitor/activation/other marker	$HLA\text{-}DR^+$
Adhesion marker	$CD11b^-$, $CD44^+$

Genetic characterization[1,3,5]

Cytogenetic karyotype	46, XY, −7, ?7p−
Receptor gene rearrangement	*IGH* RG, *IGL* R, *TCRD* R
Unique gene alteration	*P53* mutation

Functional characterization[3,4]

Cytokine production	IL-1β $mRNA^+$
Cytokine response	growth stimulation: EPO, IL-1β, IL-3, IL-4, IL-6, IL-7, GM-CSF, SCF; growth inhibition: IFN-γ, TNF-α

Comments

- Precursor B-cell line of undetermined type.
- Rare case of jCML in blast crisis.
- Allegedly proliferatively responsive to various cytokines.

References

[1] Lau, R.C. et al. (1994) Leukemia 8, 903–908.
[2] Brown, G.A. et al. (1995) Cancer Res. 55, 78–82.
[3] Freedman, M.H. et al. (1993) Blood 81, 3068–3075.
[4] Attias, D. et al. (1995) Leukemia 9, 884–888.
[5] Lam, V. et al. (1999) Leukemia Res. 23, 871–880.

Culture characterization[1]

Culture medium	90% RPMI 1640 + 10% FBS
Doubling time	24–30 h
Viral status	EBV⁻
Authentication	no
Primary reference	Estrov et al.[1]
Availability of cell line	original authors

Clinical characterization[1]

Patient	53-year-old female
Disease diagnosis	pre B-ALL (L2)
Treatment status	refractory
Specimen site	bone marrow
Year of establishment	1990

Immunophenotypic characterization[1]

T-/NK cell marker	$CD2^-$, $CD3^-$, $CD4^-$, $CD5^+$, $CD7^-$, $CD8^-$, $CD56^-$
B-cell marker	$CD10^+$, $CD19^+$, $CD20^+$, $CD22^+$, IgM^-
Myelomonocytic marker	$CD13^-$, $CD14^-$, $CD15^+$, $CD33^+$
Erythroid-megakaryocytic marker	$CD41^-$
Progenitor/activation/other marker	$CD34^-$, $CD38^+$, HLA-DR^+, TdT^+
Adhesion marker	$CD11b^-$
Cytokine receptor	$CD25^-$, $CD116^+$

Genetic characterization[1,2]

Cytogenetic karyotype	45, XX, −7, t(8;10)(q24;q24), t(9;22)(q34;q11.2), der(12), t(7;12)(q11;p21)
Unique translocation/fusion gene	Ph^+ t(9;22)(q34;q11) → *BCR-ABL* e1-a2 fusion gene
Receptor gene rearrangement	*IGH* R

Functional characterization[1]

Colony formation	in methylcellulose
Cytochemistry	$ANAE^-$, CAE^-, MPO^-, PAS^-
Cytokine production	G-CSF, GM-CSF, IL-1β, IL-6, TGF-β, TNF-α
Cytokine response	growth stimulation: GM-CSF

Comments

- Precursor B-cell line of undetermined type.
- Carries Ph chromosome leading to *BCR-ABL* e1-a2 fusion gene.
- Co-expression of myeloid surface markers.

References

[1] Estrov, Z. et al. (1996) Leukemia 10, 1534–1543.
[2] Drexler, H.G. et al. (1999) Leukemia Res. 23, 207–215.

Z-119

Culture characterization[1]

Culture medium	90% RPMI 1640 + 10% FBS
Doubling time	20–30 h
Viral status	EBV$^-$
Authentication	yes (by *IGH* rearrangement)
Primary reference	Estrov et al.[1]
Availability of cell line	original authors

Clinical characterization[1]

Patient	25-year-old female
Disease diagnosis	pre B-ALL (L2)
Treatment status	at relapse
Specimen site	bone marrow
Year of establishment	1990

Immunophenotypic characterization[1]

T-/NK cell marker	CD2$^-$, CD3$^-$, CD4$^-$, CD5$^-$, CD7$^-$, CD8$^-$, CD56$^-$
B-cell marker	CD10$^+$, CD19$^+$, CD20$^+$, CD22$^+$, sIg$^-$
Myelomonocytic marker	CD13$^+$, CD14$^-$, CD15$^-$, CD33$^+$
Erythroid-megakaryocytic marker	CD41$^-$
Progenitor/activation/other marker	CD34$^+$, CD38$^-$, HLA-DR$^+$, TdT$^-$
Adhesion marker	CD11a$^-$, CD11b$^+$
Cytokine receptor	CD25$^-$, CD116$^-$

Genetic characterization[1,2]

Cytogenetic karyotype	45, X, −?X/Y, der(1)t(1;?8)(q42;?q13), t(7;12)(q11.2;q24.3), t(9;22)(q34;q11.2)
Unique translocation/fusion gene	Ph$^+$ t(9;22)(q34;q11) → *BCR-ABL* e1-a2 fusion gene
Receptor gene rearrangement	*IGH* R

Functional characterization[1]

Colony formation	in methylcellulose
Cytochemistry	ANAE$^-$, CAE$^-$, MPO$^-$, PAS$^-$
Cytokine production	G-CSF, GM-CSF, IL-1β

Comments

- Precursor B-cell line of undetermined type.
- Carries Ph chromosome leading to *BCR-ABL* e1-a2 fusion gene.

References

[1] Estrov, Z. et al. (1996) J. Cell. Physiol. 166, 618–630.
[2] Drexler, H.G. et al. (1999) Leukemia Res. 23, 207–215.

Z-181

Culture characterization[1]

Culture medium	90% α-MEM + 10% FBS
Doubling time	30–40 h
Viral status	EBV⁻
Authentication	yes (by *IGH* rearrangement)
Primary reference	Estrov et al.[1]
Availability of cell line	original authors

Clinical characterization[1]

Patient	33-year-old male
Disease diagnosis	pre B-ALL (L2)
Treatment status	at relapse
Specimen site	bone marrow
Year of establishment	1991

Immunophenotypic characterization[1]

T-/NK cell marker	$CD2^-$, $CD3^-$, $CD4^-$, $CD5^-$, $CD7^-$, $CD8^-$, $CD56^-$
B-cell marker	$CD10^+$, $CD19^+$, $CD20^+$, $CD22^+$, sIg^-
Myelomonocytic marker	$CD13^+$, $CD14^-$, $CD15^{(+)}$, $CD33^-$
Erythroid-megakaryocytic marker	$CD41^-$
Progenitor/activation/other marker	$CD34^+$, $CD38^-$, $HLA\text{-}DR^+$, TdT^+
Adhesion marker	$CD11a^-$, $CD11b^-$
Cytokine receptor	$CD25^-$, $CD116^+$

Genetic characterization[1,2]

Cytogenetic karyotype	45, X −?X/Y, +8, t(6;[9];22)(q25;[q34];q11.2), del(7)(p11.2p22), der(9)t(6;[9];22), Ph^+
Unique translocation/fusion gene	Ph^+ t(9;22)(q34;q11) → *BCR-ABL* e1-a2 fusion gene
Receptor gene rearrangement	*IGH* R

Functional characterization[1]

Colony formation	in methylcellulose
Cytochemistry	$ANAE^-$, CAE^-, MPO^-, PAS^-
Cytokine production	G-CSF, GM-CSF, IL-1β

Comments

- Precursor B-cell line of undetermined type.
- Carries Ph chromosome leading to *BCR-ABL* e1-a2 fusion gene.
- Co-expression of myeloid surface antigens.

References

[1] Estrov, Z. et al. (1996) J. Cell. Physiol. 166, 618–630.

[2] Drexler, H.G. et al. (1999) Leukemia Res. 23, 207–215.

Part 2
Mature B-Cell Lines

A3/Kawakami

Culture characterization[1]

Culture medium	90% RPMI 1640 + 10% FBS
Doubling time	20–22 h
Viral status	EBNA$^-$
Authentication	no
Primary reference	Hirose et al.[1]
Availability of cell line	JCRB 0101

Clinical characterization[1]

Patient	68-year-old female
Disease diagnosis	NHL (diffuse large cell) of stomach
Treatment status	terminal
Specimen site	ascites
Year of establishment	1980

Immunophenotypic characterization[1,2]

T-/NK cell marker	CD1$^-$, CD2$^-$, CD3$^-$, cyCD3$^-$, CD4$^+$, CD5$^-$, CD7$^-$, CD8$^-$, CD28$^-$, CD57$^-$, TCR$\alpha\beta^-$, TCRδ^-, cyTCRα^-, cyTCRβ^-, cyTCRγ^-
B-cell marker	CD9$^-$, CD10$^-$, CD19$^-$, CD20$^-$, CD21$^-$, CD22$^-$, CD24$^-$, CD79$^-$, cyIg$^-$, sIg$^-$
Myelomonocytic marker	CD13$^-$, CD14$^-$, CD15$^-$, CD33$^-$
Erythroid-megakaryocytic marker	CD41a$^-$, CD61$^-$
Progenitor/activation/other marker	CD34$^-$, CD38$^+$, CD71$^+$, HLA-DR$^-$, TdT$^-$
Adhesion marker	CD11b$^-$, CD29$^+$
Cytokine receptor	CD25$^-$

Genetic characterization[1-3]

Cytogenetic karyotype	48, X, −X, −1, −4, −5, −10, −12, −13, −14, −22, +4mar, +der(1)t (1;?)(1pter→1q11::?), dup(1)(1qter→1p36::1p31→1pter), inv(3) (q21q26), +der(4)t(4;?)(4pter→4q3?3::?), +der(5)t(1;5)(1qter→1q21::5p15→5qter), +der(10)t(1;10)(1qter→1q11::10q26→10pter), +der(11)?, +der(14)t(14;?)(14pter→14q32::?), del(16)(q2?2:), +der(22)t(13;22)(13qter→13q14::22q13→22pter)
Unique gene alteration	*P53* mutation

Functional characterization[1]

Cytochemistry	ACP$^+$, PAS$^+$

Comments

- Described as mature EBV^- B-cell line derived from B-NHL (DLCL) of the stomach, but all specific or associated B-cell markers investigated are negative (further analysis for unequivocal cell lineage assignment is required).
- Carries alteration at 14q32 possibly involving the *IGH* gene.
- Available from cell line bank.

References

[1] Hirose, M. et al. (1983) Gann 74, 106–115.
[2] Minowada, J. and Matsuo, Y. (1999) unpublished data.
[3] Cheng, J. and Haas, M. (1990) Mol. Cell. Biol. 10, 5502–5509.

Culture characterization[1]

Culture medium	90% RPMI 1640 + 10% FBS
Doubling time	20–22 h
Viral status	EBNA$^-$
Authentication	no
Primary reference	Hirose et al.[1]
Availability of cell line	JCRB 0097

Clinical characterization[1]

Patient	52-year-old female
Disease diagnosis	B-NHL (DLCL) (double cancer of colon: adenocarcinoma + lymphoma)
Treatment status	post-colon operation
Specimen site	ascites
Year of establishment	1980

Immunophenotypic characterization[1]

T-/NK cell marker	CD1a$^-$, CD2$^-$, CD3$^-$, CD4$^+$, CD5$^{(+)}$, CD8$^-$
B-cell marker	CD10$^-$, CD20$^-$, cyIg$^-$, sIg$^-$
Progenitor/activation/other marker	HLA-DR$^+$
Adhesion marker	CD11a$^-$, CD11b$^-$

Genetic characterization[1]

Cytogenetic karyotype	49, XX, +7, +10, +15, −17, t(2;14)(2pter→2q24::14q22→14qter; 14qter→14q22::2q31→2qter), t(3;6)(3pter→3q22::6q27→6qter; 6pter→6q27::3q23→3qter), +del(6)(qter→p21:), t(8;14)(8pter→8q23::14q22-14qter;14pter→14q22::8q23→8qter), t(13;?)

Functional characterization[1]

Colony formation	in agar
Cytochemistry	ACP$^+$, PAS$^{(+)}$
Heterotransplantation	into newborn hamster or nude mice
Special features	secretion: IgM

Comments

- Mature EBV$^-$ B-cell line derived from B-NHL (DLCL-type).
- Available from cell line bank.

Reference

[1] Hirose, M. et al. (1983) Gann 74, 106–115.

BAL-KHs

Culture characterization[1]

Sister cell line	**BAL-KHc** (simultaneous sister cell line – from same peripheral blood sample – some immunoprofile differences)
Culture medium	90% RPMI 1640 + 10% FBS
Doubling time	32 h
Viral status	EBNA$^-$
Authentication	no
Primary reference	Hirata et al.[1]
Availability of cell line	original authors

Clinical characterization[1]

Patient	71-year-old female
Disease diagnosis	B-ALL
Treatment status	at diagnosis
Specimen site	peripheral blood
Year of establishment	1986

Immunophenotypic characterization[1,2]

T-/NK cell marker	CD1$^-$, CD2$^-$, CD3$^-$, CD4$^-$, CD5$^-$, CD7$^-$, CD8$^-$, CD28$^-$, CD57$^-$, TCRαβ$^-$, TCRγδ$^-$
B-cell marker	CD9$^+$, CD10$^+$, CD19$^-$, CD20$^+$, CD21$^-$, CD23$^-$, CD24$^+$, FMC-7$^-$, NU-B1$^{(+)}$, cyIgGκ$^+$, sIgGκ$^+$
Myelomonocytic marker	CD13$^-$, CD14$^-$, CD15$^-$, CD16$^-$, CD33$^-$
Erythroid-megakaryocytic marker	CD41a$^-$, CD61$^-$
Progenitor/activation/other marker	CD34$^-$, CD71$^+$, HLA-DR$^+$, TdT$^-$
Adhesion marker	CD11b$^-$
Cytokine receptor	CD25$^{(+)}$

Genetic characterization[1]

Cytogenetic karyotype	47, XX, −1, −12, −14, −15, +2mar, +der(1)t(1;?)(p36;?), t(8;14) (q24;q32), +der(12)t(12;?)(q13;?)
Unique translocation/fusion gene	t(8;14)(q24;q32) → *MYC-IGH* genes altered?

Functional characterization[1]

Cytochemistry	ACP$^+$, ANAE$^-$, CAE$^-$, MPO$^-$, PAS$^-$, SBB$^-$

Comments

- Mature EBV$^-$ B-cell line derived from B-ALL.
- Carries t(8;14)(q24;q32) which may lead to alteration of the *MYC-IGH* genes.

References

[1] Hirata, T. et al. (1989) Leukemia Res. 13, 203–212.
[2] Minowada, J. and Matsuo, Y. (1999) unpublished data.

BALL-1

Culture characterization[1,2]

Culture medium	80% RPMI 1640 + 20% FBS
Doubling time	66 h
Viral status	EBV⁻
Authentication	no
Primary reference	Miyoshi et al.[1]
Availability of cell line	JCRB 0071, RCB 0256

Clinical characterization[1,2]

Patient	75-year-old male
Disease diagnosis	ALL
Treatment status	at diagnosis (pre-terminal)
Specimen site	peripheral blood
Year of establishment	1976

Immunophenotypic characterization[1-4]

T-/NK cell marker	$CD1^-$, $CD2^-$, $CD3^-$, $CD4^+$, $CD5^+$, $CD7^+$, $CD8^-$, $CD28^-$, $CD57^-$, $TCR\alpha\beta^-$, $TCR\gamma\delta^-$
B-cell marker	$CD9^-$, $CD10^-$, $CD19^+$, $CD20^+$, $CD21^-$, $CD22^-$, $CD24^-$, FMC-7^+, cyIgMDλ^+, sIgMDλ^+
Myelomonocytic marker	$CD13^-$, $CD14^-$, $CD15^-$, $CD33^-$
Erythroid-megakaryocytic marker	$CD41a^-$, $CD61^-$
Progenitor/activation/other marker	$CD34^-$, $CD38^+$, $CD71^+$, HLA-DR^+, TdT^-
Adhesion marker	$CD11b^-$
Cytokine receptor	$CD25^+$

Genetic characterization[3,5]

Cytogenetic karyotype	47, X, −Y, −1, −3, −4, −6, +7, −9, −11, −11, −13, −14, −15, −15, −18, +22, +2mar, del(2)(q33), del(14)(q22q24), del(7)(p11), del(8)(q24), +der(1)t(?;1;?)(?::1p36→1q12::?), +der(3)t(15;3;?)(15qter→15q1?1::3p13→3q21::?), +der(4)t(1;4)(1qter→1q12::4p14→4qter), +der(6)t(6;11)(6q22→6p23::11q13→11qter), +dic(9)t (9;13)(9qter→9p2?1::13q1?1→13q12::13q14→13qter), +der(11)t (11;?)(11pter→11q23::?), +der(13)t(13;?)(13q21::?), +der(14)t(1;11;14)(1q42→1q25::11q25→11q13::14q32→14pter), +der(14)t(14;8;?)(14pter→14q32::8?q2?4→8?qter::?), +der(18)t(18;?)(18pter→18q2?3::?), +der(21)t(21;?)(21qter→21pter::?)

Unique translocation/fusion gene	t(8;14)(q24;q32) → *MYC-IGH* genes altered?
Unique gene alteration	*P16INK4A* deletion

Functional characterization[1]

Heterotransplantation	into newborn hamster

Comments

- Mature EBV⁻ B-cell line derived from B-ALL.
- Carries t(8;14)(q24;q32) which may lead to alteration of the *MYC-IGH* genes.
- Available from cell line banks.
- Cells grow very well.

References

[1] Miyoshi, I. et al. (1977) Nature 267, 843–844.
[2] Hiraki, S. et al. (1977) J. Natl. Cancer Inst. 59, 93–94.
[3] Minowada, J. and Matsuo, Y. (1999) unpublished data.
[4] Inoue, K. et al. (1997) Blood 89, 1405–1412.
[5] Ogawa, M. et al. (1994) Blood 84, 2431–2435.

BALL-2

Culture characterization[1]

Culture medium	80% RPMI 1640 + 20% FBS
Doubling time	36 h
Viral status	EBV$^+$
Authentication	yes (by cytogenetics)
Primary reference	Kubonishi et al.[1]
Availability of cell line	restricted

Clinical characterization[1]

Patient	14-year-old male
Disease diagnosis	B-ALL (L2)
Treatment status	at diagnosis (pre-terminal)
Specimen site	peripheral blood
Year of establishment	1987

Immunophenotypic characterization[1]

T-/NK cell marker	CD1a$^-$, CD2$^-$, CD3$^-$, CD4$^-$, CD5$^-$, CD8$^-$, CD57$^-$
B-cell marker	CD10$^-$, CD19$^+$, CD20$^+$, CD21$^+$, cyIgMκ^+, sIgMDκ^+
Myelomonocytic marker	CD13$^-$, CD14$^-$
Progenitor/activation/other marker	CD38$^+$, HLA-DR$^+$, TdT$^-$
Cytokine receptor	CD25$^-$

Genetic characterization[1]

Cytogenetic karyotype	46, XY, t(8;14)(q24;q32), del(10)(q24), der(13)t(13;1)(q34;q11)
Unique translocation/fusion gene	t(8;14)(q24;q32) → *MYC-IGH* genes altered
Receptor gene rearrangement	*IGH* RR, *TCRB* G
Unique gene alteration	*MYC* rearrangement

Functional characterization[1]

Cytochemistry	ANAE$^-$, MPO$^-$

Comments

- Mature EBV$^+$ B-cell line derived from B-ALL.
- Cellular derivation and neoplastic nature verified.
- Carries t(8;14)(q24;q32) causing alteration of the *MYC-IGH* genes.

Reference

[1] Kubonishi, I. et al. (1991) Am. J. Hematol. 37, 179–185.

BALM-1

Culture characterization[1]

Sister cell line	**BALM-2** (simultaneous sister cell line – from the same peripheral blood sample – identical features)
Culture medium	90% RPMI 1640 + 10% FBS
Viral status	EBNA$^+$
Authentication	yes (by cytogenetics)
Primary reference	Minowada et al.[1]
Availability of cell line	original authors

Clinical characterization[1]

Patient	56-year-old male
Disease diagnosis	B-ALL
Treatment status	at diagnosis
Specimen site	peripheral blood
Year of establishment	1976

Immunophenotypic characterization[1,2]

T-/NK cell marker	CD1$^-$, CD2$^-$, CD3$^-$, CD4$^-$, CD5$^-$, CD7$^-$, CD8$^-$, CD28$^-$, CD57$^-$, TCR$\alpha\beta^-$
B-cell marker	CD9$^+$, CD10$^-$, CD19$^+$, CD20$^+$, CD21$^+$, CD24$^+$, FMC-7$^{(+)}$, PCA-1$^-$, cyIgMDκ^+, sIgMDκ^+
Myelomonocytic marker	CD13$^-$, CD14$^-$, CD15$^-$, CD33$^-$
Erythroid-megakaryocytic marker	CD41a$^-$, CD61$^-$
Progenitor/activation/other marker	CD34$^-$, CD38$^+$, HLA-DR$^+$, TdT$^-$
Adhesion marker	CD11b$^-$
Cytokine receptor	CD25$^-$

Genetic characterization[1]

Cytogenetic karyotype	47, 14q+, t(12;17;?)

Functional characterization[1,3]

Colony formation	in agar
Heterotransplantation	into nude mice

Comments

- Mature EBV$^+$ B-cell line derived from B-ALL.
- Authentic derivation and neoplastic nature verified.
- Cells grow very well.

References

[1] Minowada, J. et al. (1977) Cancer Res. 37, 3096–3099.
[2] Minowada, J. and Matsuo, Y. (1999) unpublished data.
[3] Kubonishi, I. et al. (1980) Cancer 45, 2324–2329.

BALM-3

Culture characterization[1]

Sister cell lines	**BALM-4, BALM-5** (simultaneous sister cell lines – from the same pleural effusion sample – nearly identical features)
Culture medium	90% RPMI 1640 + 10% FBS
Doubling time	60–72 h
Viral status	EBNA⁻
Authentication	no
Primary reference	Lok et al.[1]
Availability of cell line	original authors

Clinical characterization[1]

Patient	63-year-old female
Disease diagnosis	B-NHL (poorly differentiated diffuse, small non-cleaved cell type)
Treatment status	at diagnosis
Specimen site	pleural effusion
Year of establishment	1978

Immunophenotypic characterization[1,2]

T-/NK cell marker	$CD1^-$, $CD2^-$, $CD3^-$, $CD4^-$, $CD5^-$, $CD7^-$, $CD8^-$, $CD57^-$
B-cell marker	$CD9^-$, $CD10^+$, $CD19^+$, $CD20^{(+)}$, $CD21^-$, $CD24^+$, FMC-7^+, cyIgGκ^+, sIgGκ^+
Myelomonocytic marker	$CD13^-$, $CD14^-$, $CD15^-$, $CD33^-$
Progenitor/activation/other marker	$CD34^-$, $CD38^+$, HLA-DR^+, TdT^-
Adhesion marker	$CD11b^-$
Cytokine receptor	$CD25^-$

Genetic characterization[1]

Cytogenetic karyotype	52, XX, +7, +8, +8, +19, +2mar, t(14;18)(q32;q21)
Unique translocation/fusion gene	t(14;18)(q32;q21) → *IGH-BCL2* genes altered?

Functional characterization[1]

Cytochemistry	$ANAE^-$, $CAE^{(+)}$, MPO^-, PAS^+, SBB^-

Comments

- Mature EBV^- B-cell line derived from B-NHL.
- Carries t(14;18)(q32;q21) possibly leading to alteration of the *IGH-BCL2*.

References

[1] Lok, M.S. et al. (1979) Int. J. Cancer 24, 572–578.
[2] Minowada, J. and Matsuo, Y. (1999) unpublished data.

BALM-7

Culture characterization[1]

Sister cell lines	**BALM-6, BALM-8** (simultaneous sister cell lines – from the same bone marrow sample – similar features)
Culture medium	90% RPMI 1640 + 10% FBS
Doubling time	60 h
Authentication	yes (by cytogenetics)
Primary reference	Matsuo et al.[1]
Availability of cell line	original authors

Clinical characterization[1]

Patient	38-year-old male
Disease diagnosis	ALL
Treatment status	at diagnosis
Specimen site	bone marrow
Year of establishment	1989

Immunophenotypic characterization[1-3]

T-/NK cell marker	CD1a⁻, CD2⁻, CD3⁻, CD4⁻, CD5⁻, CD7⁻, CD8⁻, CD28⁻, CD57⁻, TCRαβ⁻, TCRγδ⁻
B-cell marker	CD9⁻, CD10⁺, CD19⁺, CD20⁺, CD21⁺, CD22⁽⁺⁾, CD24⁺, NU-B1⁺, cyIgMλ⁺, sIgMλ⁺
Myelomonocytic marker	CD13⁻, CD14⁻, CD15⁻, CD16⁻, CD33⁻
Erythroid-megakaryocytic marker	CD41a⁻, CD61⁻
Progenitor/activation/other marker	CD34⁻, CD71⁺, HLA-DR⁺, TdT⁻
Adhesion marker	CD11b⁻
Cytokine receptor	CD25⁺

Genetic characterization[1]

Cytogenetic karyotype	48, XY, −14, −19, +21?, +1mar, t(8;14)(q24;q32), der/der?(1) 7(p?), +der(14), t(14;?)(14pter→14q32::?), +der(19)t(1;19)(1qter→1q11orq12::19p13→19qter)
Unique translocation/fusion gene	t(8;14)(q24;q32) → *MYC-IGH* genes altered?
Receptor gene rearrangement	*IGH* RG, *TCRB* GG, *TCRG* GG, *TCRD* DG (BALM-8 mRNA: IGH⁺, TCRA⁻, TCRB⁻, TCRG⁻, TCRD⁻)

Functional characterization[1]

Cytochemistry	MPO⁻, NBT⁻

Comments

- Mature B-cell line established from B-ALL.
- Carries t(8;14)(q24;q32) possibly causing alteration of the *MYC-IGH* genes.

References

[1] Matsuo, Y. et al. (1992) Human Cell 5, 187–190.
[2] Minowada, J. and Matsuo, Y. (1999) unpublished data.
[3] Stöckbauer, P. et al. (1994) Human Cell 7, 40–46.

BALM-9

Culture characterization[1,2]

Sister cell lines	**BALM-10, BALM-11, BALM-12** (simultaneous sister cell lines – from the same blood sample – similar features)
Subclones	**BALM-9K, BALM-9KL, BALM-9L, BALM-9N** (subclones with slightly different immunoprofiles: Igκ$^+$λ$^-$, Igκ$^+$λ$^+$, Igκ$^-$λ$^+$, Igκ$^-$λ$^-$)
Culture medium	90% RPMI 1640 + 10% FBS
Viral status	EBV$^-$
Authentication	no
Primary reference	Matsuo et al.[1]
Availability of cell line	original authors

Clinical characterization[1]

Patient	12-year-old female
Disease diagnosis	B-ALL (L3)
Treatment status	refractory, terminal
Specimen site	peripheral blood
Year of establishment	1993

Immunophenotypic characterization[1,2]

T-/NK cell marker	CD1a$^-$, CD2$^-$, CD3$^-$, CD4$^+$, CD5$^-$, CD7$^-$, CD8$^-$, CD28$^-$, CD57$^-$, TCRαβ$^-$, TCRγδ$^-$
B-cell marker	CD10$^+$, CD19$^+$, CD20$^+$, CD21$^-$, CD23$^-$, CD24$^+$, CD40$^+$, CD72$^+$, CD73$^-$, CD74$^+$, CD76$^-$, CD77$^{(+)}$, CD78$^+$, CD79b$^+$, CD80$^{(+)}$, CD81$^+$, CD83$^-$, CD84$^-$, CD85$^+$, CD86$^+$, NU-B1$^+$, cyIgMDκλ$^+$, sIgMDκλ$^+$
Myelomonocytic marker	CD13$^-$, CD14$^-$, CD15$^-$, CD33$^-$, CD64$^-$, CD65$^-$, CD68$^-$
Erythroid-megakaryocytic marker	CD41a$^-$, CD61$^-$
Progenitor/activation/other marker	CD30$^-$, CD34$^-$, CD70$^+$, CD71$^+$, CD95$^-$, HLA-DR$^+$, TdT$^-$
Adhesion marker	CD11b$^-$, CD58$^-$, CD62E$^-$, CD62L$^-$
Cytokine receptor	CD25$^+$

Genetic characterization[1]

Cytogenetic karyotype	55–56, XX, +1, +7, +10, +11, +16, +17, +20, +add(X) (q22), +add(1)(p11), add(4)(q25), del(8)(q24), inv(9)(p12q13), add(14)(q32), add(21)(q22), +add(21)(q22)
Receptor gene rearrangement	*IGH* R, *IGK* R, *IGL* R

Comments

- Mature EBV$^-$ B-cell line derived from B-ALL (L3).
- Various subclones and sister cell lines established.

References

[1] Matsuo, Y. et al. (1996) Leukemia 10, 700–706.
[2] Minowada, J. and Matsuo, Y. (1999) unpublished data.

BALM-13

Culture characterization[1]

Sister cell line	**BALM-14** (simultaneous EBV$^-$ sister cell line – from the same bone marrow sample – intraclonal immunophenotyping differences)
Culture medium	90% RPMI 1640 + 10% FBS
Viral status	EBNA$^+$
Authentication	no
Primary reference	Matsuo and Ariyasu[1]
Availability of cell line	original authors

Clinical characterization[1]

Patient	13-year-old male
Disease diagnosis	B-ALL (L3)
Treatment status	at diagnosis
Specimen site	bone marrow
Year of establishment	1994

Immunophenotypic characterization[1,2]

T-/NK cell marker	CD1a$^-$, CD2$^-$, CD3$^-$, CD4$^-$, CD5$^{(+)}$, CD7$^-$, CD8$^-$, CD28$^-$, CD56$^-$, CD57$^-$, TCR$\alpha\beta^-$, TCR$\gamma\delta^-$
B-cell marker	CD9$^+$, CD10$^-$, CD19$^+$, CD20$^+$, CD21$^+$, CD22$^+$, CD23$^+$, CD24$^+$, CD37$^+$, CD40$^+$, CD72$^{(+)}$, CD73$^{(+)}$, CD76$^+$, CD77$^+$, CD78$^+$, CD79b$^+$, CD80$^+$, CD81$^+$, CD83$^+$, CD84$^-$, CD85$^+$, FMC-7$^+$, NU-B1$^+$, cyIgMDλ^+, sIgMDλ^+
Myelomonocytic marker	CD13$^-$, CD15$^-$, CD16$^{(+)}$, CD32$^{(+)}$, CD33$^-$, CD35$^+$, CD64$^-$, CD65$^-$, CD68$^-$
Erythroid-megakaryocytic marker	CD41a$^-$, CD42a$^-$, CD61$^-$
Progenitor/activation/other marker	CD30$^+$, CD34$^-$, CD38$^+$, CD45$^+$, CD45RA$^+$, CD45RO$^+$, CD70$^+$, CD71$^+$, CD95$^+$, HLA-DR$^+$, TdT$^{(+)}$
Adhesion marker	CD11a$^+$, CD18$^+$, CD29$^+$, CD43$^+$, CD44$^+$, CD49b$^-$, CD49d$^+$, CD49f$^-$, CD50$^+$, CD51$^-$, CD58$^+$, CD62E$^-$
Cytokine receptor	CD25$^+$, CD115$^+$, CD116$^{(+)}$, CD117$^+$, CD119$^+$, CD120a$^+$, CD120b$^+$, CD121a$^+$, CD121b$^+$, CD122$^+$, CD124$^+$, CD126$^+$, CD127$^+$, CD128$^+$, CD130$^+$

Genetic characterization[1]

Cytogenetic karyotype	46, XY, invdup(1)(q12q25), del(8)(q24), add(13)(q22), add(14) (q32), add(16)(q22)
Unique translocation/fusion gene	t(8;14)(q24;q32) → *MYC-IGH* genes altered?

Comments

- Mature EBV+ B-cell line derived from B-ALL (L3).
- Neoplastic nature verified.
- Carries t(8;14)(q24;q32) possibly leading to alteration of the *MYC* and *IGH* genes.
- Extensive immunoprofile.
- EBV+ and EBV− neoplastic sister cell lines established.

References

[1] Matsuo, Y. and Ariyasu, T. (1996) Human Cell 9, 57–62.
[2] Minowada, J. and Matsuo, Y. (1999) unpublished data.

BALM-16

Culture characterization[1]

Culture medium	90% RPMI 1640 + 10% FBS
Doubling time	32 h
Viral status	EBNA$^-$
Authentication	yes (by cytogenetics)
Primary reference	Matsuo et al.[1]
Availability of cell line	original authors

Clinical characterization[1]

Patient	42-year-old male
Disease diagnosis	B-ALL (L3)
Treatment status	at 2nd relapse
Specimen site	peripheral blood
Year of establishment	1996

Immunophenotypic characterization[1]

T-/NK cell marker	CD3$^-$, CD4$^-$, CD5$^-$, CD7$^-$, CD28$^-$, CD56$^-$, CD57$^-$
B-cell marker	CD9$^+$, CD10$^+$, CD19$^+$, CD20$^-$, CD21$^-$, CD22$^-$, cyCD22$^+$, CD23$^-$, CD24$^+$, CD39$^+$, CD40$^+$, CD72$^+$, CD76$^+$, CD77$^-$, cyCD79a$^-$, cyCD79b$^+$, CD80$^-$, CD86$^-$, FMC-7$^-$, cyIg$^-$, sIg$^-$
Myelomonocytic marker	CD13$^-$, CD14$^-$, CD15$^-$, CD33$^-$
Progenitor/activation/other marker	CD30$^-$, CD34$^-$, CD38$^+$, CD45$^+$, CD45RO$^+$, CD45RA$^+$, CD71$^+$, HLA-DR$^+$, TdT$^-$
Adhesion marker	CD44$^-$
Cytokine receptor	CD25$^+$

Genetic characterization[1]

Cytogenetic karyotype	47<2n>XY, +18, t(1;4)(q32.2;q35), t(8;22)(q24;q11), add(11) (q13), del(13)(q13q14.2)
Unique translocation/fusion gene	t(8;22)(q24;q11) → *MYC-IGL* genes altered?
Receptor gene rearrangement	*IGH* RR, *IGK* DD, *IGL* GG

Functional characterization[1]

Special features	parathryoid hormone-related peptide mRNA$^+$

Comments

- Mature EBV$^-$ B-cell line derived from B-ALL (L3).
- Carries t(8;22)(q24;q11) possibly leading to alteration of *MYC* and *IGL* genes.

Reference

[1] Matsuo, Y. et al. (1997) Leukemia 11, 2168–2174.

BALM-18

Culture characterization[1]

Establishment	initially on feeder layer (bone marrow stromal cells)
Culture medium	90% RPMI 1640 + 10% FBS
Doubling time	36 h
Viral status	EBV⁻
Authentication	no
Primary reference	Matsuo et al.[1]
Availability of cell line	original authors

Clinical characterization[1]

Patient	35-year-old male
Disease diagnosis	B-ALL (L3)
Treatment status	at diagnosis
Specimen site	peripheral blood
Year of establishment	1997

Immunophenotypic characterization[1]

T-/NK cell marker	$CD3^-$, $CD4^{(+)}$, $CD5^-$, $CD7^-$
B-cell marker	$CD9^{(+)}$, $CD10^+$, $CD19^+$, $CD20^+$, $CD22^+$, $CD23^-$, $CD24^+$, $CD39^-$, $CD77^+$, $CD79a^-$, $cyCD79a^+$, $CD79b^+$, $CD80^+$, $CD86^+$, FMC-$7^{(+)}$, NU-$B1^+$, cyIgMDκ$^+$, sIgMDκ$^+$
Myelomonocytic marker	$CD13^-$, $CD33^-$
Progenitor/activation/other marker	$CD34^-$, $CD38^+$, $CD70^+$, $CD71^+$, $CD95^+$, HLA-DR^+, TdT^-
Adhesion marker	$CD11a^+$, $CD49d^-$
Cytokine receptor	$CD25^+$

Genetic characterization[1]

Cytogenetic karyotype	46<2n>XY, +1, der(1;18)(q10;q10), t(8;14)(q24.1;q32)
Unique translocation/fusion gene	t(8;14)(q24.1;q32) → *MYC-IGH* genes altered?

Functional characterization[1]

Special features	resistant to FAS-induced apoptosis

Comments

- Mature EBV⁻ B-cell line derived from B-ALL (L3).
- Carries t(8;14)(q24.1;q32) which may cause alteration of *MYC-IGH* genes.

Reference

[1] Matsuo, Y. et al. (1999) Leukemia Res. 23, 559–568.

Bay91

Culture characterization[1]

Culture medium	90% RPMI 1640 + 10% FBS
Viral status	EBNA$^-$
Authentication	no
Primary reference	Nakamura et al.[1]
Availability of cell line	not known

Clinical characterization[1]

Patient	24-year-old female
Disease diagnosis	B-ALL (L2)
Treatment status	at diagnosis
Specimen site	peripheral blood
Year of establishment	1991

Immunophenotypic characterization[1]

T-/NK cell marker	CD2$^-$, CD5$^-$, CD7$^-$
B-cell marker	CD10$^+$, CD19$^+$, CD20$^-$, CD21$^-$, CD22$^+$, CD24$^+$, sIgGκ^+
Myelomonocytic marker	CD13$^-$, CD33$^-$
Progenitor/activation/other marker	CD34$^-$, HLA-DR$^+$, TdT$^+$
Adhesion marker	CD11b$^-$
Cytokine receptor	CD25$^-$

Genetic characterization[1]

Cytogenetic karyotype	46, X, der(X)t(X;1)(q22;p12), t(1;7)(q42;p15), der(1)del(p12) dup(1)(q21q44), del(9)(p13p22), t(14;18)(q32;q21)
Unique translocation/fusion gene	t(14;18)(q32;q21) → *IGH-BCL2* genes altered

Functional characterization[1]

Special features	RAG1 mRNA$^+$

Comments

- Mature EBV$^-$ B-cell line derived from B-ALL (L2).
- Carries t(14;18)(q32;q21) leading to alteration of the *IGH* and *BCL2* genes.

Reference

[1] Nakamura, F. et al. (1996) Leukemia 10, 1159–1163.

BC-1

Culture characterization[1–3]

Sister cell line	**HBL-6** (independently established sister cell line – from ascites – some immunoprofile, cytogenetic and functional differences)
Establishment	initially on feeder layer (heterologous peripheral blood cells)
Culture medium	90% RPMI 1640 + 10% FBS
Doubling time	48–72 h
Viral status	EBV$^+$, HHV-8$^+$, HIV$^-$, HTLV-I/II$^-$
Authentication	yes (by *IGH* rearrangement, EBV sequencing)
Primary reference	Cesarman et al.[1]
Availability of cell line	ATCC CRL 2230

Clinical characterization[1–5]

Patient	46-year-old male (HIV$^+$)
Disease diagnosis	PEL + AIDS
Treatment status	at diagnosis (stage IE-B)
Specimen site	ascites
Year of establishment	1992

Immunophenotypic characterization[1–3]

T-/NK cell marker	CD3$^-$, CD5$^-$
B-cell marker	CD19$^-$, CD20$^-$, CD21$^-$, CD22$^-$, CD23$^-$, CD138$^+$, sIg$^-$
Progenitor/activation/other marker	CD30$^+$, CD45$^+$, EMA$^+$

Genetic characterization[1–3,6]

Cytogenetic karyotype	49, XY, +7, der(3)t(1;3)(q11;q28)del(1)(q42), +inv(8)(p23q13), add(9)(q34), +der(12)t(9;12)(?;q22), der(15)t(4;15)(?q28;p13), der(16)t(15;16)(q15;p13)
Receptor gene rearrangement	*IGH* R
Unique gene alteration	*BCL6* mutation

Functional characterization[3,7–9]

Colony formation	HBL-6: in agar
Cytokine production	IL-6, IL-10
Heterotransplantation	HBL-6: into NOD/SCID mice
Special features	EBNA2$^+$, LMP1$^-$

Comments

- First B-cell line derived from patient with AIDS-associated PEL.
- Cells are HHV-8$^+$ and EBV$^+$.
- Authentic derivation and neoplastic nature verified.

- Sister cell line HBL-6 established independently by other investigators.
- More extensive immunoprofile on HBL-6 [2,8].
- Available from cell line bank.

References

[1] Cesarman, E. et al. (1995) Blood 86, 2708–2714.
[2] Drexler, H.G. et al. (1998) Leukemia 12, 1507–1517. (review on PEL cell lines)
[3] Gaidano, G. et al. (1996) Leukemia 10, 1237–1240.
[4] Chadburn, A. et al. (1993) Cancer 72, 3078–3090. (case report)
[5] Cesarman, E. et al. (1995) New Engl. J. Med. 332, 1186–1191. (case report)
[6] Gaidano, G. et al. (1999) Genes Chromosomes Cancer 24, 16–23.
[7] Drexler, H.G. et al. (1999) Leukemia 13, 634–640.
[8] Boshoff, C. et al. (1998) Blood 91, 1671–1679.
[9] Asou, H. et al. (1998) Blood, 2475–2481.

BC-2

Culture characterization[1,2]

Establishment	initially on feeder layer (heterologous peripheral blood cells)
Culture medium	90% RPMI 1640 + 10% FBS
Doubling time	48–72 h
Viral status	EBV^+, $HHV\text{-}8^+$, HIV^-, $HTLV\text{-}I/II^-$
Authentication	yes (by EBV sequencing)
Primary reference	Cesarman et al.[1]
Availability of cell line	ATCC CRL 2231

Clinical characterization[1-4]

Patient	31-year-old male (HIV^+)
Disease diagnosis	PEL + AIDS
Treatment status	at diagnosis
Specimen site	pleural effusion
Year of establishment	1992

Immunophenotypic characterization[1,2]

T-/NK cell marker	$CD3^-$, $CD5^-$
B-cell marker	$CD19^-$, $CD20^-$, $CD21^-$, $CD22^-$, $CD23^-$, $CD138^+$, sIg^-
Progenitor/activation/other marker	$CD30^-$, $CD45^+$, EMA^+

Genetic characterization[1,2,5]

Cytogenetic karyotype	49–51, Y, −2, −7, +12, −14, +19, +3-5mar, ?del(X)(q24), add(9) (q34), add(11)(q25), add(12)(q24), der(14)add(14)(p13), add(14) (q32), ?del(17)(q23), +?del(22)(q12)
Receptor gene rearrangement	*IGH* R
Unique gene alteration	*BCL6* mutation

Functional characterization[6,7]

Cytokine production	IL-6, IL-10
Special features	$EBNA2^+$, $LPM1^-$

Comments

- Mature B-cell line derived from patient with AIDS-associated PEL.
- Cells are $HHV\text{-}8^+$ and EBV^+.
- Authentic derivation and neoplastic nature verified.
- Available from cell line bank.

References

[1] Cesarman, E. et al. (1995) Blood 86, 2708–2714.
[2] Drexler, H.G. et al. (1998) Leukemia 12, 1507–1517. (review on PEL cell lines)
[3] Cesarman, E. et al. (1995) New Engl. J. Med. 332, 1186–1191. (case report)
[4] Knowles, D.M. et al. (1989) Blood 73, 792–799. (case report)
[5] Gaidano, G. et al. (1999) Genes Chromosomes Cancer 24, 16–23.
[6] Drexler, H.G. et al. (1999) Leukemia 13, 634–640.
[7] Asou, H. et al. (1998) Blood, 2475–2481.

BC-3

Culture characterization[1-3]

Sister cell line	**KS-1** (independently established sister cell line – from pleural effusion – similar profile)
Establishment	initially on feeder layer (heterologous peripheral blood cells)
Culture medium	80% RPMI 1640 + 20% FBS
Doubling time	KS-1: 48–72 h
Viral status	CMV$^-$, EBV$^-$, HHV-8$^+$, HIV$^-$, HSV-1$^-$, HSV-2$^-$
Authentication	yes (by *IGH* rearrangement)
Primary reference	Arvanitakis et al.[1]
Availability of cell line	ATCC CRL 2277

Clinical characterization[1-4]

Patient	85-year-old male (HIV$^-$)
Disease diagnosis	PEL (without AIDS)
Treatment status	at diagnosis
Specimen site	pleural effusion
Year of establishment	1995

Immunophenotypic characterization[1-3]

T-/NK cell marker	CD2$^-$, CD3$^-$, CD4$^-$, CD5$^-$, CD8$^-$
B-cell marker	CD19$^-$, CD20$^-$, CD21$^-$, CD22$^-$, CD138$^+$, sIg$^-$
Progenitor/activation/other marker	CD30$^+$, CD38$^+$, CD45$^+$, CD71$^+$, EMA$^+$, HLA-DR$^+$
Adhesion marker	CD54$^+$

Genetic characterization[1,2,5]

Cytogenetic karyotype	47, X, −Y, +7, +8, +12, −13, add(1)(q44), ?t(12;14)(p11;q11), ?del(16)(q13)
Receptor gene rearrangement	*IGH* R, *IGK* R, *IGL* R, *TCRB* G
Unique gene alteration	*BCL6* mutation

Functional characterization[3,6]

Cytokine production	KS-1: IL-6, IL-10
Heterotransplantation	KS-1: into BNX mice
Special features	KS-1: EBER$^-$, LMP1$^-$

Comments

- Mature EBV$^-$ HHV-8$^+$ B-cell line derived from non-AIDS PEL.
- Sister cell line KS-1 established independently by other investigators.
- Available from cell line bank.

References

[1] Arvanitakis, L. et al. (1996) Blood 88, 2648–2654.
[2] Drexler, H.G. et al. (1998) Leukemia 12, 1507–1517.
[3] Said, J.W. et al. (1996) Blood 87, 4937–4943.
[4] Nador, R.G. et al. (1995) New Engl. J. Med. 333, 943. (case report)
[5] Gaidano, G. et al. (1999) Genes Chromosomes Cancer 24, 16–23.
[6] Asou, H. et al. (1998) Blood 91, 2475–2481.

BCBL-1

Culture characterization[1,2]

Culture medium	90% RPMI 1640 + 10% FBS
Viral status	EBV⁻, HHV-8⁺, HIV⁻, HTLV-I/II⁻
Authentication	no
Primary reference	Renne et al.[1]
Availability of cell line	NIH AIDS Program (no. 3233)

Clinical characterization[1,2]

Patient	40-year-old male (HIV⁺)
Disease diagnosis	Kaposi's sarcoma → PEL + AIDS
Specimen site	ascites

Immunophenotypic characterization[2]

B-cell marker	CD19⁽⁺⁾, CD138⁺
Progenitor/activation/other marker	CD38⁺, CD45⁺

Genetic characterization[2–4]

Cytogenetic karyotype	48, X, −Y, +7, +12, −14, −18, −22, +3mar, dup(1)(q12q32), +add(2)(q37), del(4)(?q27), add(6)(p25), del(9)(q12q21), add(11) (q25), add(14)(q32), add(15)(q24)
Receptor gene rearrangement	*IGH* R, *TCRB* G
Unique gene alteration	*BCL6* mutation

Functional characterization[1,2]

Heterotransplantation	into SCID mice
Special features	TPA → production of HHV-8 (lysis); LMP1⁻

Comments

- Mature EBV⁻ HHV-8⁺ B-cell line derived from AIDS-associated PEL.
- Freely available from National Institutes of Health, Bethesda, MD, USA.

References

[1] Renne, R. et al. (1996) Nature Medicine 2, 342–346.
[2] Drexler, H.G. et al. (1998) Leukemia 12, 1507–1517.
[3] Gaidano, G. et al. (1999) Genes Chromosomes Cancer 24, 16–23.
[4] Katano, H. et al. (1999) J. Med. Virol. 58, 394–401.

BCP-1

Culture characterization[1,2]

Culture medium	80% RPMI 1640 + 20% FBS
Viral status	CMV$^-$, EBV$^-$, HHV-6$^-$, HHV-8$^+$, HIV$^-$
Authentication	yes (by *IGH* rearrangement)
Primary reference	Boshoff et al.[1]
Availability of cell line	ATCC CRL 2294

Clinical characterization[1-3]

Patient	94-year-old male (HIV$^-$)
Disease diagnosis	Kaposi's sarcoma → PEL (without AIDS)
Treatment status	at diagnosis
Specimen site	peripheral blood
Year of establishment	1995

Immunophenotypic characterization[1,2]

T-/NK cell marker	CD3$^-$, CD5$^-$, CD7$^-$
B-cell marker	CD10$^-$, CD19$^-$, CD20$^-$, CD22$^-$, CD23$^+$, CD72$^-$, CD79a$^-$, CD79b$^-$, cyIgG$^+$, sIg$^-$
Progenitor/activation/other marker	CD30$^-$, CD34$^-$, CD38$^-$, CD39$^-$, CD45$^-$, EMA$^+$, HLA-DR$^-$, TdT$^-$
Adhesion marker	CD11a$^-$, CD11b$^+$, CD11c$^-$, CD18$^-$, CD29$^+$, CD31$^-$, CD43$^+$, CD44$^-$, CD49d$^-$, CD49e$^-$, CD54$^-$, CD58$^-$, CD62E$^-$, CD62L$^-$
Cytokine receptor	CD25$^+$

Genetic characterization[1,2]

Cytogenetic karyotype	der(X)t(X;12;14)(q12;q11:q22-23;q23-24)
Receptor gene rearrangement	*IGH* R, *IGK* R

Functional characterization[1,2]

Colony formation	in agar
Heterotransplantation	into NOD/SCID mice

Comments

- Mature EBV$^-$ HHV-8$^+$ B-cell line derived from non-AIDS PEL.
- Available from cell line bank.

References

[1] Boshoff, C. et al. (1998) Blood 91, 1671–1679.

[2] Drexler, H.G. et al. (1998) Leukemia 12, 1507–1517.

[3] Strauchen, J.A. et al. (1997) Ann. Int. Med. 125, 822–825. (case report)

BEVA

Culture characterization[1]

Other name of cell line	**BEVA-FBS**
Sister cell line	**EBV-Ev** (= EBV$^+$ B-LCL)
Culture medium	90% RPMI 1640 + 10% human serum or 10% FBS
Doubling time	in human serum: 72 h; in FBS: 96 h
Viral status	EBV$^+$
Authentication	yes (by HLA typing, *BCL2* breakpoint)
Primary reference	De Kroon et al.[1]
Availability of cell line	original authors

Clinical characterization[1]

Patient	48-year-old male
Disease diagnosis	AML M5b + (subclinical) B-NHL (follicular)
Treatment status	refractory
Specimen site	bone marrow
Year of establishment	1991

Immunophenotypic characterization[1,2]

T-/NK cell marker	CD7$^-$
B-cell marker	CD10$^-$, CD19$^+$, CD20$^+$, CD21$^+$, CD22$^+$, CD23$^+$, CD24$^+$, sIgGκ^+
Myelomonocytic marker	CD13$^-$, CD15$^+$, CD33$^-$, CD65$^+$
Progenitor/activation/other marker	CD34$^-$, HLA-DR$^+$
Adhesion marker	CD11a$^{(+)}$, CD11b$^+$, CD11c$^-$, CD44$^+$, CD49d$^+$, CD54$^+$

Genetic characterization[1,2]

Cytogenetic karyotype	48, XY, +8, +der(2)t(2;?)(q;?), t(3;17)(q29;q24), der(5)t(1;5) (q32;p15), der(11)t(4;11)(q21;q23), t(14;18)(q32;q21), trp(14) (q22;q32), der(15)t(15;15)(p11;q11)
Unique translocation/fusion gene	(1) t(14;18)(q32;q21) → *IGH-BCL2* genes altered (MBR) (2) t(4;11)(q21;q23) → no *MLL-AF4* fusion gene
Receptor gene rearrangement	*IGH* R

Functional characterization[1,2]

Colony formation	in methylcellulose
Cytochemistry	ANAE$^{(+)}$, MPO$^-$
Cytokine production	mRNA$^+$: IL-1β, IL-6
Cytokine response	growth stimulation: IL-6; growth inhibition: IL-4
Heterotransplantation	into SCID mice

Comments

- Mature EBV+ B-cell line derived from AML M5 with subclinical follicular B-NHL.
- Neoplastic nature verified.
- No clonal relationship between primary AML cells and BEVA, but between B-NHL and BEVA.
- Carries t(14;18)(q32;q21) leading to alteration of the *IGH* and *BCL2* genes (major breakpoint region).
- Co-expression of myeloid and lymphoid immunomarkers.

References

[1] De Kroon, J.F.E.M. et al. (1997) Exp. Hematol. 25, 1062–1068.
[2] De Kroon, J.F.E.M. et al. (1992) Blood 80, 436a.

BHL-89

Culture characterization[1]

Culture medium	85% RPMI 1640 or Ham's F12 + 15% FBS
Doubling time	48 h
Viral status	EBNA⁻
Authentication	no
Primary reference	Pályi et al.[1]
Availability of cell line	not known

Clinical characterization[1]

Patient	56-year-old female
Disease diagnosis	B-NHL (diffuse, small cleaved and large cell, intermediate grade)
Treatment status	at diagnosis
Specimen site	lymph node
Year of establishment	1981

Immunophenotypic characterization[1]

T-/NK cell marker	$CD1^-$, $CD2^-$, $CD3^-$
B-cell marker	$CD9^{(+)}$, $CD10^-$, $CD19^-$, $CD20^-$, $CD21^-$, $cyCD22^+$, $CD23^-$, $CD24^+$, $CD80^+$, $sIgM^+$
Myelomonocytic marker	$CD13^-$, $CD14^-$, $CD16^-$, $CD33^-$
Progenitor/activation/other marker	$CD71^-$, $HLA\text{-}DR^-$, TdT^+
Adhesion marker	$CD11a^-$

Genetic characterization[1]

Cytogenetic karyotype	47, +1, 14q+, −15

Functional characterization[1]

Colony formation	in agar
Cytochemistry	ACP^-, $ANAE^+$, $CAE^{(+)}$, MPO^-, $PAS^{(+)}$
Heterotransplantation	into nude mice
Inducibility of differentiation	resistant to TPA

Comments

- Mature EBV⁻ B-cell line derived from B-NHL.

Reference

[1] Pályi, I. et al. (1995) Haematologica 80, 206–211.

BNBH-1

Culture characterization[1]

Culture medium	90% RPMI 1640 + 10% FBS
Doubling time	51 h
Viral status	EBNA$^+$
Authentication	yes (by *IGH* rearrangement)
Primary reference	Tokumine et al.[1]
Availability of cell line	not known

Clinical characterization[1]

Patient	40-year-old male
Disease diagnosis	HCL
Treatment status	stable phase, post-splenectomy
Specimen site	peripheral blood
Year of establishment	1985

Immunophenotypic characterization[1]

T-/NK cell marker	CD2$^-$, CD5$^-$
B-cell marker	CD20$^+$, CD21$^+$, FMC-7$^-$, PCA-1$^+$, cyIgG$^+$, sIgGκ$^+$
Progenitor/activation/other marker	HLA-DR$^+$
Adhesion marker	CD11b$^-$, CD11c$^+$
Cytokine receptor	CD25$^+$

Genetic characterization[1]

Cytogenetic karyotype	46, XY, −6, −14, +der(6)t(3;6)(q11;q27), +der(14)t(14;?) (q32.1;?); sideline with +der(6)t(6;8)(q27;q22)
Receptor gene rearrangement	*IGH* RR

Functional characterization[1]

Cytochemistry	ACP$^+$, ANBE$^+$, CAE$^-$, MPO$^-$, PAS$^-$, TRAP$^+$
Inducibility of differentiation	TPA → spreading, adherence

Comments

- Mature EBV$^+$ B-cell line derived from HCL.
- Authentic derivation and neoplastic nature verified.

Reference

[1] Tokumine, Y. et al. (1988) Int. J. Cancer 42, 99–103.

BONNA-12

Culture characterization[1,2]

Culture medium	90% RPMI 1640 + 10% FBS
Doubling time	36–60 h
Viral status	EBV+, HBV−, HCV−, HHV-8−, HIV−, HTLV-I/II−
Authentication	yes (by HLA typing, minisatellite DNA anaylsis)
Primary reference	Wientjens et al.[1]
Availability of cell line	DSM ACC 150

Clinical characterization[1]

Patient	46-year-old male
Disease diagnosis	HCL
Treatment status	at diagnosis
Specimen site	spleen
Year of establishment	1988

Immunophenotypic characterization[1,2]

T-/NK cell marker	CD3−, CD5−
B-cell marker	CD9+, CD10−, CD19+, CD20+, CD21+, CD22+, CD23+, CD24+, CD37+, CD103−, FMC-7+, cyIgMκ+, sIgMκ+
Myelomonocytic marker	CD13−
Progenitor/activation/other marker	HLA-DR+
Adhesion marker	CD11c+
Cytokine receptor	CD25−

Genetic characterization[1,2]

Cytogenetic karyotype	hyperdiploid, 4% polyploidy; 47(45–47)<2n>X, −Y, +12, t(9;10)(p12/13;q12); supernumerary ring or acf present in most cells
Receptor gene rearrangement	*IGH* R, *IGK* R

Functional characterization[1]

Colony formation	in methylcellulose
Cytochemistry	ACP+, ANAE−, ANBE−, MPO−, PAS−, TRAP+

Comments

- Mature EBV+ B-cell derived from HCL.
- Cellular derivation and neoplastic nature verified.
- Available from cell line bank.

References

[1] Wientjens, G.J.H.M. et al. (1991) Leukemia Lymphoma 5, 415–422.

[2] Drexler, H.G. et al. (1999) DSMZ Catalogue of Cell Lines, 7th edn, Braunschweig, Germany.

Ci-1

Culture characterization[1]

Culture medium	90% RPMI 1640 + 10% FBS
Viral status	EBNA$^-$
Authentication	no
Primary reference	Th'ng et al.[1]
Availability of cell line	original authors

Clinical characterization[1]

Patient	38-year-old female
Disease diagnosis	B-NHL (lymphoblastic, non-convoluted cell type)
Treatment status	at diagnosis
Specimen site	ascites
Year of establishment	1981

Immunophenotypic characterization[1,2]

T-/NK cell marker	CD3$^-$, CD5$^-$, CD56$^-$
B-cell marker	CD10$^-$, CD19$^+$, CD20$^+$, CD21$^+$, CD22$^+$, CD23$^+$, CD24$^+$, CD37$^+$, FMC-7$^+$, cyIgλ$^+$, sIg$^-$
Myelomonocytic marker	CD13$^-$, CD15$^-$, CD33$^-$
Erythroid-megakaryocytic marker	CD31$^-$, CD36$^-$, CD41a$^+$, CD61$^+$
Progenitor/activation/other marker	CD34$^-$, CD38$^+$, HLA-DR$^+$
Adhesion marker	CD11a$^+$, CD11b$^-$, CD11c$^-$, CD18$^+$, CD29$^+$, CD44$^+$, CD49a$^+$, CD49b$^+$, CD49c$^+$, CD49d$^+$, CD49e$^+$, CD49f$^-$, CD51$^+$, CD54$^+$, CD58$^+$
Cytokine receptor	CD25$^+$

Genetic characterization[1]

Cytogenetic karyotype	46, XX, t(2;8)(p11;q24), t(14;22)(q32;q11)
Unique translocation/fusion gene	t(2;8)(p11;q24) → *IGK-MYC* genes altered?
Receptor gene rearrangement	*IGH* R, *IGK* RR, *IGL* RG

Comments

- Mature EBV$^-$ B-cell line derived from lymphoblastic B-NHL.
- Carries t(2;8)(p11;q32) which may lead to alteration of the *IGK-MYC* genes.

References

[1] Th'ng, K.H. et al. (1987) Int. J. Cancer 39, 89–93.
[2] Segat, D. et al. (1994) Blood 83, 1586–1594.

CRO-AP/2

Culture characterization[1-3]

Culture medium	90% RPMI 1640 + 10% FBS
Doubling time	20–40 h
Viral status	EBV+, HBV−, HCV−, HHV-8+, HIV−, HTLV-I/II−
Authentication	yes (by *IGH* rearrangement)
Primary reference	Carbone et al.[1]
Availability of cell line	DSM ACC 48

Clinical characterization[1,2]

Patient	49-year-old male (HIV+)
Disease diagnosis	Kaposi's sarcoma → PEL + AIDS
Treatment status	at diagnosis
Specimen site	pleural effusion
Year of establishment	1996

Immunophenotypic characterization[1-3]

T-/NK cell marker	CD3−, CD5−
B-cell marker	CD10−, CD19−, CD20+, CD22−, CD23+, CD37+, CD40−, CD74+, CD79a−, CD80−, CD138+, sIg−
Myelomonocytic marker	CD13−
Progenitor/activation/other marker	CD30+, CD34−, CD38+, CD39+, CD45+, CD45RA−, CD45RO−, CD70+, EMA+, HLA-DR+
Adhesion marker	CD43−

Genetic characterization[1-4]

Cytogenetic karyotype	hyperdiploid, 12% polyploidy; 48(47–48)<2n>X, +X, −Y, +12, add(?Y)(p11), add(1)(q43.1), der(15)t(8;15)(q11;q25), der(22)t(1;22)(q21;q13); sideline with add(9)(q13)
Receptor gene rearrangement	*IGH* R, *IGK* R
Unique gene alteration	*BCL6* mutation

Functional characterization[1]

Special features	EBNA2−, LMP1+

Comments

- Mature EBV+ HHV-8+ B-cell line derived from AIDS-associated PEL.
- Authentic derivation and neoplastic nature verified.
- Available from cell line bank.

References
[1] Carbone, A. et al. (1997) Int. J. Cancer 73, 562–569.
[2] Drexler, H.G. et al. (1998) Leukemia 12, 1507–1517. (review on PEL cell lines).
[3] Drexler, H.G. et al. (1999) DSMZ Catalogue of Cell Lines, 7th edn, Braunschweig, Germany.
[4] Gaidano, G. et al. (1999) Genes Chromosomes Cancer 24, 16–23.

Culture characterization[1-3]

Culture medium 90% RPMI 1640 + 10% FBS
Doubling time 50 h
Viral status EBV⁻, HBV⁻, HCV⁻, HHV-8⁺, HIV⁻, HTLV-I/II⁻
Authentication yes (by *IGH* rearrangement)
Primary reference Carbone et al.[1]
Availability of cell line DSM ACC 275

Clinical characterization[1,2]

Patient 42-year-old male (HIV⁺)
Disease diagnosis PEL + AIDS
Treatment status at diagnosis
Specimen site ascites
Year of establishment 1997

Immunophenotypic characterization[1-3]

T-/NK cell marker CD3⁻, CD4⁻, CD5⁻
B-cell marker CD10⁻, CD19⁻, CD20⁻, CD22⁻, CD23⁺, CD37⁻, CD40⁻, CD74⁺, CD79a⁻, CD80⁻, CD138⁺, cyIg⁻, sIg⁻
Myelomonocytic marker CD13⁻
Progenitor/activation/other marker CD30⁺, CD34⁻, CD38⁺, CD45⁺, CD45RA⁻, CD45RO⁻, CD70⁺, EMA⁺, HLA-DR⁺
Adhesion marker CD11a⁺, CD11b⁻, CD11c⁻, CD39⁻, CD43⁺, CD44⁺, CD49a⁻, CD49d⁺, CD49f⁺, CD54⁺, CD56⁻, CD58⁺, CD103⁻

Genetic characterization[1-4]

Cytogenetic karyotype flat-moded, hypotetraploid; 80–88<4n>XX, −Y, −Y, −1, −2, +6, −7, −10, −13, −13, −22, −22, +6mar, i(1q), add(2)(p2?), add(3) (q2?7), der(4;9)(q10;q10) × 1–2, del(5)(p13p14) ×1–2, del(5)(q2?1) ×1–2, der(6)t(6;7)(q15;q1)add(7)(q3?5), del(8)(p11), del(9)(p21), add(9)(p11), del(11)(q2?2q2?3) × 2, der(12)t(2;12)(p12;q24.32), add(13)(p11), t(14;19)(q11;q13) × 2, der(15)t(1;?)(?15)(q21;) (?;q11), ider(?)t(8;?)(q11;?); extensive subclonal rearrangements present
Receptor gene rearrangement *IGH* R, *IGK* R
Unique gene alteration *BCL6* mutation

Functional characterization[1]

Special features EBNA2⁻, LMP1⁻

Comments

- Mature EBV⁻ HHV-8⁺ B-cell line derived from AIDS-associated PEL.
- Available from cell line bank.

References

[1] Carbone, A. et al. (1998) Brit. J. Haematol. 102, 1081–1089.
[2] Drexler, H.G. et al. (1998) Leukemia 12, 1507–1517.
[3] Drexler, H.G. et al. (1999) DSMZ Catalogue of Cell Lines, 7th edn, Braunschweig, Germany.
[4] Gaidano, G. et al. (1999) Genes Chromosomes Cancer 24, 16–23.

Culture characterization[1-3]

Culture medium	90% RPMI 1640 + 10% FBS
Doubling time	35–40 h
Viral status	EBV+, HBV−, HCV−, HHV-8+, HIV−, HTLV-I/II−
Authentication	yes (by *IGK* rearrangement)
Primary reference	Carbone et al.[1]
Availability of cell line	DSM ACC 215

Clinical characterization[1,2]

Patient	35-year-old male (HIV+)
Disease diagnosis	Kaposi's sarcoma → PEL + AIDS
Treatment status	progressive, terminal
Specimen site	pleural effusion
Year of establishment	1998

Immunophenotypic characterization[1-3]

T-/NK cell marker	CD3−, CD5−
B-cell marker	CD10−, CD19−, CD20−, CD22−, CD23+, CD37+, CD40−, CD74+, CD79a−, CD80−, CD138+, sIg−
Myelomonocytic marker	CD13−
Progenitor/activation/other marker	CD30+, CD34−, CD38+, CD45+, CD45RA−, CD45RO+, CD70+, EMA+, HLA-DR+
Adhesion marker	CD11a+, CD11b−, CD11c−, CD39+, CD43−, CD44+, CD49a−, CD49d+, CD49f+, CD54+, CD56−, CD58+, CD103−

Genetic characterization[1-3]

Cytogenetic karyotype	hyperdiploid, 12% polyploidy; 44–49<2n>X, −Y, +8, +2mar, del(4)(q25), add(8)(p11), dup(12)(q11q14), del(14)(q24); sideline with i(8q)
Receptor gene rearrangement	*IGH* R, *IGK* R
Unique gene alteration	*BCL6* mutation

Functional characterization[1]

Special features	EBNA2−, LMP1−

Comments

- Mature EBV+ HHV-8+ B-cell line derived from AIDS-associated PEL.
- Authentic derivation and neoplastic nature verified.
- Available from cell line bank.

References

[1] Carbone, A. et al. (1998) Brit. J. Haematol. 102, 1081–1089.

[2] Drexler, H.G. et al. (1998) Leukemia 12, 1507–1517. (review on PEL cell lines).

[3] Drexler, H.G. et al. (1999) DSMZ Catalogue of Cell Lines, 7th edn, Braunschweig, Germany.

CTB-1

Culture characterization[1]

Culture medium	90% RPMI 1640 + 10% FBS
Doubling time	36 h
Authentication	yes (by cytogenetics)
Primary reference	Uchida et al.[1]
Availability of cell line	RCB 1316

Clinical characterization[1]

Patient	70-year-old male
Disease diagnosis	B-NHL (diffuse large cell)
Treatment status	stage IV, refractory, progressive
Specimen site	pericardial effusion
Year of establishment	1994

Immunophenotypic characterization[1]

T-/NK cell marker	$CD3^-$, $CD4^-$, $CD5^-$, $CD8^-$
B-cell marker	$CD10^+$, $CD19^+$, $CD20^+$, $CD21^+$, $sIgG\kappa^+$
Progenitor/activation/other marker	$CD38^+$, $CD95^+$, HLA-DR^+

Genetic characterization[1]

Cytogenetic karyotype	48, XY, +7, −16, del(1)(p13), +add(1)(p36), +add(12)(p13), t(14;22)(q32;q11)

Functional characterization[1]

Heterotransplantation	into nude mice
Proto-oncogene	BCL2 overexpression
Special features	resistant to FAS-induced apoptosis

Comments

- Mature B-cell line derived from B-NHL (DLCL).
- Available from cell line bank.

Reference

[1] Uchida, Y. et al. (1997) Int. J. Oncol. 10, 1103–1107.

DEAU

Culture characterization[1]

Culture medium	90% IMDM + 10% FBS
Viral status	EBNA$^-$
Authentication	yes (by cytogenetics)
Primary reference	Al Saati et al.[1]
Availability of cell line	not known

Clinical characterization[1]

Patient	42-year-old male
Disease diagnosis	B-NHL (diffuse large cell, centroblastic, intermediate grade)
Specimen site	lymph node

Immunophenotypic characterization[1]

T-/NK cell marker	CD3$^-$, CD4$^-$, CD5$^-$, CD7$^-$, CD8$^-$
B-cell marker	CD10$^+$, CD19$^+$, CD21$^+$, CD22$^+$, sIg$^-$
Progenitor/activation/other marker	HLA-DR$^+$

Genetic characterization[1]

Cytogenetic karyotype	88, XX, −Y, del(X)(q24), −1, −2, −3, −3, −7, −7, −8, −10, −10, −14, −14, −15, −17, −17, −17, +21, +2mar, +iso(3p), del(4)(q23), +der(7)t(7;?)(q33;?), +der(8)t(8;?)(q24;?), +der(8)t(8;?)(q24;?), +der(8)t(8;?)(p23;?), del(9)(q33), +der(10)t(9;10)(q12;p15), inv(11)(p15q12), del(14)(q24), +iso(17q), +iso(17q), +iso(17p), iso(17p)

Functional characterization[1]

Heterotransplantation	into nude mice

Comments

- Mature EBV$^-$ B-cell line derived from B-NHL (DLCL).

Reference

[1] Al Saati, T. et al. (1989) Blood 74, 2476–2485.

DoHH2

Culture characterization[1,2]

Culture medium 90–95% RPMI 1640 + 5–10% FBS
Doubling time 40–48 h
Viral status EBV$^-$, HBV$^-$, HCV$^-$, HHV-8$^-$, HIV$^-$, HTLV-I/II$^-$
Authentication yes (by *IGH* rearrangement, HLA typing, *BCL2* breakpoint)
Primary reference Kluin-Nelemans et al.[1]
Availability of cell line DSM ACC 47

Clinical characterization[1]

Patient 60-year-old male
Disease diagnosis B-NHL (follicular centroblastic/centrocytic → immunoblastic)
Treatment status refractory, progressive
Specimen site pleural effusion
Year of establishment 1990

Immunophenotypic characterization[1–4]

T-/NK cell marker CD3$^-$, CD5$^-$, CD7$^-$
B-cell marker CD10$^+$, CD19$^+$, CD20$^+$, CD22$^-$, CD23$^-$, CD24$^+$, CD37$^+$, CD138$^-$, cyIgGλ$^+$, sIgGλ$^+$
Myelomonocytic marker CD13$^-$, CD15$^-$, CD33$^-$, CD65$^-$
Progenitor/activation/other marker CD34$^-$, HLA-DR$^+$
Adhesion marker CD11a$^{(+)}$, CD11b$^-$, CD11c$^-$, CD18$^{(+)}$, CD44$^+$, CD49a$^-$, CD49b$^-$, CD49d$^+$, CD49e$^-$, CD49f$^-$, CD54$^+$

Genetic characterization[1,2,5,6]

Cytogenetic karyotype 47(43–48)<2n>XY, +7, der(8)t(8;18)(q24;q21), der(14)t(8;14)(q24;q32), der(18)t(14;18)(q32;q21)
Unique translocation/fusion gene (1) t(8;14)(q24;q32) → *MYC-IGH* genes altered
(2) t(14;18)(q32;q21) → *IGH-BCL2* genes altered (MBR)
Receptor gene rearrangement *IGH* RR, *IGK* D, *IGL* RR
Unique gene alteration *MYC* rearrangement, *P16INK4A* deletion

Functional characterization[1,3–5]

Colony formation in methylcellulose
Cytochemistry ANAE$^-$
Cytokine response growth stimulation: IL-7; growth inhibition: IFN-α
Heterotransplantation into SCID mice

Comments

- Mature EBV⁻ B-cell derived from follicular B-NHL (reference cell line).
- Carries both t(8;14)(q24;q32) and t(14;18)(q32;q21) leading to alterations of the *MYC-IGH* genes and of the *IGH-BCL2* genes (major breakpoint region), respectively.
- Available from cell line bank.

References

[1] Kluin-Nelemans, H.C. et al. (1991) Leukemia 5, 221–224.
[2] Drexler, H.G. et al. (1999) DSMZ Catalogue of Cell Lines, 7th edn, Braunschweig, Germany.
[3] De Kroon, J.F.E.M. et al. (1994) Leukemia 8, 1385–1391.
[4] De Kroon, J.F.E.M. et al. (1997) Exp. Hematol. 25, 1062–1068.
[5] Stranks, G. et al. (1995) Blood 85, 893–901.
[6] Dyer, M.J.S. et al. (1996) Leukemia 10, 1198–1208.

Culture characterization[1]

Culture medium serum-free medium or 90% IMDM + 10% FBS
Authentication yes (by cytogenetics)
Primary reference Lambrechts et al.[1]
Availability of cell line original authors

Clinical characterization[1]

Patient 70-year-old female
Disease diagnosis B-NHL (immunoblastic large cell)
Treatment status at diagnosis
Specimen site bone marrow

Immunophenotypic characterization[1]

T-/NK cell marker CD2⁻, CD3⁻
B-cell marker CD10⁺, CD19⁺, CD20⁻, CD22⁻, CD37⁺, cyIg⁻, sIg⁻
Myelomonocytic marker CD13⁻, CD15⁻, CD33⁻, CD65⁻
Progenitor/activation/other marker CD45⁺, HLA-DR⁺, TdT⁻

Genetic characterization[1]

Cytogenetic karyotype 48, XX, t(1;6)(q12;q26), add(2)(q37), +del(7)(q21q32), t(8;14) (q24;q32), +inv(12)(p12.3q24), der(13)t(1;13)(q12;p12), t(14;18) (q32;q21), del(20)(p11p13), +ace
Unique translocation/fusion gene (1) t(8;14)(q24;q32) → *MYC-IGH* genes altered
(2) t(14;18)(q32;q21) → *IGH-BCL2* genes altered (MBR)
Receptor gene rearrangement *IGH* R, *IGK* D, *IGL* R

Functional characterization[1]

Proto-oncogene mRNA⁺: BCL2, MYC (truncated)

Comments

- Mature B-cell line derived from immunoblastic B-NHL.
- Carries both t(8;14)(q24;q32) and t(14;18)(q32;q21) leading to alterations of the *MYC-IGH* genes and of the *IGH-BCL2* genes (major breakpoint region), respectively.

References

[1] Lambrechts, A.C. et al. (1994) Leukemia 8, 1164–1171.
[2] van Ooteghem, R.B.C. et al. (1994) Cancer Genet. Cytogenet. 74, 87–94.

DS-1

Culture characterization[1]

Culture medium	90% RPMI 1640 + 10% FBS + 10 U/ml IL-6
Viral status	EBV⁻
Authentication	yes (by cytogenetics)
Primary reference	Bock et al.[1]
Availability of cell line	not known

Clinical characterization[1]

Patient	29-year-old female
Disease diagnosis	B-NHL (immunodeficiency with intestinal lymphangiectasia → immunoblastic lymphoma)
Treatment status	at diagnosis
Specimen site	pleural effusion

Immunophenotypic characterization[1]

T-/NK cell marker	$CD2^-$, $CD3^-$, $CD4^-$, $CD5^-$, $CD7^-$, $CD8^-$
B-cell marker	$CD19^-$, $CD20^-$, $PCA\text{-}1^+$, $sIgG\kappa^+$
Progenitor/activation/other marker	$CD30^-$, $CD38^-$, $CD45^-$, $HLA\text{-}DR^+$
Cytokine receptor	$CD25^-$, $CD126^+$, $CD130^+$

Genetic characterization[1]

Cytogenetic karyotype	t(7;14)(q32;q32), t(8;22)(q24;q11)
Unique translocation/fusion gene	t(8;22)(q24;q11) → *MYC-IGL* genes altered?
Receptor gene rearrangement	*IGH* R

Functional characterization[1]

Cytokine production	IL-6 $mRNA^+$
Cytokine response	IL-6-dependent
Proto-oncogene	MYC mRNA overexpression
Special features	secretion: IgG

Comments

- Mature EBV⁻ B-cell line derived from immunoblastic B-NHL.
- Possibly PEL cell line, but HHV-8 status not known.
- Carries t(8;22)(q24;q11) which may lead to alteration of the *MYC-IGL* genes.

Reference

[1] Bock, G.H. et al. (1993) Cytokine 5, 480–489.

Culture characterization[1,2]

Culture medium	90% RPMI 1640 + 10% FBS
Doubling time	48–72 h
Viral status	EBV$^+$, HBV$^-$, HCV$^-$, HHV-8$^-$, HIV$^-$, HTLV-I/II$^-$
Authentication	yes (by *IGH* rearrangements)
Primary reference	Saltman et al.[1]
Availability of cell line	DSM ACC 67

Clinical characterization[1]

Patient	69-year-old female
Disease diagnosis	B-CLL
Treatment status	prior to any treatment
Specimen site	peripheral blood
Year of establishment	1988

Immunophenotypic characterization[1,2]

T-/NK cell marker	CD3$^-$, CD5$^+$
B-cell marker	CD19$^+$, CD37$^+$, sIgMκ$^+$
Myelomonocytic marker	CD13$^-$
Progenitor/activation/other marker	HLA-DR$^+$

Genetic characterization[1,2]

Cytogenetic karyotype	hypodiploid, 1.6% polyploidy; 45(42–46)<2n>X, −X; no structural rearrangements
Receptor gene rearrangement	*IGH* RR, *IGK* RG, *IGL* GG

Comments

- Mature EBV$^+$ B-cell line derived from B-CLL.
- Cellular derivation and neoplastic nature verified.
- Cells carry normal karyotype without any structural rearrangements/ abnormalities.
- Available from cell line bank.

References

[1] Saltman, D. et al. (1990) Leukemia Res. 14, 381–387.

[2] Drexler, H.G. et al. (1999) DSMZ Catalogue of Cell Lines, 7th edn, Braunschweig, Germany.

Culture characterization[1]

Culture medium 90% RPMI 1640 + 10% FBS
Doubling time 28 h
Viral status EBNA⁺
Authentication no
Primary reference Harvey et al.[1]
Availability of cell line original authors

Clinical characterization[1]

Patient 69-year-old male
Disease diagnosis HCL
Specimen site peripheral blood

Immunophenotypic characterization[1]

T-/NK cell marker $CD5^{+}$
B-cell marker $CD19^{+}$, $CD20^{+}$, $CD21^{+}$, $CD22^{-}$, $CD24^{+}$, FMC-7^{+}, HC-2^{+}, PCA-1^{+}, $sIg\lambda^{+}$
Progenitor/activation/other marker HLA-DR^{-}
Adhesion marker $CD11a^{+}$, $CD11b^{(+)}$, $CD11c^{-}$, $CD18^{+}$
Cytokine receptor $CD25^{+}$

Genetic characterization[2]

Cytogenetic karyotype 46,XY, del(4)(p14), der(6)t(4;6)(p14;p21.3), del(7)(q32q36.1)

Functional characterization[1,2]

Cytochemistry $TRAP^{+}$
Inducibility of differentiation IFN-$\alpha \rightarrow$ differentiation
Proto-oncogene $mRNA^{+}$: FOS, HRAS, MYB, MYC, RAF2

Comments

- Mature EBV^{+} B-cell line derived from HCL.
- Neoplastic nature of cells is verified, but derivation not authenticated.

References

[1] Harvey, W. et al. (1991) Leukemia Res. 15, 733–744.
[2] Harvey, W. et al. (1994) Leukemia Res. 18, 577–585.

Farage

Culture characterization[1]

Culture medium	90% RPMI 1640 + 10% FBS
Doubling time	48 h
Viral status	EBNA$^+$
Authentication	no
Primary reference	Ben-Bassat et al.[1]
Availability of cell line	not known

Clinical characterization[1]

Patient	70-year-old female
Disease diagnosis	B-NHL (diffuse large cell)
Treatment status	at diagnosis
Specimen site	lymph node

Immunophenotypic characterization[1]

T-/NK cell marker	CD1a$^-$, CD2$^-$, CD3$^-$, CD4$^-$, CD8$^-$, CD57$^-$
B-cell marker	CD10$^+$, CD19$^+$, CD20$^+$, CD22$^+$, CD23$^+$, CD39$^+$, CD103$^-$, HC-2$^-$, cyIgMκ$^+$, sIg$^-$
Myelomonocytic marker	CD16$^-$
Progenitor/activation/other marker	CD38$^+$, CD71$^+$, HLA-DR$^+$, TdT$^+$
Adhesion marker	CD11b$^-$
Cytokine receptor	CD25$^-$

Genetic characterization[1,2]

Cytogenetic karyotype	47, XX, +11
Receptor gene rearrangement	*IGH* RG, *IGK* R, *TCRB* G, *TCRG* RD

Functional characterization[1]

Cytochemistry	ACP$^+$, ANAE$^-$, MPO$^-$, SBB$^-$

Comments

- Mature EBV$^+$ B-cell line derived from B-NHL (DLCL).
- Neoplastic nature confirmed, but derivation not authenticated.

References

[1] Ben-Bassat, H. et al. (1992) Leukemia Lymphoma 6, 513–521.
[2] Gabay, C. et al. (1999) Leukemia Lymphoma (in press).

FL-18

Culture characterization[1,2]

Sister cell line	**FL-18EB** (simultaneous EBV$^+$ sister cell line with same karyotype and clonal identity)
Culture medium	90% RPMI 1640 + 10% FBS
Doubling time	36–48 h
Viral status	EBNA$^-$
Authentication	yes (by cytogenetics)
Primary reference	Ohno et al.[1]
Availability of cell line	not known

Clinical characterization[1,2]

Patient	69-year-old male
Disease diagnosis	B-NHL (follicular small cleaved cell)
Treatment status	progressive, terminal
Specimen site	pleural effusion
Year of establishment	1984

Immunophenotypic characterization[1–4]

T-/NK cell marker	CD2$^-$, CD3$^-$, CD4$^-$, CD5$^-$, CD7$^-$, CD8$^-$
B-cell marker	CD10$^+$, CD19$^+$, CD20$^+$, CD21$^-$, CD22$^+$, CD23$^-$, CD24$^+$, PCA-1$^-$, cyIg$^-$, sIgGκ$^+$
Myelomonocytic marker	CD13$^-$
Progenitor/activation/other marker	CD95$^-$, HLA-DR$^+$, TdT$^-$
Cytokine receptor	CD25$^-$

Genetic characterization[1–3]

Cytogenetic karyotype	49, XY, +7, +12, −17, +del(X)(p11), del(6)(p11), dirins(13;8) (q14;q22q24), t(8;22)(q24;q13), t(14;18)(q32.3;q21.3), +der(17)t (17;?)(q23;?)
Unique translocation/fusion gene	t(14;18)(q32;q21) → *IGH-BCL2* genes altered?
Receptor gene rearrangement	*IGH* RR, *IGK* R

Functional characterization[2]

Cytochemistry	ANAE$^-$, CAE$^-$, MPO$^-$, PAS$^-$

Comments

- Mature EBV$^-$ B-cell line derived from B-NHL (FCL).
- Carries t(14;18)(q32;q21) possibly leading to alteration of the *IGH-BCL2*.

References

[1] Ohno, H. et al. (1985) Jpn. J. Cancer Res. 76, 563–566.
[2] Doi, S. et al. (1987) Blood 70, 1619–1623.
[3] Ohno, H. et al. (1987) Int. J. Cancer 39, 785–788.
[4] Tani, A. et al. (1996) Leukemia 10, 1592–1603.

FL-218

Culture characterization[1]

Culture medium	80% RPMI 1640 + 20% FBS
Doubling time	36–48 h
Viral status	EBNA⁻
Authentication	yes (by cytogenetics, *BCL2-IGH* rearrangement)
Primary reference	Amakawa et al.[1]
Availability of cell line	restricted

Clinical characterization[1]

Patient	73-year-old female
Disease diagnosis	B-NHL (follicular, small cell, cleaved → diffuse mixed small/large)
Treatment status	refractory, terminal
Specimen site	ascites
Year of establishment	1986

Immunophenotypic characterization[1,2]

T-/NK cell marker	$CD2^-$, $CD3^-$, $CD4^-$, $CD5^-$, $CD7^-$, $CD8^-$
B-cell marker	$CD10^-$, $CD19^+$, $CD20^+$, $CD21^-$, $CD22^+$, $CD23^-$, $CD24^+$, $sIgG\lambda^+$
Myelomonocytic marker	$CD13^-$
Progenitor/activation/other marker	$CD95^+$, HLA-DR^+, TdT^-
Cytokine receptor	$CD25^-$

Genetic characterization[1]

Cytogenetic karyotype	47, XX, +12, −18, del(10)(q23.2q24.3), t(14;18)(q32.3;q21.3), +18q−
Unique translocation/fusion gene	t(14;18)(q32;q21) → *IGH-BCL2* genes altered (MBR)
Receptor gene rearrangement	*IGH* RR
Unique gene alteration	*BCL2* amplification

Functional characterization[1,3]

Special features	mRNA overexpression: BCL2, BCL6

Comments

- Mature EBV⁻ B-cell line derived from B-NHL (follicular transformed into DLCL).
- Carries t(14;18)(q32;q21) leading to alteration of the *IGH* and *BCL2* genes (major breakpoint region).

References

[1] Amakawa, R. et al. (1990) Cancer Res. 50, 2423–2428.
[2] Tani, A. et al. (1996) Leukemia 10, 1592–1603.
[3] Yonetani, N. et al. (1998) Oncogene 17, 971–979.

FL-318

Culture characterization[1]

Culture medium	80% RPMI 1640 + 20% FBS
Doubling time	36–48 h
Viral status	EBNA⁻
Authentication	yes (by cytogenetics, *BCL2-IGH* rearrangement)
Primary reference	Amakawa et al.[1]
Availability of cell line	restricted

Clinical characterization[1]

Patient	67-year-old male
Disease diagnosis	B-NHL (follicular, small cell, cleaved → DLCL)
Treatment status	refractory, terminal
Specimen site	ascites
Year of establishment	1987

Immunophenotypic characterization[1,2]

T-/NK cell marker	$CD2^-$, $CD3^-$, $CD4^-$, $CD5^-$, $CD7^-$, $CD8^-$
B-cell marker	$CD10^+$, $CD19^+$, $CD20^+$, $CD21^-$, $CD22^-$, $CD23^-$, $CD24^+$, $sIgM\kappa^+$
Myelomonocytic marker	$CD13^-$
Progenitor/activation/other marker	$CD95^+$, $HLA\text{-}DR^+$, TdT^-
Cytokine receptor	$CD25^-$

Genetic characterization[1,3]

Cytogenetic karyotype	51, XY, +7, +13, +15, del(2)(p21p23), t(14;18)(q32.3;q21.3), del(15)(q13q15), +18q−, +i(18q−)
Unique translocation/fusion gene	t(14;18)(q32;q21) → *IGH-BCL2* genes altered (MBR)
Receptor gene rearrangement	*IGH* RR
Unique gene alteration	*BCL2* amplification

Comments

- Mature EBV⁻ B-cell line derived from B-NHL (follicular transformed into DLCL).
- Carries t(14;18)(q32;q21) leading to alteration of the *IGH* and *BCL2* genes (major breakpoint region).

References

[1] Amakawa, R. et al. (1990) Cancer Res. 50, 2423–2428.
[2] Tani, A. et al. (1996) Leukemia 10, 1592–1603.
[3] Kadowaki, N. et al. (1995) Leukemia 9, 1139–1143.

FMC-Hu-1-B

Culture characterization[1]

Establishment	initially on normal bone marrow stromal cells + PHA-LCM
Culture medium	95% McCoy's 5A + 5% FBS
Doubling time	36 h
Viral status	EBNA$^-$
Authentication	no
Primary reference	Seshadri et al.[1]
Availability of cell line	not known

Clinical characterization[1]

Patient	10-year-old male
Disease diagnosis	B-NHL (diffuse lymphoblastic)
Treatment status	at relapse
Specimen site	abdominal mass

Immunophenotypic characterization[1]

T-/NK cell marker	CD2$^-$, CD3$^-$, CD4$^-$, CD5$^-$, CD7$^-$, CD8$^-$
B-cell marker	CD9$^{(+)}$, CD10$^{(+)}$, CD20$^+$, CD21$^-$, CD24$^{(+)}$, CD37$^+$, FMC-7$^+$, cyIg$^-$, sIgMκ$^+$
Myelomonocytic marker	CD14$^-$
Progenitor/activation/other marker	CD45$^+$, CD45RA$^{(+)}$, CD71$^+$, HLA-DR$^+$

Genetic characterization[1]

Cytogenetic karyotype	49, +t(3;12)(pter→p13;p13), +del(6q)(q15), +t(7;dirdup(13q))(p15;q12), +t(6;10)(pter→q13;q26), +t(8;14)(q24;q32), +t(12;?) (qter→cen;?), +dirdup(13q)(q34;q21→qter)×2, +der(16)(q24+)
Unique translocation/fusion gene	t(8;14)(q24;q32) → *MYC-IGH* genes altered?

Functional characterization[1]

Colony formation	in soft agar
Special features	secretion: IgM

Comments

- Mature EBV$^-$ B-cell line derived from B-NHL.
- Carries t(8;14)(q24;q32) which may lead to alteration of the *MYC-IGH* genes.

Reference

[1] Seshadri, R. et al. (1985) Leukemia Res. 9, 97–111.

Granta 519

Culture characterization[1,2]

Culture medium 90% DMEM + 10% FBS
Doubling time 49 h
Viral status EBV⁺, HBV⁻, HCV⁻, HHV-8⁻, HIV⁻, HTLV-I/II⁻
Authentication yes (by cytogenetics)
Primary reference Jadayel et al.[1]
Availability of cell line DSM ACC 342

Clinical characterization[1]

Patient 58-year-old female
Disease diagnosis B-NHL (leukemic transformation of mantle cell NHL)
Treatment status refractory, stage IV
Specimen site peripheral blood
Year of establishment 1991

Immunophenotypic characterization[1,2]

T-/NK cell marker $CD3^-$, $CD5^-$
B-cell marker $CD10^-$, $CD19^+$, $CD20^+$, $CD22^-$, $CD23^+$, $CD24^+$, $CD37^+$, FMC-7^+, cyIgMλ^+, sIgDMλ^+
Progenitor/activation/other marker $CD30^{(+)}$, HLA-DR^+

Genetic characterization[1-4]

Cytogenetic karyotype hypodiploid, 8% polyploidy; 44(39–44)<2n>XX, −12, −17, −18, +mar, add(1)(p22), del(3)(p14p23), i(8p), i(8q), add(9)(p22)×1–2, t(11;14)(q13;q32), add(13)(p12), add(18)(q21); sideline with two copies of der(14) and der(9)
Unique translocation/fusion gene t(11;14)(q13;q32) → *BCL1-IGH* genes altered
Unique gene alteration *P15INK4B* deletion, *P16INK4A* deletion

Functional characterization[1]

Special features overexpression: CDK4, cyclin D1

Comments

- Mature EBV⁺ B-cell line derived from B-NHL (mantle cell lymphoma).
- Carries t(11;14)(q13;q32) leading to alteration of the *BCL1* and *IGH* genes.
- Available from cell line bank.
- Cells grow well.

References

[1] Jadayel, D.M. et al. (1997) Leukemia 11, 64–72.
[2] Drexler, H.G. et al. (1999) DSMZ Catalogue of Cell Lines, 7th edn, Braunschweig, Germany.
[3] Stranks, G. et al. (1995) Blood 85, 893–901.
[4] Willis, T.G. et al. (1997) Blood 90, 2456–2464.

Hair-M

Culture characterization[1,2]

Culture medium 90% RPMI 1640 + 10% FBS
Doubling time 48 h
Viral status EBNA$^-$
Authentication no
Primary reference Matsuo et al.[1]
Availability of cell line original authors

Clinical characterization[1,2]

Patient 86-year-old male
Disease diagnosis HCL
Treatment status terminal
Specimen site peripheral blood
Year of establishment 1981

Immunophenotypic characterization[1-3]

T-/NK cell marker CD1a$^-$, CD2$^-$, CD3$^-$, CD4$^-$, CD5$^-$, CD7$^-$, CD8$^-$, CD28$^+$, CD56$^-$, CD57$^-$, TCR$\alpha\beta^-$, TCR$\gamma\delta^-$
B-cell marker CD9$^+$, CD10$^-$, CD19$^+$, CD20$^+$, CD21$^+$, CD22$^+$, CD24$^-$, FMC-7$^+$, NU-B1$^+$, cyIgGκ^+, sIgGκ^+
Myelomonocytic marker CD13$^-$, CD14$^-$, CD15$^+$, CD16$^-$, CD33$^-$
Erythroid-megakaryocytic marker CD41a$^-$, CD61$^-$
Progenitor/activation/other marker CD34$^-$, CD38$^+$, CD71$^+$, HLA-DR$^+$, TdT$^-$
Adhesion marker CD11b$^{(+)}$
Cytokine receptor CD25$^+$

Genetic characterization[1,3]

Cytogenetic karyotype 46, XY, −11, −14
Receptor gene rearrangement *TCRG* GG, *TCRD* GG

Functional characterization[1]

Cytochemistry ACP$^+$, ALP$^-$, ANAE$^-$, ANBE$^-$, CAE$^-$, GLC$^+$, MPO$^-$, PAS$^+$, SBB$^-$, TRAP$^-$
Special features phagocytosis$^{(+)}$

Comments

- Mature EBV$^-$ B-cell line derived from HCL.

References

[1] Matsuo, Y. et al. (1986) J. Natl. Cancer Inst. 76, 207–216.
[2] Matsuo, Y. et al. (1983) Acta Haematol. Jap. 46, 1222–1232.
[3] Minowada, J. and Matsuo, Y. (1999) unpublished data.

HBL-1

Culture characterization[1,2]

Culture medium	90% RPMI 1640 + 10% FBS
Doubling time	48 h
Viral status	EBNA$^-$
Authentication	yes (by cytogenetics)
Primary reference	Abe et al.[1]
Availability of cell line	original authors

Clinical characterization[1,2]

Patient	65-year-old male
Disease diagnosis	B-NHL (diffuse large cell)
Treatment status	progressive, terminal
Specimen site	pleural effusion
Year of establishment	1985

Immunophenotypic characterization[1,2]

T-/NK cell marker	CD1$^-$, CD2$^-$, CD3$^-$, CD4$^-$, CD5$^-$, CD8$^-$, CD57$^-$
B-cell marker	CD10$^-$, CD20$^+$, CD21$^-$, CD24$^+$, NU-B1$^-$, cyIgMκ$^+$, sIgMκ$^+$
Myelomonocytic marker	CD15$^-$
Progenitor/activation/other marker	CD30$^-$, HLA-DR$^+$
Cytokine receptor	CD25$^-$

Genetic characterization[1,2]

Cytogenetic karyotype	44, X, −Y, −6, −7, −12, −13, −14, −16, −17, +der(3)t(3;13)(p11.2;p12), del(4)(q31.1), +der(6)t(6;14;16)(p21.3;q24.3; p13.1), psudic(12)t(12;7)(p13.3;p11.2), +der(14)t(14;17) (q24.3;p11.2), +psudic(16)t(16;13)(p13.1;p12), +der(17)t(6;17) (p21.3;p11.2), i(18q)

Functional characterization[1,2]

Cytochemistry	ACP$^+$, ALP$^-$, ANBE$^-$, CAE$^-$, MPO$^-$, PAS$^-$, SBB$^-$
Heterotransplantation	into nude mice

Comments

- Mature EBV$^-$ B-cell line derived from B-NHL (DLCL).
- Note that there is another mature B-cell line termed **HBL-1** which was derived independently from an unrelated patient in New York, NY, USA [reference: Gaidano, G. et al. (1993) Leukemia 7, 1621–1629].

References

[1] Abe, M. et al. (1988) Cancer 61, 483–490.
[2] Nozawa, Y. et al. (1988) Tohoku J. Exp. Med. 156, 319–330.

Culture characterization[1]

Culture medium	90% IMDM + 10% FBS
Viral status	EBV+, HIV−, HTLV-I−
Authentication	yes (by *IGH, IGK* rearrangement, *P53* mutation)
Primary reference	Gaidano et al.[1]
Availability of cell line	original authors

Clinical characterization[1]

Patient	32-year-old female
Disease diagnosis	B-NHL (small non-cleaved cell) + AIDS
Treatment status	stage IV, at diagnosis, prior to therapy
Specimen site	peripheral blood

Immunophenotypic characterization[1]

T-/NK cell marker	CD3−, CD4−, CD5−, CD8−
B-cell marker	CD9+, CD10+, CD19+, CD20+, CD21−, CD22+, CD23+, CD24+, sIgMDκ+
Progenitor/activation/other marker	CD30−, CD38+, CD71+, HLA-DR+, TdT−
Cytokine receptor	CD25+

Genetic characterization[1]

Cytogenetic karyotype	46, XX, t(8;14)(q24;q32), der(13)t(1;13)(q25;q32)
Unique translocation/fusion gene	t(8;14)(q24;q32) → *MYC-IGH* genes altered?
Receptor gene rearrangement	*IGH* R, *IGK* R
Unique gene alteration	*MYC* germline, *P53* mutation

Comments

- Mature EBV+ HIV− B-cell line derived from AIDS-NHL.
- Cellular derivation and neoplastic nature verified.
- Carries t(8;14)(q24;q32) possibly causing alteration of the *MYC-IGH* genes.
- Note that there is another mature B-cell line termed **HBL-1** which was derived independently from an unrelated patient in Fukushima, Japan in 1985 [reference: Abe, M. et al. (1988) Cancer 61, 483–490].

Reference

[1] Gaidano, G. et al. (1993) Leukemia 7, 1621–1629.

HBL-2

Culture characterization[1]

Culture medium	90% RPMI 1640 + 10% FBS
Doubling time	36 h
Viral status	EBNA⁻
Authentication	no
Primary reference	Abe et al.[1]
Availability of cell line	not known

Clinical characterization[1]

Patient	84-year-old male
Disease diagnosis	B-NHL (diffuse large cell)
Treatment status	at diagnosis
Specimen site	lymph node
Year of establishment	1985

Immunophenotypic characterization[1,2]

T-/NK cell marker	CD1⁻, CD3⁻, CD5⁻, CD57⁻
B-cell marker	CD10⁻, CD20⁺, CD21⁻, NU-B1⁻, sIgMDλ⁺
Myelomonocytic marker	CD15⁻
Progenitor/activation/other marker	CD30⁻, HLA-DR⁺
Cytokine receptor	CD25⁻

Genetic characterization[1-3]

Cytogenetic karyotype	46, X, −Y, +7, der(?)t(3;?)(p13;?), der(4)t(1;4)(q21;p15.2), t(11;6;9)(q11;p23q21;p22), add(8)(p23), der(9)t(3;9)(q21;p22), del(9)(q32), add(11)(q23), der(11)t(?;11;14)(?;q13;q32), add(13)(q23), der(14)t(11;14)(q12.2;q32.3), t(11;14)(q32.3;q15), add(15) (p12), add(17)(p13), add(18)(q23), der(22)t(9;22)(q11;p12)
Unique translocation/fusion gene	t(11;14)(q13;q32) → *BCL1-IGH* genes altered
Unique gene alteration	*BCL2* amplification on add(18)(q23)

Functional characterization[1,4]

Cytochemistry	ACP⁻, ALP⁻, ANBE⁻, CAE⁻, MPO⁻
Heterotransplantation	into nude mice
Proto-oncogene	mRNA overexpression: BCL2, BCL6

Comments

- Mature EBV⁻ B-cell line derived from B-NHL (DLCL).
- Carries t(11;14)(q13;q32) leading to alteration of the *BCL1* and *IGH* genes.
- Note that there is another mature B-cell line termed **HBL-2** which was derived independently from an unrelated patient in New York, NY, USA [reference: Gaidano, G. et al. (1993) Leukemia 7, 1621–1629].

References
[1] Abe, M. et al. (1988) Cancer 61, 483–490.
[2] Taniwaki, M. et al. (1995) Blood 85, 3223–3228.
[3] Taniwaki, M. et al. (1995) Blood 86, 1481–1486.
[4] Yonetani, N. et al. (1998) Oncogene 17, 971–979.

HBL-2

Culture characterization[1]

Culture medium	90% IMDM + 10% FBS
Viral status	EBV$^-$, HIV$^-$, HTLV-I$^-$
Authentication	yes (by *IGH, IGK* rearrangement, *P53* mutation)
Primary reference	Gaidano et al.[1]
Availability of cell line	original authors

Clinical characterization[1]

Patient	34-year-old male
Disease diagnosis	B-NHL (small non-cleaved cell) + AIDS
Treatment status	stage IV, at diagnosis, prior to therapy
Specimen site	pleural effusion

Immunophenotypic characterization[1]

T-/NK cell marker	CD3$^-$, CD4$^-$, CD5$^-$, CD8$^-$
B-cell marker	CD9$^-$, CD10$^+$, CD19$^+$, CD20$^+$, CD21$^-$, CD22$^+$, CD23$^-$, CD24$^+$, sIgMλ^+
Progenitor/activation/other marker	CD30$^-$, CD38$^+$, CD71$^+$, HLA-DR$^+$, TdT$^-$
Cytokine receptor	CD25$^-$

Genetic characterization[1]

Cytogenetic karyotype	47, XY, +21, t(8;14)(q24;q32)
Unique translocation/fusion gene	t(8;14)(q24;q32) → *MYC-IGH* genes altered
Receptor gene rearrangement	*IGH* R, *IGK* R
Unique gene alteration	*P53* mutation

Comments

- Mature B-cell line derived from AIDS-NHL.
- Cells are EBV$^-$, HIV$^-$.
- Carries t(8;14)(q24;q32) which leads to alteration of the *MYC-IGH* genes.
- Note that there is another mature B-cell line termed **HBL-2** which was derived independently from an unrelated patient in Fukushima, Japan, in 1985 [reference: Abe, M. et al. (1988) Cancer 61, 483–490].

Reference

[1] Gaidano, G. et al. (1993) Leukemia 7, 1621–1629.

HBL-3

Culture characterization[1]

Culture medium	90% IMDM + 10% FBS
Viral status	EBV$^-$, HIV$^-$, HTLV-I$^-$
Authentication	yes (by *IGH, IGK* rearrangement, *P53* mutation)
Primary reference	Gaidano et al.[1]
Availability of cell line	original authors

Clinical characterization[1]

Patient	36-year-old male
Disease diagnosis	B-NHL (small non-cleaved cell) + AIDS
Treatment status	stage III, at diagnosis, prior to therapy
Specimen site	liver

Immunophenotypic characterization[1]

T-/NK cell marker	CD3$^-$, CD4$^-$, CD5$^-$, CD8$^-$
B-cell marker	CD9$^-$, CD10$^+$, CD19$^+$, CD20$^+$, CD21$^-$, CD22$^+$, CD23$^-$, CD24$^-$, sIgMDλ$^+$
Progenitor/activation/other marker	CD30$^-$, CD38$^-$, CD71$^+$, HLA-DR$^+$, TdT$^-$
Cytokine receptor	CD25$^-$

Genetic characterization[1]

Cytogenetic karyotype	49, XY, +7, +12, del(2)(q22orq24), t(8;22)(q24;q11), +add(11) (p15)
Unique translocation/fusion gene	t(8;22)(q24;q11) → *MYC-IGL* genes altered
Receptor gene rearrangement	*IGH* R, *IGK* R
Unique gene alteration	*P53* mutation

Comments

- Mature B-cell line derived from AIDS-NHL.
- Cells are EBV$^-$, HIV$^-$.
- Carries t(8;22)(q24;q11) leading to alteration of *MYC* and *IGL* genes.
- Note that there is a precursor B-cell line also termed **HBL-3** which was derived independently from an unrelated patient in Fukushima, Japan, in 1985 [reference: Abe, M. et al. (1990) Virchows Archiv B Cell. Pathol. 59, 107–113].

Reference

[1] Gaidano, G. et al. (1993) Leukemia 7, 1621–1629.

HC1

Culture characterization[1,2]

Culture medium	80% RPMI 1640 + 20% FBS
Doubling time	40–45 h
Viral status	EBV⁺, HBV⁻, HCV⁻, HHV-8⁻, HIV⁻, HTLV-I/II⁻
Authentication	no
Primary reference	Schiller et al.[1]
Availability of cell line	DSM ACC 301

Clinical characterization[1]

Patient	56-year-old male
Disease diagnosis	HCL
Treatment status	at diagnosis
Specimen site	peripheral blood

Immunophenotypic characterization[1,2]

T-/NK cell marker	$CD2^-$, $CD3^-$, $CD4^-$, $CD8^-$
B-cell marker	$CD10^-$, $CD19^+$, $CD20^+$, $CD22^+$, $CD37^+$, FMC-7^+, PCA-1^+, $sIgA\kappa^+$
Myelomonocytic marker	$CD13^-$
Progenitor/activation/other marker	HLA-DR^+
Adhesion marker	$CD11c^+$
Cytokine receptor	$CD25^{(+)}$, $CD118^+$

Genetic characterization[1,2]

Cytogenetic karyotype	near-tetraploid, 3% polyploidy; 92(89–92)<4n>XXYY, add(2)(p13→23); differs from originally published karyotype
Receptor gene rearrangement	*IGH* R, *IGK* R

Functional characterization[1]

Colony formation	in soft agar
Cytochemistry	$TRAP^-$
Cytokine response	growth inhibition: IFN-α, IFN-β
Heterotransplantation	into nude mice

Comments

- Mature EBV^+ B-cell line derived from HCL.
- Neoplastic nature confirmed, but derivation not authenticated.
- Available from cell line bank.

References

[1] Schiller, J.H. et al. (1991) Leukemia 5, 399–407.

[2] Drexler, H.G. et al. (1999) DSMZ Catalogue of Cell Cultures, 7th edn, Braunschweig, Germany.

HF-1

Culture characterization[1]

Establishment	initially frozen sample rescued on feeder cells
Culture medium	90% DMEM + 10% FBS
Authentication	yes (by cytogenetics)
Primary reference	Knuutila et al.[1]
Availability of cell line	not known

Clinical characterization[1]

Patient	64-year-old female
Disease diagnosis	B-NHL (nodular and diffuse large non-cleaved, centroblastic lymphoma)
Treatment status	stage IVB, in partial remission
Specimen site	lymph node
Year of establishment	1990

Immunophenotypic characterization[1]

B-cell marker	$CD19^+$, $CD20^+$, $CD22^+$, $CD39^+$, $sIgG\kappa^+$
Progenitor/activation/other marker	$CD45^+$, HLA-DR^+

Genetic characterization[1]

Cytogenetic karyotype	52, XX, +X, +12, +21, t(2;8)(p12;q24), +?i(5)(p10), +?i(5) (p10), del(7)(q35), +add(7)(q32), add(9)(q34), t(14;18)(q32;q21)
Unique translocation/fusion gene	t(14;18)(q32;q21) → *IGH-BCL2* genes altered (MBR)

Functional characterization[1]

Proto-oncogene	mRNA overexpression: BCL2, MYC

Comments

- Mature B-cell line derived from B-NHL.
- Carries t(14;18)(q32;q21) leading to alteration of the *IGH* and *BCL2* genes (major breakpoint region).

Reference

[1] Knuutila, S. et al. (1994) Eur. J. Haematol. 52, 65–72.

HF-4a

Culture characterization[1]

Sister cell line — **HF-4b** (serial sister cell line – from lymph node at less aggressive disease stage – similar features, some cytogenetic differences)
Establishment — initially frozen samples rescued on feeder cells
Culture medium — 90% DMEM + 10% FBS
Authentication — yes (by cytogenetics)
Primary reference — Knuutila et al.[1]
Availability of cell line — original authors

Clinical characterization[1]

Patient — 62-year-old female
Disease diagnosis — B-NHL (mantle cell lymphoma, anaplastic variant)
Treatment status — progressive, terminal
Specimen site — lymph node
Year of establishment — 1990

Immunophenotypic characterization[1]

B-cell marker — CD19+, CD20+, CD22+, CD39+, sIgMDκ+
Progenitor/activation/other marker — CD45+, HLA-DR+

Genetic characterization[1]

Cytogenetic karyotype — 46, X, +7, −13, −15, add(X)(q22), t(1;6)(1qter→1p32::6q15→6qter;6pter→6p15), t(1;8)(p21;q24), t(3;19)(q28;p13.2), inv(4) (p14p16), del(5)(q13q22), del(10)(q24), +der(14)t(14;18) (q32;q21), der(18)t(14;18)(q32;q21), +der(19)t(3;19)(q28;p13.2), der(22)t(13;22)(q12or14;q11or13)
Unique translocation/fusion gene — t(14;18)(q32;q21)→*IGH-BCL2* genes altered (MBR)

Functional characterization[1]

Proto-oncogene — mRNA overexpression: BCL2, MYC

Comments

- Mature B-cell line derived from B-NHL (MCL).
- Carries t(14;18)(q32;q21) leading to alteration of the *IGH* and *BCL2* genes (major breakpoint region).

Reference

[1] Knuutila, S. et al. (1994) Eur. J. Haematol. 52, 65–72.

Culture characterization[1,2]

Establishment	by serial xenotransplantation into mice
Culture medium	90% RPMI 1640 + 10% FBS
Doubling time	24 h
Viral status	EBNA$^-$
Authentication	no
Primary reference	Kopper et al.[1]
Availability of cell line	original authors

Clinical characterization[1]

Patient	12-year-old male
Disease diagnosis	B-NHL (diffuse centroblastic)
Treatment status	at diagnosis
Specimen site	intestinal tumor tissue
Year of establishment	1984

Immunophenotypic characterization[1,2]

T-/NK cell marker	CD3$^-$
B-cell marker	CD10$^+$, CD19$^+$, CD20$^+$, CD21$^+$, CD22$^+$, CD24$^+$, CD37$^+$, sIgMλ$^+$
Myelomonocytic marker	CD14$^-$
Progenitor/activation/other marker	CD38$^+$, CD71$^+$, HLA-DR$^+$
Cytokine receptor	CD25$^-$

Genetic characterization[2]

Cytogenetic karyotype	46, XY, −1, −2, −3, −14, +der(1), +der(2), +der(3), +der(14), ins(1;3)(q11;p25−p26), t(2;14)(p13;q24)

Functional characterization[1,2]

Heterotransplantation	into nude mice
Special features	secretion: IgM, chondroitin sulfate proteoglycans

Comments

- Mature EBV$^-$ B-cell line derived from B-NHL.
- First established as serially transplantable tumor in mice.

References

[1] Kopper, L. et al. (1987) Anticancer Res. 7, 193–198.
[2] Kopper, L. et al. (1991) Anticancer Res. 11, 1645–1650.

JeKo-1

Culture characterization[1]

Culture medium	80% RPMI 1640, Ham's F12, Hanks balanced salt solution (2:1:1) + 20% FBS
Doubling time	33 h
Viral status	EBV⁻
Authentication	no
Primary reference	Jeon et al.[1]
Availability of cell line	original authors

Clinical characterization[1]

Patient	78-year-old female
Disease diagnosis	B-NHL (mantle cell lymphoma, large cell variant in leukemic conversion)
Specimen site	peripheral blood

Immunophenotypic characterization[1]

T-/NK cell marker	CD3⁻, CD4⁻, CD5⁺, CD8⁻, CD56⁻
B-cell marker	CD10⁻, CD19⁺, CD20⁺, CD22⁺, CD23⁻, cyIgMDκ⁺, sIgMDκ⁺
Myelomonocytic marker	CD14⁻
Progenitor/activation/other marker	CD45RA⁺, HLA-DR⁺

Genetic characterization[1]

Cytogenetic karyotype	40–41, X0, −2, −6, −8, −9, −12, −13, −14, −14, −16, −16, −20, −22, +6mar, +add(1)(p13), add(1)(q12), add(3)(q27), add(5)(p13), add(6)(q12), add(7)(q22), add(9)(q34), add(10)(p15), add(11) (p11), add(21)(p13)
Unique translocation/fusion gene	*BCL1-IGH* genes altered without t(11;14)(q13;q32)

Functional characterization[1]

Heterotransplantation	into SCID mice
Special features	overexpression: BCL2, cyclin D1, MYC, RB1

Comments

- Mature EBV⁻ B-cell line derived from B-NHL (MCL).
- Carries *BCL1-IGH* gene rearrangement without apparent t(11;14)(q13;q32).
- Cells are difficult to culture.

Reference

[1] Jeon, H.J. et al. (1998) Brit. J. Haematol. 102, 1323–1326.

JVM-2

Culture characterization[1,2]

Establishment	by exposure to EBV and TPA
Culture medium	90% RPMI 1640 + 10% FBS
Doubling time	50–72 h
Viral status	EBV$^+$, HBV$^-$, HCV$^-$, HHV-8$^-$, HIV$^-$, HTLV-I/II$^-$
Authentication	yes (by *IGH*, *IGL* rearrangement; cytogenetics)
Primary reference	Melo et al.[1]
Availability of cell line	DSM ACC 12

Clinical characterization[1]

Patient	63-year-old female
Disease diagnosis	B-PLL
Treatment status	at diagnosis
Specimen site	peripheral blood
Year of establishment	1984

Immunophenotypic characterization[1,2]

T-/NK cell marker	CD3$^-$, CD5$^+$
B-cell marker	CD9$^+$, CD10$^-$, CD19$^+$, CD20$^+$, CD21$^+$, CD37$^+$, CD138$^+$, FMC-7$^+$, cyIgMλ$^+$, sIgMλ$^+$
Myelomonocytic marker	CD13$^-$
Progenitor/activation/other marker	CD30$^+$, CD34$^-$, CD38$^+$, CD71$^+$, HLA-DR$^+$
Cytokine receptor	CD25$^+$

Genetic characterization[1–3]

Cytogenetic karyotype	pseudodiploid, 1.8% polyploidy; 46(42–46)<2n>XX, t(1;15)(q32.3;q22.1), der(8)t(3;8)(q13;p21), t(11;14)(q13;q32), dup(16) (p11p13.2)
Unique translocation/fusion gene	t(11;14)(q13;q32) → *BCL1-IGH* genes altered (MTC)
Receptor gene rearrangement	*IGH* RR, *IGK* GG, *IGL* R

Functional characterization[1]

Special features	secretion: IgM

Comments

- Mature EBV$^+$ B-cell line derived from B-PLL.
- Carries t(11;14)(q13;q32) leading to alteration of the *BCL1* and *IGH* (in the major translocation cluster).
- Cellular derivation and neoplastic nature verified.
- Available from cell line bank.

References

[1] Melo, J.V. et al. (1986) Int. J. Cancer 38, 531–538.
[2] Drexler, H.G. et al. (1999) DSMZ Catalogue of Cell Lines, 7th edn, Braunschweig, Germany.
[3] Melo, J.V. et al. (1988) Clin. Exp. Immunol. 73, 23–28.

JVM-3

Culture characterization[1,2]

Establishment	by exposure to EBV and TPA
Culture medium	90% RPMI 1640 + 10% FBS
Doubling time	60–72 h
Viral status	EBV⁺, HBV⁻, HCV⁻, HHV-8⁻, HIV⁻, HTLV-I/II⁻
Authentication	yes (by *IGH, IGK* rearrangement; cytogenetics)
Primary reference	Melo et al.[1]
Availability of cell line	DSM ACC 18

Clinical characterization[1,3]

Patient	73-year-old male
Disease diagnosis	B-PLL
Treatment status	at diagnosis
Specimen site	peripheral blood
Year of establishment	1984

Immunophenotypic characterization[1,2]

T-/NK cell marker	$CD3^-$, $CD5^+$
B-cell marker	$CD9^+$, $CD10^-$, $CD19^+$, $CD20^+$, $CD21^{(+)}$, $CD37^+$, FMC-7^+, cyIgMDκ^+, sIgMDκ^+
Myelomonocytic marker	$CD13^-$
Progenitor/activation/other marker	$CD30^+$, $CD38^+$, $CD71^+$, HLA-DR^+
Cytokine receptor	$CD25^+$

Genetic characterization[1,2,4]

Cytogenetic karyotype	near tetraploid, no clear mode; 87–93<4n>XXYY, +3, −7, −8, −10, +12, +12, −14, −19, −19, −20, −20
Receptor gene rearrangement	*IGH* RR, *IGK* R

Functional characterization[1]

Special features	secretion: IgM

Comments

- Mature EBV⁺ B-cell line derived from B-PLL.
- Cellular derivation and neoplastic nature verified.
- Available from cell line bank.

References

[1] Melo, J.V. et al. (1986) Int. J. Cancer 38, 531–538.

[2] Drexler, H.G. et al. (1999) DSMZ Catalogue of Cell Lines, 7th edn, Braunschweig, Germany.

[3] Robinson, D.S.F. et al. (1985) J. Clin. Pathol. 38, 897–903. (case report)

[4] Melo, J.V. et al. (1988) Clin. Exp. Immunol. 73, 23–28.

Culture characterization[1,2]

Establishment	by exposure to EBV and TPA
Culture medium	90% RPMI 1640 + 10% FBS
Doubling time	60–72 h
Viral status	EBV^+, HBV^-, HCV^-, $HHV\text{-}8^-$, HIV^-, $HTLV\text{-}I/II^-$
Authentication	yes (by *IGH, IGL* rearrangement)
Primary reference	Melo et al.[1]
Availability of cell line	DSM ACC 19

Clinical characterization[1]

Patient	male
Disease diagnosis	B-PLL
Treatment status	at diagnosis
Specimen site	peripheral blood
Year of establishment	1985

Immunophenotypic characterization[1,2]

T-/NK cell marker	$CD3^-$, $CD5^-$
B-cell marker	$CD9^+$, $CD10^-$, $CD19^+$, $CD20^+$, $CD21^+$, $CD37^+$, $FMC\text{-}7^+$, $HC\text{-}2^-$, $cyIgM\lambda^+$, $sIgMD\lambda^+$
Myelomonocytic marker	$CD13^-$
Progenitor/activation/other marker	$CD38^+$, $CD71^+$, $HLA\text{-}DR^+$
Cytokine receptor	$CD25^-$

Genetic characterization[1-3]

Cytogenetic karyotype	hyperdiploid, 4% tetraploidy; 47(44–48)<2n>X, +8, −13, t(2;?)(q35;?), t(8;13)(p12;q13), t(8;?)(p12;?)
Receptor gene rearrangement	*IGH* RR, *IGL* R
Unique gene alteration	*P15INK4B* deletion

Comments

- Mature EBV^+ B-cell line derived from B-PLL.
- Cellular derivation and neoplastic nature verified.
- Available from cell line bank.

References

[1] Melo, J.V. et al. (1988) Clin. Exp. Immunol. 73, 23–28.

[2] Drexler, H.G. et al. (1999) DSMZ Catalogue of Cell Lines, 7th edn, Braunschweig, Germany.

[3] Aguiar, R.C.T. et al. (1997) Leukemia 11, 233–238.

KAL-1

Culture characterization[1]

Culture medium	90% IMDM + 10% FBS or serum-free
Doubling time	24 h
Viral status	EBV$^-$
Authentication	no
Primary reference	Ichinose et al.[1]
Availability of cell line	not known

Clinical characterization[1]

Patient	37-year-old male
Disease diagnosis	B-NHL (ileocecal, diffuse large cell)
Treatment status	refractory, terminal
Specimen site	pleural effusion
Year of establishment	1988

Immunophenotypic characterization[1]

T-/NK cell marker	CD2$^-$, CD3$^-$, CD4$^-$, CD5$^-$, CD7$^-$, CD8$^-$
B-cell marker	CD10$^+$, CD19$^+$, CD20$^+$, CD21$^-$, sIgMλ$^+$
Progenitor/activation/other marker	HLA-DR$^+$, TdT$^-$

Genetic characterization[1]

Cytogenetic karyotype	46, XY, dup(1)(q21→q32), t(8;22)(q24;q11)
Unique translocation/fusion gene	t(8;22)(q24;q11) → *MYC-IGL* genes altered?
Receptor gene rearrangement	*IGH* R, *IGL* R, *TCRB* G, *TCRG* G

Functional characterization[1]

Cytochemistry	ACP$^-$, ALP$^-$, MPO$^-$, PAS$^+$
Heterotransplantation	into nude mice
Proto-oncogene	MYC mRNA overexpression

Comments

- Mature EBV$^-$ B-cell line derived from B-NHL (DLCL).
- Carries t(8;22)(q24;q11) possibly leading to alteration of *MYC* and *IGL* genes.

Reference

[1] Ichinose, I. et al. (1991) Cancer Res. 51, 5392–5397.

Karpas 231

Culture characterization[1]

Culture medium	90% RPMI 1640 + 10% FBS
Viral status	EBV$^-$
Authentication	yes (by *IGH* rearrangement)
Primary reference	Nacheva et al.[1]
Availability of cell line	original authors

Clinical characterization[1]

Patient	71-year-old female
Disease diagnosis	B-ALL (L3)
Treatment status	at diagnosis
Specimen site	peripheral blood
Year of establishment	1988

Immunophenotypic characterization[1]

T-/NK cell marker	CD2$^-$, CD3$^-$, CD5$^-$
B-cell marker	CD10$^-$, CD19$^+$, CD20$^+$, CD22$^+$, FMC-7$^-$, sIgMλ$^+$
Myelomonocytic marker	CD13$^-$, CD33$^-$
Progenitor/activation/other marker	CD30$^-$, CD34$^-$, CD45$^+$, HLA-DR$^+$, TdT$^+$
Cytokine receptor	CD25$^-$

Genetic characterization[1–4]

Cytogenetic karyotype	51, XX, +7, +12, +20, t(1;3;11)(q42.3;q27.1;q23.1), der(8)t(8;9)(q24.2;p13.3), der(8)t(8;9)(q24.2;p13.3), der(9)t(8;9)(q24.2;p13.3)×2, intdel13q(q12.3q21.2), t(14;18)(q32.3;q21.3), der(17)t(17;?)(p13;?)
Unique translocation/fusion gene	(1) t(14;18)(q32;q21) → *IGH-BCL2* genes altered (MBR) (2) t(3;11)(q27;q23.1) → *LAZ3/BCL6-BOB1/OBF1* fusion gene
Receptor gene rearrangement	*IGH* RR, *IGK* DD, *IGL* RR, *TCRB* GG, *TCRG* GG, *TCRD* DD
Unique gene alteration	*MYC* amplification

Functional characterization[1]

Proto-oncogene	MYC mRNA overexpression; BCL2$^+$

Comments

- Mature EBV$^-$ B-cell line derived from B-ALL (L3).
- Carries t(14;18)(q32;q21) leading to alteration of the *IGH* and *BCL2* genes (major breakpoint region).
- Carries also t(3;11)(q27;q23.1) leading to *LAZ3/BCL6-BOB1/OBF1* fusion gene.

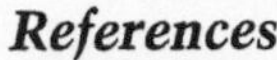

References

[1] Nacheva, E. et al. (1993) Blood 82, 231–240.
[2] Galiègue-Zouitina, S. et al. (1996) Leukemia 10, 579–587.
[3] Yuille, M.A.R. et al. (1996) Leukemia 10, 1492–1496.
[4] Willis, T.G. et al. (1997) Blood 90, 2456–2464.

Karpas 353

Culture characterization[1]

Culture medium	90% RPMI 1640 + 10% FBS
Viral status	EBV$^-$
Authentication	yes (by *IGH* rearrangement)
Primary reference	Nacheva et al.[1]
Availability of cell line	original authors

Clinical characterization[1]

Patient	60-year-old female
Disease diagnosis	B-ALL (L2)
Treatment status	at diagnosis
Specimen site	peripheral blood
Year of establishment	1988

Immunophenotypic characterization[1]

T-/NK cell marker	CD2$^-$, CD3$^-$, CD5$^-$, CD7$^-$
B-cell marker	CD10$^+$, CD19$^+$, CD20$^+$, CD22$^-$, FMC-7$^-$, sIgMλ^+
Myelomonocytic marker	CD13$^-$, CD33$^-$
Progenitor/activation/other marker	CD30$^-$, CD34$^-$, CD45$^+$, HLA-DR$^+$, TdT$^+$
Cytokine receptor	CD25$^-$

Genetic characterization[1–3]

Cytogenetic karyotype	47, +12, t(1;3;7)(p32.1;q21.1;q22.1), t(8;9)(q24.2;p13.3), t(14;18)(q32.3;q21.3)
Unique translocation/fusion gene	t(14;18)(q32;q21) → *IGH-BCL2* genes altered (MBR)
Receptor gene rearrangement	*IGH* RR, *IGK* DD, *IGL* RG, *TCRB* GG, *TCRG* GG, *TCRD* GG
Unique gene alteration	*P16INK4A* deletion

Functional characterization[1]

Proto-oncogene	MYC mRNA overexpression; BCL2$^+$

Comments

- Mature EBV$^-$ B-cell line derived from B-ALL.
- Carries t(14;18)(q32;q21) leading to alteration of the *IGH* and *BCL2* genes (major breakpoint region).

References

[1] Nacheva, E. et al. (1993) Blood 82, 231–240.
[2] Stranks, G. et al. (1995) Blood 85, 893–901.
[3] Willis, T.G. et al. (1997) Blood 90, 2456–2464.

Karpas 422

Culture characterization[1,2]

Other name of cell line	**K422**
Culture medium	80–90% RPMI 1640 + 10–20% FBS
Doubling time	60–90 h
Viral status	EBNA⁻, EBV⁺, HBV⁻, HCV⁻, HHV-8⁻, HIV⁻, HTLV-I/II⁻
Authentication	yes (by cytogenetics)
Primary reference	Dyer et al.[1]
Availability of cell line	DSM ACC 32

Clinical characterization[1]

Patient	73-year-old female
Disease diagnosis	B-NHL (intraabdominal; diffuse large cell)
Treatment status	refractory, terminal
Specimen site	pleural effusion
Year of establishment	1987

Immunophenotypic characterization[1–3]

T-/NK cell marker	CD1⁻, CD2⁻, CD3⁻, cyCD3⁻, CD4⁻, CD5⁻, CD7⁻, CD8⁻
B-cell marker	CD10⁺, CD19⁺, CD20⁺, CD37⁺, cyIgM⁺, sIgMGκ⁽⁺⁾
Myelomonocytic marker	CD13⁻, CD14⁻, CD15⁻, CD33⁺
Erythroid-megakaryocytic marker	CD41b⁻
Progenitor/activation/other marker	CD34⁻, CD38⁻, CD45⁺, CD45RC⁺, HLA-DR⁺
Cytokine receptor	CD25⁻

Genetic characterization[1–6]

Cytogenetic karyotype	hyperdiploid, 10% polyploidy; 47(44–48)<2n>XX, +14, t(2;10)(p23;q22), t(4;11)(q21;q24), t(4;16)(q21;p13), der(14)t(14;18)(q32;q21)×2; FISH: t(4;16;10)(q21;p13;q23), no t(4;11)(q21;q23)
Unique translocation/fusion gene	t(14;18)(q32;q21) → *IGH-BCL2* genes altered (MBR)
Receptor gene rearrangement	*IGH* RD, *IGK* RR, *TCRB* RG, *TCRG* G

Functional characterization[1]

Cytochemistry	MPO⁻

Comments

- Mature B-cell line derived from B-NHL (DLCL).
- Carries t(14;18)(q32;q21) leading to alteration of the *IGH* and *BCL2* genes (major breakpoint region).
- Available from cell line bank.
- Cells are difficult to culture.

References

[1] Dyer, M.J.S. et al. (1990) Blood 75, 709–714.

[2] Drexler, H.G. et al. (1999) DSMZ Catalogue of Cell Lines, 7th edn, Braunschweig, Germany.

[3] Iida, S. et al. (1992) Leukemia Res. 16, 1155–1163.

[4] Kobayashi, H. et al. (1993) Blood 81, 3027–3033.

[5] Kearney, L. et al. (1992) Blood 80, 1659–1665.

[6] Farrugia, M.M. et al. (1994) Blood 83, 191–198.

Karpas 1106P

Culture characterization[1]

Sister cell line	**Karpas 1106A** (simultaneous sister cell line – from ascites – clonal identity confirmed – identical features)
Culture medium	90% RPMI 1640 + 10% FBS
Authentication	no
Primary reference	Nacheva et al.[1]
Availability of cell line	original authors

Clinical characterization[1]

Patient	23-year-old female
Disease diagnosis	B-NHL (mediastinal lymphoblastic)
Treatment status	at relapse
Specimen site	pleural effusion
Year of establishment	1984

Immunophenotypic characterization[1]

T-/NK cell marker	CD5⁻
B-cell marker	CD10⁻, CD19⁺, CD22⁺, CD23⁻, CD37⁺, FMC-7⁺, sIgGλ⁺

Genetic characterization[1,2]

Cytogenetic karyotype	49, X, −20, del(2)(p11.2p13.3), der(3)t(2;3)(p13.3;p25.1), +i(9p), ins(12;?)(q13.1q13.3), del(14)(q11.2q13.1), del(15)(q11.2q15.3), der(18)t(X;13;18)(q28;q12.1;q21.3), del(20)(q13.1q13.3)×2, der(X)t(X;13;18)(q28;q12.1;q21.3), +i(Xp)
Unique translocation/fusion gene	no *BCL2* rearrangement at 18q21
Receptor gene rearrangement	*IGH* RR, *IGK* DD, *IGL* RR
Unique gene alteration	*P16INK4A* deletion

Comments

- Mature B-cell line derived from mediastinal lymphoblastic B-NHL.
- Carries unusual three-way translocation t(X;13;18)(q28;q12.1;q21.3) not involving *BCL2* gene.
- Sister cell lines from pleural effusion and ascites established.

References

[1] Nacheva, E. et al. (1994) Blood 84, 3422–3428.
[2] Stranks, G. et al. (1995) Blood 85, 893–901.

KHM-2B

Culture characterization[1]

Culture medium	90% RPMI 1640 + 10% FBS
Authentication	yes (by *BCL2, IGH, MYC* rearrangements)
Primary reference	Matsuzaki et al.[1]
Availability of cell line	original authors

Clinical characterization[1]

Patient	62-year-old male
Disease diagnosis	B-ALL (L3)
Treatment status	at diagnosis
Specimen site	peripheral blood
Year of establishment	1988

Immunophenotypic characterization[1]

T-/NK cell marker	CD2⁻, CD3⁻, CD4⁻, CD8⁻
B-cell marker	CD10⁺, CD19⁺, CD20⁺, sIgMλ⁺
Progenitor/activation/other marker	HLA-DR⁺

Genetic characterization[1,2]

Cytogenetic karyotype	46, XY, −21, t(8;14)(q24;q32), t(14;18)(q32;q21), +der(21) t(21;?)(p12;?)
Unique translocation/fusion gene	(1) t(8;14)(q24;q32) → *MYC-IGH* genes altered (2) t(14;18)(q32;q21) → *IGH-BCL2* genes altered
Receptor gene rearrangement	*IGH* RR

Functional characterization[1]

Proto-oncogene	mRNA⁺: BCL2, MYC

Comments

- Mature B-cell line established from B-ALL (L3).
- Carries both t(8;14)(q24;q32) and t(14;18)(q32;q21) leading to alterations of the *MYC-IGH* genes and *IGH-BCL2* genes.

References

[1] Matsuzaki, H. et al. (1990) Acta Haematol. 84, 156–161.
[2] van Ooteghem, R.B.C. et al. (1994) Cancer Genet. Cytogenet. 74, 87–94.

KHM-10B

Culture characterization[1]

Culture medium	90% RPMI 1640 + 10% FBS
Doubling time	24 h
Viral status	EBV⁻
Authentication	yes (by *IGH* rearrangement)
Primary reference	Sonoki et al.[1]
Availability of cell line	not known

Clinical characterization[1]

Patient	41-year-old male
Disease diagnosis	B-ALL (L3)
Treatment status	at diagnosis
Specimen site	peripheral blood
Year of establishment	1991

Immunophenotypic characterization[1]

T-/NK cell marker	$CD2^+$, $CD3^-$
B-cell marker	$CD10^+$, $CD19^+$, $CD20^-$, $sIgM\lambda^+$
Myelomonocytic marker	$CD13^-$
Progenitor/activation/other marker	$HLA\text{-}DR^+$
Cytokine receptor	$CD25^-$

Genetic characterization[1]

Cytogenetic karyotype	47, XY, +1, −11, +mar, der(10)t(10;?)(q26;?), der(13)t(13;?)(q34;?); FISH: t(8;22)(q24;q11)
Unique translocation/fusion gene	t(8;22)(q24;q11) → *MYC-IGL* genes altered?
Receptor gene rearrangement	*IGH* RR

Functional characterization[1]

Cytochemistry	$ANBE^-$, CAE^-, MPO^-, $PAS^{(+)}$
Proto-oncogene	$mRNA^+$: MAX, MYC

Comments

- Mature EBV⁻ B-cell line derived from B-ALL (L3).
- Carries t(8;22)(q24;q11) possibly leading to alteration of *MYC* and *IGL* genes.

Reference

[1] Sonoki, T. et al. (1995) Leukemia 9, 2093–2099.

KIS-1

Culture characterization[1]

Culture medium	90% RPMI 1640 + 10% FBS
Doubling time	24 h
Viral status	EBNA$^-$
Authentication	yes (by cytogenetics, *IGH* rearrangement)
Primary reference	Kamesaki et al.[1]
Availability of cell line	original authors

Clinical characterization[1]

Patient	53-year-old male
Disease diagnosis	B-NHL (diffuse large cell)
Treatment status	at presentation
Specimen site	ascites
Year of establishment	1985

Immunophenotypic characterization[1]

T-/NK cell marker	CD1$^-$, CD2$^-$, CD3$^-$, CD4$^-$, CD5$^-$, CD8$^-$, CD57$^-$
B-cell marker	CD10$^-$, CD19$^+$, CD20$^-$, cyIgλ$^+$
Myelomonocytic marker	CD13$^-$, CD14$^-$, CD15$^-$
Progenitor/activation/other marker	CD30$^+$, CD45$^+$, HLA-DR$^+$
Adhesion marker	CD11b$^-$
Cytokine receptor	CD25$^-$

Genetic characterization[1–3]

Cytogenetic karyotype	50, X, Y, −1, −14, −17, −22, +mar, +min, +der(X)t(X;?) (p11.4;?), +1p+(HSR), t(9;14)(p13;q32.3), t(12;13)(q13.1;p12), +der(14)t(9;14;?)(?::14p12→14q32.3::9p13→9pter)×2, del(16) (q22), +der(17)t(17;?)(p11.2;?), +der(22)t(1;22)(q11;q13)
Unique translocation/fusion gene	t(9;14)(p13;q32) → *PAX5-IGH* genes altered
Receptor gene rearrangement	*IGH* R, *IGK* RD, *IGL* RR, *TCRB* G

Functional characterization[1,4]

Cytochemistry	ACP$^{(+)}$, ALP$^-$, ANAE$^{(+)}$, CAE$^-$, MPO$^-$
Proto-oncogene	mRNA overexpression: BCL2, PAX5
Special features	secretion: Igλ

Comments

- Mature EBV$^-$ B-cell line derived from B-NHL (DLCL).
- Carries t(9;14)(p13;q32) leading to alteration of the *PAX5* and *IGH* genes.
- Cells grow well.

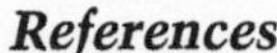

References

[1] Kamesaki, H. et al. (1988) Jpn. J. Cancer Res. 79, 1193–1200.
[2] Ohno, H. et al. (1990) Proc. Natl. Acad. Sci. USA 87, 628–632.
[3] Iida, S. et al. (1996) Blood 88, 4110–4117.
[4] Yonetani, N. et al. (1998) Oncogene 17, 971–979.

KML-1

Culture characterization[1]

Culture medium	90% RPMI 1640 + 10% FBS
Authentication	no
Primary reference	Ikezoe et al.[1]
Availability of cell line	not known

Clinical characterization[1]

Patient	28-year-old female
Disease diagnosis	B-NHL (diffuse large cell)
Treatment status	refractory, terminal
Specimen site	pleural effusion
Year of establishment	1987

Immunophenotypic characterization[1]

T-/NK cell marker	CD1$^-$, CD2$^-$, CD3$^-$, CD4$^-$, CD5$^-$, CD7$^-$, CD8$^-$
B-cell marker	CD10$^+$, CD19$^+$, CD20$^+$, cyIgMκ^+, sIgκ^+
Myelomonocytic marker	CD13$^+$, CD14$^-$, CD33$^+$
Progenitor/activation/other marker	CD30$^-$, CD34$^-$, CD45RO$^-$, HLA-DR$^+$

Genetic characterization[1,2]

Cytogenetic karyotype	43–45, X, −X, −4, −10, −13, −14, −18, +5mar, del(6)(p22), add(10)(q2?5), der(12)t(12;?)(p12;?), der(17)t(17;?)(p11;?), add(18)(q22), add(19)(p13)
Receptor gene rearrangement	*IGH* R, *IGK* DD, *TCRB* G

Functional characterization[1]

Cytochemistry	ANBE$^-$, MPO$^-$
Cytokine production	TGF-β mRNA$^+$
Heterotransplantation	into nude mice
Transcription factor	DCC mRNA decreased

Comments

- Mature B-cell line derived from B-NHL (DLCL).

References

[1] Ikezoe, T. et al. (1995) Am. J. Hematol. 50, 124–132.
[2] Miyagi, T. et al. (1993) Leukemia 7, 970–977.

MC116

Culture characterization[1,2]

Culture medium	80% RPMI 1640 + 20% FBS
Doubling time	28 h
Viral status	EBV$^-$, HBV$^-$, HCV$^-$, HHV-8$^-$, HIV$^-$, HTLV-I/II$^-$
Authentication	no
Primary reference	Magrath et al.[1]
Availability of cell line	ATCC CRL 1649, DSM ACC 82

Clinical characterization[1]

Disease diagnosis	B-NHL (undifferentiated)
Specimen site	pleural effusion

Immunophenotypic characterization[2,3]

T-/NK cell marker	CD3$^-$
B-cell marker	CD10$^{(+)}$, CD19$^+$, CD20$^+$, CD37$^+$, CD138$^-$, cyIgMλ$^+$, sIgMλ$^+$
Myelomonocytic marker	CD13$^-$
Progenitor/activation/other marker	CD34$^-$, HLA-DR$^+$

Genetic characterization[1,2,4]

Cytogenetic karyotype	near diploid, 1.5% polyploidy; 46(37–46)<2n>XY, dup(1) (q21q32), t(8;14)(q24;q32), del(10)(q23)
Unique translocation/fusion gene	t(8;14)(q24;q32) → *MYC-IGH* genes altered?
Unique gene alteration	*P53* mutation

Functional characterization[1]

Special features	secretion: IgMλ

Comments

- Mature EBV$^-$ B-cell line derived from undifferentiated B-NHL.
- Carries t(8;14)(q24;q32) possibly leading to alterations of the *MYC* and *IGH* genes.
- Available from cell line banks.

References

1 Magrath, I.T. et al. (1980) J. Natl. Cancer Inst. 64, 465–476.

2 Drexler, H.G. et al. (1999) DSMZ Catalogue of Cell Lines, 7th edn, Braunschweig, Germany.

3 Magrath, I.T. et al. (1980) J. Natl. Cancer Inst. 64, 477–483.

4 Gaidano, G. et al. (1991) Proc. Natl. Acad. Sci. USA 88, 5413–5417.

MEC1

Culture characterization[1]

Sister cell line	**MEC2** (serial sister cell line – cytogenetic differences)
Culture medium	90% IMDM + 10% FBS
Doubling time	40 h
Viral status	EBV$^+$
Authentication	yes (by *IGH* rearrangement/sequencing)
Primary reference	Stacchini et al.[1]
Availability of cell line	original authors

Clinical characterization[1]

Patient	61-year-old male
Disease diagnosis	B-CLL → prolymphocytoid transformation
Treatment status	at diagnosis, prior to treatment
Specimen site	peripheral blood
Year of establishment	1993

Immunophenotypic characterization[1]

T-/NK cell marker	CD2$^-$, CD3$^-$, CD4$^-$, CD5$^-$, CD7$^-$, CD8$^-$, CD28$^-$, CD56$^-$, CD57$^-$
B-cell marker	CD10$^-$, CD19$^+$, CD20$^+$, CD21$^+$, CD22$^+$, CD23$^+$, CD40$^+$, CD80$^+$, CD86$^+$, FMC-7$^{(+)}$, sIgMDκ$^+$
Myelomonocytic marker	CD13$^-$, CD14$^-$, CD15$^-$, CD33$^-$
Progenitor/activation/other marker	CD30$^+$, CD34$^-$, CD38$^+$, CD45$^+$, CD95$^+$, HLA-DR$^+$
Adhesion marker	CD11a$^+$, CD11b$^-$, CD11c$^+$, CD18$^+$, CD29$^+$, CD44$^+$, CD49c$^-$, CD49d$^+$, CD54$^+$
Cytokine receptor	CD25$^{(+)}$

Genetic characterization[1]

Cytogenetic karyotype	46, XY, −2, −10, +mar, t(1;6)(q23;p23), der(7), +der(10)t(2;?;10)(q23;?;q22.1), del(12)(p12p13)(q13q15), del(17)(p11.2pter)
Receptor gene rearrangement	*IGH* R

Functional characterization[1]

Proto-oncogene	mRNA$^+$: BAK, BAX, BCL2, BCLXL, BCLXS

Comments

- Mature EBV$^+$ B-cell line derived from B-CLL in transformation.
- Cellular derivation and neoplastic nature verified.

• Serial sister cell lines established prior to treatment but during transformation.
• Cells grow very well.

Reference

[1] Stacchini, A. et al. (1999) Leukemia Res. 23, 127–136.

MN-60

Culture characterization[1,2]

Culture medium	90% Ham's F10 + 10% FBS
Doubling time	25 h
Viral status	EBV⁻, HBV⁻, HCV⁻, HHV-8⁻, HIV⁻, HTLV-I/II⁻
Authentication	yes (by cytogenetics)
Primary reference	Roos et al.[1]
Availability of cell line	DSM ACC 138

Clinical characterization[1]

Patient	20-year-old male
Disease diagnosis	B-ALL (L3)
Treatment status	in partial remission
Specimen site	peripheral blood
Year of establishment	1981

Immunophenotypic characterization[1,2]

T-/NK cell marker	$CD2^-$, $CD3^-$
B-cell marker	$CD10^+$, $CD19^+$, $CD37^+$, $CD77^-$, $cyIgM^+$, $sIgM\lambda^+$
Myelomonocytic marker	$CD13^-$
Erythroid-megakaryocytic marker	$GlyA^-$
Progenitor/activation/other marker	$HLA\text{-}DR^+$

Genetic characterization[1,2]

Cytogenetic karyotype	46(45–47)<2n>XY, dup(1)(q21q41), del(6)(q21), t(8;14) (q24;q32), i(13q)
Unique translocation/fusion gene	t(8;14)(q24;q32) → *MYC-IGH* genes altered?

Functional characterization[1]

Colony formation	in agar
Cytochemistry	$ACP^{(+)}$, $ANAE^{(+)}$, $CAE^{(+)}$, $Lysozyme^-$, MPO^-
Heterotransplantation	into nude mice
Inducibility of differentiation	DMSO, IFN-α, TPA → growth arrest, no differentiation

Comments

- Mature EBV⁻ B-cell line derived from B-ALL (L3).
- Carries t(8;14)(q24;q32) which may lead to alteration of the *MYC-IGH* genes.
- Available from cell line bank.

References

[1] Roos, G. et al. (1982) Leukemia Res. 6, 685–693.

[2] Drexler, H.G. et al. (1999) DSMZ Catalogue of Cell Lines, 7th edn, Braunschweig, Germany.

Culture characterization[1]

Culture medium	85% RPMI 1640 + 10% FBS
Doubling time	72–96 h
Viral status	EBNA$^+$
Authentication	yes (by cytogenetics, *IGH* rearrangement)
Primary reference	Saltman et al.[1]
Availability of cell line	original authors

Clinical characterization[1]

Patient	57-year-old male
Disease diagnosis	B-NHL (diffuse centroblastic-centrocytic)
Specimen site	peripheral blood
Year of establishment	1987

Immunophenotypic characterization[1]

T-/NK cell marker	CD5$^+$
B-cell marker	CD19$^+$, CD20$^+$, CD21$^+$, CD23$^+$, cyIgMλ$^+$, sIg$^-$
Progenitor/activation/other marker	HLA-DR$^+$, TdT$^-$
Cytokine receptor	CD25$^+$

Genetic characterization[1–3]

Cytogenetic karyotype	43, X, −Y, −12, −15, −17, +mar, der(5)t(5;12)(p14;q12), der(8)t(8;?)(p11.2;?), t(9;?;13)(q32;?;q14), der(10)t(10;15)(q24;q15), t(11;14)(q13;q32)
Unique translocation/fusion gene	t(11;14)(q13;q32) → *BCL1-IGH* genes altered
Receptor gene rearrangement	*IGH* RR

Functional characterization[1]

Cytochemistry	ANAE$^-$, PAS$^+$
Special features	cyclin D1 mRNA overexpression

Comments

- Mature EBV$^+$ B-cell line derived from B-NHL (diffuse centroblastic-centrocytic).
- Cellular derivation and neoplastic nature verified.
- Carries t(11;14)(q13;q32) leading to alteration of the *BCL1* and *IGH*.

References

[1] Saltman, D.L. et al. (1988) Blood 72, 2026–2030.
[2] Brown, A.G. et al. (1993) Nature Genetics 3, 67–72.
[3] Willis, T.G. et al. (1997) Blood 90, 2456–2464.

OCI-Ly18

Culture characterization[1]

Culture medium	80% IMDM + 20% human plasma
Doubling time	36 h
Viral status	EBV⁻
Authentication	yes (by *IGH* rearrangement)
Primary reference	Chang et al.[1]
Availability of cell line	not known

Clinical characterization[1]

Patient	56-year-old male
Disease diagnosis	B-NHL (diffuse large cell, immunoblastic)
Treatment status	stage IIIB, at relapse
Specimen site	pleural effusion
Year of establishment	1988

Immunophenotypic characterization[1]

T-/NK cell marker	$CD3^-$
B-cell marker	$CD10^+$, $CD19^+$, $CD20^+$, $CD21^-$, $CD23^-$, $cyIgM\lambda^+$, $sIgM\lambda^+$
Progenitor/activation/other marker	$CD34^-$, $HLA\text{-}DR^+$, TdT^-

Genetic characterization[1,2]

Cytogenetic karyotype	52, X, −?Y, +7, +9, +18, +18, +20, +21, t(8;18;14)(q24.1; q21.1;q32.2), ins(12;?)(q13;?), i(17q), +der(18)t(8;18)(q24.1; q21.2)
Unique translocation/fusion gene	t(8;14)(q24;q32) → *MYC-IGH* genes altered (*BCL2* in germline)
Receptor gene rearrangement	*IGH* RR, *IGK* DD, *IGL* R
Unique gene alteration	*P53* deletion

Functional characterization[1,3,4]

Cytochemistry	ACP^+, $ANAE^-$, MPO^-, PAS^+
Cytokine production	IL-6
Cytokine response	growth stimulation: IL-6
Heterotransplantation	into SCID mice

Comments

- Mature EBV⁻ B-cell line derived from B-NHL (DLCL).
- Carries three-way t(8;18;14)(q24;q21.1;q32) which led to alteration of the *MYC-IGH* genes (*BCL2* apparently in germline).

References

1 Chang, H. et al. (1992) J. Clin. Invest. 89, 1014–1020.
2 Chang, H. et al. (1995) Leukemia Lymphoma 19, 165–171.
3 Chang, H. et al. (1992) Leukemia Lymphoma 8, 129–136.
4 Yee, C. et al. (1992) Leukemia Lymphoma 7, 123–129.

ONHL-1

Culture characterization[1]

Culture medium	90% RPMI 1640 + 10% FBS
Doubling time	25 h
Viral status	EBNA$^-$
Authentication	yes (by *IGH, IGK, IGL* rearrangement)
Primary reference	Matsumura et al.[1]
Availability of cell line	not known

Clinical characterization[1]

Patient	58-year-old male
Disease diagnosis	B-NHL (follicular, small cleaved cell type)
Treatment status	refractory, terminal
Specimen site	bone marrow
Year of establishment	1987

Immunophenotypic characterization[1]

T-/NK cell marker	CD2$^-$
B-cell marker	CD10$^-$, CD19$^-$, CD20$^+$, CD24$^+$, PCA-1$^-$, sIgMDκ$^+$
Progenitor/activation/other marker	CD38$^-$, HLA-DR$^+$

Genetic characterization[1]

Cytogenetic karyotype	46, X, −Y, −3, −8, −9, −10, −11, −13, −14, −15, −16, −18, −20, +2mar, t(2;12)(p11.2;q24.1), +der(3)t(3;9;10)(p21.1;p22;q22), del(5)(q22q31.1), del(6)(q15q22.2), +der(8)t(8;?)(p23.1;?), +der (9)t(3;9;10)(p21.1;p22;q22), +der(11)t(11;?)(q25;?), +der(14)t (14;?)(q32.3;?), t(15;16)(q21.1;q22), +der(15)t(15;?)(p13;?), +der(16)t(16;?)(p13.1;?), t(17;17)(p13;q21.1), +der(18)t(18;?) (q21;?), +der(20)t(20;?)(q13.3;?), +der(21)t(8;21)(q13;p13)
Unique translocation/fusion gene	t(14;18)(q32;q21) → *IGH-BCL2* genes altered
Receptor gene rearrangement	*IGH* RR, *IGK* RG, *IGL* RR

Functional characterization[1]

Cytochemistry	ACP$^-$, MPO$^-$, PAS$^-$
Inducibility of differentiation	TPA → terminal B-cell differentiation
Proto-oncogene	mRNA$^+$: BCL2, MYC

Comments

- Mature EBV⁻ B-cell line established from B-NHL (FCL).
- Carries t(14;18)(q32;q21) leading to alteration of the *IGH* and *BCL2* genes.

Reference

[1] Matsumura, I. et al. (1990) Int. J. Cancer 46, 1107–1111.

OPL-1

Culture characterization[1]

Culture medium	85% RPMI 1640 + 15% FBS
Doubling time	48 h
Viral status	EBV^+ (EBV type B)
Authentication	yes (by EBV clonality)
Primary reference	Kanno et al.[1]
Availability of cell line	original authors

Clinical characterization[1]

Patient	76-year-old male
Disease diagnosis	B-NHL (diffuse immunoblastic, pyothorax-associated lymphoma)
Treatment status	prior to chemotherapy
Specimen site	biopsy specimen
Year of establishment	1993

Immunophenotypic characterization[1,2]

T-/NK cell marker	$CD3^-$, $CD4^-$, $CD5^-$, $CD8^-$, $CD57^-$
B-cell marker	$CD10^-$, $CD19^-$, $CD20^+$, $CD21^+$, $CD22^+$, $CD23^+$, $CD24^+$, $CD74^+$, $CD75^-$, $sIgG\kappa^+$
Myelomonocytic marker	$CD15^-$
Progenitor/activation/other marker	$CD30^-$
Adhesion marker	$CD11a^+$, $CD11c^-$, $CD54^+$
Cytokine receptor	$CD126^+$

Genetic characterization[1]

Cytogenetic karyotype	46–207, multiple numerical and structural rearrangements

Functional characterization[1,2]

Cytokine production	IL-6; IL-10 $mRNA^+$
Cytokine response	growth stimulation: IL-6
Proto-oncogene	$mRNA^+$: BCL2, FGR, MYC
Special features	$EBNA2^+$, $LMP1^-$; $mRNA^+$: EBNA1, LMP1

Comments

- Mature EBV^+ B-cell line derived from B-NHL (pyothorax-associated lymphoma).
- Cellular derivation and neoplastic nature verified.

References

[1] Kanno, H. et al. (1996) Int. J. Cancer 67, 86–94.

[2] Kanno, H. et al. (1996) Lab. Invest. 75, 167–173.

OPL-2

Culture characterization[1]

Culture medium	85% RPMI 1640 + 15% FBS
Doubling time	24 h
Viral status	EBV^+ (EBV type A)
Authentication	yes (by *IGH* rearrangement, EBV clonality)
Primary reference	Kanno et al.[1]
Availability of cell line	original authors

Clinical characterization[1]

Patient	67-year-old male
Disease diagnosis	B-NHL (diffuse immunoblastic, pyothorax-associated lymphoma)
Treatment status	prior to chemotherapy
Specimen site	biopsy specimen
Year of establishment	1994

Immunophenotypic characterization[1,2]

T-/NK cell marker	$CD4^-$, $CD5^-$, $CD8^-$, $CD57^-$
B-cell marker	$CD10^-$, $CD19^-$, $CD20^+$, $CD21^+$, $CD22^+$, $CD23^+$, $CD24^+$, $CD74^+$, $CD75^+$, $sIgG\lambda^+$
Myelomonocytic marker	$CD15^-$
Progenitor/activation/other marker	$CD30^+$
Adhesion marker	$CD11a^-$, $CD11c^-$, $CD54^+$
Cytokine receptor	$CD126^+$

Genetic characterization[1]

Cytogenetic karyotype	53–129, del(1)(q21)×1–2, add(2)(p11.1), add(7)(q32)×2, der(11)t(11;14)(p15;q11)×2, add(15)(p13), der(19)t(1;19)(q21;p13.3)×2
Receptor gene rearrangement	*IGH* R

Functional characterization[1,2]

Colony formation	in soft agar
Cytokine response	growth stimulation: IL-6
Proto-oncogene	$mRNA^+$: BCL2, MYC
Special features	$mRNA^+$: EBNA1, LMP1

Comments

- Mature EBV^+ B-cell line derived from B-NHL (pyothorax-associated lymphoma).
- Cellular derivation and neoplastic nature verified.

References

[1] Kanno, H. et al. (1996) Int. J. Cancer 67, 86–94.

[2] Kanno, H. et al. (1996) Lab. Invest. 75, 167–173.

PER-377

Culture characterization[1]

Sister cell line	**PER-365** (serial sister cell line – at 1st relapse – similar features)
Culture medium	90% RPMI 1640 + 10% FBS
Doubling time	3.6 d
Viral status	EBV$^-$
Authentication	yes (by *IGH*, *MLL* rearrangements)
Primary reference	Kees et al.[1]
Availability of cell line	not known

Clinical characterization[1]

Patient	19-month-old male
Disease diagnosis	AML → ALL (1st relapse) → ALL L2 (2nd relapse)
Treatment status	at 2nd relapse, terminal
Specimen site	bone marrow
Year of establishment	1988

Immunophenotypic characterization[1]

T-/NK cell marker	CD2$^-$, CD3$^-$, cyCD3$^-$, CD7$^+$
B-cell marker	CD10$^-$, CD19$^+$, CD22$^+$, cyIgM$^+$, sIgMλ$^+$
Myelomonocytic marker	CD13$^+$, CD15$^+$, CD33$^+$
Progenitor/activation/other marker	HLA-DR$^+$, TdT$^-$

Genetic characterization[1,2]

Cytogenetic karyotype	46, XY, t(2;13)(p12;q34), del(7)(q11q21), inv(14)(q11q23), der(17)t(8;17)(q11;p11), t(1;20)(q32;q13); SKY: t(2;13)(p12;q32)
Receptor gene rearrangement	*IGH* R, *IGK* D, *TCRB* G
Unique gene alteration	*MLL* partial, non-tandem duplication

Comments

- Mature EBV$^-$ B-cell line established from B-ALL (L2).
- Carries alterations of *MLL* gene.

References

[1] Kees, U.R. et al. (1995) Genes Chromosomes Cancer 12, 201–208.
[2] Whitman, S.P. et al. (1998) Blood 92, 71a.

RC-K8

Culture characterization[1,2]

Culture medium	85% RPMI 1640 + 15% FBS
Doubling time	48–60 h
Viral status	EBNA⁻
Authentication	no
Primary reference	Kubonishi et al.[1]
Availability of cell line	restricted

Clinical characterization[1,2]

Patient	55-year-old male
Disease diagnosis	true histiocytic lymphoma
Treatment status	refractory, terminal
Specimen site	peritoneal effusion
Year of establishment	1984

Immunophenotypic characterization[1-3]

T-/NK cell marker	$CD2^-$, $CD3^-$, $cyCD3^-$, $CD4^-$, $CD5^-$, $CD7^-$, $CD8^-$, $CD57^-$
B-cell marker	$CD10^-$, $CD19^+$, $CD20^+$, $cyIg^-$, sIg^-
Myelomonocytic marker	$CD13^-$, $CD14^-$, $CD15^-$, $CD33^-$
Erythroid-megakaryocytic marker	$CD41b^-$
Progenitor/activation/other marker	$HLA\text{-}DR^+$, TdT^-

Genetic characterization[1-4]

Cytogenetic karyotype	46, XY, −8, −14, +mar, t(Y;7)(q12;q32), t(2;2)(p25;p23), t(3;4)(q29;q31), +der(8)t(8;8)(p22;q11), t(10;15)(p11;p13), t(11;14) (q23;q32), t(13;20)(q12;q13)
Unique translocation/fusion gene	t(11;14)(q23;q32) → *BCL1-IGH* genes altered
Receptor gene rearrangement	*IGH* RR, *IGK* RR, *TCRB* G, *TCRG* RG

Functional characterization[1,2]

Cytochemistry	ACP^+, ALP^-, $ANBE^+$ (NaF inhibitable), CAE^-, $Lysozyme^-$, MPO^-, PAS^+, SBB^-
Heterotransplantation	into newborn hamster
Special features	secretion: plasminogen activator

Comments

- Mature EBV^- B-cell line derived from B-NHL (allegedly but unlikely 'true' histiocytic lymphoma).
- Carries t(11;14)(q13;q32) leading to alteration of the *BCL1* and *IGH* genes.

References

1 Kubonishi, I. et al. (1985) Jpn. J. Cancer Res. 76, 12–15.
2 Kubonishi, I. et al. (1986) Cancer 58, 1453–1460.
3 Iida, S. et al. (1992) Leukemia Res. 16, 1155–1163.
4 Kobayashi, H. et al. (1993) Blood 81, 3027–3033.

Culture characterization[1]

Other name of cell line	Riva
Culture medium	90% RPMI 1640 + 10% FBS
Viral status	EBNA⁻
Authentication	no
Primary reference	Th'ng et al.[1]
Availability of cell line	original authors

Clinical characterization[1]

Patient	48-year-old female
Disease diagnosis	B-NHL (lymphocytic, small cell type)
Treatment status	progressive, refractory, terminal
Specimen site	peripheral blood
Year of establishment	1977

Immunophenotypic characterization[1-3]

T-/NK cell marker	$CD1^-$, $CD2^-$, $CD3^-$, $CD4^-$, $CD5^-$, $CD7^-$, $CD8^-$, $CD56^-$
B-cell marker	$CD9^+$, $CD10^-$, $CD19^+$, $CD20^+$, $CD21^-$, $CD22^+$, $CD23^-$, $CD24^+$, $CD37^+$, FMC-7^+, NU-$B1^+$, PCA-1^-, cyIgMκ$^+$, sIgMκ$^+$
Myelomonocytic marker	$CD13^-$, $CD15^-$, $CD33^-$
Erythroid-megakaryocytic marker	$CD31^-$, $CD36^-$, $CD41a^+$, $CD61^+$
Progenitor/activation/other marker	$CD34^-$, $CD38^+$, HLA-DR^+, TdT^-
Adhesion marker	$CD11a^-$, $CD11b^-$, $CD11c^-$, $CD18^+$, $CD29^+$, $CD44^+$, $CD49a^-$, $CD49b^-$, $CD49c^+$, $CD49d^+$, $CD49e^+$, $CD49f^-$, $CD51^-$, $CD54^+$, $CD58^+$
Cytokine receptor	$CD25^{(+)}$

Genetic characterization[1,4,5]

Cytogenetic karyotype	46–49, XX, +7, +19, −16, −18, +2mar, ?inv(1)(?p31;p36.2), del(3)(?q12/13), 3q+, t(4;8)(q12;q22), del(6)(q13), del(13)? t(13;?)(q12/13;?), ?14, 15p+, del(19)(q13.3), der(19)t(19;?), der(22)t(22;?)
Receptor gene rearrangement	*IGH* R, *IGK* RG, *IGL* GG
Unique gene alteration	*P15INK4B* deletion, *P16INK4A* deletion, *RB1* deletion/mutation

Comments

- Mature EBV⁻ B-cell line derived from lymphocytic B-NHL.

References

[1] Th'ng, K.H. et al. (1987) Int. J. Cancer 39, 89–93.
[2] Segat, D. et al. (1994) Blood 83, 1586–1594.
[3] Minowada, J. and Matsuo, Y. (1999) unpublished data.
[4] Weide, R. et al. (1991) Brit. J. Haematol. 78, 500–505.
[5] Aguiar, R.C.T. et al. (1997) Leukemia 11, 233–238.

Culture characterization[1–3]

Subclone RL-7
Culture medium 90% RPMI 1640 + 10% FBS
Viral status EBV⁻
Authentication yes (by cytogenetics)
Primary reference Beckwith et al.[1]
Availability of cell line ATCC CRL 2261

Clinical characterization[1]

Patient 52-year-old male
Disease diagnosis B-NHL (diffuse undifferentiated, small non-cleaved, large cell)
Specimen site ascites
Year of establishment 1983

Immunophenotypic characterization[1]

T-/NK cell marker $TCR\alpha\beta^-$
B-cell marker $CD19^+$, $CD20^+$, $CD21^+$, $CD22^+$, $sIgMD\lambda^+$
Progenitor/activation/other marker $CD45^+$, HLA-DR^+
Cytokine receptor $CD25^-$

Genetic characterization[1–4]

Cytogenetic karyotype RL-7: 51, XY, +7, +8, +21, +del(12)(q21;q24.3), 14q+, t(14;18)(q32;q21), t(17;19)(q11;q13.3)
Unique translocation/fusion gene t(14;18)(q32;q21) → *IGH-BCL2* genes altered (MBR)
Receptor gene rearrangement *IGH* R, *TCRB* G
Unique gene alteration *P53* mutation

Functional characterization[1]

Special features secretion: IgM

Comments

- Mature EBV⁻ B-cell line derived from B-NHL (DLCL).
- Carries t(14;18)(q32;q21) leading to alteration of the *IGH* and *BCL2* genes (major breakpoint region).
- Available from cell line bank.

References

[1] Beckwith, M. et al. (1990) J. Natl. Cancer Inst. 82, 501–509.
[2] Li, C.C. et al. (1995) Leukemia 9, 650–655.
[3] Lipford, E. et al. (1987) Blood 70, 1816–1823.
[4] Widmer, L. et al. (1996) Blood 88, 3166–3175.

ROS-50

Culture characterization[1]

Culture medium 90% RPMI 1640 + 10% FBS
Authentication no
Primary reference van Ooteghem et al.[1]
Availability of cell line original authors

Clinical characterization[1]

Patient 69-year-old male
Disease diagnosis B-ALL (L3)
Treatment status at relapse
Specimen site peripheral blood

Immunophenotypic characterization[1]

B-cell marker $CD10^+$, $CD19^+$, $CD20^+$, $CD21^+$, $CD24^+$, $CD37^+$, FMC-7^+, sIgMDκ^+
Progenitor/activation/other marker $CD34^+$, $CD71^+$, HLA-DR^+, TdT^-

Genetic characterization[1-3]

Cytogenetic karyotype 47, XY, der(8), +der(8), add(14)(q32), der(7;17)(q10;q10), del(18)(q21); FISH: t(8;14;18)(q24;q32;q32)
Unique translocation/fusion gene (1) t(8;14)(q24;q32) → *MYC-IGH* genes altered
(2) t(14;18)(q32;q21) → *IGH-BCL2* genes altered (mcr)
Receptor gene rearrangement *IGH* RR

Functional characterization[1,2]

Proto-oncogene overexpression: BCL2, MYC

Comments

- Mature B-cell line established from B-ALL (L3).
- Carries both t(8;14)(q24;q32) and t(14;18)(q32;q21) leading to alterations of the *MYC-IGH* and *IGH-BCL2* (minor cluster region) genes.

References

[1] van Ooteghem, R.B.C. et al. (1994) Cancer Genet. Cytogenet. 74, 87–94.
[2] Dyer, M.J.S. et al. (1996) Leukemia 10, 1198–1208.
[3] Lambrechts, A.C. et al. (1994) Leukemia 8, 1164–1171.

Sc-1

Culture characterization[1]

Other name of cell line	**Scott**
Culture medium	90% RPMI 1640 + 10% FBS
Viral status	EBNA$^-$
Authentication	no
Primary reference	Th'ng et al.[1]
Availability of cell line	original authors

Clinical characterization[1]

Patient	67-year-old male
Disease diagnosis	B-NHL (follicular, small cleaved cell type)
Treatment status	at diagnosis
Specimen site	ascites
Year of establishment	1977

Immunophenotypic characterization[1-3]

T-/NK cell marker	CD1$^-$, CD2$^-$, CD3$^-$, CD4$^-$, CD5$^-$, CD7$^-$, CD8$^-$, CD56$^-$
B-cell marker	CD9$^-$, CD10$^+$, CD19$^+$, CD20$^+$, CD21$^+$, CD22$^+$, CD23$^+$, CD24$^+$, CD37$^+$, FMC-7$^-$, NU-B1$^+$, cyIgλ$^+$, sIg$^-$
Myelomonocytic marker	CD13$^-$, CD15$^-$, CD33$^-$
Erythroid-megakaryocytic marker	CD31$^-$, CD36$^-$, CD41a$^+$, CD61$^+$
Progenitor/activation/other marker	CD34$^-$, CD38$^+$, HLA-DR$^+$, TdT$^-$
Adhesion marker	CD11a$^+$, CD11b$^-$, CD11c$^-$, CD18$^+$, CD29$^+$, CD44$^+$, CD49a$^+$, CD49b$^+$, CD49c$^+$, CD49d$^+$, CD49e$^+$, CD49f$^-$, CD51$^+$, CD54$^+$, CD58$^+$
Cytokine receptor	CD25$^{(+)}$

Genetic characterization[1]

Cytogenetic karyotype	49, XY, +3, +7, +8, t(14;17)(q32;q21)
Receptor gene rearrangement	*IGH* R, *IGK* DD, *IGL* RG

Comments

- Mature EBV$^-$ B-cell line derived from follicular B-NHL.

References

1 Th'ng, K.H. et al. (1987) Int. J. Cancer 39, 89–93.
2 Segat, D. et al. (1994) Blood 84, 1586–1594.
3 Minowada, J. and Matsuo, Y. (1999) unpublished data.

Culture characterization[1]

Establishment	initially on feeder layer (adult skin fibroblasts or glial cells)
Culture medium	90–98% RPMI 1640 or Ham's F10 + 2–10% FBS
Doubling time	48 h
Viral status	EBNA⁻
Authentication	no
Primary reference	Nilsson et al.[1]
Availability of cell line	not known

Clinical characterization[1]

Patient	male
Disease diagnosis	B-NHL (diffuse histiocytic)
Treatment status	at presentation
Specimen site	pleural effusion
Year of establishment	1977

Immunophenotypic characterization[1,2]

T-/NK cell marker	$CD1^-$, $CD2^-$, $CD3^-$, $CD5^-$, $CD7^-$, $CD8^-$
B-cell marker	$CD9^-$, $CD10^-$, $CD19^+$, $CD20^+$, $CD21^+$, $CD24^+$, $CD39^+$, $CD80^+$, FMC-7^+, NU-$B1^+$, PCA-1^+, $sIgM\kappa^+$
Myelomonocytic marker	$CD15^-$, $CD87^-$
Progenitor/activation/other marker	$CD38^+$, HLA-DR^+
Adhesion marker	$CD11b^-$
Cytokine receptor	$CD25^+$

Genetic characterization[1]

Cytogenetic karyotype	46–49(49), −Y, +9, −10, −11, −12, +13, −14, −16, −19, t(6;6)(q;q), t(?;2)(?;q), t(6;1)(p;q), t(1;2)(p;p), t(?;14)(?;q), t(?;7)(?;q), 7p+, t(19;?)(p;?), 10q+, 11q−, t(8;11)(−;q)

Functional characterization[1]

Colony formation	in agarose
Cytochemistry	ACP^-, ALP^-, $ANAE^+$, CAE^+, GLC^-, MPO^-, PAS^-

Comments

- Mature EBV⁻ B-cell line derived from B-NHL ('diffuse histiocytic'/large cell).

References

[1] Nilsson, K. et al. (1983) Hematol. Oncol. 1, 277–295.
[2] Minowada, J. and Matsuo, Y. (1999) unpublished data.

SP-53

Culture characterization[1-3]

Sister cell lines (1) **SP-49** (EBNA⁻) and **SP-50B** (EBNA⁺) (simultaneous sister cell lines – both from peripheral blood – identical features)
(2) **SP-52EB** (simultaneous sister cell line = EBV⁺ B-LCL with normal karyotype)
Culture medium 80% RPMI 1640 + 20% FBS
Doubling time 60 h
Viral status EBNA⁻, HTLV-I/II⁻
Authentication no
Primary reference Daibata et al.[1]
Availability of cell line not known

Clinical characterization[1,2]

Patient 58-year-old female
Disease diagnosis B-NHL (intermediate lymphocytic)
Treatment status leukemic conversion, refractory, terminal
Specimen site peripheral blood
Year of establishment 1986

Immunophenotypic characterization[1,2]

T-/NK cell marker CD1a⁻, CD2⁻, CD4⁻, CD5⁻, CD8⁻, CD57⁻
B-cell marker CD10⁺, CD19⁺, cyIgMλ⁺, sIgMλ⁺
Myelomonocytic marker CD13⁻, CD14⁻
Progenitor/activation/other marker CD38⁻, HLA-DR⁺

Genetic characterization[1,2]

Cytogenetic karyotype 46, XX, 1p−, 2q−, del(4)(p14), del(6)(q21), del(7)(q32), t(11;14)(q13;q32)
Unique translocation/fusion gene t(11;14)(q13;q32) → *BCL1-IGH* genes altered?

Comments

- Mature EBV⁻ B-cell line derived from B-NHL (MCL).
- Carries t(11;14)(q13;q32) possibly leading to alteration of the *BCL1* and *IGH* genes.
- Neoplastic nature of sister cell line SP-50B (EBV⁺) confirmed.
- Sister cell line SP-52EB is an EBV⁺ B-LCL with normal karyotype.

References

1 Daibata, M. et al. (1989) Cancer 64, 1248–1253.
2 Daibata, M. et al. (1987) Jpn. J. Cancer Res. 78, 1182–1185.
3 Kubonishi, I. et al. (1990) Am. J. Hematol. 35, 62–64.

SU-DHL-4

Culture characterization[1,2]

Establishment	initially on feeder layer (skin fibroblasts)
Culture medium	80–85% RPMI 1640 + 10–15% FBS
Viral status	EBNA⁻
Authentication	no
Primary reference	Epstein et al.[1]
Availability of cell line	restricted

Clinical characterization[1,2]

Patient	38-year-old male
Disease diagnosis	B-NHL (diffuse large cell, cleaved cell type)
Specimen site	peritoneal effusion
Year of establishment	1975

Immunophenotypic characterization[1-4]

T-/NK cell marker	$CD1^-$, $CD2^-$, $CD3^-$, $CD4^-$, $CD5^-$, $CD8^-$
B-cell marker	$CD9^{(+)}$, $CD10^{(+)}$, $CD19^+$, $CD20^+$, $CD21^-$, $CD24^+$, $CD75^+$, FMC-7^+, NU-$B1^+$, PCA-1^-, cyIgGκ^+, sIgGκ^+
Myelomonocytic marker	$CD13^-$, $CD14^-$, $CD15^-$, $CD33^-$
Progenitor/activation/other marker	$CD34^-$, $CD38^+$, HLA-DR^+, TdT^-
Adhesion marker	$CD11b^-$
Cytokine receptor	$CD25^-$

Genetic characterization[4-6]

Cytogenetic karyotype	47, XY, +8, t(3;?), t(14;18)(q32;q21)
Unique translocation/fusion gene	t(14;18)(q32;q21) → *IGH-BCL2* genes altered
Unique gene alteration	*P53* mutation

Functional characterization[1,2,7-9]

Colony formation	in agar
Cytochemistry	ACP^+, ALP^-, $ANAE^+$, $ANBE^+$, CAE^-, GLC^-, MPO^-, Oil Red O^-, PAS^+, SBB^-, $TRAP^-$
Heterotransplantation	into nude or SCID mice

Comments

- Mature EBV⁻ B-cell line established from B-NHL (DLCL).
- Carries t(14;18)(q32;q21) leading to alteration of the *IGH* and *BCL2* genes.

References

[1] Epstein, A.L. et al. (1978) Cancer 42, 2379–2391.
[2] Winter, J.N. et al. (1984) Blood 63, 140–146.
[3] Minowada, J. and Matsuo, Y. (1999) unpublished data.
[4] Kaiser-McCaw Hecht, B. et al. (1985) Cancer Genet. Cytogenet. 14, 205–218.
[5] Farrugia, M.M. et al. (1994) Blood 83, 191–198.
[6] Li, C.C. et al. (1995) Leukemia 9, 650–655.
[7] Schmidt-Wolf, I.G.H. et al. (1991) J. Exp. Med. 174, 139–149.
[8] Epstein, A.L. et al. (1976) Cancer 37, 2158–2186.
[9] Epstein, A.L. et al. (1979) Cancer Res. 39, 1748–1759.

SU-DHL-5

Culture characterization[1,2]

Establishment	initally on feeder layer (skin fibroblasts)
Culture medium	80–85% RPMI 1640 + 10–15% FBS
Viral status	EBNA⁻
Authentication	no
Primary reference	Epstein et al.[1]
Availability of cell line	not known

Clinical characterization[1,2]

Patient	17-year-old female
Disease diagnosis	B-NHL (diffuse large cell, noncleaved cell type)
Specimen site	lymph node

Immunophenotypic characterization[1–3]

T-/NK cell marker	$CD1a^-$, $CD2^-$, $CD3^-$, $CD4^-$, $CD8^-$
B-cell marker	$CD9^-$, $CD10^-$, $CD20^+$, $CD21^-$, $CD24^+$, $CD75^+$, $cyIgM\lambda^+$, $sIgM\lambda^+$
Progenitor/activation/other marker	$HLA\text{-}DR^+$, TdT^-
Adhesion marker	$CD11b^-$

Genetic characterization[3]

Cytogenetic karyotype	47, XX, +12, 2q−, 6q−

Functional characterization[1,2,4,5]

Colony formation	in agar
Cytochemistry	ACP^+, ALP^-, $ANAE^{(+)}$, $ANBE^{(+)}$, CAE^-, GLC^-, MPO^-, Oil Red O^+, PAS^+, SBB^-, $TRAP^-$
Heterotransplantation	into nude mice

Comments

- Mature EBV^- B-cell line derived from B-NHL (DLCL).

References

[1] Epstein, A.L. et al. (1978) Cancer 42, 2379–2391.

[2] Winter, J.N. et al. (1984) Blood 63, 140–146.

[3] Kaiser-McCaw Hecht, B. et al. (1985) Cancer Genet. Cytogenet. 14, 205–218.

[4] Epstein, A.L. et al. (1976) Cancer 37, 2158–2176.

[5] Epstein, A.L. et al. (1979) Cancer Res. 39, 1748–1759.

SU-DHL-6

Culture characterization[1,2]

Establishment	initially on feeder layer (skin fibroblasts)
Culture medium	80–85% RPMI 1640 + 10–15% FBS
Viral status	EBNA⁻
Authentication	no
Primary reference	Epstein et al.[1]
Availability of cell line	not known

Clinical characterization[1,2]

Patient	43-year-old male
Disease diagnosis	B-NHL (diffuse, mixed small and large cell type)
Specimen site	peritoneal effusion

Immunophenotypic characterization[1-3]

T-/NK cell marker	CD1a⁻, CD2⁻, CD3⁻, CD4⁻, CD8⁻
B-cell marker	CD9⁻, CD10⁻, CD20⁺, CD21⁻, CD24⁺, CD75⁺, cyIgMκorλ⁺, sIgMκorλ⁺
Progenitor/activation/other marker	HLA-DR⁺, TdT⁻
Adhesion marker	CD11b⁻

Genetic characterization[3]

Cytogenetic karyotype	47, X, −4, +6, +7, −8, −17, −22, +3mar, +i(17q), 6p−, 7q−, 9p−, 18q−, t(11;?)(q25;?), t(14;18)(q32;q21)
Unique translocation/fusion gene	t(14;18)(q32;q21) → *IGH-BCL2* genes altered?

Functional characterization[1,2]

Cytochemistry	ACP⁺, ALP⁺, ANAE⁽⁺⁾, ANBE⁽⁺⁾, CAE⁻, GLC⁺, MPO⁻, Oil Red O⁺, PAS⁺, SBB⁻, TRAP⁻

Comments

- Mature EBV⁻ B-cell line derived from B-NHL.
- Carries t(14;18)(q32;q21) possibly causing rearrangement of the *IGH-BCL2* genes.
- Discrepancy on immunoglobulin isotype (sIgMκ or sIgMλ?).

References

1. Epstein, A.L. et al. (1978) Cancer 42, 2379–2391.
2. Winter, J.N. et al. (1984) Blood 63, 140–146.
3. Kaiser-McGaw Hecht, B. et al. (1985) Cancer Genet. Cytogenet. 14, 205–218.

SU-DHL-7

Culture characterization[1,2]

Establishment	initially on feeder layer (skin fibroblasts)
Culture medium	80–85% RPMI 1640 + 10–15% FBS
Viral status	EBNA⁻
Authentication	no
Primary reference	Epstein et al.[1]
Availability of cell line	not known

Clinical characterization[1,2]

Patient	47-year-old female
Disease diagnosis	B-NHL (follicular large cell)
Specimen site	pleural effusion

Immunophenotypic characterization[1-3]

T-/NK cell marker	CD1a⁻, CD2⁻, CD3⁻, CD4⁻, CD8⁻
B-cell marker	CD9⁻, CD10⁻, CD20⁺, CD21⁻, CD24⁺, CD75⁺, cyIgAGκorλ⁺, sIgAGκorλ⁺
Progenitor/activation/other marker	HLA-DR⁺, TdT⁻
Adhesion marker	CD11b⁻

Genetic characterization[3,4]

Cytogenetic karyotype	44, XX, −10, −12, −13, −18, −21, −22, +4mar, 2q−, 6q−, t(1;X)(p11;q28), t(14;?)(q32;?)
Unique gene alteration	*P53* mutation

Functional characterization[1,2]

Cytochemistry	ACP⁺, ALP⁻, ANAE⁺, ANBE⁽⁺⁾, CAE⁻, GLC⁺, MPO⁻, Oil Red O⁺, PAS⁺, SBB⁻ TRAP⁻

Comments

- Mature EBV⁻ B-cell line derived from B-NHL (FCL).
- Discrepancy on immunoglobulin isotype (sIgAGκ or sIgAGκ?).

References

[1] Epstein, A.L. et al. (1978) Cancer 42, 2379–2391.
[2] Winter, J.N. et al. (1984) Blood 63, 140–146.
[3] Kaiser-McCaw Hecht, B. et al. (1985) Cancer Genet. Cytogenet. 14, 205–218.
[4] Li, C.C. et al. (1995) Leukemia 9, 650–655.

SU-DHL-8

Culture characterization[1,2]

Establishment	initially on feeder layer (skin fibroblasts)
Culture medium	80–85% RPMI 1640 + 10–15% FBS
Viral status	EBNA⁻
Authentication	no
Primary reference	Epstein et al.[1]
Availability of cell line	not known

Clinical characterization[1,2]

Patient	59-year-old male
Disease diagnosis	B-NHL (diffuse large cell, non-cleaved cell type)
Specimen site	pleural effusion

Immunophenotypic characterization[1–3]

T-/NK cell marker	$CD1a^-$, $CD2^-$, $CD3^-$, $CD4^-$, $CD8^-$
B-cell marker	$CD9^-$, $CD10^-$, $CD20^+$, $CD21^-$, $CD24^+$, $CD75^+$, $cyIg\lambda^+$, sIg^-
Progenitor/activation/other marker	$HLA\text{-}DR^+$, TdT^-
Adhesion marker	$CD11b^-$

Genetic characterization[3,4]

Cytogenetic karyotype	51, X, +8, +12, +13, +20, +20, −22, +2mar, t(8;?)(q22;?)
Unique gene alteration	*P53* mutation

Functional characterization[1,2,5]

Colony formation	in agar
Cytochemistry	ACP^+, ALP^-, $ANAE^+$, $ANBE^+$, CAE^-, GLC^+, MPO^-, Oil Red O^-, PAS^-, SBB^-, $TRAP^-$

Comments

- Mature EBV^- B-cell line established from B-NHL (DLCL).

References

[1] Epstein, A.L. et al. (1978) Cancer 42, 2379–2391.
[2] Winter, J.N. et al. (1984) Blood 63, 140–146.
[3] Kaiser-McCaw Hecht, B. et al. (1985) Cancer Genet. Cytogenet. 14, 205–218.
[4] Li, C.C. et al. (1995) Leukemia 9, 650–655.
[5] Epstein, A.L. et al. (1979) Cancer Res. 39, 1748–1759.

SU-DHL-9

Culture characterization[1,2]

Establishment	initially on feeder layer (skin fibroblasts)
Culture medium	80–85% RPMI 1640 + 10–15% FBS
Viral status	EBNA⁻
Authentication	no
Primary reference	Epstein et al.[1]
Availability of cell line	not known

Clinical characterization[1,2]

Patient	64-year-old female
Disease diagnosis	B-NHL (diffuse large cell, non-cleaved cell type)
Specimen site	pleural effusion

Immunophenotypic characterization[1-3]

T-/NK cell marker	$CD1a^-$, $CD2^-$, $CD3^-$, $CD4^-$, $CD8^-$
B-cell marker	$CD9^-$, $CD10^-$, $CD20^+$, $CD21^-$, $CD24^+$, $CD75^+$, $cyIg\kappa^+$, sIg^-
Progenitor/activation/other marker	$HLA\text{-}DR^+$, TdT^-
Adhesion marker	$CD11b^-$

Genetic characterization[3,4]

Cytogenetic karyotype	51, XX, +1, +3, +7, −8, +9, +13, +20, −22, +mar, 1p−, 1q−, t(5;?)
Unique gene alteration	*P53* mutation

Functional characterization[1,2,5]

Colony formation	in agar
Cytochemistry	ACP^+, ALP^-, $ANAE^+$, $ANBE^+$, CAE^-, GLC^+, MPO^-, Oil Red O^-, PAS^+, SBB^-, $TRAP^-$

Comments

- Mature EBV^- B-cell line established from B-NHL (DLCL).

References

[1] Epstein, A.L. et al. (1978) Cancer 42, 2379–2391.
[2] Winter, J.N. et al. (1984) Blood 63, 140–146.
[3] Kaiser-McCaw Hecht, B. et al. (1985) Cancer Genet. Cytogenet. 14, 205–218.
[4] Li, C.C. et al. (1995) Leukemia 9, 650–655.
[5] Epstein, A.L. et al. (1979) Cancer Res. 39, 1748–1759.

SU-DHL-10

Culture characterization[1]

Establishment	initially on feeder layer (skin fibroblasts)
Culture medium	80% RPMI 1640 + 20% FBS
Viral status	EBV⁻
Authentication	no
Primary reference	Epstein et al.[1]
Availability of cell line	not known

Clinical characterization[1]

Patient	25-year-old male
Disease diagnosis	B-NHL (diffuse histiocytic)
Specimen site	pleural effusion

Immunophenotypic characterization[1-3]

T-/NK cell marker	$CD2^-$, $CD3^-$, $cyCD3^-$, $CD5^-$, $CD7^-$
B-cell marker	$CD10^-$, $CD19^+$, $cyIgG\lambda^+$, $sIgG\lambda^+$
Myelomonocytic marker	$CD13^-$, $CD14^-$, $CD15^-$, $CD33^-$
Erythroid-megakaryocytic marker	$CD41b^-$
Progenitor/activation/other marker	$HLA\text{-}DR^+$

Genetic characterization[2-5]

Cytogenetic karyotype	96, XXYY, −14, −14, +6mar, t(7;?), t(11;Y)(q23;q11), t(11;Y) (q23;q11), 18q−, 18q−
Receptor gene rearrangement	*IGH* RD, *IGK* RRR, *TCRB* G, *TCRG* G
Unique gene alteration	*P15INK4B* deletion, *P53* mutation

Functional characterization[5]

Cytochemistry	MPO^-

Comments

- Mature EBV⁻ B-cell line established from B-NHL ('diffuse histiocytic').

References

[1] Epstein, A.L. et al. (1978) Cancer 42, 2379–2391.
[2] Iida, S. et al. (1992) Leukemia Res. 16, 1155–1163.
[3] Kaiser-McCaw Hecht, B. et al. (1985) Cancer Genet. Cytogenet. 14, 205–218.
[4] Li, C.C. et al. (1995) Leukemia 9, 650–655.
[5] Siebert, R. et al. (1995) Brit. J. Haematol. 91, 350–354.

Tanoue

Culture characterization[1,2]

Subclones	**GR-Tanoue, GR-ST** (obtained by transfection with G-CSFR)
Culture medium	90% RPMI 1640 + 10% FBS
Doubling time	40–48 h
Viral status	EBV^-, HBV^-, HCV^-, HHV-8^-, HIV^-, HTLV-I/II^-
Authentication	yes (by cytogenetics)
Primary reference	El-Sonbaty et al.[1]
Availability of cell line	DSM ACC 399, RCB 1180

Clinical characterization[1]

Patient	11-year-old male
Disease diagnosis	B-ALL (L2)
Specimen site	peripheral blood
Year of establishment	1990

Immunophenotypic characterization[1,2]

T-/NK cell marker	$CD3^-$, $CD7^+$
B-cell marker	$CD10^+$, $CD19^+$, $CD20^+$, $CD37^{(+)}$, cyIgMλ^+, sIgMλ^+
Myelomonocytic marker	$CD13^-$, $CD14^-$, $CD15^-$, $CD33^+$
Progenitor/activation/other marker	$CD34^-$, HLA-DR^+
Adhesion marker	$CD11b^-$
Cytokine receptor	$CD114^-$

Genetic characterization[1-3]

Cytogenetic karyotype	hyperdiploid, 12% polyploidy; 47/48<2n>X, −Y/XYqh+, +7, +14, dup(1)(q21.1/21.2q23.1/23.2), t(2;4)(q2?2;q2?5), del(6) (q27), t(8;14)(q24;q32), del(13)(q34), add(22)(q13); extensive subclonal variation
Unique translocation/fusion gene	t(8;14)(q24;q32) → *MYC-IGH* genes altered?
Unique gene alteration	*P53* mutation

Functional characterization[1]

Cytochemistry	$ANAE^-$, MPO^-, $PAS^{(+)}$

Comments

- Mature EBV^- B-cell line derived from B-ALL (L2).
- Carries t(8;14)(q24;q32) which may lead to alteration of the *MYC-IGH* genes.
- Available from cell line banks.
- Cells grow well.

References
[1] El-Sonbaty, S.S. et al. (1995) Leukemia Res. 19, 249–256.
[2] Drexler, H.G. et al. (1999) DSMZ Catalogue of Cell Lines, 7th edn, Braunschweig, Germany.
[3] Drexler, H.G. et al. (2000) Leukemia 14, 198–206.

Culture characterization[1]

Culture medium	90% α-MEM + 10% FBS
Doubling time	5.3 d
Viral status	EBNA$^-$
Authentication	no
Primary reference	Yamaguchi et al.[1]
Availability of cell line	original authors

Clinical characterization[1]

Patient	80-year-old female
Disease diagnosis	B-NHL (large cell)
Specimen site	ascites
Year of establishment	1996

Immunophenotypic characterization[1]

B-cell marker	CD10$^+$, CD19$^+$, CD20$^+$
Progenitor/activation/other marker	HLA-DR$^+$

Genetic characterization[1]

Cytogenetic karyotype	69, XX, del(1)(p21), add(5)(p15), add(7)(p22), del(8)(p22), del(10)(q23), del(13)(q31), t(14;18)(q32;q21), add(17)(p22)
Unique translocation/fusion gene	t(14;18)(q32;q21) → *IGH-BCL2* genes altered (MBR)
Unique gene alteration	*P53* mutation

Functional characterization[1]

Proto-oncogene	BCL2 overexpression

Comments

- Mature EBV$^-$ B-cell line derived from B-NHL (large cell).
- Carries t(14;18)(q32;q21) leading to alteration of *BCL2-IGH* genes in the major breakpoint region.

Reference

[1] Yamaguchi, H. et al. (1999) Brit. J. Haematol. 105, 764–767.

Culture characterization[1]

Culture medium	80% α-MEM + 20% FBS + 5 ng/ml IL-3
Viral status	EBNA⁻
Authentication	yes (by *IGH* rearrangement)
Primary reference	Tohda et al.[1]
Availability of cell line	not known

Clinical characterization[1]

Patient	67-year-old male
Disease diagnosis	B-CLL
Treatment status	acute phase, lymphoblastic transformation
Specimen site	peripheral blood
Year of establishment	1990

Immunophenotypic characterization[1]

T-/NK cell marker	$CD2^-$
B-cell marker	$CD10^+$, $CD19^+$, $CD20^+$, $sIgM\kappa^+$
Myelomonocytic marker	$CD13^-$, $CD14^-$, $CD33^-$
Progenitor/activation/other marker	$HLA\text{-}DR^+$
Adhesion marker	$CD11b^-$
Cytokine receptor	$CD123^+$

Genetic characterization[1]

Cytogenetic karyotype	46, X, −Y, 5q−, 6p+, −8, 9p−, 11q+, 17p−, +2mar
Receptor gene rearrangement	*IGH* RR, *TCRB* G

Functional characterization[1,2]

Colony formation	in methylcellulose
Cytochemistry	$ANBE^-$, CAE^-, MPO^-, PAS^-
Cytokine response	IL-3-dependent

Comments

- Mature EBV^- B-cell line established from B-CLL in transformation.
- Constitutively IL-3-dependent.

References

[1] Tohda, S. et al. (1991) Blood 78, 1789–1794.
[2] Nara, N. (1992) Leukemia Lymphoma 7, 331–335. (mini-review on TMD2).

Culture characterization[1]

Culture medium	90% RPMI 1640 + 10% FBS
Doubling time	35 h
Viral status	EBNA⁻
Authentication	no
Primary reference	Nakamura et al.[1]
Availability of cell line	not known

Clinical characterization[1]

Patient	2-year-old female
Disease diagnosis	B-ALL (L3)
Treatment status	at 2nd relapse, refractory, pre-terminal
Specimen site	peripheral blood
Year of establishment	1992

Immunophenotypic characterization[1,2]

T-/NK cell marker	$CD2^-$, $CD3^-$, $CD4^-$, $CD5^-$, $CD7^-$, $CD8^-$
B-cell marker	$CD10^-$, $CD19^+$, $CD20^+$, $CD21^+$, $CD22^+$, $CD23^+$, $CD24^+$, $sIgG\kappa^+$ or $sIgM\lambda^+$
Myelomonocytic marker	$CD13^-$, $CD33^-$
Progenitor/activation/other marker	$CD34^-$, $CD95^+$, HLA-DR^+, TdT^+
Adhesion marker	$CD11b^-$
Cytokine receptor	$CD25^-$

Genetic characterization[1,2]

Cytogenetic karyotype	47, XX, dup(1)(q21q41), +der(6)(q12or21;q13), t(8;14)(q24;p32)
Unique translocation/fusion gene	t(8;14)(q24;q32) → *MYC-IGH* genes altered?

Functional characterization[1]

Special features	RAG1 $mRNA^+$

Comments

- Mature EBV^- B-cell line derived from B-ALL (L3).
- Carries t(8;14)(q24;q32) leading commonly to alterations of the *MYC* and *IGH* genes.
- Discrepancies regarding immunoglobulin expression (sIgGκ or sIgMλ).

References

[1] Nakamura, F. et al. (1996) Leukemia 10, 1159–1163.
[2] Tani, A. et al. (1996) Leukemia 10, 1592–1603.

TY-1

Culture characterization[1]

Culture medium	80% RPMI 1640 + 20% FBS
Viral status	EBV$^-$, HHV-8$^+$, HIV$^-$
Authentication	no
Primary reference	Katano et al.[1]
Availability of cell line	original authors

Clinical characterization[1]

Patient	47-year-old male (HIV$^+$)
Disease diagnosis	PEL + AIDS
Treatment status	at diagnosis
Specimen site	pericardial effusion
Year of establishment	1996

Immunophenotypic characterization[1]

T-/NK cell marker	CD3$^-$, CD4$^-$, CD5$^-$, CD8$^-$, CD56$^-$
B-cell marker	CD10$^-$, CD19$^-$, CD20$^-$, CD21$^-$, sIg$^-$
Myelomonocytic marker	CD14$^-$, CD68$^-$
Progenitor/activation/other marker	CD30$^+$, CD45$^+$, CD45RO$^+$
Adhesion marker	CD11b$^-$, CD11c$^-$

Genetic characterization[1]

Cytogenetic karyotype	50, XYq−, +7, +8, +11, +15, t(12;13), t(21;22)
Receptor gene rearrangement	*IGH* DD, *TCRB* G

Functional characterization[1]

Special features	TPA induces conversion from latent to lytic phase

Comments

- Mature EBV$^-$ HHV-8$^+$ HIV$^-$ B-cell line derived from AIDS-PEL.

Reference

[1] Katano, H. et al. (1999) J. Med. Virol. 58, 394–401.

U-698 M

Culture characterization[1,2]

Establishment	initially on allogeneic skin fibroblast feeder cells
Culture medium	90% Ham's F10 or RPMI 1640 + 10% FBS
Doubling time	48–60 h
Viral status	EBV⁻, HBV⁻, HCV⁻, HHV-8⁻, HIV⁻, HTLV-I/II⁻
Authentication	yes (by HLA typing)
Primary reference	Nilsson and Sundström[1]
Availability of cell line	DSM ACC 4

Clinical characterization[1]

Patient	7-year-old male
Disease diagnosis	lymphoblastic lymphosarcoma
Treatment status	at diagnosis, prior to therapy
Specimen site	tonsil
Year of establishment	1972

Immunophenotypic characterization[1-3]

T-/NK cell marker	CD1⁻, CD2⁻, CD3⁻, CD4⁻, CD5⁻, CD7⁻, CD8⁻
B-cell marker	CD9⁺, CD10⁺, CD19⁺, CD20⁺, CD21⁻, CD24⁺, CD37⁺, FMC-7⁺, NU-B1⁺, cyIgMDκ⁺, sIgMDκ⁺
Myelomonocytic marker	CD13⁻, CD15⁻
Progenitor/activation/other marker	CD38⁺, HLA-DR⁺, TdT⁻
Cytokine receptor	CD25⁺

Genetic characterization[1,2]

Cytogenetic karyotype	hyperdiploid, 5.5% polyploidy; 49(44–50)<2n>XY, +3, +7, −14, +mar, dup(1)(q43q21.2), der(2)t(2;3)(p16;p11), add(3)(p11), del(6)(q15q22), del(9)(p22), dup(11)(q23q13), add(13)(p12), add(16)(q24)

Functional characterization[1,4]

Cytochemistry	Alcian Blue⁻, PAS⁻
Heterotransplantation	into nude mice

Comments

- Mature EBV⁻ B-cell line derived from B-cell lymphoma.
- Available from cell line bank.

References

[1] Nilsson, K. and Sundström, C. (1974) Int. J. Cancer 13, 808–823.

[2] Drexler, H.G. et al. (1999) DSMZ Catalogue of Cell Lines, 7th edn, Braunschweig, Germany.

[3] Minowada, J. and Matsuo, Y. (1999) unpublished data.

[4] Nilsson, K. et al. (1977) Int. J. Cancer 19, 337–344.

U 2904

Culture characterization[1]

Culture medium	90% RPMI 1640 + 10% FBS
Doubling time	24 h
Viral status	EBV⁻
Authentication	yes (by *BCL2, IGH, MYC* rearrangements)
Primary reference	Sambade et al.[1]
Availability of cell line	original authors

Clinical characterization[1]

Patient	59-year-old male
Disease diagnosis	B-NHL (nodular centroblastic-centrocytic → centroblastic)
Treatment status	at relapse
Specimen site	pleural effusion
Year of establishment	1991

Immunophenotypic characterization[1]

T-/NK cell marker	$CD2^-$, $CD3^-$, $CD4^-$, $CD5^-$, $CD7^-$, $CD8^-$
B-cell marker	$CD10^{(+)}$, $CD19^+$, $CD20^+$, $sIgA\lambda^+$
Myelomonocytic marker	$CD13^-$, $CD14^-$, $CD15^-$, $CD25^-$, $CD33^-$
Progenitor/activation/other marker	$CD34^-$, $HLA\text{-}DR^+$, TdT^-
Cytokine receptor	$CD25^-$

Genetic characterization[1]

Cytogenetic karyotype	45, X, −Y, +4, −22, inv(1)(q12;q23), del(6)(q11→qtr), der(8)t (8;14)(q24;q32), der(10)t(1;10)(q1;q11), der(13)t(10;13)(q11;p11), t(14;18)(q32;q21), del(16)(q22)
Unique translocation/fusion gene	(1) t(8;14)(q24;q32) → *MYC-IGH* genes altered (2) t(14;18)(q32;q21) → *IGH-BCL2* genes altered (mcr)
Receptor gene rearrangement	*IGH* RR, *TCRB* G

Functional characterization[1]

Colony formation	in agarose
Heterotransplantation	into nude mice
Proto-oncogene	mRNA overexpression: BCL2, MYC

Comments

- Mature EBV⁻ B-cell line derived from secondary transformed centroblastic lymphoma.
- Carries both t(8;14)(q24;q32) and t(14;18)(q32;q21) leading to alterations of the *MYC-IGH* and *IGH-BCL2* (minor cluster region) genes.

Reference

[1] Sambade, C. et al. (1995) Int. J. Cancer 63, 710–715.

UoC-B2

Culture characterization[1]

Establishment	initially on feeder layer (media, agar, human serum)
Culture medium	85% McCoy's 5A + 15% FBS
Doubling time	50 h
Viral status	EBV⁻
Authentication	yes (by cytogenetics, *IGH* rearrangement)
Primary reference	Lorenzana et al.[1]
Availability of cell line	not known

Clinical characterization[1]

Patient	2-year-old male
Disease diagnosis	B-NHL (primary diffuse large cell lymphoma of skin)
Treatment status	at relapse
Specimen site	peripheral blood

Immunophenotypic characterization[1]

T-/NK cell marker	$CD1a^-$, $CD2^-$, $CD3^-$, $CD4^-$, $CD5^-$, $CD7^-$, $CD8^-$, $CD56^-$
B-cell marker	$CD10^-$, $CD19^+$, $CD20^-$, $CD21^-$, $CD22^+$, $sIgMD\lambda^+$
Myelomonocytic marker	$CD14^-$
Progenitor/activation/other marker	$CD30^-$, $CD34^+$, $CD38^+$, $CD45^+$, $CD71^-$, $HLA\text{-}DR^+$, TdT^-
Adhesion marker	$CD11c^-$
Cytokine receptor	$CD25^-$, $CD126^+$

Genetic characterization[1]

Cytogenetic karyotype	46, XY, +1, −11, +12, −13, −21, der(1)t(1;13)(p13;q11), del(6) (q2?1q2?5), add(7)(q11), t(10;12)(q11orq21;p11), +der(?)t(?;7) (?;q21)
Receptor gene rearrangement	*IGH* RR, *TCRB* G

Comments

- Mature EBV⁻ B-cell line derived from B-NHL (cutaneous DLCL).

Reference

[1] Lorenzana, A.N. et al. (1993) Cancer 72, 931–937.

UoC-B10

Culture characterization[1]

Establishment	initially on feeder layer (media, agar, human serum)
Culture medium	85% McCoy's 5A + 15% FBS
Viral status	EBV$^-$
Authentication	yes (by cytogenetics, *IGH* rearrangement)
Primary reference	Downie et al.[1]
Availability of cell line	not known

Clinical characterization[1]

Patient	26-year-old male
Disease diagnosis	mediastinal yolk sac tumor (mature teratoma) → MDS → BCP-ALL (L1/L3)
Treatment status	at diagnosis of leukemia
Specimen site	peripheral blood
Year of establishment	1991

Immunophenotypic characterization[1,2]

T-/NK cell marker	CD1a$^-$, CD2$^-$, CD3$^-$, CD4$^+$, CD5$^-$, CD7$^-$, CD8$^-$, CD56$^-$
B-cell marker	CD10$^+$, CD19$^+$, CD20$^-$, CD24$^-$, sIgDλ$^+$
Myelomonocytic marker	CD13$^-$, CD14$^-$, CD15$^-$, CD33$^-$
Erythroid-megakaryocytic marker	CD61$^-$, GlyA$^+$
Progenitor/activation/other marker	CD34$^-$, CD38$^+$, CD45$^+$, HLA-DR$^+$, TdT$^-$
Cytokine receptor	CD114$^+$

Genetic characterization[1,2]

Cytogenetic karyotype	46, XY, +11, dic(8;22)(q24;p11), del(11)(p12), del(11)(q13), +i(12)(p10), der(13;13)(q10;q10), der(20)t(17;20)(q11;q13.3)
Receptor gene rearrangement	*IGH* R

Functional characterization[1]

Cytochemistry	ANAE$^-$, MPO$^-$, PAS$^-$

Comments

- Mature EBV$^-$ B-cell line established from secondary BCP-ALL post-treatment for mediastinal germ cell tumor.

References

[1] Downie, P.A. et al. (1994) Cancer Res. 54, 4999–5004.

[2] Zhang, L.Q. et al. (1993) Leukemia 7, 1865–1874.

WSU-CLL

Culture characterization[1]

Culture medium	90% RPMI 1640 + 10% FBS
Doubling time	18 h
Viral status	EBNA$^-$
Authentication	no
Primary reference	Mohammad et al.[1]
Availability of cell line	original authors

Clinical characterization[1]

Patient	68-year-old male
Disease diagnosis	B-CLL
Treatment status	resistant, terminal
Specimen site	peripheral blood
Year of establishment	1992

Immunophenotypic characterization[1]

T-/NK cell marker	CD2$^-$, CD5$^-$, CD8$^-$
B-cell marker	CD10$^+$, CD19$^+$, CD20$^+$, CD21$^-$, CD22$^+$, CD24$^+$, CD37$^+$, cyIgGλ^+, sIgGλ^+
Myelomonocytic marker	CD13$^-$, CD14$^-$, CD33$^-$
Progenitor/activation/other marker	CD45$^+$, CD45R$^+$, HLA-DR$^+$
Adhesion marker	CD11c$^+$

Genetic characterization[1]

Cytogenetic karyotype	45, X, del(3)(p14;p24), t(4;12;12)(q31;q22;p13), t(5;12)(q31;p13), add(16)(q24)×2, t(18;21)(q12;p12)
Receptor gene rearrangement	*IGH* RR, *IGK* DD, *IGL* RD

Functional characterization[1]

Colony formation	in soft agar
Heterotransplantation	into SCID mice
Special features	sensitive to various chemotherapeutic agents

Comments

- Mature EBV$^-$ B-cell line derived from B-CLL.
- Model for xenografting of CLL cells and *in vivo* drug-sensitivity testing.

Reference

[1] Mohammad, R.M. et al. (1996) Leukemia 10, 130–137.

WSU-DLCL

Culture characterization[1]

Culture medium	90% RPMI 1640 + 10% FBS
Doubling time	20 h
Viral status	EBNA$^-$
Authentication	no
Primary reference	Mohammad et al.[1]
Availability of cell line	original authors

Clinical characterization[1]

Patient	32-year-old female
Disease diagnosis	B-NHL (diffuse large cell)
Treatment status	primary resistant, terminal
Specimen site	pleural effusion
Year of establishment	1988

Immunophenotypic characterization[1]

T-/NK cell marker	CD2$^-$, CD3$^-$, CD4$^-$, CD5$^-$, CD8$^-$
B-cell marker	CD10$^+$, CD19$^+$, CD20$^+$, CD21$^-$, CD22$^+$, CD24$^+$, CD37$^+$, PCA-1$^+$, cyIgMκ$^+$, sIgMκ$^+$
Myelomonocytic marker	CD13$^-$, CD14$^-$, CD33$^-$
Progenitor/activation/other marker	CD38$^-$, CD45R$^+$, HLA-DR$^+$
Adhesion marker	CD11b$^+$, CD11c$^-$
Cytokine receptor	CD25$^+$

Genetic characterization[1]

Cytogenetic karyotype	75, XXXXX, 14q+, 8q, 10q, t(3;4)

Functional characterization[1]

Colony formation	in soft agar
Special features	MDR1 mRNA overexpression

Comments

- Mature EBV$^-$ B-cell line derived from B-NHL (DLCL).

Reference

[1] Mohammad, R.M. et al. (1992) Cancer 69, 1468–1474.

WSU-FSCCL

Culture characterization[1]

Culture medium	90% RPMI 1640 + 10% FBS
Doubling time	26 h
Viral status	EBNA$^-$
Authentication	no
Primary reference	Mohammad et al.[1]
Availability of cell line	original authors

Clinical characterization[1]

Patient	37-year-old male
Disease diagnosis	B-NHL (low-grade follicular, small cleaved cell)
Treatment status	leukemic phase, stage IVB, refractory, terminal
Specimen site	peripheral blood
Year of establishment	1989

Immunophenotypic characterization[1]

T-/NK cell marker	CD2$^-$, CD3$^-$, CD4$^-$, CD5$^-$, CD8$^-$
B-cell marker	CD10$^+$, CD19$^+$, CD20$^+$, CD21$^-$, CD22$^+$, CD37$^+$, PCA-1$^-$, cyIgMκ$^+$, sIgMκ$^+$
Myelomonocytic marker	CD13$^-$, CD14$^-$, CD33$^-$
Progenitor/activation/other marker	CD38$^+$, HLA-DR$^+$
Adhesion marker	CD11b$^-$, CD11c$^+$

Genetic characterization[1]

Cytogenetic karyotype	47, XY, dup(1)(p11p34), +i(1)(q10q32), del(6q), t(8;11) (q24;q21), t(14;18)(q32;q21)
Unique translocation/fusion gene	t(14;18)(q32;q21) → *IGH-BCL2* genes altered (MBR)
Unique gene alteration	*MYC* rearrangement

Comments

- Mature EBV$^-$ B-cell line derived from follicular B-NHL.
- Carries t(14;18)(q32;q21) leading to alteration of the *IGH* and *BCL2* genes (major breakpoint region).
- Both *BCL2* and *MYC* genes are rearranged.

Reference

[1] Mohammad, R.M. et al. (1993) Cancer Genet. Cytogenet. 70, 62–67.

WSU-NHL

Culture characterization[1,2]

Culture medium	90% RPMI 1640 + 10% FBS
Doubling time	45–57 h
Viral status	EBV^-, HBV^-, HCV^-, $HHV\text{-}8^-$, HIV^-, $HTLV\text{-}I/II^-$
Authentication	no
Primary reference	Mohamed and Al-Katib[1]
Availability of cell line	DSM ACC 58

Clinical characterization[1]

Patient	48-year-old female
Disease diagnosis	B-NHL (nodular histiocytic, follicular, large cell)
Treatment status	refractory, progressive
Specimen site	pleural effusion
Year of establishment	1986

Immunophenotypic characterization[1,2]

T-/NK cell marker	$CD2^-$, $CD3^-$, $CD4^-$, $CD5^-$, $CD8^-$
B-cell marker	$CD10^+$, $CD19^+$, $CD20^+$, $CD21^-$, $CD24^+$, $CD37^-$, $PCA\text{-}1^-$, $cyIgG\lambda^+$, $sIgG\lambda^+$
Myelomonocytic marker	$CD13^-$, $CD14^-$, $CD33^-$
Progenitor/activation/other marker	$HLA\text{-}DR^+$
Adhesion marker	$CD11b^-$, $CD11c^-$

Genetic characterization[1–3]

Cytogenetic karyotype	hypodiploid, 6% polyploidy; 45(42–45)<2n>XX, −4, −15, +mar, der(1)t(1;4)(p36;q11), der(1)t(1;?7)(p34;q31.2)t(1;16)(q22;q22), del(5)(q32q35), der(6)t(1;6)(q22;p21), t(9;13)(p22;q13–14), der(13)t(13;?16)(p11;p11), t(14;18)(q32;q21), der(16)t(16;?)(q22;?)t(?;5)(?;q31), add(17)(p1?), add(19)(q13.2)
Unique translocation/fusion gene	t(14;18)(q32;q21) → *IGH-BCL2* genes altered (MBR)

Functional characterization[1]

Inducibility of differentiation	IFN-γ, TPA → differentiation

Comments

- Mature EBV^- B-cell line derived from follicular B-NHL.
- Carries t(14;18)(q32;q21) leading to alteration of the *IGH* and *BCL2* genes (major breakpoint region).
- Available from cell line bank.

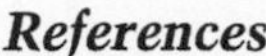

References

[1] Mohamed, A.N. and Al-Katib, A. (1988) Leukemia Res. 12, 833–843.
[2] Drexler, H.G. et al. (1999) DSMZ Catalogue of Cell Lines, 7th edn, Braunschweig, Germany.
[3] Mohammad, R.M. et al. (1993) Cancer Genet. Cytogenet. 70, 62–67.

WSU-WM

Culture characterization[1]

Culture medium	90% RPMI 1640 + 10% FBS
Doubling time	16 h
Viral status	EBNA$^-$
Authentication	no
Primary reference	Al-Katib et al.[1]
Availability of cell line	original authors

Clinical characterization[1]

Patient	60-year-old male
Disease diagnosis	Waldenström's macroglobulinemia (IgMκ)
Treatment status	refractory, pre-terminal
Specimen site	pleural effusion
Year of establishment	1990

Immunophenotypic characterization[1]

T-/NK cell marker	CD4$^-$, CD5$^-$
B-cell marker	CD10$^+$, CD19$^+$, CD20$^+$, CD21$^-$, CD22$^+$, CD37$^+$, cyIgMλ$^+$, sIgMλ$^+$
Myelomonocytic marker	CD13$^-$, CD14$^-$, CD33$^-$
Progenitor/activation/other marker	HLA-DR$^+$
Adhesion marker	CD11c$^-$
Cytokine receptor	CD25$^-$

Genetic characterization[1]

Cytogenetic karyotype	46–54, t(8;14)(q24;q32), t(12;17)(q24;q21), 10p−, t(Y;8)
Unique translocation/fusion gene	t(8;14)(q24;q32) → *MYC-IGH* genes altered
Receptor gene rearrangement	*IGK* D, *IGL* R

Functional characterization[1]

Colony formation	in soft agar
Heterotransplantation	into nude or SCID mice
Proto-oncogene	MYC mRNA overexpression
Special features	secretion: IgMλ

Comments

- Mature EBV$^-$ B-cell line derived from Waldenström's macroglobulinemia.
- Carries t(8;14)(q24;q32) causing alteration of the *MYC-IGH* genes.
- Discrepancy between IgMκ of patient's primary cells and IgMλ of WSU-WM: authentication of proper derivation and exclusion of cross-contamination with another cell line is strongly recommended.
- Cells grow very well.

Reference

[1] Al-Katib, A. et al. (1993) Blood 81, 3034–3042.

Culture characterization[1]

Culture medium 90% RPMI 1640 + 10% FBS
Doubling time 36–48 h
Authentication no
Primary reference Yonetani et al.[1]
Availability of cell line original authors

Clinical characterization[1]

Patient 61-year-old male
Disease diagnosis B-NHL (DLCL, immunoblastic variant)
Treatment status stage III, refractory
Specimen site ascites
Year of establishment 1992

Immunophenotypic characterization[1]

B-cell marker CD10⁻, CD19⁺, CD20⁺, sIgMDκ⁺
Progenitor/activation/other marker HLA-DR⁺

Genetic characterization[1–3]

Cytogenetic karyotype 46, X, +X, −Y, −2, −2, −4, −7, −8, −18, +6mar, add(1)(p11), t(2;18)(p11;q21), t(3;16)(q27;p11.2), add(13)(q22), der(14)dir ins(14;?)(q11;?), add(17)(p11), add(21)(p11)
Unique translocation/fusion gene (1) t(2;18)(p11;q21) → *IGK-BCL2* genes altered
(2) t(3;16)(q27;p11.2) → *BCL6*-? genes altered

Functional characterization[1]

Proto-oncogene mRNA overexpression: BCL2, BCL6

Comments

- Mature B-cell line derived from B-NHL (DLCL).
- Cells carry both *BCL2* and *BCL6* rearrangements deregulated by t(2;18)(p11;q21) and t(3;16)(q27;p11.2).

References

[1] Yonetani, N. et al. (1998) Oncogene 17, 971–979.
[2] Muramatsu, M. et al. (1996) Brit. J. Haematol. 93, 911–920.
[3] Ohno, H. et al. (1994) Jpn. J. Cancer Res. 85, 592–600.

Z-138

Culture characterization[1]

Culture medium	90% RPMI 1640 + 10% FBS
Doubling time	18–24 h
Viral status	EBV⁻
Authentication	yes (by cytogenetics, *IGH* rearrangement)
Primary reference	Estrov et al.[1]
Availability of cell line	not known

Clinical characterization[1]

Patient	70-year-old male
Disease diagnosis	B-CLL → B-ALL (L2)
Treatment status	at relapse, progressive
Specimen site	bone marrow
Year of establishment	1990

Immunophenotypic characterization[1]

T-/NK cell marker	CD2⁻, CD3⁻, CD4⁽⁺⁾, CD5⁻, CD7⁻, CD8⁻, CD56⁻
B-cell marker	CD10⁻, CD19⁺, CD20⁺, CD22⁺, sIgMDλ⁺
Myelomonocytic marker	CD13⁻, CD14⁻, CD33⁻
Progenitor/activation/other marker	CD34⁻, CD38⁺, HLA-DR⁺, TdT⁻
Adhesion marker	CD11b⁻
Cytokine receptor	CD25⁻

Genetic characterization[1]

Cytogenetic karyotype	48, XY, +12, +13, der(8)t(8;14;?)(q24;q32;?), der(9)t(9;?)(q34;?), del(11)(q13q25), del(12)(q22q24.1), der(14)t(14;18;?)(q32;q21;?), der(18)t(14;18;?)(q32;q21;?)
Unique translocation/fusion gene	(1) t(8;14)(q24;q32) → *MYC-IGH* genes altered? (*MYC* in germline) (2) t(14;18)(q32;q21) → *IGH-BCL2* genes altered?
Receptor gene rearrangement	*IGH* RD

Functional characterization[1]

Cytochemistry	CAE⁻, MPO⁻, PAS⁽⁺⁾
Cytokine production	G-CSF, GM-CSF
Proto-oncogene	cyclin D1 mRNA overexpression

Comments

- Mature EBV⁻ B-cell line derived from B-CLL transformed into B-ALL.
- Carries t(8;14)(q24;q32), del(11)(q13q25), and t(14;18)(q32;q21) possibly leading to alterations of the *MYC, BCL1, IGH* and *BCL2* genes.

Reference

[1] Estrov, Z. et al. (1998) Leukemia Res. 22, 341–353.

Section III

MYELOMA AND PLASMA CELL LEUKEMIA CELL LINES

ACB-885

Culture characterization[1]

Sister cell line	**ACB-1085** (simultaneous sister cell line – from same pleural effusion sample – identical features)
Culture medium	80% IMDM + 5% FBS + 5% PHA-LCM + 10% human plasma
Doubling time	30–35 h
Viral status	EBV⁻
Authentication	yes (by cytogenetics)
Primary reference	Brox et al.[1]
Availability of cell line	not known

Clinical characterization[1]

Patient	39-year-old male
Disease diagnosis	multiple myeloma (IgGκ)
Treatment status	refractory
Specimen site	pleural effusion

Immunophenotypic characterization[1]

T-/NK cell marker	$CD2^-$, $CD3^-$, $CD4^-$, $CD8^-$
B-cell marker	$CD10^-$, $CD20^-$, $CD21^-$, $cyIgG\kappa^+$
Progenitor/activation/other marker	$CD38^+$, HLA-DR^-

Genetic characterization[1]

Cytogenetic karyotype	46, XY, −1, −2, −6, −7, −8, −10, −12, −13, +21, −22, +8mar, t(1;10)(1qter→1q22::10p11.2→10qter), t(2;10)(2qter→2p11.2::10p11.2→10pter), t(1;9)(1pter→1p32::9p21→9p22), t(1;9)(1qter→ 1q11::9p13→9p22), t(2;8)(2pter→2p11.2::8q24→8pter)

Functional characterization[1]

Cytochemistry	MGP^+, PAS^+

Comments

- Myeloma cell line of IgGκ-type.
- Cell lines possibly lost due to retirement of principal investigator.

Reference

[1] Brox, L.W. et al. (1987) Cancer Genet. Cytogenet. 27, 135–144.

AMO1

Culture characterization[1]

Establishment	initially on macrophage feeder layer
Culture medium	93% RPMI 1640 + 7% FBS
Doubling time	24 h
Viral status	EBV$^-$
Authentication	no
Primary reference	Shimizu et al.[1]
Availability of cell line	original authors

Clinical characterization[1]

Patient	64-year-old female
Disease diagnosis	plasmacytoma (in duodenum; IgAκ)
Treatment status	at presentation
Specimen site	ascites
Year of establishment	1984

Immunophenotypic characterization[1]

T-/NK cell marker	CD2$^-$, CD3$^-$, CD4$^+$, CD5$^-$, CD8$^-$
B-cell marker	CD10$^-$, CD19$^-$, CD20$^-$, CD21$^-$, PCA-1$^+$, cyIgAκ$^+$
Myelomonocytic marker	CD13$^-$, CD33$^-$
Progenitor/activation/other marker	CD34$^-$, CD38$^+$
Adhesion marker	CD11b$^-$, CD11c$^-$

Genetic characterization[1]

Cytogenetic karyotype	46, X, −X, +12, 2p+, 8q+, 10q+, 12q+, 12p−, 14q+, 15p+, der(12)t(12;14)(p11;q32), der(14)t(14;?)(q32;?), der(15)t(12;15)(p11;p13)
Receptor gene rearrangement	*IGH* R, *IGK* R, *TCRB* G, *TCRG* G

Functional characterization[1]

Cytochemistry	ACP$^+$, ANAE$^{(+)}$, MPO$^-$

Comments

- Plasmacytoma cell line of IgAκ-type.
- Expresses also T-cell marker CD4.

Reference

[1] Shimizu, S. et al. (1993) Leukemia 7, 274–280.

ANBL-6

Culture characterization[1]

Sister cell line	**ANBM-6** (simultaneous sister cell line – from bone marrow – phenotypically indistinguishable)
Culture medium	90% RPMI 1640 + 10% FBS + 1 ng/ml IL-6
Viral status	EBNA$^-$
Authentication	no
Primary reference	Jelinek et al.[1]
Availability of cell line	original authors

Clinical characterization[1]

Patient	67-year-old female
Disease diagnosis	multiple myeloma
Treatment status	at 4th relapse
Specimen site	peripheral blood
Year of establishment	1992

Immunophenotypic characterization[1]

T-/NK cell marker	CD2$^-$, CD3$^-$, CD5$^-$
B-cell marker	CD10$^+$, CD19$^-$, CD20$^-$, CD21$^-$, CD22$^-$, CD23$^-$, CD40$^+$, CD73$^-$, cyIgλ^+, sIg$^-$
Myelomonocytic marker	CD32$^+$
Progenitor/activation/other marker	CD38$^+$, CD45$^-$, CD95$^-$, HLA-DR$^+$
Adhesion marker	CD11a$^-$, CD11b$^-$, CD18$^-$, CD29$^+$, CD44$^+$, CD49d$^+$, CD49e$^-$, CD54$^+$, CD56$^-$
Cytokine receptor	CD25$^+$

Genetic characterization[1,2]

Cytogenetic karyotype	87–88, XX, −X, −X, −1, −1, −5, −9, −9, −13, −13, −14, −14, −14, −15, −17, −17, −22, −22, +15-17mar, der(6)t(1;6)(q11;q13)×2, der(?8)t(8;?)(?p11.1;?)×2, der(10)t(9;10)(q13;q26)×2, t(14;16)(q32;q23), der(15)t(15;?)(q1?5;?); near-diploid sideline with similar structural changes
Receptor gene rearrangement	*IGH* R, *IGK* R

Functional characterization[1–3]

Cytokine response	IL-6-dependent; growth inhibition: IFN-α, IFN-γ, TGF-β, TNF-α
Proto-oncogene	MAF mRNA$^+$
Special features	resistant to FAS-induced apoptosis

Comments

- Myeloma cell line of Igλ type.
- Cells described to be IL-6-dependent, but resistant to FAS-induced apoptosis.

References

[1] Jelinek, D.F. et al. (1993) Cancer Res. 53, 5320–5327.
[2] Chesi, M. et al. (1998) Blood 91, 4457–4463.
[3] Westendorf, J.J. et al. (1995) Blood 85, 3566–3576.

Culture characterization[1]

Culture medium	90% RPMI 1640 + 10% FBS
Doubling time	40 h
Viral status	EBNA$^-$
Authentication	no
Primary reference	Ishii et al.[1]
Availability of cell line	restricted

Clinical characterization[1]

Patient	73-year-old male
Disease diagnosis	myeloma (testicular tumor; IgDλ)
Treatment status	refractory, terminal
Specimen site	ascites
Year of establishment	1986

Immunophenotypic characterization[1]

T-/NK cell marker	CD2$^-$, CD4$^{(+)}$, CD5$^-$, CD8$^-$
B-cell marker	CD10$^-$, CD19$^-$, PCA-1$^-$, cyIgDλ$^+$, sIg$^-$
Myelomonocytic marker	CD13$^-$
Progenitor/activation/other marker	CD30$^+$, CD38$^+$, HLA-DR$^-$

Genetic characterization[1]

Cytogenetic karyotype	46, X, −Y, −1, −2, −8, −10, −11, −11, −12, −14, +6mar, +der(1)t(1;?)(p34;?), +der(2)t(2;?)(p23;?), +der(14)t(14;?)(q32;?)

Functional characterization[1]

Special features	secretion: IgDλ

Comments

- Myeloma cell line of rare IgDλ-type.

Reference

[1] Ishii, K. et al. (1992) Am. J. Hematol. 41, 218–224.

DOBIL-6

Culture characterization[1]

Culture medium	90% RPMI 1640 + 10% FBS
Doubling time	36 h
Viral status	EBV⁻
Authentication	yes (by cytogenetics)
Primary reference	Ohmori et al.[1]
Availability of cell line	original authors

Clinical characterization[1]

Patient	65-year-old female
Disease diagnosis	myeloma (non-secretory)
Treatment status	relapse, refractory
Specimen site	bone marrow
Year of establishment	1995

Immunophenotypic characterization[1]

T-/NK cell marker	$CD2^-$, $CD3^-$, $CD4^-$, $CD5^-$, $CD7^-$, $CD8^-$
B-cell marker	$CD10^-$, $CD19^+$, $CD20^+$, $CD21^-$, $CD40^+$, PCA-1^+, $cyIgG^+$, sIg^-
Myelomonocytic marker	$CD13^-$, $CD33^-$
Progenitor/activation/other marker	$CD38^+$, $CD45RA^-$, $CD45RO^+$, HLA-DR^+
Adhesion marker	$CD11a^-$, $CD18^-$, $CD29^+$, $CD44^+$, $CD49d^+$, $CD49e^-$, $CD54^+$, $CD56^-$
Cytokine receptor	$CD126^+$

Genetic characterization[1]

Cytogenetic karyotype	47, X, −X, +9, +18, der(1)t(1;3)(p36.1;p25)t(1;22)(q21;q13.1), der(2)t(2;10)(p11.2;p11.2), del(2)(q32.2q35), inv(3)(q25q29), add(3)(p25), add(6)(p21.3), der(10)t(2;10)(p11.2;p11.2), der(11)t(6;11)(p21.3;q13.1), t(12;16)(p11.2;q12), der(14)t(11;14)(q13; q32.1), der(17)t(14;17)(q24;q21), i(17)(q10), add(19)(p13.3), der(19)t(17;19)(q21;q13.3), del(20)(q11.2q13.1), der(22)t(1;22)(q21;q13.1)
Unique translocation/fusion gene	no *PRAD1-IGH* fusion gene
Receptor gene rearrangement	*IGH* RR, *IGK* RR, *IGL* GG, *TCRB* RG

Functional characterization[1]

Cytokine response	growth stimulation: IL-6
Heterotransplantation	into nude mice
Special features	secretion: IL-6, parathyroid hormone-related protein

Comments

- Myeloma cell line of IgG-type.
- Cells secrete IL-6 and parathyroid hormone-related protein.

Reference

[1] Ohmori, M. et al. (1998) Brit. J. Haematol. 101, 688–693.

Culture characterization[1,2]

Culture medium 90% IMDM + Ham's F12 (1:1) + 10% FBS
Doubling time 60–72 h
Viral status EBV⁻
Authentication no
Primary reference Hamilton et al.[1]
Availability of cell line original authors

Clinical characterization[1]

Patient 58-year-old female
Disease diagnosis myeloma (IgGλ)
Treatment status refractory, terminal
Specimen site peritoneal effusion (ascites)
Year of establishment 1988

Immunophenotypic characterization[1]

B-cell marker $CD9^+$, $CD20^+$, $CD75^+$, $CD78^+$, $cyIgG\lambda^+$, sIg^-
Myelomonocytic marker $CD32^+$
Progenitor/activation/other marker $CD38^+$, $CD71^+$, HLA-DR^+
Adhesion marker $CD44^+$, $CD54^+$, $CD56^-$

Genetic characterization[1,3,4]

Cytogenetic karyotype 48, X, +1, −4, +5, −13, −13, −14, −15, −15, −16, −16, −18, +20, −21, −22, +7mar, del(3q), del(6)(q21), +9p, +9p, +11p, +11p, +12q, iso(14q), +18p
Unique gene alteration *P53* mutation

Functional characterization[1,5]

Cytochemistry $ANAE^-$, PAS^-, SBB^-
Cytokine production TGF-β $mRNA^+$
Cytokine response growth stimulation: IL-6; growth inhibition: IFN-α
Special features secretion: IgGλ

Comments

- Myeloma cell line of IgGλ-type.

References

[1] Hamilton, M.S. et al. (1990) Brit. J. Haematol. 75, 378–384.
[2] Georgii-Hemming, P. et al. (1999) Blood 93, 1724–1731.
[3] Mazars, G.R. et al. (1992) Oncogene 7, 1015–1018.
[4] Kuipers, J. et al. (1999) Cancer Genet. Cytogenet. 109, 99–107.
[5] Portier, M. et al. (1993) Brit. J. Haematol. 85, 514–520.

FLAM-76

Culture characterization[1]

Culture medium	90% RPMI 1640 + 10% FBS + 4 ng/ml IL-6
Viral status	EBNA⁻
Authentication	no
Primary reference	Kubonishi et al.[1]
Availability of cell line	original authors

Clinical characterization[1]

Patient	77-year-old male
Disease diagnosis	plasma cell leukemia (Igκ) (aggressive, non-secretory)
Treatment status	refractory, terminal
Specimen site	peripheral blood
Year of establishment	1989

Immunophenotypic characterization[1]

T-/NK cell marker	$CD2^-$, $CD3^-$, $CD4^-$, $CD5^-$, $CD8^-$
B-cell marker	$CD10^-$, $CD19^-$, $CD20^-$, PCA-1^-, cyIgκ^+, sIg^-
Myelomonocytic marker	$CD13^-$, $CD15^-$, $CD32^-$, $CD33^-$
Progenitor/activation/other marker	$CD30^-$, $CD38^+$, HLA-DR^-
Adhesion marker	$CD11a^-$, $CD18^-$, $CD44^+$, $CD54^+$

Genetic characterization[1–3]

Cytogenetic karyotype	43, X, −Y, −8, −13, −14, −21, −22, +2mar, t(11;14)(q13;q32), +der(14)t(11;14)(q13;q32)
Unique translocation/fusion gene	*BCL1* not rearranged
Receptor gene rearrangement	*IGH* R, *IGK* R
Unique gene alteration	*P16INK4A* deletion, *P18INK4C* deletion

Functional characterization[1]

Cytochemistry	ACP^+, ALP^-, $ANBE^{(+)}$, CAE^-, MPO^-, PAS^-, SBB^-
Cytokine response	IL-6-dependent; growth inhibition: IFN-α
Transcription factor	cyclin D1 overexpression

Comments

- Plasma cell line of Igκ-type.
- Constitutively IL-6-dependent.

References

[1] Kubonishi, I. et al. (1992) Cancer 70, 1528–1535.
[2] Williams, M.E. et al. (1996) Blood 88, 102a.
[3] Kuipers, J. et al. (1999) Cancer Genet. Cytogenet. 109, 99–107.

Culture characterization[1]

Subclone	**AD3** (with immunological and cytogenetic differences)
Culture medium	90% RPMI 1640 + 10% FBS
Doubling time	131 h
Viral status	EBV$^-$
Authentication	no
Primary reference	Tagawa et al.[1]
Availability of cell line	restricted

Clinical characterization[1]

Patient	69-year-old male
Disease diagnosis	plasmacytoma (IgAκ)
Treatment status	at diagnosis
Specimen site	ascites
Year of establishment	1986

Immunophenotypic characterization[1]

B-cell marker	CD10$^+$, CD19$^{(+)}$, CD20$^{(+)}$, PCA-1$^+$, cyIgAκ$^+$, sIg$^-$
Progenitor/activation/other marker	CD38$^+$, HLA-DR$^+$

Genetic characterization[1-4]

Cytogenetic karyotype	67, X, −Y, +3, +7, +9, +11, +11, +12, −13, −13, −14, − 15, +16, +17, +18, +19, +20, +22, +22, +3mar, del(1)(q21)×2, +der(1)t(1;7)(q11;p11.2), +der(1)t(?;1)(1;?)(?;p32q22;?), del(5)(q22q35), +i(6q), +der(8)dic(1;8)(p13;q24.1), +der(8)t(8;?)(q24;?), +der(13)t(1;13)(q11;p11)×2, der(14)t(8;14)(q24.1;q32)×2, +der(15)t(9;15)(q13;p11), i(21q); SKY: t(8;14)(q24;q32), add(14)(q32)
Unique translocation/fusion gene	t(6;14)(p25;q32) → *MUM1/IRF4-IGH* genes altered
Unique gene alteration	*MYC* G

Functional characterization[1]

Cytochemistry	ACP$^+$, ANBE$^+$
Proto-oncogene	MYC mRNA$^+$
Special features	secretion: IgA, amylase

Comments

- Plasma cell line of IgAκ-type.
- Carries t(6;14)(p25;q32) leading to alteration of *MUM1/IRF4-IGH* genes.
- Cells secrete amylase and IgA.

References

[1] Tagawa, S. et al. (1990) Leukemia 4, 600–605.
[2] Taniwaki, M. et al. (1994) Blood 83, 2962–2969.
[3] Rao, P.H. et al. (1998) Blood 92, 1743–1748.
[4] Yoshida, S. et al. (1999) Leukemia 13, 1812–1816.

Fravel

Culture characterization[1]

Other name of cell line	Fr
Culture medium	85% RPMI 1640 + 15% FBS
Doubling time	62 h
Viral status	EBNA$^-$
Authentication	no
Primary reference	Miller et al.[1]
Availability of cell line	restricted

Clinical characterization[1]

Patient	male
Disease diagnosis	multiple myeloma (IgGκ)
Specimen site	bone marrow
Year of establishment	1978

Immunophenotypic characterization[1,2]

T-/NK cell marker	CD2$^-$, CD3$^-$, CD4$^-$, CD5$^-$, CD7$^-$, CD8$^-$
B-cell marker	CD9$^-$, CD10$^+$, CD19$^+$, CD20$^+$, CD21$^+$, CD22$^+$, CD23$^-$, CD24$^-$, CD37$^+$, PCA-1$^-$, cyIg$^-$, sIg$^-$
Myelomonocytic marker	CD15$^-$, CD16$^-$
Progenitor/activation/other marker	CD30$^-$, CD45$^+$, CD71$^+$, EMA$^+$, HLA-DR$^+$
Adhesion marker	CD11$^-$, CD18$^+$, CD56$^-$
Cytokine receptor	CD25$^-$

Genetic characterization[1,3]

Cytogenetic karyotype	46–49, XY, −7, −8, del(7)(q32)×2, 14q+

Functional characterization[1,4]

Colony formation	in agar or methylcellulose
Cytokine production	TNF-β
Special features	secretion: IgGκ

Comments

- Myeloma cell line of IgGκ-type.

References

[1] Miller, C.H. et al. (1982) Cancer 49, 2091–2096.
[2] Duperray, C. et al. (1989) Blood 73, 566–572.
[3] Kuipers, J. et al. (1999) Cancer Genet. Cytogenet. 109, 99–107.
[4] Bataille, R. et al. (1989) Cancer 63, 877–880.

HL461

Culture characterization[1]

Other name of cell line	**UCD-HL461**
Establishment	on bone marrow stroma feeder cells, then independent
Culture medium	90% RPMI 1640 + 10% FBS
Doubling time	50 h
Viral status	EBNA$^-$
Authentication	yes (by cytogenetics)
Primary reference	Scibienski et al.[1]
Availability of cell line	original authors

Clinical characterization[1]

Disease diagnosis	multiple myeloma (IgGκ)
Treatment status	stage IIIB, at diagnosis
Specimen site	bone marrow
Year of establishment	1986

Immunophenotypic characterization[1]

T-/NK cell marker	CD1$^-$, CD2$^-$, CD3$^-$, CD4$^-$, CD5$^+$, CD7$^-$, CD8$^-$, CD57$^-$
B-cell marker	CD9$^-$, CD10$^+$, CD19$^+$, CD20$^+$, CD21$^-$, CD22$^+$, CD24$^+$, CD35$^-$, PCA-1$^+$, cyIgM$^-$, sIgκ$^+$
Myelomonocytic marker	CD13$^-$, CD14$^-$, CD16$^-$, CD33$^-$
Progenitor/activation/other marker	CD38$^+$, HLA-DR$^+$
Adhesion marker	CD11$^-$, CD56$^-$
Cytokine receptor	CD25$^+$

Genetic characterization[1]

Cytogenetic karyotype	80–83, multiple abnormalities

Functional characterization[1,2]

Cytokine receptor	RT-PCR$^+$: IL-6R
Special features	secretion: $IgG_4κ$

Comments

- Myeloma cell line of IgGκ-type.

References

[1] Scibienski, R.J. et al. (1990) Leukemia 4, 775–780.
[2] Scibienski, R.J. et al. (1992) Leukemia 6, 940–947.

HSM-2

Culture characterization[1]

Subclone	**HSM-2.3** (IL-3/IL-6-dependent)
Culture medium	80% IMDM + 20% FBS + 10 U/ml IL-6
Authentication	no
Primary reference	Kobayashi et al.[1]
Availability of cell line	not known

Clinical characterization[1]

Patient	90-year-old male
Disease diagnosis	plasma cell leukemia
Treatment status	at diagnosis
Specimen site	bone marrow
Year of establishment	1989

Immunophenotypic characterization[1]

T-/NK cell marker	$CD1^-$, $CD2^-$, $CD7^-$
B-cell marker	$CD10^-$, $CD19^-$, $CD20^-$, $CD24^-$, PCA-1^+, $cyIgM^+$, $sIg\kappa^+$
Myelomonocytic marker	$CD13^-$, $CD14^-$, $CD33^-$
Progenitor/activation/other marker	$CD38^+$, HLA-DR^-

Genetic characterization[1]

Cytogenetic karyotype	45(44–47), −5, −9, −9, −13, −13, −15, −15, −22, −Y, 1q+, 3q−, 4q−, 5q−, 7p+, 10p−, 10q+, 19p+, 22p+

Functional characterization[1]

Cytochemistry	$ANBE^-$, CAE^-, MPO^-, PAS^+, SBB^-
Cytokine response	IL-6-dependent

Comments

- Plasma cell line of IgMκ-type.
- Constitutively IL-6-dependent.

Reference

[1] Kobayashi, M. et al. (1991) Brit. J. Haematol. 78, 217–221.

JJN-2

Culture characterization[1]

Subclones	**JJN-1** (parental cell line, hypodiploid); **JJN-3** (subclone of JJN-1, IL-6-independent)
Culture medium	40% IMDM + 40% Ham's F12 + 20% FBS + IL-6
Doubling time	3–4 d
Viral status	EBNA$^-$
Authentication	no
Primary reference	Jackson et al.[1]
Availability of cell line	original authors

Clinical characterization[1]

Patient	57-year-old female
Disease diagnosis	plasma cell leukemia ($IgA_1\kappa$)
Treatment status	at diagnosis
Specimen site	bone marrow
Year of establishment	1987

Immunophenotypic characterization[1]

B-cell marker	CD9$^-$, CD10$^-$, CD19$^-$, CD20$^-$, CD21$^-$, CD22$^-$, CD23$^-$, CD24$^+$, CD37$^-$, CD39$^+$, CD40$^-$, cyIgAκ$^+$, sIgAκ$^{(+)}$
Myelomonocytic marker	CD32$^+$
Progenitor/activation/other marker	CD38$^+$, CD45RA$^-$, HLA-DR$^-$
Cytokine receptor	CD25$^-$

Genetic characterization[1-3]

Cytogenetic karyotype	75(61–82)XXX, −1, −2, −3, −7, −7, −9, −9, −10, −10, −11, −11, −12, −12, −13, −13, −14, −14, −15, −16, −16, −17, −20, +4mar, 1q+, 1q−, del(3)(q23), 5p+, 5p+, del(6)(q15)×2, +der(7)t(7;11)(q32;q12), del(8)(p21)×2, 14q+, 14q+, del(20)(p11) JJN-3: t(14;16)(q32;q23)

Functional characterization[1,4]

Cytokine production	JJN-3: TGF-β mRNA$^+$
Cytokine response	IL-6-dependent
Proto-oncogene	JJN-3: MAF mRNA$^+$
Special features	secretion: IgAκ

Comments

- Plasma cell line of IgAκ-type.
- Constitutively IL-6-dependent.
- Various subclones available.

References

[1] Jackson, N. et al. (1989) Clin. Exp. Immunol. 75, 93–99.
[2] Chesi, M. et al. (1998) Blood 91, 4457–4463.
[3] Kuipers, J. et al. (1999) Cancer Genet. Cytogenet. 109, 99–107.
[4] Portier, M. et al. (1993) Brit. J. Haematol. 85, 514–520.

Karpas 620

Culture characterization[1]

Other name of cell line	**K620**
Culture medium	80% RPMI 1640 + 20% FBS
Viral status	EBNA$^-$
Authentication	yes (by cytogenetics)
Primary reference	Nacheva et al.[1]
Availability of cell line	original authors

Clinical characterization[1]

Patient	77-year-old female
Disease diagnosis	plasma cell leukemia (IgGκ)
Treatment status	at diagnosis
Specimen site	peripheral blood
Year of establishment	1987

Immunophenotypic characterization[1]

T-/NK cell marker	CD1a$^-$, CD2$^-$, CD3$^-$, CD4$^-$, CD5$^-$, CD7$^-$, CD8$^-$, CD57$^-$
B-cell marker	CD10$^-$, CD20$^-$, CD37$^+$, cyIg$^+$, sIg$^{(+)}$
Myelomonocytic marker	CD14$^-$, CD15$^-$, CD33$^-$
Progenitor/activation/other marker	CD38$^-$, CD45$^-$, CD52$^-$, HLA-DR$^+$

Genetic characterization[1]

Cytogenetic karyotype	82<4n>X-XXX, −2, −3, −4, −5, −6, −9, −12, −22, t(1;14) (q11;q32.3)×2, t(1;17)(q11;p13.1)×2, t(1;11)(q32.1;q13.3)×2, t(8;11)(q24.22;q13.3)×2, t(8;14)(q24.1;q32.3)×2, t(11;13) (q13.3;q14.3)×2
Unique translocation/fusion gene	t(8;14)(q24;q32) → *MYC-IGH* genes altered
Unique gene alteration	*BCL1* rearrangement

Functional characterization[1]

Cytochemistry	ACP$^+$, ANBE$^-$, CAE$^-$, SBB$^-$
Special features	secretion: Igκ

Comments

- Plasma cell line of IgGκ-type.
- Carries t(8;14)(q24;q32) affecting the *MYC-IGH* genes.
- *BCL1* gene on 11q13 is rearranged.

Reference

[1] Nacheva, E. et al. (1990) Brit. J. Haematol. 74, 70–76.

Karpas 707

Culture characterization[1,2]

Sister cell line	two identical serial sister cell lines from bone marrow and peripheral blood (same cell line name)
Culture medium	90% RPMI 1640 + 10% FBS
Doubling time	48–72 h
Viral status	EBNA⁻
Authentication	no
Primary reference	Karpas et al.[1]
Availability of cell line	original authors

Clinical characterization[1,2]

Patient	53-year-old male
Disease diagnosis	multiple myeloma (IgGλ)
Treatment status	at relapse, refractory, terminal
Specimen site	peripheral blood and bone marrow
Year of establishment	1981

Immunophenotypic characterization[1,2]

T-/NK cell marker	$CD2^-$
B-cell marker	$cyIg\lambda^+$, sIg^-

Genetic characterization[1,2]

Cytogenetic karyotype	45, XY, −5, −6, +9, −12, −13, −16, −17, 4p+, +6p, +t(6q;7q), Ph

Functional characterization[1,2]

Cytochemistry	ACP^+, ALP^-, CAE^+, PAS^-, SBB^-
Special features	secretion: Igλ

Comments

- Myeloma cell line of IgGλ-type.
- Unlikely to carry Philadelphia chromosome t(9;22).

References

[1] Karpas, A. et al. (1982) Lancet i, 931–933.
[2] Karpas, A. et al. (1982) Science 216, 997–999.

KMM-1

Culture characterization[1–3]

Culture medium	90% RPMI 1640 + 10% FBS (or serum-free 50% IMDM + 50% Ham's F12 + additives)
Doubling time	36–40 h
Viral status	EBNA$^-$
Authentication	no
Primary reference	Togawa et al.[1]
Availability of cell line	RCB 0193

Clinical characterization[1–3]

Patient	62-year-old male
Disease diagnosis	multiple myeloma (Igλ)
Treatment status	refractory, terminal
Specimen site	subcutaneous plasmacytoma
Year of establishment	1980

Immunophenotypic characterization[1–6]

T-/NK cell marker	CD1b$^-$, CD2$^-$, CD3$^-$, CD4$^-$, CD5$^-$, CD7$^-$, CD8$^-$, CD28$^+$, CD57$^-$, TCRαβ$^-$
B-cell marker	CD9$^-$, CD10$^-$, CD19$^-$, CD20$^-$, CD21$^-$, FMC-7$^-$, NU-B1$^-$, PCA-1$^+$, cyIgλ$^+$, sIg$^-$
Myelomonocytic marker	CD13$^-$, CD15$^-$
Erythroid-megakaryocytic marker	CD41a$^-$, CD61$^-$
Progenitor/activation/other marker	CD38$^+$, CD71$^+$, CD95$^+$, HLA-DR$^-$, TdT$^-$
Cytokine receptor	CD25$^-$, CD114$^-$, CD117$^-$, CD119$^+$, CD120a$^-$, CD120b$^+$, CD122$^+$, CD124$^-$, CD126$^-$, CD127$^-$, CD128$^-$, CD130$^+$

Genetic characterization[1]

Cytogenetic karyotype	47(40–58), X, −Y, 1q+, −2, +t(1;2)(cen;cen), +7, 12q+, 14q+, +mar

Functional characterization[1,2,3,5]

Cytochemistry	CAE$^-$, MGP$^+$, MPO$^-$, PAS$^-$
Proto-oncogene	MYC mRNA overexpression
Special features	secretion: Igλ, amylase

Comments

- Myeloma cell line of Igλ-type.
- Available from cell line bank.
- Cells grow very well.

References

[1] Togawa, A. et al. (1982) Int. J. Cancer 29, 495–500.
[2] Namba, M. et al. (1989) In Vitro Cell. Dev. Biol. 25, 723–729.
[3] Ohtsuki, T. et al. (1991) Brit. J. Haematol. 77, 172–179.
[4] Shima, Y. et al. (1995) Blood 85, 757–764.
[5] Shirato, L. et al. (1996) Cancer Letters 107, 131–136.
[6] Minowada, J. and Matsuo, Y. (1999) unpublished data.

KMS-11

Culture characterization[1,2]

Culture medium	90% RPMI 1640 + 10% FBS (or serum-free in 50% IMDM + 50% Ham's F12 + additives)
Doubling time	36 h
Viral status	EBNA⁻
Authentication	no
Primary reference	Namba et al.[1]
Availability of cell line	original authors

Clinical characterization[1,2]

Patient	67-year-old female
Disease diagnosis	multiple myeloma (IgGκ)
Specimen site	pleural effusion
Year of establishment	1987

Immunophenotypic characterization[1-3]

B-cell marker	CD20$^{(+)}$, CD21$^{(+)}$, PCA-1^{+}, cyIgκ^{+}, sIgκ^{+}
Progenitor/activation/other marker	CD38$^{(+)}$, CD71^{+}, HLA-DR^{-}

Genetic characterization[4-8]

Cytogenetic karyotype	hyperdiploid, t(4;14)(p16.3;q32.3), t(14;16)(q32;q23)
Unique translocation/fusion gene	t(4;14)(p16.3;q32.3) → *IGH-FGFR3-MMSET* genes altered

Functional characterization[1-3]

Colony formation	in agar
Cytochemistry	CAE^{-}, MGP^{+}, MPO^{-}, PAS^{-}
Cytokine response	growth stimulation: IL-6
Transcription factor	mRNA overexpression: MAF, MYC
Special features	secretion: Igκ

Comments

- Myeloma cell line of IgGκ-type.
- Carries t(4;14)(p16.3;q32.3) resulting in *IGH-MMSET* hybrid transcripts.

References

[1] Namba, M. et al. (1989) In Vitro Cell. Dev. Biol. 25, 723–729.
[2] Ohtsuki, T. et al. (1991) Brit. J. Haematol. 77, 172–179.
[3] Shirato, L. et al. (1996) Cancer Letters 107, 131–136.
[4] Chesi, M. et al. (1997) Nature Genetics 16, 260–265.
[5] Richelda, R. et al. (1997) Blood 90, 4062–4070.
[6] Chesi, M. et al. (1998) Blood 91, 4457–4463.
[7] Chesi, M. et al. (1998) Blood 92, 3025–3034.
[8] Finelli, P. et al. (1999) Blood 94, 724–732.

KMS-12-PE

Culture characterization[1-3]

Sister cell line	**KMS-12-BM** (serial sister cell line – from bone marrow at terminal stage – immunological, cytogenetic, functional differences)
Culture medium	90% RPMI 1640 + 10% FBS (or serum-free in 50% IMDM + 50% Ham's F12 + additives)
Doubling time	62 h
Viral status	EBNA⁻
Authentication	partly (by amylase isozyme analysis)
Primary reference	Ohtsuki et al.[1]
Availability of cell line	KMS-12-PE: JCRB 0430; KMS-12-BM: JCRB 0429

Clinical characterization[1-3]

Patient	64-year-old female
Disease diagnosis	multiple myeloma
Treatment status	at relapse
Specimen site	pleural effusion
Year of establishment	1987

Immunophenotypic characterization[1-3]

T-/NK cell marker	$CD3^-$
B-cell marker	$CD10^-$, $CD19^-$, $CD20^-$, $CD21^-$, PCA-1^+, $cyIg^-$, sIg^-
Progenitor/activation/other marker	$CD38^+$, $CD71^+$, HLA-DR^-
Adhesion marker	$CD11b^-$
Cytokine receptor	$CD25^-$

Genetic characterization[1-4]

Cytogenetic karyotype	41, t(9;11)(q34;q13), del(1p22→1pter), t(11;14)(q13;q32)
Unique translocation/fusion gene	t(11;14)(q13;q32) → *BCL1-IGH* genes altered?
Unique gene alteration	*P16INK4A* deletion, *P18INK4C* deletion

Functional characterization[1-3,5,6]

Cytochemistry	CAE^-, MGP^+, MPO^-, PAS^-
Cytokine response	growth inhibition: IL-6
Proto-oncogene	MYC mRNA overexpression
Transcription factor	cyclin D1 overexpression
Special features	production: amylase

Comments

- Myeloma cell line of undetermined Ig-type.
- Carries t(11;14)(q13;q32) possibly leading to alteration of the *BCL1* and *IGH* genes.

- Serial sister cell lines established.
- Available from cell line bank.

References

[1] Ohtsuki, T. et al. (1989) Brit. J. Haematol. 73, 199–204.
[2] Namba, M. et al. (1989) In Vitro Cell. Dev. Biol. 25, 723–729.
[3] Ohtsuki, T. et al. (1991) Brit. J. Haematol. 77, 172–179.
[4] Williams, M.E. et al. (1996) Blood 88, 102a.
[5] Vaandrager, J.W. et al. (1997) Blood 89, 349–350.
[6] Shirato, L. et al. (1996) Cancer Letters 107, 131–136.

KMS-18

Culture characterization[1]

Culture medium	90% RPMI 1640 + 10% FBS
Doubling time	72 h
Viral status	EBV$^-$
Authentication	yes (by cytogenetics)
Primary reference	Ohtsuki et al.[1]
Availability of cell line	original authors

Clinical characterization[1]

Patient	60-year-old male
Disease diagnosis	multiple myeloma (IgAλ)
Treatment status	at relapse, leukemic phase, terminal
Specimen site	peripheral blood
Year of establishment	1996

Immunophenotypic characterization[1]

T-/NK cell marker	CD1$^-$, CD2$^-$, CD3$^-$, CD4$^-$, CD5$^-$, CD7$^-$, CD8$^-$
B-cell marker	CD10$^-$, CD19$^-$, CD20$^-$, CD23$^+$
Myelomonocytic marker	CD13$^-$, CD14$^-$, CD33$^-$
Erythroid-megakaryocytic marker	CD36$^-$
Progenitor/activation/other marker	CD30$^-$, CD34$^-$, CD38$^-$, HLA-DR$^+$

Genetic characterization[1,2]

Cytogenetic karyotype	42, add(1)(q32), t(4;14)(p16.3;q32.3), add(10)(q24), add(17)(p11)
Unique translocation/fusion gene	t(4;14)(p16;q32) → *FGFR3-IGH* genes altered

Functional characterization[1]

Special features	secretion: Igλ, ammonia

Comments

- Myeloma cell line of IgAλ-type.
- Carries t(4;14)(p16;q32) leading to alteration of the *FGFR3* and *IGH* genes.

References

[1] Ohtsuki, T. et al. (1998) Int. J. Oncol. 12, 545–552.
[2] Richelda, R. et al. (1997) Blood 97, 4062–4070.

KPMM2

Culture characterization[1]

Culture medium	80% RPMI 1640 + 20% FBS + 4 ng/ml IL-6
Doubling time	48 h
Viral status	EBV^-
Authentication	no
Primary reference	Goto et al.[1]
Availability of cell line	restricted

Clinical characterization[1]

Patient	77-year-old female
Disease diagnosis	multiple myeloma (IgGλ)
Treatment status	at relapse, terminal
Specimen site	ascites
Year of establishment	1991

Immunophenotypic characterization[1]

T-/NK cell marker	$CD3^-$, $CD4^-$, $CD5^-$, $CD7^-$, $CD8^-$
B-cell marker	$CD10^-$, $CD19^-$, $CD20^-$, $CD22^-$, $PCA\text{-}1^+$, $cyIgG\lambda^+$, $sIgG\lambda^+$
Myelomonocytic marker	$CD13^-$, $CD14^-$, $CD33^-$
Erythroid-megakaryocytic marker	$CD42b^-$, $CD63^+$, $GlyA^-$
Progenitor/activation/other marker	$CD34^-$, $CD38^+$, $CD45^+$, $CD45RA^-$, $CD45RO^-$, $CD69^-$, $CD71^+$, $HLA\text{-}DR^-$
Adhesion marker	$CD11a^-$, $CD11b^-$, $CD11c^-$, $CD18^-$, $CD29^+$, $CD44^+$, $CD49d^+$, $CD49e^-$, $CD54^+$, $CD56^+$, $CD58^+$, $CD61^-$, $CD62L^-$
Cytokine receptor	$CD25^-$, $CD126^+$

Genetic characterization[1]

Cytogenetic karyotype	46, XX, −4, +16, der(1;19)(q10;q10), t(3;14)(q21;q32), t(6;11) (p12;p15), der(10)add(10)(p13)dic(9;10)(q10;q26)
Receptor gene rearrangement	*IGH* R, *IGK* G, *IGL* R

Functional characterization[1,2]

Cytochemistry	ACP^+, ALP^-, $ANBE^{(+)}$, CAE^-, MPO^-, PAS^-
Cytokine production	IL-6
Cytokine response	IL-6-dependent; growth inhibition: INF-α, IFN-γ
Heterotransplantation	into SCID mice
Special features	secretion: IgG

Comments

- Myeloma cell line of IgGλ-type.
- Constitutively IL-6-dependent.

References

[1] Goto, H. et al. (1995) Leukemia 9, 711–718.
[2] Tsunenari, T. et al. (1997) Blood 90, 2437–2444.

Culture characterization[1,2]

Culture medium 85% RPMI 1640 + 15% FBS
Doubling time 65–75 h
Viral status EBV⁻, HBV⁻, HCV⁻, HHV-8⁻, HIV⁻, HTLV-I/II⁻
Authentication no
Primary reference Diehl et al.[1]
Availability of cell line DSM ACC 49

Clinical characterization[1]

Patient 36-year-old female
Disease diagnosis plasma cell leukemia (IgG)
Specimen site peripheral blood
Year of establishment 1977

Immunophenotypic characterization[1–5]

T-/NK cell marker $CD2^-$, $CD3^-$, $CD4^-$, $CD5^-$, $CD7^-$, $CD8^-$, $CD28^+$
B-cell marker $CD9^-$, $CD10^-$, $CD19^-$, $CD23^-$, $CD24^-$, $CD37^-$, $CD40^-$, $CD138^+$, PCA-1^-, sIg^-
Myelomonocytic marker $CD13^-$, $CD14^-$, $CD15^-$, $CD16^-$
Progenitor/activation/other marker $CD30^-$, $CD38^+$, $CD45^+$, $CD71^+$, EMA^-, HLA-DR^+
Adhesion marker $CD11a^-$, $CD11b^-$, $CD18^+$, $CD49e^-$, $CD56^+$
Cytokine receptor $CD25^-$

Genetic characterization[2,6–8]

Cytogenetic karyotype hyperdiploid, 11% polyploidy; 49(45–50)<2n>X, −X, −X, −6, +7, +8, +11, +11, +14, +19, −22, t(5;8)(q12/13;q24.32), der(8)t(5;8)(q12/13;q24.32), der(8)t(7;8)(q11.2;p22), del(11) (q13)×2, add(14)(q32), del(17)(p12), add(19)(q13); sideline with add(13)(q34)
Unique gene alteration *P53* mutation

Functional characterization[1,6,9]

Cytochemistry ACP^+, $ANAE^-$, GLC^+, MPO^-, PAS^+
Cytokine production TNF-β
Heterotransplantation into nude mice

Comments

- One of the oldest plasma cell lines (IgG-type) (reference cell line).
- Available from cell line bank.
- Cells grow well.

References

[1] Diehl, V. et al. (1978) Blut 36, 331–338.
[2] Drexler, H.G. et al. (1999) DSMZ Catalogue of Cell Lines, 7th edn, Braunschweig, Germany.
[3] Duperray, C. et al. (1989) Blood 73, 566–572.
[4] Pellat-Deceunynck, C. et al. (1994) Blood 84, 2597–2603.
[5] Pellat-Deceunynck, C. et al. (1995) Blood 86, 4001–4002.
[6] Kirchner, H.H. et al. (1981) Blut 43, 93–97.
[7] Egle, A. et al. (1997) Brit. J. Haematol. 97, 418–428.
[8] Kuipers, J. et al. (1999) Cancer Genet. Cytogenet. 109, 99–107.
[9] Bataille, R. et al. (1989) Cancer 63, 877–880.

LB 84-1

Culture characterization[1]

Culture medium	85% L-15 + 15% FBS
Doubling time	34 h
Viral status	EBV⁻
Authentication	yes (by cytogenetics)
Primary reference	Durie et al.[1]
Availability of cell line	restricted

Clinical characterization[1]

Patient	45-year-old female
Disease diagnosis	multiple myeloma (IgAκ)
Treatment status	at relapse, terminal
Specimen site	bone marrow
Year of establishment	1984

Immunophenotypic characterization[1,2]

T-/NK cell marker	$CD1^-$, $CD2^-$, $CD3^-$, $CD4^-$, $CD5^-$, $CD7^-$, $CD8^-$, $CD57^-$
B-cell marker	$CD9^-$, $CD10^+$, $CD19^+$, $CD20^+$, $CD21^+$, $CD22^+$, $CD23^-$, $CD24^-$, $CD37^+$, PC-1^+, PCA-1^-, cyIgAλ^+
Myelomonocytic marker	$CD13^-$, $CD14^-$, $CD15^-$, $CD16^-$
Progenitor/activation/other marker	$CD34^-$, $CD38^+$, $CD45^+$, $CD71^+$, EMA^+, HLA-DR^+
Adhesion marker	$CD11b^-$, $CD18^+$, $CD56^-$, $CD62L^-$
Cytokine receptor	$CD25^-$

Genetic characterization[1]

Cytogenetic karyotype	hyperdiploid, −X, +2, −4, −14, +17, +18, +19, +20, +21, +22, del(1)(p36), t(2;?)(q37;?), del(3)(q22), dup(3)(q26q29), dup(3), del(5)(p14), t(5;?)(q35;?), del(6)(q15), del(6)(q21), del(7)(q31)
Receptor gene rearrangement	*IGH* D, *IGK* R, *IGL* R

Functional characterization[1,3]

Cytochemistry	$ANBE^+$, CAE^+
Cytokine production	TNF-β
Special features	constitutive drug-resistance

Comments

- Myeloma cell line of IgAλ-type.
- Originally described as 'myelomonocytic myeloma cell line'; but presumably only non-specific antibody binding.
- Discrepancy between IgAκ of patient and IgAλ of cell line.

References

[1] Durie, B.G.M. et al. (1989) Blood 73, 770–776.
[2] Duperray, C. et al. (1989) Blood 73, 566–572.
[3] Bataille, R. et al. (1989) Cancer 63, 877–880.

LB-831

Culture characterization[1]

Sister cell line	**LB-832** (serial sister cell line – also from pleural effusion – cytogenetic differences)
Establishment	initially on autologous feeder layer
Culture medium	85% L-15 + 15% FBS
Doubling time	38 h
Viral status	EBV⁻
Authentication	yes (by cytogenetics)
Primary reference	Durie et al.[1]
Availability of cell line	restricted

Clinical characterization[1]

Patient	36-year-old female
Disease diagnosis	multiple myeloma (IgGκ)
Treatment status	refractory, terminal
Specimen site	pleural effusion
Year of establishment	1983

Immunophenotypic characterization[1]

B-cell marker	CD10$^+$, CD20$^+$, CD21$^-$, cyIgGκ$^+$
Progenitor/activation/other marker	HLA-DR$^+$, TdT$^-$

Genetic characterization[1,2]

Cytogenetic karyotype	69, XX-X, +2, +3, +4, +5, +6, +8, −9, +10, +11, +14, −16, −17, +18, +19, +20, −22, +6mar, t(1;?)(q32;?), t(1;15) (p11;p11), del(3)(p21p25), t(5;?)(q35;?), del(7)(p15:), t(7;?) (p22;?), ?HSR7(p22), i(8q), inv(11)(p11;q13), t(13;15)(p12;p12)
Unique gene alteration	*P53* mutation

Functional characterization[1,3]

Cytochemistry	ACP$^+$, GLC$^+$
Cytokine production	TGF-β mRNA$^+$
Special features	secretion: β2-microglobulin, IgGκ, osteoclast activating factor

Comments

- Myeloma cell line of IgGκ-type.
- Serial sister cell lines established.

References

[1] Durie, B.G.M. et al. (1985) Blood 66, 548–555.
[2] Mazars, G.R. et al. (1992) Oncogene 7, 1015–1018.
[3] Portier, M. et al. (1993) Brit. J. Haematol. 85, 514–520.

LOPRA-1

Culture characterization[1]

Subclones	LOPRA-1/4, LOPRA-1/5 (differences in Ig production)
Culture medium	80% IMDM + 20% FBS (or serum-free in HB 104)
Doubling time	30 h
Viral status	EBNA$^-$
Authentication	no
Primary reference	Lohmeyer et al.[1]
Availability of cell line	not known

Clinical characterization[1]

Patient	63-year-old female
Disease diagnosis	multiple myeloma ($IgA_2\kappa$)
Treatment status	at relapse, terminal
Specimen site	ascites
Year of establishment	1982

Immunophenotypic characterization[1]

T-/NK cell marker	CD1$^-$, CD2$^-$, CD3$^-$, CD4$^-$, CD5$^-$, CD6$^-$, CD7$^-$, CD8$^-$, CD28$^+$
B-cell marker	CD10$^-$, CD19$^-$, CD20$^-$, CD21$^-$, CD22$^-$, CD23$^-$, CD24$^{(+)}$, CD37$^-$, CD39$^-$, PCA-1$^+$, cyIgAκ^+, sIg$^-$
Myelomonocytic marker	CD13$^-$
Progenitor/activation/other marker	CD30$^-$, CD38$^+$, CD45$^-$, CD71$^+$, HLA-DR$^-$
Cytokine receptor	CD25$^-$

Genetic characterization[1]

Cytogenetic karyotype	70(66–71)XX, −X, −1, −4, −6, +7, −8, −8, −13, −16, +18, +21, +i(1q), +i(1q), +6q−, +3mar

Functional characterization[1]

Cytochemistry	ACP$^+$, ALP$^-$, ANAE$^{(+)}$, GLC$^+$, MPO$^-$, PAS$^{(+)}$
Special features	secretion: IgAκ

Comments

- Myeloma cell line of IgAκ-type.

Reference

[1] Lohmeyer, J. et al. (1988) Brit. J. Haematol. 69, 335–343.

LP-1

Culture characterization[1,2]

Culture medium	80–90% IMDM + 10–20% FBS
Doubling time	50 h
Viral status	EBV$^-$, HBV$^-$, HCV$^-$, HHV-8$^-$, HIV$^-$, HTLV-I/II$^-$
Authentication	yes (by cytogenetics)
Primary reference	Pegoraro et al.[1]
Availability of cell line	DSM ACC 41

Clinical characterization[1]

Patient	57-year-old female
Disease diagnosis	multiple myeloma (IgGλ)
Treatment status	leukemic phase, refractory, terminal
Specimen site	peripheral blood
Year of establishment	1986

Immunophenotypic characterization[1–4]

T-/NK cell marker	CD2$^-$, CD3$^-$, CD28$^+$
B-cell marker	CD10$^-$, CD19$^-$, CD20$^-$, CD21$^-$, CD37$^-$, CD40$^+$, CD138$^+$, PC-1$^-$, PCA-1$^-$, cyIgGλ$^+$, sIgλ$^+$
Myelomonocytic marker	CD13$^-$, CD16$^-$, CD35$^-$
Progenitor/activation/other marker	CD38$^+$, HLA-DR$^+$
Adhesion marker	CD11a$^-$, CD11b$^-$, CD49e$^-$, CD56$^+$
Cytokine receptor	CD25$^-$

Genetic characterization[1,2,5–7]

Cytogenetic karyotype	flat-moded, aneuploid; 79(72–79)<3n>XX, −X, −1, +3, +4, +5, +6, +8, +11, +12, −13, +14, +15, +16, +18, −19, +20, +21, +22, +dmin×1–4, add(1)(p11-21), del(3)(q22), der(4) add(4)(p1?)t(4;11)(11;?)(q33;q13)(q24;?)×2, der(5)t(1;5) (p11;q35), del(6)(q15)×1–2, del(8)(p21)×1–2, add(9)(q34-35)×1–2, t(12;13)(p12;q11)×2
Unique translocation/fusion gene	t(4;14)(p16.3;q32.3) → *IGH-MMSET* genes altered
Unique gene alteration	*P53* mutation

Functional characterization[1]

Cytochemistry	ACP$^+$, PAS$^-$
Inducibility of differentiation	PWM, TPA → terminal B-cell differentiation
Proto-oncogene	MYC mRNA$^+$
Special features	secretion: IgGλ

Comments

- Myeloma cell line of IgGλ-type.
- Carries t(4;14)(p16.3;q32.3) leading to alteration of the *IGH-MMSET* genes.
- Available from cell line bank.
- Cells are difficult to culture.

References

[1] Pegoraro, L. et al. (1989) Blood 73, 1020–1027.
[2] Drexler, H.G. et al. (1999) DSMZ Catalogue of Cell Lines, 7th edn, Braunschweig, Germany.
[3] Pellat-Deceunynck, C. et al. (1994) Blood 84, 2597–2603.
[4] Pellat-Deceunynck, C. et al. (1995) Blood 86, 4001–4002.
[5] Chesi, M. et al. (1998) Blood 91, 4457–4463.
[6] Chesi, M. et al. (1998) Blood 92, 3025–3034.
[7] Egle, A. et al. (1997) Brit. J. Haematol. 97, 418–428.

MEF-1

Culture characterization[1]

Culture medium	80% IMDM + 20% FBS
Doubling time	36 h
Viral status	EBV⁻
Authentication	yes (by *IGH* rearrangement, *P53* mutation)
Primary reference	Yufu et al.[1]
Availability of cell line	original authors

Clinical characterization[1]

Patient	67-year-old female
Disease diagnosis	multiple myeloma (Igκ)
Treatment status	at diagnosis
Specimen site	bone marrow

Immunophenotypic characterization[1]

T-/NK cell marker	$CD2^-$, $CD5^-$, $CD7^-$, $CD28^+$
B-cell marker	$CD10^-$, $CD19^-$, $CD20^-$, $CD21^-$, $CD22^-$, $CD23^-$, $CD24^-$, PCA-1^-, sIg^-
Myelomonocytic marker	$CD13^-$, $CD14^-$
Erythroid-megakaryocytic marker	$CD41a^-$
Progenitor/activation/other marker	$CD34^-$, $CD38^-$, $CD45^-$, $CD71^+$, HLA-DR^-
Adhesion marker	$CD44^+$, $CD54^+$, $CD56^-$

Genetic characterization[1]

Cytogenetic karyotype	50, X, −X, +5, +8, +der(3)add(3)(p22)add(3)(q13), t(4;6)(q21;p23), add(7)(q36), +add(7)(q36), t(11;14)(q13;q32), +der(14)t(11;14)(q13;q32), der(19)dic(1;19)(p13;p13.3)
Unique translocation/fusion gene	t(11;14)(q13;q32) → *BCL1-IGH* genes altered?
Receptor gene rearrangement	*IGH* R
Unique gene alteration	*P53* mutation

Functional characterization[1]

Cytokine production	IL-6
Transcription factor	cyclin D1 $mRNA^+$
Special features	production: Igκ

Comments

- Myeloma cell line of Igκ-type.
- Carries t(11;14)(q13;q32) possibly leading to alteration of the *BCL1* and *IGH* genes.

Reference

[1] Yufu, Y. et al. (1999) Cancer 85, 1750–1757.

MM.1

Culture characterization[1]

Culture medium	90% RPMI 1640 + 10% FBS
Doubling time	72 h
Viral status	EBV⁻
Authentication	yes (by HLA typing, *IGL* rearrangement)
Primary reference	Goldman-Leikin et al.[1]
Availability of cell line	original authors

Clinical characterization[1]

Patient	45-year-old female
Disease diagnosis	multiple myeloma (IgAλ)
Treatment status	progressive, terminal
Specimen site	peripheral blood
Year of establishment	1986

Immunophenotypic characterization[1]

T-/NK cell marker	$CD2^-$
B-cell marker	$CD10^-$, $CD19^-$, $CD20^-$, $CD75^+$, $PCA\text{-}1^+$, $cyIg\lambda^+$, sIg^-
Progenitor/activation/other marker	$CD38^+$, $CD71^+$, $HLA\text{-}DR^+$

Genetic characterization[1–4]

Cytogenetic karyotype	44, XX, −8, −13, −14, −16, −21, del(1)(p13p22), t(2;?)(q37;?), t(3;?)(p25;?), t(6;?)(q22;?), t(12;14)(q24.3;q32.3), +der(8)t(8;13)(q21;q22), +der(16)t(8;16)(p21.1;q12), +der(21)t(1;21)(q12;p13); t(14;16)(q32;q23)
Receptor gene rearrangement	*IGL* R
Unique gene alteration	*P18INK4C* deletion

Functional characterization[1,2]

Cytochemistry	ACP^+, $ANAE^{(+)}$, $ANBE^{(+)}$, MGP^+, MPO^-, PAS^-, $TRAP^-$
Proto-oncogene	MAF $mRNA^+$
Special features	secretion: Igλ

Comments

- Myeloma cell line of IgAλ-type.

References

[1] Goldman-Leikin, R.E. et al. (1989) J. Lab. Clin. Med. 113, 335–345.
[2] Chesi, M. et al. (1998) Blood 91, 4457–4463.
[3] Williams, M.E. et al. (1996) Blood 88, 102a.
[4] Kuipers, J. et al. (1999) Cancer Genet. Cytogenet. 109, 99–107.

MOLP-5

Culture characterization[1]

Sister cell line	**B407** (= EBV$^+$ B-LCL; simultaneous sister cell line)
Establishment	on bone marrow stroma feeder layer
Culture medium	90% RPMI 1640 + 10% FBS + bone marrow stroma feeder cells
Doubling time	48 h
Viral status	EBV$^-$, HBV$^-$, HCV$^-$, HHV-8$^-$, HIV$^-$, HTLV-I/II$^-$
Authentication	no
Primary reference	Matsuo et al.[1]
Availability of cell line	original authors

Clinical characterization[1]

Patient	71-year-old male
Disease diagnosis	multiple myeloma (Igκ; with hyperammonemia/hypercalcemia)
Treatment status	stage IIIB, at relapse, terminal
Specimen site	peripheral blood
Year of establishment	1997

Immunophenotypic characterization[1]

T-/NK cell marker	CD2$^-$, CD3$^-$, CD4$^-$, CD5$^-$, CD7$^-$, CD8$^-$, CD28$^+$
B-cell marker	CD9$^-$, CD10$^-$, CD19$^-$, CD20$^-$, CD21$^-$, CD22$^-$, CD23$^-$, CD40$^+$, CD79a$^-$, CD79b$^-$, CD138$^+$, NU-B1$^-$, PCA-1$^+$, cyIgκ$^+$, sIg$^-$
Myelomonocytic marker	CD13$^-$, CD14$^-$, CD15$^-$, CD33$^-$
Erythroid-megakaryocytic marker	CD61$^-$, CD62P$^-$
Progenitor/activation/other marker	CD34$^-$, CD38$^+$, CD39$^-$, CD71$^+$, HLA-DR$^-$, TdT$^-$
Adhesion marker	CD11a$^-$, CD11b$^-$, CD18$^-$, CD29$^+$, CD44$^+$, CD49a$^-$, CD49b$^-$, CD49c$^-$, CD49d$^+$, CD49e$^-$, CD49f$^-$, CD51$^-$, CD54$^+$, CD56$^+$, CD58$^+$
Cytokine receptor	CD116$^-$, CD117$^-$, CD119$^-$, CD120a$^-$, CD120b$^-$, CD123$^-$, CD124$^-$, CD126$^-$, CD127$^-$, CD130$^-$, CD131$^{(+)}$, CD132$^-$, CD135$^-$

Genetic characterization[1]

Cytogenetic karyotype	45, X, −Y, +1, +5, −14, −15, +1mar, dic(1;?13)(p?22;q?14), add(3)(p11), add(5)(q?11.2), t(11;14)(q13;q32), del(17)(p11.2)
Unique translocation/fusion gene	t(11;14)(q13;q32) → *BCL1-IGH* genes altered?
Receptor gene rearrangement	*IGH* R, *IGK* R, *IGL* G

Functional characterization[1]

Cytochemistry	ACP+, ALP−, ANBE−, CAE−, Fe−, MPO−, PAS−, TRAP−
Cytokine receptor	RT-PCR+: IL-6R
Cytokine response	growth stimulation: IL-6, IL-10; growth inhibition: IFN-α, IFN-β, TGF-β
Special features	RT-PCR+: parathyroid hormone-related protein

Comments

- Myeloma cell line of Igκ-type.
- Carries t(11;14)(q13;q32) possibly leading to alteration of the *BCL1* and *IGH* genes.
- Cells are continuously stroma-dependent.

Reference

[1] Matsuo, Y. et al. (2000) Brit. J. Haematol. 109, 54–63.

Culture characterization[1]

Culture medium	80% RPMI 1640 + 20% FBS
Doubling time	40–50 h
Viral status	EBV⁻
Authentication	yes (by *IGH* rearrangement)
Primary reference	Donelli et al.[1]
Availability of cell line	not known

Clinical characterization[1]

Patient	51-year-old female
Disease diagnosis	plasma cell leukemia (IgAκ)
Treatment status	at diagnosis
Specimen site	peripheral blood
Year of establishment	1986

Immunophenotypic characterization[1]

T-/NK cell marker	CD1a⁻, CD3⁻, CD7⁻
B-cell marker	CD10⁽⁺⁾, CD19⁻, CD20⁻, CD21⁻, PCA-1⁺, cyIgAκ⁺, sIgAκ⁺
Myelomonocytic marker	CD13⁻, CD14⁻, CD33⁻
Progenitor/activation/other marker	CD38⁺, CD71⁺, HLA-DR⁽⁺⁾, TdT⁻

Genetic characterization[1]

Cytogenetic karyotype	hypertriploid, 74, +1, −2, +3, +6, +7, +8, +11, +12, +13, +16, +17, +18, +19, +20, +21, +22, +3mar, 4q+, 13q−, 14q+, 22q−, +3mar
Receptor gene rearrangement	*IGH* R, *IGK* R, *IGL* G

Functional characterization[1]

Cytochemistry	ACP⁺, GLC⁺, PAS⁻
Special features	secretion: IgAκ

Comments

- Plasma cell line of IgAκ-type.

Reference

[1] Donelli, A. et al. (1987) Int. J. Cancer 40, 383–388.

NCI-H929

Culture characterization[1,2]

Culture medium	90% RPMI 1640 + 10% FBS
Doubling time	50 h
Viral status	EBV⁻, HBV⁻, HCV⁻, HHV-8⁻, HIV⁻, HTLV-I/II⁻
Authentication	yes (by *IGH* rearrangement)
Primary reference	Gazdar et al.[1]
Availability of cell line	ATCC CRL 9068, DSM ACC 163

Clinical characterization[1]

Patient	62-year-old female
Disease diagnosis	myeloma (IgAκ)
Treatment status	at relapse
Specimen site	pleural effusion
Year of establishment	1984

Immunophenotypic characterization[1–3]

T-/NK cell marker	$CD3^-$, $CD4^-$, $CD5^-$, $CD7^-$, $CD8^-$, $CD28^+$, $CD57^-$
B-cell marker	$CD10^-$, $CD19^-$, $CD20^-$, $CD21^-$, $CD37^-$, $CD138^+$, PCA-1^+, cyIgκ^+, sIgAκ^+
Myelomonocytic marker	$CD13^-$, $CD14^-$, $CD15^-$, $CD16^-$
Progenitor/activation/other marker	$CD38^+$, $CD71^+$, HLA-DR^-, TdT^-
Adhesion marker	$CD11a^-$, $CD49e^-$, $CD62L^-$
Cytokine receptor	$CD25^-$

Genetic characterization[1,2,4–6]

Cytogenetic karyotype	hypodiploid, 16% polyploidy; 45(43–46)<2n>X, −X, +8, −10, −13, +mar, dup(1)(q11q25), del(1)(p11p25), del(6)(q25), der(8)t (8;?)(q23;?)×2, del(9)(p13), del(12)(p12)
Unique translocation/fusion gene	t(4;14)(p16.3;q32.3) → *IGH-FGFR3-MMSET* genes altered
Receptor gene rearrangement	*IGH* R, *IGK* R
Unique gene alteration	*MYC* rearrangement

Functional characterization[1,7]

Cytochemistry	$ANAE^+$, GLC^+, MGP^+, PAS^-
Proto-oncogene	$mRNA^+$: MAF, MYC
Special features	secretion: IgAκ

Comments

- Myeloma cell line of IgAκ-type (reference cell line).
- Carries t(4;14)(p16.3;q32.3) leading to alterations of the *IGH, FGFR3* and *MMSET* genes.
- Cells are difficult to culture.
- Available from cell line banks.

References

[1] Gazdar, A.F. et al. (1986) Blood 67, 1542–1549.
[2] Drexler, H.G. et al. (1999) DSMZ Catalogue of Cell Lines, 7th edn, Braunschweig, Germany.
[3] Pellat-Deceunynck, C. et al. (1995) Blood 86, 4001–4002.
[4] Chesi, M. et al. (1997) Nature Genetics 16, 260–265.
[5] Chesi, M. et al. (1998) Blood 92, 3025–3034.
[6] Kuipers, J. et al. (1999) Cancer Genet. Cytogenet. 109, 99–107.
[7] Chesi, M. et al. (1998) Blood 91, 4457–4463.

NOP-2

Culture characterization[1]

Culture medium	90% RPMI 1640 + 10% FBS
Doubling time	48–72 h
Viral status	EBNA⁻
Authentication	yes (by cytogenetics)
Primary reference	Nagai et al.[1]
Availability of cell line	not known

Clinical characterization[1]

Patient	50-year-old male
Disease diagnosis	multiple myeloma (Igλ)
Treatment status	at relapse
Specimen site	plasmacytoma (chest wall)
Year of establishment	1986

Immunophenotypic characterization[1]

T-/NK cell marker	$CD2^-$, $CD3^-$, $CD7^-$
B-cell marker	$CD10^-$, $CD19^-$, $CD20^-$, $CD21^-$, $CD24^-$, PCA-1^+, cyIgλ^+, sIg$^-$
Myelomonocytic marker	$CD14^-$, $CD15^-$
Progenitor/activation/other marker	$CD38^-$, HLA-DR^-
Adhesion marker	$CD11b^-$
Cytokine receptor	$CD25^-$

Genetic characterization[1]

Cytogenetic karyotype	47, X, −Y, +6, +7, −15, inv dup del(1)(p13-q21;q21), t(8;22) (q24;q11), t(11;14)(q13;q32), +der(15)t(15;1)(p11;p22)
Unique translocation/fusion gene	t(11;14)(q13;q32) → *BCL1-IGH* genes altered?
Receptor gene rearrangement	*IGH* R, *IGL* R

Functional characterization[1,2]

Proto-oncogene	MYC mRNA$^+$
Special features	secretion: Igλ

Comments

- Myeloma cell line of Igλ-type.
- Carries t(11;14)(q13;q32) which may lead to alteration of the *BCL1* and *IGH* genes.

References

[1] Nagai, T. et al. (1991) Int. J. Hematol. 54, 141–149.
[2] Mahmoud, M.S. et al. (1996) Blood 87, 4311–4315.

OPM-2

Culture characterization[1,2]

Sister cell line	**OPM-1** (simultaneous sister cell line – different karyotype)
Culture medium	90% RPMI 1640 + 10% FBS
Doubling time	30–60 h
Viral status	EBV⁻, HBV⁻, HCV⁻, HHV-8⁻, HIV⁻, HTLV-I/II⁻
Authentication	no
Primary reference	Katagiri et al.[1]
Availability of cell line	DSM ACC 50

Clinical characterization[1]

Patient	56-year-old female
Disease diagnosis	multiple myeloma (Igλ)
Treatment status	at relapse, leukemic phase, terminal
Specimen site	peripheral blood
Year of establishment	1982

Immunophenotypic characterization[1–3]

T-/NK cell marker	CD1a⁻, CD2⁻, CD3⁻, CD4⁻, CD8⁻, CD28⁺
B-cell marker	CD9⁻, CD10⁻, CD19⁻, CD20⁻, CD24⁻, CD37⁻, CD138⁺, cyIgλ⁺, sIg⁻
Myelomonocytic marker	CD13⁻
Progenitor/activation/other marker	CD38⁺, HLA-DR⁻
Adhesion marker	CD11a⁻, CD49e⁻

Genetic characterization[1,2,4–8]

Cytogenetic karyotype	78(77–81)<3n>XX, −X, +1, +6, −8, +11, +8mar, del(1) (p13)×2, der(1)t(1;2)(p13;p13)×2, del(7) (p15), del(10)(q22), der(22)t(1;22)(q12;p13)×2; FISH: t(4;14)(p16.3;q32.3)
Unique translocation/fusion gene	t(4;14)(p16.3;q32.3) → *IGH-FGFR3-MMSET* genes altered
Unique gene alteration	*P53* mutation

Functional characterization[1]

Cytochemistry	ACP⁺, MPO⁻, PAS⁽⁺⁾

Comments

- Myeloma cell line of Igλ-type (reference cell line).
- Carries t(4;14)(p16.3;q32.3) resulting in *IGH-MMSET* hybrid transcripts.
- Available from cell line bank.

References

[1] Katagiri, S. et al. (1985) Int. J. Cancer 34, 241–246.
[2] Drexler, H.G. et al. (1999) DSMZ Catalogue of Cell Lines, 7th edn, Braunschweig, Germany.
[3] Pellat-Deceunynck, C. et al. (1995) Blood 86, 4001–4002.
[4] Chesi, M. et al. (1997) Nature Genetics 16, 260–265.
[5] Chesi, M. et al. (1998) Blood 92, 3025–3034.
[6] Egle, A. et al. (1997) Brit. J. Haematol. 97, 418–428.
[7] Finelli, P. et al. (1999) Blood 94, 724–732.
[8] Kuipers, J. et al. (1999) Cancer Genet. Cytogenet. 109, 99–107.

Culture characterization[1]

Culture medium	80% IMDM + 20% FBS + 30 U/ml IL-6
Doubling time	40–50 h
Viral status	EBV$^-$
Authentication	yes (by cytogenetics)
Primary reference	Takahira et al.[1]
Availability of cell line	not known

Clinical characterization[1]

Patient	63-year-old female
Disease diagnosis	multiple myeloma (IgGλ)
Treatment status	at relapse, terminal
Specimen site	peripheral blood
Year of establishment	1992

Immunophenotypic characterization[1]

B-cell marker	CD10$^-$, CD19$^-$, PCA-1$^-$, cyIgGλ$^+$
Myelomonocytic marker	CD13$^-$
Progenitor/activation/other marker	CD38$^+$, HLA-DR$^-$

Genetic characterization[1]

Cytogenetic karyotype	45(44–47)<2n>X, −X, 1p+, 1p−, +1q−, t(2;8)(q23;p23), 3p−, 6q−, 12q−, 14q+, −16, −17, +mar

Functional characterization[1]

Cytokine response	IL-6-dependent; growth inhibition: IFN-α
Special features	secretion: IgGλ

Comments

- Myeloma cell line of IgGλ-type.
- Constitutively IL-6-dependent.

Reference

[1] Takahira, H. et al. (1994) Exp. Hematol. 22, 261–266.

RPMI 8226

Culture characterization[1,2]

Culture medium	90% RPMI 1640 + 10% FBS
Doubling time	60–70 h
Viral status	EBV$^-$, HBV$^-$, HCV$^-$, HIV$^-$, HTLV-I/II$^-$
Authentication	no
Primary reference	Matsuoka et al.[1]
Availability of cell line	ATCC CCL 155, DSM ACC 402, IFO 50013, JCRB 0034

Clinical characterization[1]

Patient	61-year-old male
Disease diagnosis	multiple myeloma (IgGλ)
Treatment status	at diagnosis
Specimen site	peripheral blood
Year of establishment	1966

Immunophenotypic characterization[1-8]

T-/NK cell marker	CD1$^-$, CD2$^-$, CD3$^-$, CD4$^-$, CD5$^-$, CD7$^-$, CD8$^-$, CD28$^+$, CD57$^-$
B-cell marker	CD9$^+$, CD10$^-$, CD19$^-$, CD20$^-$, CD21$^{(+)}$, CD22$^-$, CD23$^+$, CD24$^-$, CD37$^+$, CD40$^+$, CD138$^+$, FMC-7$^-$, NU-B1$^-$, PCA-1$^+$, cyIgλ$^+$, sIg$^-$
Myelomonocytic marker	CD13$^-$, CD14$^-$, CD15$^-$, CD16$^-$
Progenitor/activation/other marker	CD30$^+$, CD34$^-$, CD38$^+$, CD45$^{(+)}$, CD71$^+$, CD95$^+$, HLA-DR$^+$, TdT$^-$
Adhesion marker	CD11a$^{(+)}$, CD11b$^-$, CD18$^-$, CD49e$^+$, CD56$^+$
Cytokine receptor	CD25$^-$, CD116$^-$, CD117$^-$, CD119$^+$, CD120a$^-$, CD120b$^-$, CD122$^-$, CD124$^-$, CD126$^+$, CD127$^-$, CD128$^-$, CD130$^+$

Genetic characterization[2,9-13]

Cytogenetic karyotype	flat-moded hypotriploid, 7.5% polyploidy; 62–67<3n>XXY, −1, +5, +6, +7, −8, −9, −10, +11, −12, −13, −14, +16, +18, −19, +21, −22, der(1)t(1;1)(p36.3;q11-12), add(3)(q27-29)×2, del(4) (p14p15.4), der(5;6)(q10;p10)×1–2, del(5)(q32), del(6)(q15), add(9)(p24), add(11)(p15), del(11)(q25), der(14)t(1;14)(q11-12;q32)×2, del(15)(q11.2q14)×2, add(16)(q23-24), der(17)t(?8;17) (q21.2;q25), del(22)(q13.2); FISH: t(1;14)(p13;q32), t(16;22) (q23;q11)
Unique gene alteration	*P16INK4A* methylation, *P53* mutation

Functional characterization[1,9,14,15]

Colony formation	in agar
Cytokine production	TGF-β mRNA⁺
Heterotransplantation	into nude mice or SCID mice
Proto-oncogene	MAF mRNA⁺
Special features	secretion: Igλ

Comments

- First myeloma cell line established in 1966 (reference cell line).
- Myeloma cell line of IgGλ-type.
- Available from cell line banks.
- Cells grow well.

References

[1] Matsuoka, Y. et al. (1967) Proc. Soc. Exp. Biol. Med. 125, 1246–1250.
[2] Drexler, H.G. et al. (1999) DSMZ Catalogue of Cell Lines, 7th edn, Braunschweig, Germany.
[3] Minowada, J. and Matsuo, Y. (1999) unpublished data.
[4] Duperray, C. et al. (1989) Blood 73, 566–572.
[5] Pellat-Deceunynck, C. et al. (1994) Blood 84, 2597–2603.
[6] Pellat-Deceunynck, C. et al. (1995) Blood 86, 4001–4002.
[7] Shima, Y. et al. (1995) Blood 85, 757–764.
[8] Westendorf, J.J. et al. (1995) Blood 85, 3566–3576.
[9] Chesi, M. et al. (1998) Blood 91, 4457–4463.
[10] Egle, A. et al. (1997) Brit. J. Haematol. 97, 418–428.
[11] Mazars, G.R. et al. (1992) Oncogene 7, 1015–1018.
[12] Raynaud, S. et al. (1993) Genes Chromosomes Cancer 8, 80–87.
[13] Kuipers, J. et al. (1999) Cancer Genet. Cytogenet. 109, 99–107.
[14] Portier, M. et al. (1993) Brit. J. Haematol. 85, 514–520.
[15] Nilsson, K. et al. (1977) Int. J. Cancer 19, 337–344.

SK-MM-1

Culture characterization[1]

Culture medium	90% RPMI 1640 + 10% FBS
Doubling time	32 h
Viral status	EBNA$^-$
Authentication	no
Primary reference	Eton et al.[1]
Availability of cell line	original authors

Clinical characterization[1]

Patient	55-year-old male
Disease diagnosis	plasma cell leukemia (Igκ)
Treatment status	at relapse, terminal
Specimen site	bone marrow
Year of establishment	1981

Immunophenotypic characterization[1]

T-/NK cell marker	CD2$^-$, CD5$^-$
B-cell marker	CD10$^-$, CD19$^-$, CD20$^+$, CD21$^-$, PCA-1$^-$, cyIg$^-$, sIg$^-$
Myelomonocytic marker	CD33$^-$
Progenitor/activation/other marker	CD38$^-$, HLA-DR$^-$

Genetic characterization[1-4]

Cytogenetic karyotype	33-45, −X, −9, −10, −15, −18, −20, 1p−, 6q−, t(4;19), t(8q;10q), +11p, t(11q;15q), t(13p;1q), 14p+, 14q+; FISH: t(6;14) (p25;q32), add(14)(q32), t(14;20)(q32;q11)
Unique translocation/fusion gene	t(6;14)(p25;q32) → *MUM1/IRF4-IGH* genes altered
Receptor gene rearrangement	*IGH* RR, *IGK* RG, *TCRB* G, *TCRG* G

Functional characterization[1]

Cytochemistry	ACP$^+$, MGP$^+$, SBB$^-$
Special features	secretion: Igκ

Comments

- Plasma cell line of Igκ-type.
- Carries t(6;14)(p25;q32) leading to alteration of *MUM1/IRF4-IGH* genes.
- Secretion of Igκ.

References

[1] Eton, O. et al. (1989) Leukemia 3, 729–735.
[2] Iida, S. et al. (1997) Nature Genetics 17, 226–230.
[3] Rao, P.H. et al. (1998) Blood 92, 1743–1748.
[4] Yoshida, S. et al. (1999) Leukemia 13, 1812–1816.

SK-MM-2

Culture characterization[1,2]

Culture medium	90% RPMI 1640 + 10% FBS
Doubling time	60–100 h
Viral status	EBV⁻, HBV⁻, HCV⁻, HHV-8⁻, HIV⁻, HTLV-I/II⁻
Authentication	no
Primary reference	Eton et al.[1]
Availability of cell line	DSM ACC 430

Clinical characterization[1]

Patient	54-year-old male
Disease diagnosis	plasma cell leukemia (Igκ)
Treatment status	refractory, terminal
Specimen site	peripheral blood
Year of establishment	1982

Immunophenotypic characterization[1,2]

T-/NK cell marker	$CD2^-$, $CD3^-$, $CD5^-$
B-cell marker	$CD10^-$, $CD19^-$, $CD20^+$, $CD21^-$, $CD37^-$, $CD79a^-$, $cyCD79a^+$, $CD80^-$, $CD138^{(+)}$, PCA-1^+, $cyIg\kappa^+$, sIg^-
Myelomonocytic marker	$CD13^-$, $CD33^-$
Progenitor/activation/other marker	$CD34^-$, $CD38^+$, HLA-DR^-

Genetic characterization[1–5]

Cytogenetic karyotype	69–84, +1, +1, +3, +5, +9, +10, +11, +12, −14, +15, −17, +19, +20, +21, +13−15mar, +2q+, 2q+, +6q−, +6q−, 14q+; FISH: t(11;14)(q13;q32), add(14)(q32)
Unique translocation/fusion gene	t(11;14)(q13;q32) → *BCL1-IGH* genes altered?
Receptor gene rearrangement	*IGH* RD, *IGK* RG, *TCRB* G, *TCRG* G
Unique gene alteration	*P16INK4A* deletion, *P18INK4C* deletion

Functional characterization[1,5]

Cytochemistry	ACP^+, MGP^+, SBB^-
Transcription factor	cyclin D1 overexpression
Special features	secretion: Igκ

Comments

- Plasma cell line of Igκ-type.
- Carries t(11;14)(q13;q32) possibly leading to alteration of the *BCL1* and *IGH* genes.
- Secretion of Igκ.
- Available from cell line bank.
- Cells grow well.

References
[1] Eton, O. et al. (1989) Leukemia 3, 729–735.
[2] Drexler, H.G. et al. (1999) DSMZ Catalogue of Cell Lines, 7th edn, Braunschweig, Germany.
[3] Rao, P.H. et al. (1998) Blood 92, 1743–1748.
[4] Williams, M.E. et al. (1996) Blood 88, 102a.
[5] Vandraager, J.W. et al. (1997) Blood 89, 349–350.

Culture characterization[1]

Culture medium	90% IMDM + 10% FBS
Doubling time	24 h
Viral status	$EBNA^-$
Authentication	no
Primary reference	Weinreich et al.[1]
Availability of cell line	not known

Clinical characterization[1]

Patient	32-year-old male
Disease diagnosis	multiple myeloma
Treatment status	refractory
Specimen site	tumor

Immunophenotypic characterization[1]

T-/NK cell marker	$CD5^-$, $CD6^-$, $CD28^+$
B-cell marker	$CD9^+$, $CD10^-$, $CD19^-$, $CD20^-$, $CD22^-$, $CD23^-$, $CD24^-$, $CD40^-$, PCA-1^+, $cyIgG_3\kappa^+$, $sIg\kappa^{(+)}$
Progenitor/activation/other marker	$CD34^-$, $CD38^+$, $CD45^-$, $HLA\text{-}DR^+$
Cytokine receptor	$CD25^-$

Genetic characterization[1,2]

Cytogenetic karyotype	69–75, XX, −Y, 14q+, der(1)t(1;9)(1qter→1p11::9p21→qter), del(2)(p24)×2, der(5)t(5;18)(p15;q23), inv(8)(p23q13), der(10)t (10;?)(q26;?), der(13)t(13;21)(p11;q11)×2, add(18q21), add(18q23)
Unique translocation/fusion gene	no t(14;18) *BCL2-IGH* fusion gene
Receptor gene rearrangement	*IGH* RD, *IGK* RG

Functional characterization[1]

Cytokine response	growth stimulation: IL-6
Special features	secretion: $IgG_3\kappa$

Comments

- Myeloma cell line of $IgG_3\kappa$-type.

References

[1] Weinreich, S.S. et al. (1991) Brit. J. Haematol. 79, 226–234.
[2] Kuipers, J. et al. (1999) Cancer Genet. Cytogenet. 109, 99–107.

U-266

Culture characterization[1,2]

Other name of cell line	originally **266 Bl**
Sister cell lines	(1) **U-268** (serial sister cell line – from spleen post-mortem) (2) **U-255** (= EBV^+ B-LCL – serial sister cell line)
Establishment	initially as grid organ culture, then on feeder layer (adult allogeneic skin fibroblasts)
Culture medium	originally 90% F-10 + 10% NCS; now 90% RPMI 1640 + 10% FBS
Doubling time	55 h
Viral status	EBV^-, HBV^-, HCV^-, HIV^-, HTLV-I/II^-
Authentication	no
Primary reference	Nilsson et al.[1]
Availability of cell line	ATCC TIB 196, DSM ACC 9

Clinical characterization[1]

Patient	53-year-old male
Disease diagnosis	myeloma (IgEλ)
Treatment status	refractory, terminal
Specimen site	peripheral blood
Year of establishment	1968

Immunophenotypic characterization[2–8]

T-/NK cell marker	$CD1^-$, $CD2^-$, $CD3^-$, $CD4^-$, $CD5^-$, $CD7^-$, $CD8^-$, $CD28^+$, $TCR\alpha\beta^-$
B-cell marker	$CD9^-$, $CD10^-$, $CD19^-$, $CD20^-$, $CD21^-$, $CD22^-$, $CD23^-$, $CD24^-$, $CD37^-$, $CD40^-$, $CD138^+$, FMC-7^-, NU-$B1^-$, PCA-1^+, cyIgGMκλ^-, sIgGMκλ^-
Myelomonocytic marker	$CD13^-$, $CD14^-$, $CD15^-$, $CD16^-$, $CD33^-$
Erythroid-megakaryocytic marker	$CD41a^-$, $CD61^-$
Progenitor/activation/other marker	$CD30^+$, $CD34^-$, $CD38^+$, $CD45^{(+)}$, $CD71^+$, $CD95^+$, EMA^+, HLA-DR^+, TdT^-
Adhesion marker	$CD11a^{(+)}$, $CD11b^-$, $CD18^-$, $CD29^+$, $CD49b^-$, $CD49d^+$, $CD49e^{(+)}$, $CD49f^+$, $CD54^+$, $CD56^{(+)}$, $CD58^+$, $CD106^+$
Cytokine receptor	$CD25^-$, $CD116^-$, $CD117^-$, $CD119^+$, $CD120a^-$, $CD120b^-$, $CD122^-$, $CD124^-$, $CD126^+$, $CD127^-$, $CD128^-$, $CD130^+$

Genetic characterization[2,9–11]

Cytogenetic karyotype	hypodiploid, 6.5% polyploidy; 44(40–46)<2n>XY, −8, −10, −13, −15, +2mar, t(1;11)(p33;q13), add(3)(q27), t(4;11)(q?21;q23), add(7)(q32), add(8)(q24), add(9)(q34), add(10)(p14), add(14)(p11), add(17)(p11), add(18)(p12), der(22)t(15;22)(q21;q13); SKY: der(1)t(1;3), der(3)t(3;16), der(8)t(1;8;11), der(8)t(7;8), der(11)t(1;8;11), der(13)t(2;13), ins(14;10), t(1;13)(?;q14.2)
Unique gene alteration	*P16INK4A* methylation, *P53* mutation

Functional characterization[1,12,13]

Colony formation	in agar
Cytokine production	TNF-β; IL-6 mRNA⁺
Special features	secretion: IgEλ

Comments

- Most widely distributed myeloma cell line (reference cell line).
- Myeloma cell line of IgEλ-type.
- Available from cell line banks.
- Cells grow well.

References

1 Nilsson, K. et al. (1970) Clin. Exp. Immunol. 7, 477–489.
2 Drexler, H.G. et al. (1999) DSMZ Catalogue of Cell Lines, 7th edn, Braunschweig, Germany.
3 Duperray, C. et al. (1989) Blood 73, 566–572.
4 Minowada, J. and Matsuo, Y. (1999) unpublished data.
5 Pellat-Deceunynck, C. et al. (1994) Blood 84, 2597–2603.
6 Shima, Y. et al. (1995) Blood 85, 757–764.
7 Westendorf, J.J. et al. (1995) Blood 85, 3566–3576.
8 Pellat-Deceunynck, C. et al. (1995) Blood 86, 4001–4002.
9 Mazars, G.R. et al. (1992) Oncogene 7, 1015–1018.
10 Rao, P.H. et al. (1998) Blood 92, 1743–1748.
11 Kuipers, J. et al. (1999) Cancer Genet. Cytogenet. 109, 99–107.
12 Nilsson, K. et al. (1977) Int. J. Cancer 19, 337–344.
13 Bataille, R. et al. (1989) Cancer 63, 877–880.

UTMC-2

Culture characterization[1]

Culture medium	90% RPMI 1640 + 10% FBS
Doubling time	48 h
Viral status	EBV⁻
Authentication	yes (by cytogenetics)
Primary reference	Ozaki et al.[1]
Availability of cell line	original authors

Clinical characterization[1]

Patient	72-year-old female
Disease diagnosis	multiple myeloma (IgAκ)
Treatment status	refractory, terminal
Specimen site	pleural effusion

Immunophenotypic characterization[1]

T-/NK cell marker	$CD3^-$, $CD4^-$, $CD5^-$, $CD8^-$
B-cell marker	$CD10^-$, $CD19^-$, $CD20^-$, $cyIgA\kappa^+$, sIg^-
Myelomonocytic marker	$CD13^-$
Progenitor/activation/other marker	$CD38^+$, HLA-DR^-
Adhesion marker	$CD11a^-$, $CD18^-$, $CD29^+$, $CD54^+$, $CD56^+$
Cytokine receptor	$CD126^+$

Genetic characterization[1–3]

Cytogenetic karyotype	43, t(1;5), t(2;15); FISH: t(4;14)(p16.3;q32.3)
Unique translocation/fusion gene	t(4;14)(p16.3;q32.3) → *IGH-FGFR3-MMSET* genes altered

Functional characterization[1]

Cytokine production	RT-PCR^+: IL-6
Cytokine response	growth stimulation/increase in IgAκ production by IL-6
Special features	secretion: IgAκ

Comments

- Myeloma cell line of IgAκ-type.
- Carries t(4;14)(p16.3;q32.3) resulting in *IGH-MMSET* hybrid transcripts.

References

[1] Ozaki, S. et al. (1994) Leukemia 8, 2207–2213.
[2] Chesi, M. et al. (1997) Nature Genetics 16, 260–265.
[3] Chesi, M. et al. (1998) Blood 92, 3025–3034.

XG-1

Culture characterization[1]

Culture medium	90% RPMI 1640 + 10% FBS + 1 ng/ml IL-6
Viral status	EBV⁻
Authentication	yes (by *IGH* rearrangement)
Primary reference	Zhang et al.[1]
Availability of cell line	restricted

Clinical characterization[1]

Patient	male
Disease diagnosis	multiple myeloma (IgAκ)
Treatment status	secondary plasma cell leukemia, terminal
Specimen site	peripheral blood

Immunophenotypic characterization[1–3]

T-/NK cell marker	$CD2^-$, $CD3^-$, $CD5^-$, $CD28^+$, $CD57^-$
B-cell marker	$CD10^+$, $CD19^-$, $CD20^-$, $CD21^+$, $CD23^+$, $CD24^-$, $CD37^+$, $CD40^-$, $CD72^-$, $CD77^+$, $CD80^+$, $CD138^+$, cyIgAκ$^+$, sIgκ$^{(+)}$
Myelomonocytic marker	$CD13^-$, $CD14^-$, $CD15^-$, $CD16^-$, $CD32^-$, $CD33^-$, $CD65^-$
Progenitor/activation/other marker	$CD30^+$, $CD38^+$, $CD71^+$, HLA-DR$^-$
Adhesion marker	$CD11a^-$, $CD11b^-$, $CD11c^-$, $CD18^-$, $CD44^+$, $CD49b^-$, $CD49d^+$, $CD49e^-$, $CD49f^+$, $CD54^+$, $CD56^+$, $CD58^+$
Cytokine receptor	$CD25^-$, $CD126^+$, $CD130^+$, IL-11R$^-$, LIFR$^-$

Genetic characterization[1,4–8]

Cytogenetic karyotype	hypodiploid, −13, add(8q24), t(11;14;?)(q13;q32;?); SKY: t(11;14)(q13;q32), t(17;19)(q24-25;?)
Unique translocation/fusion gene	t(11;14)(q13;q32) → *BCL1-IGH* genes altered?
Receptor gene rearrangement	*IGH* R
Unique gene alteration	*P53* mutation

Functional characterization[1,3,8–13]

Cytokine production	OSM; mRNA$^+$: IL-6, TGF-β
Cytokine response	IL-6-dependent; growth stimulation: GM-CSF, IFN-α, IL-3, TNF-α; growth inhibition: IFN-γ
Heterotransplantation	into SCID mice
Special features	production: IgAκ

Comments

- Myeloma cell line of IgAκ-type.
- Constitutively IL-6-dependent.
- Carries t(11;14)(q13;q32) possibly leading to alteration of the *BCL1* and *IGH* genes.

References

[1] Zhang, X.G. et al. (1994) Blood 83, 3654–3663.
[2] Pellat-Deceunynck, C. et al. (1995) Blood 86, 4001–4002.
[3] Gu, Z.J. et al. (1996) Blood 88, 3972–3986.
[4] Mazars, G.R. et al. (1992) Oncogene 7, 1015–1018.
[5] Raynaud, S.D. et al. (1993) Genes Chromosomes Cancer 8, 80–87.
[6] Rao, P.H. et al. (1998) Blood 92, 1743–1748.
[7] Kuipers, J. et al. (1999) Cancer Genet. Cytogenet. 109, 99–107.
[8] Zhang, X.G. et al. (1990) Blood 76, 2599–2605.
[9] Klein, B. et al. (1990) Eur. Cytokine Net. 1, 193–201.
[10] Jourdan, M. et al. (1991) J. Immunol. 147, 4402–4407.
[11] Portier, M. et al. (1993) Brit. J. Haematol. 85, 514–520.
[12] Zhang, X.G. et al. (1994) J. Exp. Med. 177, 1337–1342.
[13] Rebouissou, C. et al. (1998) Blood 91, 4727–4737.

XG-2

Culture characterization[1]

Culture medium	90% RPMI 1640 + 10% FBS + 1 ng/ml IL-6
Viral status	EBV⁻
Authentication	yes (by *IGH* rearrangement)
Primary reference	Zhang et al.[1]
Availability of cell line	restricted

Clinical characterization[1]

Patient	female
Disease diagnosis	multiple myeloma (IgGλ)
Treatment status	stage III, terminal
Specimen site	pleural effusion

Immunophenotypic characterization[1–3]

T-/NK cell marker	$CD2^-$, $CD3^-$, $CD5^-$, $CD28^+$, $CD57^-$
B-cell marker	$CD10^-$, $CD19^-$, $CD20^-$, $CD21^+$, $CD23^+$, $CD24^-$, $CD37^-$, $CD40^+$, $CD72^-$, $CD77^+$, $CD80^-$, $CD138^+$, $cyIgG\lambda^+$, $sIg\lambda^{(+)}$
Myelomonocytic marker	$CD13^-$, $CD14^-$, $CD15^-$, $CD16^-$, $CD33^+$, $CD65^-$
Progenitor/activation/other marker	$CD30^-$, $CD38^+$, $CD71^+$, $HLA\text{-}DR^+$
Adhesion marker	$CD11a^+$, $CD11b^-$, $CD11c^-$, $CD18^+$, $CD44^+$, $CD49b^-$, $CD49d^+$, $CD49e^{(+)}$, $CD49f^+$, $CD54^+$, $CD56^-$, $CD58^+$
Cytokine receptor	$CD25^-$, $CD126^+$, $CD130^+$, $IL\text{-}11R^-$, $LIFR^-$

Genetic characterization[1,4–6]

Cytogenetic karyotype	hyperdiploid, t(5;11;14)(q31;q13;q32); SKY: t(12;14)(q24:q32), add(14)(q32), der(19)t(1;7;19)
Unique translocation/fusion gene	t(11;14)(q13;q32) → *BCL1-IGH* genes altered?
Receptor gene rearrangement	*IGH* R
Unique gene alteration	*P53* mutation

Functional characterization[1,3,7–11]

Cytokine production	OSM; TGF-β $mRNA^+$
Cytokine receptor	IL-10R $mRNA^+$
Cytokine response	IL-6-dependent; growth stimulation: IFN-α; growth inhibition: IFN-γ
Heterotransplantation	into SCID mice

Comments

- Myeloma cell line of IgGλ-type.
- Constitutively IL-6-dependent.
- Carries t(11;14)(q13;q32) possibly leading to alteration of the *BCL1* and *IGH* genes.

References

[1] Zhang, X.G. et al. (1994) Blood 83, 3654–3663.
[2] Pellat-Deceunynck, C. et al. (1995) Blood 86, 4001–4002.
[3] Gu, Z.J. et al. (1996) Blood 88, 3972–3986.
[4] Mazars, G.R. et al. (1992) Oncogene 7, 1015–1018.
[5] Raynaud, S.D. et al. (1993) Genes Chromosomes Cancer 8, 80–87.
[6] Rao, P.H. et al. (1998) Blood 92, 1743–1748.
[7] Klein, B. et al. (1990) Eur. Cytokine Net. 1, 193–201.
[8] Jourdan, M. et al. (1991) J. Immunol. 147, 4402–4407.
[9] Portier, M. et al. (1993) Brit. J. Haematol. 85, 514–520.
[10] Zhang, X.G. et al. (1994) J. Exp. Med. 177, 1337–1342.
[11] Rebouissou, C. et al. (1998) Blood 91, 4727–4737.

XG-3P

Culture characterization[1]

Sister cell line	**XG-3E** (serial sister cell line – from pleural effusion at relapse – similar immunological features)
Culture medium	90% RPMI 1640 + 10% FBS + 1 ng/ml IL-6
Authentication	yes (by *IGH* rearrangement)
Primary reference	Zhang et al.[1]
Availability of cell line	restricted

Clinical characterization[1]

Patient	female
Disease diagnosis	plasma cell leukemia (Igλ)
Treatment status	at diagnosis
Specimen site	peripheral blood

Immunophenotypic characterization[1]

T-/NK cell marker	$CD2^-$, $CD3^-$, $CD5^-$, $CD28^+$, $CD57^+$
B-cell marker	$CD10^+$, $CD19^-$, $CD20^-$, $CD21^+$, $CD23^-$, $CD24^-$, $CD37^+$, $CD40^+$, $CD72^-$, $CD77^+$, $CD80^-$, $CD138^+$, $cyIg\lambda^+$, sIg^-
Myelomonocytic marker	$CD13^-$, $CD14^-$, $CD15^-$, $CD16^-$, $CD33^+$, $CD65^-$
Progenitor/activation/other marker	$CD30^-$, $CD38^+$, $CD71^+$, $HLA\text{-}DR^-$
Adhesion marker	$CD11a^-$, $CD11b^-$, $CD11c^-$, $CD44^+$, $CD49b^-$, $CD49d^+$, $CD49e^+$, $CD49f^+$, $CD54^+$, $CD56^+$, $CD58^+$
Cytokine receptor	$CD25^-$, $CD126^+$, $CD130^+$

Genetic characterization[1,2]

Cytogenetic karyotype	hyperdiploid, add(8q24), der(14)t(14;?)(q32;?)
Receptor gene rearrangement	*IGH* R

Functional characterization[1–4]

Cytokine production	TGF-β $mRNA^+$
Cytokine response	IL-6-dependent; growth stimulation: GM-CSF, IL-3; growth inhibition: IFN-γ

Comments

- Plasma cell line of Igλ-type.
- Constitutively IL-6-dependent.

References

[1] Zhang, X.G. et al. (1994) Blood 83, 3654–3663.
[2] Mazars, G.R. et al. (1992) Oncogene 7, 1015–1018.
[3] Klein, B. et al. (1990) Eur. Cytokine Net. 1, 193–201.
[4] Portier, M. et al. (1993) Brit. J. Haematol. 85, 514–520.

XG-4

Culture characterization[1]

Culture medium 90% RPMI 1640 + 10% FBS + 1 ng/ml IL-6
Viral status EBV$^-$
Authentication yes (by *IGH* rearrangement)
Primary reference Zhang et al.[1]
Availability of cell line restricted

Clinical characterization[1]

Patient male
Disease diagnosis multiple myeloma (IgGκ)
Treatment status secondary plasma cell leukemia, terminal
Specimen site peripheral blood

Immunophenotypic characterization[1,2]

T-/NK cell marker CD2$^-$, CD3$^-$, CD5$^-$, CD28$^+$, CD57$^+$
B-cell marker CD10$^-$, CD19$^-$, CD20$^-$, CD21$^-$, CD23$^-$, CD24$^-$, CD37$^+$, CD40$^+$, CD72$^-$, CD77$^+$, CD80$^-$, CD138$^+$, cyIgκ$^+$, sIg$^-$
Myelomonocytic marker CD13$^-$, CD14$^-$, CD15$^-$, CD16$^-$, CD32$^-$, CD33$^-$, CD65$^-$
Progenitor/activation/other marker CD30$^-$, CD38$^+$, CD71$^+$, HLA-DR$^-$
Adhesion marker CD11a$^-$, CD11b$^-$, CD11c$^-$, CD18$^+$, CD44$^+$, CD49b$^-$, CD49d$^+$, CD49e$^-$, CD49f$^-$, CD54$^+$, CD56$^-$, CD58$^+$
Cytokine receptor CD25$^-$, CD126$^+$, CD130$^+$, IL-11R$^+$, LIFR$^+$

Genetic characterization[1,3,4]

Cytogenetic karyotype hyperdiploid, add(8q24); SKY: t(15;17)(q?;q24-25), der(16)t(1;6), der(16)t(15;16), der(17)t(15;17), der(20)t(10;20)
Receptor gene rearrangement *IGH* R
Unique gene alteration *P53* mutation

Functional characterization[1–3,5–8]

Cytokine production IL-10; mRNA$^+$: IL-6, TGF-β
Cytokine receptor IL-10R mRNA$^+$
Cytokine response IL-6-dependent; growth stimulation: CNTF, IFN-α, IL-11, LIF, OSM; growth inhibition: IFN-γ

Comments

- Myeloma cell line of IgGκ-type.
- Constitutively IL-6-dependent.

References

[1] Zhang, X.G. et al. (1994) Blood 83, 3654–3663.
[2] Gu, X.G. et al. (1996) Blood 88, 3972–3986.
[3] Mazars, G.R. et al. (1992) Oncogene 7, 1015–1018.
[4] Rao, P.H. et al. (1998) Blood 92, 1743–1748.
[5] Klein, B. et al. (1990) Eur. Cytokine Net. 1, 193–201.
[6] Portier, M. et al. (1993) Brit. J. Haematol. 85, 514–520.
[7] Jourdan, M. et al. (1991) J. Immunol. 147, 4402–4407.
[8] Zhang, X.G. et al. (1994) J. Exp. Med. 177, 1377–1342.

XG-5

Culture characterization[1]

Culture medium	90% RPMI 1640 + 10% FBS + 1 ng/ml IL-6
Authentication	no
Primary reference	Zhang et al.[1]
Availability of cell line	restricted

Clinical characterization[1]

Patient	female
Disease diagnosis	multiple myeloma (Igλ)
Treatment status	secondary plasma cell leukemia, terminal
Specimen site	peripheral blood

Immunophenotypic characterization[1]

T-/NK cell marker	CD2$^-$, CD3$^-$, CD5$^-$, CD28$^+$, CD57$^-$
B-cell marker	CD10$^+$, CD19$^-$, CD20$^-$, CD21$^+$, CD23$^-$, CD24$^-$, CD37$^+$, CD40$^-$, CD72$^-$, CD77$^+$, CD80$^+$, CD138$^+$, cyIgλ$^+$, sIg$^-$
Myelomonocytic marker	CD13$^-$, CD14$^-$, CD15$^-$, CD16$^-$, CD32$^-$, CD33$^+$, CD65$^-$
Progenitor/activation/other marker	CD30$^+$, CD38$^+$, CD71$^+$, HLA-DR$^-$
Adhesion marker	CD11a$^-$, CD11b$^-$, CD11c$^-$, CD18$^-$, CD44$^+$, CD49b$^-$, CD49d$^+$, CD49e$^+$, CD49f$^-$, CD54$^+$, CD56$^-$, CD58$^+$
Cytokine receptor	CD25$^-$, CD126$^+$, CD130$^+$

Genetic characterization[1-4]

Cytogenetic karyotype	hypodiploid, t(8;14)(q24;q32), t(11;14)(q13;q32); SKY: der(3)t (3;5), der(5)t(3;5), t(11;14)(q13;q32), t(15;17)(q?;q24-25), der(17)t(3;6;17), der(17)t(15;17)
Unique translocation/fusion gene	t(11;14)(q13;q32) → *BCL1-IGH* genes altered?
Receptor gene rearrangement	*IGH* R
Unique gene alteration	*P53* mutation

Functional characterization[1,2,5-7]

Cytokine production	TGF-β mRNA$^+$
Cytokine response	IL-6-dependent; growth stimulation: GM-CSF, IFN-α, IL-3; growth inhibition: IFN-γ

Comments

- Myeloma cell line of Igλ-type.
- Constitutively IL-6-dependent.
- Carries t(11;14)(q13;q32) possibly leading to alteration of the *BCL1* and *IGH* genes.

References

[1] Zhang, X.G. et al. (1994) Blood 83, 3654–3663.
[2] Mazars, G.R. et al. (1992) Oncogene 7, 1015–1018.
[3] Rao, P.H. et al. (1998) Blood 92, 1743–1748.
[4] Raynaud, S.D. et al. (1993) Genes Chromosomes Cancer 8, 80–87.
[5] Klein, B. et al. (1990) Eur. Cytokine Net. 1, 193–201.
[6] Jourdan, M. et al. (1991) J. Immunol. 147, 4402–4407.
[7] Portier, M. et al. (1993) Brit. J. Haematol. 85, 514–520.

XG-6

Culture characterization[1]

Culture medium	90% RPMI 1640 + 10% FBS + 1 ng/ml IL-6
Viral status	EBV⁻
Authentication	yes (by *IGH* rearrangement)
Primary reference	Zhang et al.[1]
Availability of cell line	restricted

Clinical characterization[1]

Patient	female
Disease diagnosis	multiple myeloma (IgGλ)
Treatment status	secondary plasma cell leukemia, terminal
Specimen site	peripheral blood

Immunophenotypic characterization[1,2]

T-/NK cell marker	$CD2^-$, $CD3^-$, $CD5^-$, $CD28^+$, $CD57^-$
B-cell marker	$CD10^-$, $CD19^-$, $CD20^-$, $CD21^+$, $CD23^-$, $CD24^-$, $CD37^+$, $CD40^+$, $CD72^-$, $CD77^+$, $CD80^-$, $CD138^+$, cyIgGλ$^+$, sIgλ$^{(+)}$
Myelomonocytic marker	$CD13^-$, $CD14^-$, $CD15^-$, $CD16^-$, $CD32^-$, $CD33^-$, $CD65^-$
Progenitor/activation/other marker	$CD38^{(+)}$, $CD71^+$, HLA-DR$^-$
Adhesion marker	$CD11a^-$, $CD11b^-$, $CD11c^-$, $CD18^-$, $CD44^+$, $CD49b^-$, $CD49d^+$, $CD49e^{(+)}$, $CD49f^-$, $CD54^+$, $CD56^-$, $CD58^+$
Cytokine receptor	$CD25^-$, $CD126^+$, $CD130^+$, IL-11R$^+$, LIFR$^+$

Genetic characterization[1,3]

Cytogenetic karyotype	hypodiploid, 13q−, der(14)t(11;14)(q13;q32); SKY: der(11)t(1;3;11), der(16)t(10;16), der(16)t(16;22), t(17;19)(q24-25;?), der(17)t(17;19)
Unique translocation/fusion gene	t(11;14)(q13;q32) → *BCL1-IGH* genes altered?
Receptor gene rearrangement	*IGH* R

Functional characterization[1,4,5]

Cytokine production	IL-10
Cytokine receptor	IL-10R mRNA$^+$
Cytokine response	IL-6-dependent; growth stimulation: IL-11, LIF, OSM

Comments

- Myeloma cell line of IgGλ-type.
- Constitutively IL-6-dependent.
- Carries t(11;14)(q13;q32) possibly leading to alteration of the *BCL1* and *IGH* genes.

References

[1] Zhang, X.G. et al. (1994) Blood 83, 3654–3663.
[2] Pellat-Deceunynck, C. et al. (1995) Blood 86, 4001–4002.
[3] Rao, P.H. et al. (1998) Blood 92, 1743–1748.
[4] Zhang, X.G. et al. (1994) J. Exp. Med. 177, 1337–1342.
[5] Gu, X.G. et al. (1996) Blood 88, 3972–3986.

XG-7

Culture characterization[1]

Culture medium	90% RPMI 1640 + 10% FBS + 1 ng/ml IL-6
Viral status	EBV⁻
Authentication	yes (by *IGH* rearrangement)
Primary reference	Zhang et al.[1]
Availability of cell line	restricted

Clinical characterization[1]

Patient	female
Disease diagnosis	multiple myeloma (IgAκ)
Treatment status	secondary plasma cell leukemia, terminal
Specimen site	peripheral blood

Immunophenotypic characterization[1]

T-/NK cell marker	$CD2^-$, $CD3^-$, $CD5^-$, $CD28^+$, $CD57^-$
B-cell marker	$CD10^-$, $CD19^-$, $CD20^-$, $CD23^-$, $CD24^-$, $CD37^-$, $CD40^-$, $CD72^-$, $CD77^+$, $CD80^-$, $CD138^+$, cyIgAκ$^+$, sIg$^-$
Myelomonocytic marker	$CD13^-$, $CD14^-$, $CD15^-$, $CD16^-$, $CD32^-$, $CD65^-$
Progenitor/activation/other marker	$CD38^+$, $CD71^+$, HLA-DR$^-$
Adhesion marker	$CD11a^-$, $CD11b^-$, $CD11c^-$, $CD18^-$, $CD44^+$, $CD49b^-$, $CD49d^+$, $CD49e^+$, $CD49f^+$, $CD54^+$, $CD56^+$, $CD58^+$
Cytokine receptor	$CD25^-$, $CD126^+$, $CD130^+$, LIFR$^+$

Genetic characterization[1-4]

Cytogenetic karyotype	hypodiploid, −13; FISH: t(6;14)(p25;q32); SKY: der(6)t(3;6), der(8)t(8;13), der(10)t(2;10), der(16)t(1;16), der(22)t(2;22)
Unique translocation/fusion gene	t(6;14)(p25;q32) → *IGH-MUM1/IRF4* fusion gene
Receptor gene rearrangement	*IGH* R

Functional characterization[1,5]

Cytokine production	IL-6, IL-10, OSM
Cytokine receptor	IL-11R mRNA$^+$
Cytokine response	IL-6-dependent

Comments

- Myeloma cell line of IgAκ-type.
- Constitutively IL-6-dependent.
- Carries t(6;14)(p25;q32) leading to *IGH-MUM1/IRF4* fusion gene.

References
[1] Zhang, X.G. et al. (1994) Blood 83, 3654–3663.
[2] Iida, S. et al. (1997) Nature Genetics 17, 226–230.
[3] Rao, P.H. et al. (1998) Blood 92, 1743–1748.
[4] Yoshida, S. et al. (1999) Leukemia 13, 1812–1816.
[5] Gu, X.G. et al. (1996) Blood 88, 3972–3986.

Section IV

T-CELL LEUKEMIA AND T-CELL LYMPHOMA CELL LINES

Part 1

Immature T-Cell Lines

ALL-SIL

Culture characterization[1]

Other name of cell line	Sil-ALL
Culture medium	90% RPMI 1640 + 10% FBS
Authentication	no
Primary reference	unpublished
Availability of cell line	not known

Clinical characterization[1]

Patient	17-year-old male
Disease diagnosis	T-ALL
Treatment status	at relapse
Specimen site	peripheral blood
Year of establishment	1979

Immunophenotypic characterization[2]

T-/NK cell marker	CD1$^+$, CD2$^-$, CD3$^-$, cyCD3$^+$, CD4$^+$, CD5$^+$, CD7$^+$, CD8$^+$, CD28$^-$, CD57$^-$, TCR$\alpha\beta^-$, TCR$\gamma\delta^-$, cyTCRα^-, cyTCRβ^-, cyTCRγ^+
B-cell marker	CD9$^-$, CD10$^-$, CD19$^-$, CD20$^-$, CD21$^-$, FMC-7$^-$
Myelomonocytic marker	CD13$^-$, CD15$^-$, CD16$^-$, CD33$^-$
Erythroid-megakaryocytic marker	CD41a$^-$, CD61$^-$
Progenitor/activation/other marker	CD34$^-$, HLA-DR$^-$, TdT$^-$
Adhesion marker	CD11b$^-$
Cytokine receptor	CD25$^-$

Genetic characterization[2-7]

Cytogenetic karyotype	95, XXYY, +6, +8, +8, −17, −17, +19, −22, t(1;13)(p31;q32), t(1;13), t(10;14)(q24;q11.2), t(10;14), +der(17)t(17;?)(p12?;?), +der(17)
Unique translocation/fusion gene	t(10;14)(q24;q11) → *HOX11-TCRD* genes altered
Unique gene alteration	*IFNA* deletion, *IFNB* deletion, *P15INK4B* deletion, *P16INK4A* deletion, *RB1* deletion

Functional characterization[8]

Cytokine production	mRNA$^+$: GM-CSF, TGF-β1

Comments

- Immature T-cell line (type T-III cortical) derived from T-ALL.
- Carries t(10;14)(q24;q11) leading to alteration of the *HOX11-TCRD* genes.

References

[1] Ben-Bassat, H. (1999) personal communication.
[2] Minowada, J. and Matsuo, Y. (1999) unpublished data.
[3] Heyman, M. et al. (1994) Leukemia 8, 425–434.
[4] Borgonovo Brandter, L. et al. (1996) Eur. J. Haematol. 56, 313–318.
[5] Kawamura, M. et al. (1999) Leukemia Res. 23, 115–126.
[6] Hatano, M. et al. (1991) Science 253, 79–82.
[7] Lu, M. et al. (1991) EMBO J. 10, 2905–2910.
[8] Micallef, M. et al. (1994) Int. J. Oncol. 4, 633–638.

Be13

Culture characterization[1,2]

Culture medium	80% RPMI 1640 + 20% FBS
Doubling time	40–60 h
Viral status	EBV⁻, HBV⁻, HCV⁻, HHV-8⁻, HIV⁻, HTLV-I/II⁻
Authentication	no
Primary reference	Galili et al.[1]
Availability of cell line	DSM ACC 396

Clinical characterization[1]

Patient	11-year-old female
Disease diagnosis	T-ALL
Treatment status	at 2nd relapse
Specimen site	bone marrow

Immunophenotypic characterization[1,2]

T-/NK cell marker	$CD1^+$, $CD2^-$, $CD3^+$, $CD4^+$, $CD5^+$, $CD6^+$, $CD7^+$, $CD8^-$, $TCR\alpha\beta^-$, $TCR\gamma\delta^-$
B-cell marker	$CD19^-$, sIg^-
Myelomonocytic marker	$CD13^-$
Progenitor/activation/other marker	$CD34^-$

Genetic characterization[2]

Cytogenetic karyotype	flat-moded hypotetraploid; 81–89<4n>XXXX, +1, +4, −9, −9, −10, −16, −21, del(1)(q13.2), add(1)(p11-21), del(4)(q11q31.2) /i(4p)×2, i(4q)del(4q31.2)/i(4q)add(4)(q31)×2, del(5)(q14q21)×2, del(6)(q22)×2, del(9)(p22)×2, der(17)t(17;?21)(p11;q11)×2

Functional characterization[1,3]

Inducibility of differentiation	TPA → differentiation
Special features	highly sensitive to dexamethasone

Comments

- Immature T-cell line (type T-III cortical) derived from childhood relapsed T-ALL.
- Available from cell line bank.

References

[1] Galili, N. et al. (1981) Human Lymph. Diff. 1, 123–130.

[2] Drexler, H.G. et al. (1999) DSMZ Catalogue of Cell Lines, 7th edn, Braunschweig, Germany.

[3] Galili, U. et al. (1984) Cancer Res. 44, 4594–4599.

CCRF-CEM

Culture characterization[1,2]

Culture medium	90% Eagle's MEM (originally) or RPMI 1640 + 10% FBS
Doubling time	24 h
Viral status	EBV⁻, HBV⁻, HCV⁻, HIV⁻, HTLV-I/II⁻
Authentication	no
Primary reference	Foley et al.[1]
Availability of cell line	ATCC CCL 119/CRL 8436, DSM ACC 240, IFO 50412, JCRB 0033/9023

Clinical characterization[1]

Patient	3-year-old female
Disease diagnosis	lymphosarcoma → ALL
Treatment status	at relapse, terminal
Specimen site	peripheral blood
Year of establishment	1964

Immunophenotypic characterization[2–6]

T-/NK cell marker	CD1a⁺, CD2⁻, CD3⁺, cyCD3⁺, CD4⁺, CD5⁺, CD6⁺, CD7⁺, CD8⁻, CD27⁺, CD28⁺, CD56⁻, CD57⁻, TCRαβ⁺, TCRγδ⁻, cyTCRα⁺, cyTCRβ⁺, cyTCRγ⁻
B-cell marker	CD9⁻, CD10⁻, CD19⁻, CD20⁻, CD21⁻, CD22⁻, CD24⁻
Myelomonocytic marker	CD13⁻, CD14⁻, CD15⁺, CD16⁻, CD33⁻, CD68⁻
Erythroid-megakaryocytic marker	CD36⁻, CD41⁻, CD61⁻, GlyA⁻
Progenitor/activation/other marker	CD34⁻, CD38⁺, CD71⁺, CD90⁻, HLA-DR⁻, TdT⁺
Adhesion marker	CD11b⁻, CD29⁺
Cytokine receptor	CD25⁻, CD117⁻, CD122⁻

Genetic characterization[1,2,6–17]

Cytogenetic karyotype	near-tetraploid, extensive subclonal variation; 90(88–101) <4n>XX, −X, −X, +20, +20, t(8;9)(p11;p24)×2, der(9)del(9) (p21-22)del(9)(q11q13-21)×2; sideline with +5, +21, add(13)(q3?3), del(16)(q12)
Unique translocation/fusion gene	submicroscopic del(1)(p32) → *SIL-SCL* fusion gene
Receptor gene rearrangement	*IGH* G, *IGK* G, *TCRA* R, *TCRB* R, *TCRG* R, *TCRD* D (mRNA: TCRA⁺, TCRB⁺, TCRG⁺, TCRD⁻)
Unique gene alteration	*IFNA* deletion, *IFNB* deletion, *P16INK4A* deletion, *P53* mutation

Functional characterization[1,18–20]

Colony formation	in agar
Cytochemistry	MGP$^+$, MPO$^-$
Cytokine production	TGF-β mRNA$^+$
Proto-oncogene	mRNA$^+$: BAX, BCLXL, P53

Comments

- Immature TCRαβ$^+$ T-cell line (type T-III cortical) derived from ALL.
- First and most widely distributed T-cell line (reference cell line).
- Carries submicroscopic del(1)(p32) leading to the *SIL-SCL* fusion gene.
- Available from cell line banks.
- Cells grow very well.

References

[1] Foley, G.E. et al. (1965) Cancer 18, 522–529.
[2] Drexler, H.G. et al. (1999) DSMZ Catalogue of Cell Lines, 7th edn, Braunschweig, Germany.
[3] Minowada, J. and Matsuo, Y. (1999) unpublished data.
[4] Toba, K. et al. (1996) Exp. Hematol. 24, 894–901.
[5] Zhou, M. et al. (1995) Blood 85, 1608–1614.
[6] Burger, R. et al. (1999) Leukemia Res. 23, 19–27.
[7] Sangster, R.N. et al. (1986) J. Exp. Med. 163, 1491–1508.
[8] Heyman, M. et al. (1994) Leukemia 8, 425–434.
[9] Siebert, R. et al. (1995) Brit. J. Haematol. 91, 350–354.
[10] Zhou, M. et al. (1995) Leukemia 9, 1159–1161.
[11] Ogawa, S. et al. (1994) Blood 84, 2431–2435.
[12] Borgonovo Brandter, L. et al. (1996) Eur. J. Haematol. 56, 313–318.
[13] Otsuki, T. et al. (1995) Cancer Res. 55, 1436–1440.
[14] Cheng, J. and Haas, H. (1990) Mol. Cell. Biol. 10, 5502–5509.
[15] Kawamura, M. et al. (1999) Leukemia Res. 23, 115–126.
[16] Aplan, P.D. et al. (1990) Science 250, 1426–1429.
[17] Brown, L. et al. (1990) EMBO J. 9, 3343–3351.
[18] Micallef, M. et al. (1994) Int. J. Oncol. 4, 633–638.
[19] Findley, H.W. et al. (1997) Blood 89, 2986–2993.
[20] Nilsson, K. et al. (1977) Int. J. Cancer 19, 337–344.

CML-T1

Culture characterization[1,2]

Culture medium	90% IMDM or RPMI 1640 + 10% FBS
Doubling time	72 h
Viral status	EBV$^-$, HBV$^-$, HCV$^-$, HHV-8$^-$, HIV$^-$, HTLV-I/II$^-$
Authentication	yes (by cytogenetics)
Primary reference	Kuriyama et al.[1]
Availability of cell line	DSM ACC 7

Clinical characterization[1,3]

Patient	36-year-old female
Disease diagnosis	Ph$^-$ CML
Treatment status	in acute phase
Specimen site	peripheral blood
Year of establishment	1987

Immunophenotypic characterization[1,2]

T-/NK cell marker	CD1$^{(+)}$, CD2$^{(+)}$, CD3$^{(+)}$, CD4$^+$, CD5$^+$, CD6$^+$, CD7$^+$, CD8$^+$, CD57$^+$, TCR$\alpha\beta^-$
B-cell marker	CD10$^-$, CD19$^-$, CD20$^-$, CD21$^-$, CD24$^-$, cyIg$^-$, sIg$^-$
Myelomonocytic marker	CD13$^-$, CD14$^-$, CD16$^-$, CD33$^-$
Erythroid-megakaryocytic marker	CD41$^-$, CD42$^-$, GlyA$^-$, PPO$^-$
Progenitor/activation/other marker	CD38$^+$, CD71$^+$, HLA-DR$^-$, TdT$^-$
Adhesion marker	CD11b$^-$
Cytokine receptor	CD25$^-$

Genetic characterization[1,2,4]

Cytogenetic karyotype	near tetraploid, with diploid sideline; 92<4n>XXXX, t(6;7) (q24;q35)×2, del(11)(q22.3)×2
Unique translocation/fusion gene	no t(9;22)(q34;q11) – *BCR-ABL* b2-a2 fusion gene
Receptor gene rearrangement	*TCRB* R

Functional characterization[1]

Cytochemistry	ACP$^+$, ANAE$^+$, ANBE$^+$, CAE$^-$, GLC$^+$, MPO$^-$, PAS$^{(+)}$

Comments

- Immature T-cell line (type T-III cortical) derived from CML in acute phase.
- Does not carry Ph chromosome, but carries *BCR-ABL* b2-a2 fusion gene.
- Rare CML-derived T-cell line.
- Available from cell line bank.
- Cells grow well.

References

[1] Kuriyama, K. et al. (1989) Blood 74, 1381–1387.

[2] Drexler, H.G. et al. (1999) DSMZ Catalogue of Cell Lines, 7th edn, Braunschweig, Germany.

[3] Soda, H. et al. (1985) Brit. J. Haematol. 59, 671–679. (case report)

[4] Drexler, H.G. et al. (1999) Leukemia Res. 23, 207–215.

DND-41

Culture characterization[1,2]

Culture medium 90% RPMI 1640 + 10% FBS
Viral status EBNA⁻
Authentication no
Primary reference unpublished
Availability of cell line not known

Clinical characterization[2]

Patient 13-year-old male
Disease diagnosis ALL
Specimen site peripheral blood
Year of establishment 1977

Immunophenotypic characterization[1,2]

T-/NK cell marker $CD1^+$, $CD2^+$, $CD3^+$, $cyCD3^+$, $CD4^+$, $CD5^+$, $CD7^+$, $CD8^-$, $CD27^-$, $CD28^+$, $CD57^-$, $TCR\alpha^-$, $TCR\beta^-$, $TCR\alpha\beta^-$, $TCR\gamma\delta^+$, $cyTCR\alpha^-$, $cyTCR\beta^-$, $cyTCR\gamma^-$, $cyTCR\delta^-$
B-cell marker $CD9^-$, $CD10^+$, $CD19^-$, $CD20^-$, $CD21^+$, $CD22^-$, $CD24^-$
Myelomonocytic marker $CD13^-$, $CD14^-$, $CD15^-$, $CD16^-$, $CD33^-$
Erythroid-megakaryocytic marker $CD41a^-$, $CD61^-$
Progenitor/activation/other marker $CD34^-$, $CD38^+$, $CD71^+$, $HLA\text{-}DR^-$, TdT^+
Adhesion marker $CD11b^-$, $CD29^+$
Cytokine receptor $CD25^-$

Genetic characterization[2-6]

Cytogenetic karyotype 91, XXYY, −9, −15, +20
Receptor gene rearrangement *IGH* G, *IGK* G, *TCRA* G, *TCRB* R, *TCRG* R, *TCRD* R (mRNA: $TCRA^-$, $TCRB^+$, $TCRG^+$, $TCRD^+$)
Unique gene alteration *P15INK4B* deletion, *P16INK4A* deletion, *P53* mutation

Functional characterization[7]

Cytokine production $mRNA^+$: GM-CSF, TGF-β1

Comments

- Immature T-cell line (type T-III cortical) established from childhood ALL.

References

1 Drexler, H.G. et al. (1985) Leukemia Res. 9, 209–229.
2 Minowada, J. and Matsuo, Y. (1999) unpublished data.
3 Sangster, R.N. et al. (1986) J. Exp. Med. 163, 1491–1508.
4 Burger, R. et al. (1999) Leukemia Res. 23, 19–27.
5 Kawamura, M. et al. (1999) Leukemia Res. 23, 115–126.
6 Borgonovo Brandter, L. et al. (1996) Eur. J. Haematol. 56, 313–318.
7 Micallef, M. et al. (1994) Int. J. Oncol. 4, 633–638.

Culture characterization[1]

Culture medium	80% RPMI 1640 + 10% horse serum + 10% FBS
Doubling time	48–96 h
Authentication	yes (by cytogenetics)
Primary reference	Kurtzberg et al.[1]
Availability of cell line	original authors

Clinical characterization[1,2]

Patient	16-year-old male
Disease diagnosis	T-ALL
Treatment status	at diagnosis
Specimen site	peripheral blood

Immunophenotypic characterization[1]

T-/NK cell marker	CD1$^-$, CD2$^-$, CD3$^-$, CD4$^-$, CD7$^+$, CD8$^-$
B-cell marker	CD9$^-$, CD20$^-$, CD21$^-$, CD24$^-$
Myelomonocytic marker	CD14$^-$, CD15$^-$
Progenitor/activation/other marker	CD34$^+$, CD71$^+$, HLA-DR$^-$, TdT$^-$
Cytokine receptor	CD25$^+$, CD117$^-$

Genetic characterization[1,3–5]

Cytogenetic karyotype	46, XY, −14, +t(1;14)(p33;q11), del(1)(p33), +del(1)(q11), del(13)(q14)
Unique translocation/fusion gene	t(1;14)(p32;q11) → *SCL-TCRD* fusion gene
Receptor gene rearrangement	*TCRA* R, *TCRB* R, *TCRG* R, *TCRD* D

Functional characterization[1]

Cytochemistry	ANAE$^-$, ANBE$^+$, MPO$^-$, PAS$^{(+)}$
Inducibility of differentiation	spontaneous differentiation to T-lymphoid, monomyeloid, erythroid cells

Comments

- Immature T-cell line (type T-I pro-T) derived from pediatric T-ALL.
- Carries t(1;14)(p32;q11) leading to the *SCL-TCRD* fusion gene.
- Cells are difficult to culture.

References

[1] Kurtzberg, J. et al. (1985) J. Exp. Med. 162, 1561–1578.
[2] Hershfield, M.S. et al. (1984) Proc. Natl. Acad. Sci USA 81, 253–257. (case report)
[3] Begley, C.G. et al. (1989) Proc. Natl. Acad. Sci. USA 86, 2031–2035.
[4] Finger, L.R. et al. (1989) Proc. Natl. Acad. Sci. USA 86, 5039–5043.
[5] Begley, C.G. et al. (1989) Proc. Natl. Acad. Sci. USA 86, 10128–10132.

Culture characterization[1,2]

Culture medium 80% RPMI 1640 + 20% FBS
Doubling time 45–55 h
Viral status EBV^-
Authentication no
Primary reference Zhou et al.[1]
Availability of cell line not known

Clinical characterization[1,2]

Patient 8-year-old male
Disease diagnosis T-ALL
Treatment status at relapse
Specimen site bone marrow
Year of establishment 1993

Immunophenotypic characterization[1]

T-/NK cell marker $CD1^+$, $CD2^+$, $CD3^-$, $CD4^-$, $CD5^+$, $CD7^+$, $CD8^+$
B-cell marker $CD10^-$, $CD19^-$, $CD20^-$, $CD24^-$
Myelomonocytic marker $CD13^-$, $CD15^+$, $CD33^-$
Progenitor/activation/other marker $CD34^-$, $CD38^+$, $CD95^+$, $HLA\text{-}DR^-$, TdT^+

Genetic characterization[1-3]

Cytogenetic karyotype 46, XY, t(1;7)(p34;q34)
Unique translocation/fusion gene t(1;7)(p34;q34) → *LCK-TCRB* genes altered
Unique gene alteration *P16INK4A* deletion

Functional characterization[1,3]

Colony formation in methylcellulose
Cytochemistry $ANAE^+$, MPO^-, SBB^-
Heterotransplantation into SCID mice
Proto-oncogene $mRNA^+$: BAX, BCLXL

Comments

- Immature T-cell line (type T-III cortical) established from relapsed childhood T-ALL.
- Carries t(1;7)(p34;q34) causing alteration of the *LCK-TCRB* genes.

References

[1] Zhou, M. et al. (1995) Blood 85, 1608–1614.
[2] Zhou, M. et al. (1995) Leukemia 9, 1159–1161.
[3] Findley, H.W. et al. (1997) Blood 89, 2986–2993.

HD-Mar

Culture characterization[1]

Culture medium	80% RPMI 1640 + 20% FBS
Doubling time	24–30 h
Viral status	EBNA⁻
Authentication	no
Primary reference	Ben-Bassat et al.[1]
Availability of cell line	original authors

Clinical characterization[1]

Patient	20-year-old male
Disease diagnosis	Hodgkin's disease (mixed cellularity)? or T-lymphoblastic NHL
Treatment status	at relapse, advanced stage, refractory
Specimen site	pleural effusion
Year of establishment	1977

Immunophenotypic characterization[1,2]

T-/NK cell marker	CD1$^+$, CD2$^+$, CD3$^+$, cyCD3$^+$, CD4$^+$, CD5$^+$, CD7$^+$, CD8$^+$, CD28$^+$, CD56$^-$, CD57$^-$, CD81$^+$, CD82$^+$, CD98$^+$, CD99$^+$, CD100$^+$, TCR$\alpha\beta^+$, TCR$\gamma\delta^-$, cyTCRα^+, cyTCRβ^+, cyTCRγ^+
B-cell marker	CD9$^-$, CD10$^+$, CD19$^-$, CD20$^-$, CD21$^+$, CD22$^-$, CD23$^-$, CD24$^-$, CD37$^-$, CD39$^-$, CD40$^-$, cyIg$^-$, sIg$^-$
Myelomonocytic marker	CD13$^-$, CD14$^-$, CD15$^-$, CD16$^-$, CD32$^-$, CD33$^-$
Erythroid-megakaryocytic marker	CD41$^-$, CD61$^-$
Progenitor/activation/other marker	CD30$^-$, CD34$^-$, CD38$^+$, CD45$^+$, CD45RA$^-$, CD45RO$^+$, CD71$^+$, HLA-DR$^-$, TdT$^+$
Adhesion marker	CD11a$^+$, CD11b$^-$, CD18$^+$, CD29$^+$, CD44$^+$, CD48$^+$, CD49b$^-$, CD49d$^+$, CD49f$^-$, CD58$^+$, CD62L$^+$
Cytokine receptor	CD25$^-$, CD122$^-$, CD127$^+$

Genetic characterization[2-4]

Cytogenetic karyotype	94–96, XXYY, −2, −4, −6, −9, −10, −10, −12, −14, +16, −17, −17, −18, +15-17mar, del(2)(p22), del(9)(p22)×2
Receptor gene rearrangement	*TCRB* R (mRNA: TCRB$^+$, TCRD$^-$)
Unique gene alteration	*IFNA* deletion, *IFNB* deletion, *P15INK4B* deletion, *P16INK4A* deletion

Functional characterization[1]

Cytochemistry	ACP$^+$, ANAE$^-$, GLC$^+$, Lysozyme$^-$, MPO$^-$

Comments

- Immature TCRαβ+ T-cell line (type T-III cortical) derived from lymphoma.

References

[1] Ben-Bassat, H. et al. (1980) Int. J. Cancer 25, 583–590.
[2] Minowada, J. and Matsuo, Y. (1999) unpublished data.
[3] Heyman, M. et al. (1994) Leukemia 8, 425–434.
[4] Borgonovo Brandter, L. et al. (1996) Eur. J. Haematol. 56, 313–318.

HPB-ALL

Culture characterization[1]

Culture medium	90% RPMI 1640 + 10% FBS
Viral status	EBNA-
Authentication	no
Primary reference	Morikawa et al.[1]
Availability of cell line	not known

Clinical characterization[1]

Patient	14-year-old male
Disease diagnosis	T-ALL + thymoma
Treatment status	at diagnosis
Specimen site	peripheral blood
Year of establishment	1973

Immunophenotypic characterization[1-3]

T-/NK cell marker	CD1+, CD2+, CD3+, cyCD3+, CD4+, CD5+, CD6+, CD7+, CD8+, CD27−, CD28+, CD57−, TCRαβ+, TCRγδ−, cyTCRα+, cyTCRβ+, cyTCRγ−
B-cell marker	CD9−, CD10+, CD19−, CD20−, CD21+, CD22−, CD24−
Myelomonocytic marker	CD13−, CD14−, CD15−, CD16−, CD33−
Erythroid-megakaryocytic marker	CD41a−, CD61−
Progenitor/activation/other marker	CD34−, CD38+, CD71+, HLA-DR−, TdT+
Adhesion marker	CD11b−, CD29+
Cytokine receptor	CD25−

Genetic characterization[2-6]

Cytogenetic karyotype	94–96, XXYY, −3, −3, −9, −16, −16,− 18, +19, +4mar, t(1;5)(q23;q25)×2, del(2)(p22)×2, add(14)(q32), +del(20)(q11)×2, +del(22)(q21)×2
Receptor gene rearrangement	*IGH* R, *IGK* G, *TCRA* R, *TCRB* R, *TCRG* R, *TCRD* D (mRNA: IGH−, TCRA+, TCRB+, TCRG+, TCRD−)
Unique gene alteration	*IFNB* deletion, *P15INK4B* deletion, *P16INK4A* deletion

Functional characterization[7]

Cytokine production	mRNA+: GM-CSF, TGF-β1

Comments

- Immature TCRαβ+ T-cell line (type T-III cortical) derived from pediatric T-ALL.
- Cells grow very well.

References

[1] Morikawa, S. et al. (1978) Int. J. Cancer 21, 166–170.
[2] Minowada, J. and Matsuo, Y. (1999) unpublished data.
[3] Burger, R. et al. (1999) Leukemia Res. 23, 19–27.
[4] Heyman, M. et al. (1994) Leukemia 8, 425–434.
[5] Sangster, R.N. et al. (1986) J. Exp. Med. 163, 1491–1508.
[6] Borgonovo Brandter, L. et al. (1996) Eur. J. Haematol. 56, 313–318.
[7] Micallef, M. et al. (1994) Int. J. Oncol. 4, 633–638.

H-SB2

Culture characterization[1–4]

Other name of cell line	**SB-2** or **CCRF-H-SB2**
Sister cell line	**CCRF-SB** (= EBV$^+$ B-LCL)
Establishment	established and initially serially passaged in newborn hamster
Culture medium	90% Eagle's MEM (or RPMI 1640) + 10% FBS
Viral status	EBV$^-$
Authentication	no
Primary reference	Adams et al.[1]
Availability of cell line	H-SB2: ATCC CCL 120.1, DSM ACC 435, JCRB 0031, RCB 0016; CCRF-SB: ATCC CCL 120, IFO 50026, JCRB 0032

Clinical characterization[1–3]

Patient	11-year-old male
Disease diagnosis	lymphosarcoma → ALL
Treatment status	refractory, terminal
Specimen site	peripheral blood
Year of establishment	1966

Immunophenotypic characterization[4–6]

T-/NK cell marker	CD1$^-$, CD2$^-$, CD3$^-$, cyCD3$^+$, CD4$^-$, CD5$^+$, CD6$^+$, CD7$^+$, CD8$^-$, CD27$^+$, CD28$^-$, CD57$^+$, TCR$\alpha\beta^-$, TCR$\gamma\delta^-$, cyTCRα^-, cyTCRβ^+, cyTCRγ^+
B-cell marker	CD9$^-$, CD10$^-$, CD19$^-$, CD20$^-$, CD21$^-$, CD22$^-$, CD24$^-$
Myelomonocytic marker	CD13$^-$, CD14$^-$, CD15$^+$, CD16$^+$, CD33$^-$
Erythroid-megakaryocytic marker	CD41a$^-$, CD61$^-$
Progenitor/activation/other marker	CD34$^+$, CD38$^+$, CD71$^+$, HLA-DR$^-$, TdT$^-$
Adhesion marker	CD11b$^-$, CD29$^+$
Cytokine receptor	CD25$^-$, CD117$^-$

Genetic characterization[7–17]

Cytogenetic karyotype	46, XY, t(1;7)(p;q35)
Unique translocation/fusion gene	(1) t(1;7)(p34;q34) → *LCK-TCRB* genes altered (2) submicroscopic del(1)(p32) → *SIL-SCL* fusion gene
Receptor gene rearrangement	*IGH* R, *IGK* G, *TCRB* R, *TCRG* R, *TCRD* D
Unique gene alteration	*P15INK4B* deletion, *P16INK4A* deletion, *RB1* deletion

Functional characterization[1,18]

Cytokine production GM-CSF mRNA+
Heterotransplantation into newborn hamster

Comments

- One of oldest immature T-cell lines derived from relapsed childhood ALL (type T-II pre-T).
- Carries t(1;7)(p34;q34) involving the *LCK* and *TCRB* genes.
- Carries submicroscopic del(1)(p32) leading to the *SIL-SCL* fusion gene.
- Available from cell line banks.
- Cells grow very well.

References

1 Adams, R.A. et al. (1968) Cancer Res. 28, 1121–1125.
2 Adams, R.A. et al. (1970) Exp. Cell Res. 62, 5–10.
3 Lazarus, H. et al. (1974) In Vitro 9, 303–310.
4 Drexler, H.G. et al. (1999) DSMZ Catalogue of Cell Lines, 7th edn, Braunschweig, Germany.
5 Minowada, J. and Matsuo, Y. (1999) unpublished data.
6 Burger, R. et al. (1999) Leukemia Res. 23, 19–27.
7 Cheng, J. et al. (1990) Blood 75, 730–735.
8 Burnett, R.C. et al. (1991) Genes Chromosomes Cancer 3, 461–467.
9 Tycko, B. et al. (1991) J. Exp. Med. 174, 867–873.
10 Burnett, R.C. et al. (1994) Blood 84, 1232–1236.
11 Wright, D.D. et al. (1994) Mol. Cell. Biol. 14, 2429–2437.
12 Zhou, M. et al. (1995) Leukemia 9, 1159–1161.
13 Siebert, R. et al. (1995) Brit. J. Haematol. 91, 350–354.
14 Otsuki, T. et al. (1995) Cancer Res. 55, 1436–1440.
15 Aplan, P.D. et al. (1990) Science 250, 1426–1429.
16 Brown, L. et al. (1990) EMBO J. 9, 3343–3351.
17 Drexler, H.G. et al. (1995) Leukemia 9, 480–500. (review on gene alterations in cell lines)
18 Micallef, M. et al. (1994) Int. J. Oncol. 4, 633–638.

HT-1

Culture characterization[1]

Culture medium	85% RPMI 1640 + 15% FBS
Doubling time	51 h
Viral status	EBV$^-$
Authentication	no
Primary reference	Abe et al.[1]
Availability of cell line	not known

Clinical characterization[1]

Patient	32-year-old male
Disease diagnosis	T-lymphoblastic lymphoma
Treatment status	at diagnosis
Specimen site	pleural effusion
Year of establishment	1988

Immunophenotypic characterization[1]

T-/NK cell marker	CD1$^+$, CD2$^+$, CD3$^+$, cyCD3$^+$, CD4$^+$, CD5$^+$, CD7$^+$, CD8$^+$, CD57$^+$
B-cell marker	CD10$^-$, CD19$^-$, CD20$^-$, sIg$^-$
Myelomonocytic marker	CD13$^-$, CD14$^-$, CD16$^-$, CD33$^-$
Progenitor/activation/other marker	HLA-DR$^-$, TdT$^+$
Adhesion marker	CD62L$^-$
Cytokine receptor	CD25$^-$

Genetic characterization[1]

Cytogenetic karyotype	47, XY, del(6)(q15q23), t(9;14)(q34;q11), del(13)(q12q14), inv(14)(q11q32)
Unique translocation/fusion gene	inv(14)(q11q32) → *TCRA-IGH* genes altered?
Receptor gene rearrangement	*IGH* G, *TCRB* RR, *TCRG* RR

Functional characterization[1]

Cytochemistry	ACP$^+$, ALP$^-$, ANBE$^-$, CAE$^-$, MPO$^-$
Heterotransplantation	into nude mice

Comments

- Immature T-cell line (type T-III cortical) established from lymphoblastic T-cell lymphoma.
- Carries inv(14)(q11q32) possibly leading to alterations of the *TCRA-IGH* genes.

Reference

[1] Abe, M. et al. (1992) Cancer 69, 1235–1240.

JK-T1

Culture characterization[1]

Establishment in hypoxic environment
Culture medium 90% RPMI 1640 + 10% FBS
Doubling time 36 h
Authentication yes (by cytogenetics)
Primary reference Urashima et al.[1]
Availability of cell line not known

Clinical characterization[1]

Patient 10-year-old male
Disease diagnosis T-ALL (L2)
Treatment status at diagnosis
Specimen site bone marrow

Immunophenotypic characterization[1]

T-/NK cell marker $CD2^+$, $CD3^+$, $CD5^+$, $CD7^+$
Progenitor/activation/other marker $CD38^+$, TdT^-
Adhesion marker $CD11a^+$, $CD18^+$, $CD54^-$
Cytokine receptor $CD25^-$

Genetic characterization[1]

Cytogenetic karyotype 46, XY, del(6)(q?), t(8;14)(q24;q13), der(9)t(9;?)(q34;?)
Unique translocation/fusion gene t(8;14)(q24;q13) → *MYC-TCRA*? (*MYC* in germline)
Receptor gene rearrangement *TCRB* R

Functional characterization[1]

Colony formation in methylcellulose
Cytochemistry $ANBE^-$, CAE^-, MPO^-, PAS^-

Comments

- Immature T-cell line established from child with T-ALL.
- Carries t(8;14)(q24;q13) possibly affecting the *MYC* and *TCRA* genes.
- Immortalization of cells not confirmed.

Reference

[1] Urashima, M. et al. (1992) Cancer Genet. Cytogenet. 64, 86–90.

JURKAT

Culture characterization[1,2]

Other name of cell line	**JM** (cell line published as JM, but widely known as JURKAT – JM is described as CD4^{+} CD8^{+} and JURKAT as CD4^{+} CD8^{-})
Establishment	initially on adult allogeneic skin fibroblast feeder cells on top of Spongostan grids
Culture medium	90% RPMI 1640 + 10% FBS
Doubling time	25–35 h
Viral status	EBV^{-}, HBV^{-}, HCV^{-}, HHV-8^{-}, HIV^{-}, HTLV-I/II^{-}
Authentication	no
Primary reference	Schneider et al.[1]
Availability of cell line	DSM ACC 282, RCB 0806/0537

Clinical characterization[1]

Patient	14-year-old male
Disease diagnosis	ALL
Treatment status	at 1st relapse
Specimen site	peripheral blood
Year of establishment	1976

Immunophenotypic characterization[1–5]

T-/NK cell marker	CD1^{+}, CD2^{+}, CD3^{+}, cyCD3^{+}, CD4^{+}, CD5^{+}, CD6^{+}, CD7^{+}, CD8^{-}, CD27^{+}, CD28^{+}, CD57$^{(+)}$, TCR$\alpha\beta^{+}$, TCR$\gamma\delta^{-}$, cyTCRα^{+}, cyTCRβ^{+}, cyTCRγ^{-}
B-cell marker	CD9^{+}, CD10^{-}, CD19^{-}, CD20^{-}, CD21^{-}, CD24^{-}
Myelomonocytic marker	CD13^{-}, CD14^{-}, CD15^{-}, CD16^{-}, CD33^{-}
Erythroid-megakaryocytic marker	CD41a^{-}, CD61^{-}
Progenitor/activation/other marker	CD34$^{(+)}$, CD38^{+}, CD71^{+}, HLA-DR^{-}, TdT^{+}
Adhesion marker	CD11b^{-}, CD29^{+}
Cytokine receptor	CD25^{-}, CD117^{-}

Genetic characterization[2,3,5–12]

Cytogenetic karyotype	flat-moded hypotetraploid, 8% polypoidy; 87(78–91)<4n>XX, −Y, −Y, −5, −16, −17, −22, add(2)(p21)/del(2)(p23)×2; sideline with additional der(5)t(5;10)(q11;p15), del(9)(p11)
Receptor gene rearrangement	*IGH* R, *IGK* G, *TCRA* R, *TCRB* R, *TCRG* R, *TCRD* D (mRNA: TCRA^{+}, TCRB^{+}, TCRG^{-}, TCRD^{-})
Unique gene alteration	*IFNA* deletion, *P15INK4B* deletion, *P16INK4A* deletion, *P53* mutation

Functional characterization[13,14]

Colony formation	in agar
Cytochemistry	ACP^+, $ANAE^-$, MPO^-, PAS^-
Cytokine production	$mRNA^+$: GM-CSF, TGF-β1

Comments

- Immature $TCR\alpha\beta^+$ T-cell line (type T-III cortical) derived from relapsed childhood T-ALL.
- One of oldest T-cell lines (reference cell line).
- Available from cell line banks.
- Cells grow very well.

References

[1] Schneider, U. et al. (1977) Int. J. Cancer 19, 621–626.
[2] Drexler, H.G. et al. (1999) DSMZ Catalogue of Cell Lines, 7th edn, Braunschweig, Germany.
[3] Minowada, J. and Matsuo, Y. (1999) unpublished data.
[4] Inoue, K. et al. (1997) Blood 89, 1405–1412.
[5] Burger, R. et al. (1999) Leukemia Res. 23, 19–27.
[6] Heyman, M. et al. (1994) Leukemia 8, 425–434.
[7] Sangster, R.N. et al. (1986) J. Exp. Med. 163, 1491–1508.
[8] Cheng, J. and Haas, M. (1990) Mol. Cell. Biol. 10, 5502–5509.
[9] Ogawa, S. et al. (1994) Blood 84, 2431–2435.
[10] Siebert, R. et al. (1995) Brit. J. Haematol. 91, 350–354.
[11] Borgonovo Brandter, L. et al. (1996) Eur. J. Haematol. 56, 313–318.
[12] Kawamura, M. et al. (1999) Leukemia Res. 23, 115–126.
[13] Nilsson, K. et al. (1977) Int. J. Cancer 19, 337–344.
[14] Micallef, M. et al. (1994) Int. J. Oncol. 4, 633–638.

Karpas 45

Culture characterization[1-3]

Other name of cell line	initially only termed **45**, later also **K45**
Culture medium	90% RPMI 1640 + 10% FBS
Viral status	EBNA⁻
Authentication	no
Primary reference	Karpas et al.[1]
Availability of cell line	DSM ACC 105

Clinical characterization[1,4]

Patient	2-year-old male
Disease diagnosis	T-ALL
Treatment status	at diagnosis
Specimen site	bone marrow
Year of establishment	1972

Immunophenotypic characterization[1,5,6]

T-/NK cell marker	$CD1^-$, $CD2^+$, $CD3^-$, $cyCD3^+$, $CD4^+$, $CD5^+$, $CD6^+$, $CD7^+$, $CD8^+$, $CD27^-$, $CD28^-$, $CD57^-$, $TCR\alpha\beta^-$, $TCR\gamma\delta^-$, $cyTCR\alpha^+$, $cyTCR\beta^+$, $cyTCR\delta^-$
B-cell marker	$CD9^-$, $CD10^-$, $CD19^-$, sIg^-
Myelomonocytic marker	$CD13^-$, $CD15^-$, $CD16^-$
Progenitor/activation/other marker	$CD34^-$, $CD38^+$, $CD45^+$, HLA-DR^-, TdT^+
Adhesion marker	$CD29^+$
Cytokine receptor	$CD25^-$

Genetic characterization[3,5,7-10]

Cytogenetic karyotype	84, −Y, −Y, −2, −3, −4, +6, −9, −13, −14, +19, −20, −21, t(X;11)(q13;q23.3), der(X)t(X;11)(q13;q23.3), t(1;5)(q25;q13.1)×2, del(4) (q21.1q31.1), der(11)t(14;11)(11;X)(q11;p13)(q23.3;q13)
Unique translocation/fusion gene	t(X;11)(q13;q23) → *MLL-AFX* fusion gene
Receptor gene rearrangement	*IGH* G, *IGK* G, *TCRB* R, *TCRG* R, *TCRD* D

Functional characterization[1,2]

Cytochemistry	ACP^+, ALP^-, $ANBE^-$, CAE^-, $Lysozyme^-$, MPO^-, PAS^+, SBB^-

Comments

- Immature T-cell line (type T-II pre-T) derived from childhood T-ALL.
- Carries t(X;11)(q13;q23) leading to the *MLL-AFX* fusion gene.
- Available from cell line bank.
- Cells grow well.

References

[1] Karpas, A. et al. (1977) Leukemia Res. 1, 35–49.
[2] Karpas, A. et al. (1980) Brit. J. Haematol. 44, 415–424.
[3] Drexler, H.G. et al. (1999) DSMZ Catalogue of Cell Lines, 7th edn, Braunschweig, Germany.
[4] Smith, J.L. et al. (1973) Lancet i, 74–77. (case report)
[5] Burger, R. et al. (1999) Leukemia Res. 23, 19–27.
[6] Chou, J.L. et al. (1986) Leukemia Res. 10, 211–220.
[7] Kearney, L. et al. (1992) Blood 80, 1659–1665.
[8] Corral, J. et al. (1993) Proc. Natl. Acad. Sci. USA 90, 8538–8542.
[9] Kobayashi, H. et al. (1993) Blood 81, 3027–3033.
[10] Borkhardt, A. et al. (1997) Oncogene 14, 195–202.

KE-37

Culture characterization[1,2]

Culture medium	90% RPMI 1640 + 10% FBS
Doubling time	40 h
Viral status	EBV$^-$, HBV$^-$, HCV$^-$, HIV$^-$, HTLV-I/II$^-$
Authentication	no
Primary reference	unpublished
Availability of cell line	DSM ACC 46

Clinical characterization[1,2]

Patient	27-year-old male
Disease diagnosis	ALL
Year of establishment	1979

Immunophenotypic characterization[1-3]

T-/NK cell marker	CD1$^+$, CD2$^+$, CD3$^-$, cyCD3$^+$, CD4$^+$, CD5$^+$, CD6$^+$, CD7$^+$, CD8$^-$, TCR$\alpha\beta^-$, TCR$\gamma\delta^-$
B-cell marker	CD19$^-$
Myelomonocytic marker	CD13$^-$
Progenitor/activation/other marker	CD34$^-$, CD90$^+$, HLA-DR$^-$

Genetic characterization[2-4]

Cytogenetic karyotype	flat-moded hypotetraploid; 91(86–92)<4n>XXYY, +8, −14, −14, t(7;12)(q32-33;p12-13)×2, der(8)t(8;14)(q24;q11)×4, der(14)t (8;14)(q24;q11)×2
Unique translocation/fusion gene	t(8;14)(q24;q11) → *MYC-TCRA* genes altered?

Comments

- Immature T-cell line (type T-III cortical) derived from ALL.
- Carries t(8;14)(q24;q11) possibly leading to alteration of the *MYC-TCRA* genes.
- Available from cell line bank.
- Cells grow well.
- A KE-37 culture which is in widespread circulation was shown to be cross-contaminated (real identity of this culture: immature T-cell line CCRF-CEM)[5]. The original, correct KE-37 is available.

References

[1] Drexler, H.G. et al. (1985) Leukemia Res. 9, 209–229.

[2] Minowada, J. and Matsuo, Y. (1999) unpublished data.

[3] Drexler, H.G. et al. (1999) DSMZ Catalogue of Cell Lines, 7th edn, Braunschweig, Germany.

[4] Lange, B. et al. (1992) Leukemia 6, 613–618.

[5] Drexler, H.G. et al. (1999) Leukemia 13, 1601–1607.

KOPT-K1

Culture characterization[1]

Culture medium	90% RPMI 1640 + 10% FBS
Primary reference	unpublished
Availability of cell line	not known

Clinical characterization[1]

Patient	6-year-old male
Disease diagnosis	NHL

Immunophenotypic characterization[1,2]

T-/NK cell marker	$CD1a^+$, $CD2^+$, $CD3^-$, $cyCD3^+$, $CD4^+$, $CD5^+$, $CD7^+$, $CD8^+$, $CD28^+$, $CD57^-$, $TCR\alpha\beta^-$, $TCR\gamma\delta^-$, $cyTCR\alpha^-$, $cyTCR\beta^+$, $cyTCR\gamma^-$
B-cell marker	$CD9^-$, $CD10^-$, $CD19^-$, $CD20^-$, $CD21^-$, $CD22^-$, $CD24^-$
Myelomonocytic marker	$CD13^-$, $CD14^-$, $CD15^+$, $CD33^-$
Erythroid-megakaryocytic marker	$CD41a^-$, $CD61^-$
Progenitor/activation/other marker	$CD34^+$, $CD38^+$, $CD71^+$, HLA-DR^-, TdT^+
Adhesion marker	$CD11b^-$, $CD29^+$
Cytokine receptor	$CD25^-$

Genetic characterization[1,3–5]

Cytogenetic karyotype	95, XXY, −Y, +8, −11, −11, +12, +13, −14, −14, −15, −20, +21, +mar, del(21)(q22), +der(11)t(11;14)(p13;q11.2)×2, +der(17)t (17;?)(q21;?)×2
Unique translocation/fusion gene	t(11;14)(p13;q11) → *TTG2-TCRD* genes altered
Unique gene alteration	*P16INK4A* deletion/methylation

Functional characterization[2,6]

Cytochemistry	$ANBE^-$, MPO^-
Cytokine production	$mRNA^+$: GM-CSF, TGF-β1

Comments

- Immature T-cell line (type T-III cortical) derived from non-Hodgkin's lymphoma.
- Carries t(11;14)(p13;q11) which affects the *TTG2-TCRD* genes.
- Insufficient clinical data.

References

[1] Minowada, J. and Matsuo, Y. (1999) unpublished data.
[2] Kojika, S. et al. (1996) Leukemia 10, 994–999.
[3] Dong, W.F. et al. (1995) Leukemia 9, 1812–1817.
[4] Kawamura, M. et al. (1999) Leukemia Res. 23, 115–126.
[5] Ohnishi, H. et al. (1995) Blood 86, 1269–1275.
[6] Micallef, M. et al. (1994) Int. J. Oncol. 4, 633–638.

Culture characterization[1,2]

Establishment	initially on feeder layer (complete media, human serum, agar) in hypoxic environment
Culture medium	85% McCoy's 5A + 15% FBS
Doubling time	68 h
Viral status	EBV⁻
Authentication	no
Primary reference	Smith et al.[1]
Availability of cell line	not known

Clinical characterization[1,2]

Patient	16-year-old male
Disease diagnosis	T-ALL
Treatment status	at relapse
Specimen site	bone marrow
Year of establishment	1978

Immunophenotypic characterization[1,3]

T-/NK cell marker	$CD1^-$, $CD2^+$, $CD3^+$, $CD4^-$, $CD5^+$, $CD7^+$, $CD8^-$
B-cell marker	$CD10^-$, $CD19^-$, $cyIg^-$, sIg^-
Myelomonocytic marker	$CD15^+$
Progenitor/activation/other marker	$CD38^+$, $CD71^+$, HLA-DR^-, TdT^+

Genetic characterization[1,4]

Cytogenetic karyotype	46, XY, t(1;11)(p36.2;p13), del(2)(p16.3p22.1), del(9)(p12p21.1)
Unique gene alteration	*P15INK4B* deletion, *P16INK4A* deletion

Functional characterization[1,5]

Colony formation	in agar
Cytochemistry	ACP^+
Proto-oncogene	$mRNA^+$: BAX, BCLXL

Comments

- Immature T-cell line (type T-IV) derived from relapsed childhood T-ALL.

References

[1] Smith, S.D. et al. (1986) Blood 67, 650–656.
[2] Smith, S.D. et al. (1981) Cancer 48, 2612–2623.
[3] Zhou, M. et al. (1995) Blood 85, 1608–1614.
[4] Zhou, M. et al. (1995) Leukemia 9, 1159–1161.
[5] Findley, H.W. et al. (1997) Blood 89, 2986–2993.

L-KAW

Culture characterization[1]

Culture medium	90% RPMI 1640 + 10% FBS
Doubling time	36 h
Authentication	yes (by *TCRB* rearrangement)
Primary reference	Minegishi et al.[1]
Availability of cell line	not known

Clinical characterization[1]

Patient	6-year-old male
Disease diagnosis	T-ALL (L2)
Treatment status	at 2nd relapse
Specimen site	peripheral blood
Year of establishment	1990

Immunophenotypic characterization[1,2]

T-/NK cell marker	CD1⁻, CD2⁺, CD3⁺, CD4⁻, CD5⁺, CD7⁺, CD8⁻, CD28⁺, CD56⁻, CD57⁻, TCRαβ⁺, TCRγδ⁻
B-cell marker	CD10⁻
Progenitor/activation/other marker	CD34⁺, CD45RA⁻, CD45RO⁺, CD71⁺, CD95⁺, HLA-DR⁻
Adhesion marker	CD11a⁺, CD18⁺
Cytokine receptor	CD25⁻, CD117⁻

Genetic characterization[1,3]

Cytogenetic karyotype	46, XY, −1, −2, +6, −7, −9, −14, −18, +mar, del(3)(p23), +der(2)t(2;?)(p23;?), +der(7)t(7;7)(q12;q35), +der(9)t(9;?)(q34;?), +der(14)t(14;?)(q11;?)
Receptor gene rearrangement	*IGH* GG, *TCRB* RG, *TCRG* RD
Unique gene alteration	*P15INK4B* deletion, *P16INK4A* deletion, *P53* mutation

Functional characterization[1]

Cytochemistry	ANBE⁺, CAE⁻, MPO⁻, PAS⁺
Proto-oncogene	MYC mRNA⁺
Special features	sensitive to FAS activation and PHA (→ apoptosis)

Comments

- Immature TCRαβ⁺ T-cell line (type T-IV) derived from relapsed pediatric T-ALL.

References

[1] Minegishi, M. et al. (1995) Leukemia Res. 19, 433–442.
[2] Morita, S. et al. (1996) Leukemia 10, 102–105.
[3] Kawamura, M. et al. (1999) Leukemia Res. 23, 115–126.

Loucy

Culture characterization[1,2]

Culture medium	80–90% RPMI 1640 + 10–20% FBS
Doubling time	36–60 h
Viral status	EBV⁻, HBV⁻, HCV⁻, HHV-8⁻, HIV⁻, HTLV-I/II⁻
Authentication	no
Primary reference	Ben-Bassat et al.[1]
Availability of cell line	DSM ACC 394

Clinical characterization[1]

Patient	38-year-old female
Disease diagnosis	T-ALL (L2)
Treatment status	refractory, terminal
Specimen site	peripheral blood
Year of establishment	1987

Immunophenotypic characterization[1,2]

T-/NK cell marker	$CD1^-$, $CD2^-$, $CD3^+$, $CD4^+$, $CD5^+$, $CD6^+$, $CD7^+$, $CD8^-$, $TCR\alpha\beta^-$, $TCR\gamma\delta^+$
B-cell marker	$CD10^{(+)}$, $CD19^-$, $cyIg^-$, sIg^-
Myelomonocytic marker	$CD13^-$
Progenitor/activation/other marker	$CD34^-$, $CD45RA^-$, $HLA\text{-}DR^-$, TdT^-
Adhesion marker	$CD11b^-$, $CD29^+$

Genetic characterization[1,2]

Cytogenetic karyotype	hypodiploid, 16% polyploidy; 45<2n>X, −X, del(1)(p3?2p3?4), del(5)(q14-15q34-35), t(16;20)(p1?1;q1?3); sideline with del(6)(q23)

Functional characterization[1]

Cytochemistry	ACP^+, $ANAE^-$, MPO^-

Comments

- Immature T-cell (type T-IV) derived from T-ALL.
- Carries unique t(16;20).
- Rare $TCR\gamma\delta^+$ T-cell line.
- Available from cell line bank.

References

[1] Ben-Bassat, H. et al. (1990) Cancer Genet. Cytogenet. 49, 241–248.

[2] Drexler, H.G. et al. (1999) DSMZ Catalogue of Cell Lines, 7th edn, Braunschweig, Germany.

Culture characterization[1]

Culture medium	90% RPMI 1640 + 10% FBS
Authentication	no
Primary reference	unpublished
Availability of cell line	not known

Clinical characterization[1]

Patient	14-year-old male
Disease diagnosis	T-lymphoblastic lymphoma
Treatment status	leukemic phase, at relapse
Specimen site	peripheral blood

IMMUNOPHENOTYPICAL characterization[2]

T-/NK cell marker	$CD1^-$, $CD2^-$, $CD3^+$, $cyCD3^+$, $CD4^+$, $CD5^+$, $CD7^+$, $CD8^-$, $CD28^+$, $CD57^+$, $TCR\alpha\beta^+$, $TCR\gamma\delta^-$, $cyTCR\alpha^+$, $cyTCR\beta^+$, $cyTCR\gamma^-$
B-cell marker	$CD9^+$, $CD10^-$, $CD19^-$, $CD20^-$
Myelomonocytic marker	$CD13^-$, $CD15^-$
Erythroid-megakaryocytic marker	$CD41a^-$, $CD61^-$
Progenitor/activation/other marker	$CD34^-$, $CD71^+$, HLA-DR^-, TdT^-
Cytokine receptor	$CD25^-$

Genetic characterization[2-4]

Cytogenetic karyotype	89–90, XXYY, −3, −9, −16, −22, +2mar
Unique gene alteration	*P15INK4B* rearrangement, *P16INK4A* deletion

Comments

- Immature $TCR\alpha\beta^+$ T-cell line (type T-IV) derived from pediatric lymphoblastic T-cell lymphoma.

References

[1] Okamura, J. (1999) personal communication.
[2] Minowada, J and Matsuo, Y. (1999) unpublished data.
[3] Heyman, M. et al. (1994) Leukemia 8, 425–434.
[4] Borgonovo Brandter, L. et al. (1996) Eur. J. Haematol. 56, 313–318.

MOLT 3

Culture characterization[1,2]

Sister cell lines	**MOLT 1, MOLT 2, MOLT 4** (simultaneous sister cell lines – from same sample – MOLT 1, MOLT 2 are lost; MOLT 4 has similar features)
Subclone	**MOLT 4F** (subclone similar to MOLT 4)
Culture medium	90% RPMI 1640 + 10% FBS
Doubling time	40 h
Viral status	EBV⁻, HCV⁻, HHV-8⁻, HIV⁻, HTLV-I/II⁻
Authentication	no
Primary reference	Minowada et al.[1]
Availability of cell line	MOLT 3: ATCC CRL 1552, DSM ACC 84, JCRB 9048, RCB 1164; MOLT 4: ATCC CRL 1582, DSM ACC 362, IFO 50362, JCRB 9031, RCB 0206; MOLT 4F: JCRB 0021

Clinical characterization[1]

Patient	19-year-old male
Disease diagnosis	ALL
Treatment status	at relapse
Specimen site	peripheral blood
Year of establishment	1971

Immunophenotypic characterization[1–5]

T-/NK cell marker	$CD1^+$, $CD2^+$, $CD3^-$, $cyCD3^+$, $CD4^+$, $CD5^+$, $CD6^+$, $CD7^+$, $CD8^+$, $CD27^-$, $CD28^+$, $CD56^-$, $CD57^-$, $TCR\alpha\beta^-$, $TCR\gamma\delta^-$, $cyTCR\alpha^-$, $cyTCR\beta^+$, $cyTCR\gamma^-$
B-cell marker	$CD9^+$, $CD10^-$, $CD19^-$, $CD20^-$, $CD21^-$, $CD22^-$, $CD24^-$
Myelomonocytic marker	$CD13^-$, $CD14^-$, $CD15^-$, $CD16^-$, $CD33^-$
Erythroid-megakaryocytic marker	$CD36^-$, $CD41a^-$, $CD61^-$, $GlyA^-$
Progenitor/activation/other marker	$CD34^{(+)}$, $CD38^+$, $CD71^+$, HLA-DR^-, TdT^+
Adhesion marker	$CD11b^-$, $CD29^+$
Cytokine receptor	$CD25^-$, $CD117^-$, $CD122^-$

Genetic characterization[2,3,5–12]

Cytogenetic karyotype	98(94–101)<4n>XXYY, +6, +7, +8, +8, +17, +20, del(6) (q16)×2, der(7)t(7;7)(p15;q11)×2
Receptor gene rearrangement	*IGH* G, *IGK* G, *TCRA* R, *TCRB* R, *TCRG* G, *TCRD* D (mRNA: $TCRA^+$, $TCRB^+$)
Unique gene alteration	*NRAS* mutation, *P16INK4A* deletion

Functional characterization[13–15]

Colony formation	in agar
Cytokine production	MOLT 4: TGF-β1 mRNA$^+$
Heterotransplantation	into nude or SCID mice

Comments

- One of oldest immature T-cell lines (type T-III cortical) derived from ALL (reference cell line).
- Available from cell line banks.
- Cells grow very well.

References

1 Minowada, J. et al. (1972) J. Natl. Cancer Inst. 49, 891–895.
2 Drexler, H.G. et al. (1999) DSMZ Catalogue of Cell Lines, 7th edn, Braunschweig, Germany.
3 Minowada, J. and Matsuo, Y. (1999) unpublished data.
4 Toba, K. et al. (1996) Exp. Hematol. 24, 894–901.
5 Burger, R. et al. (1999) Leukemia Res. 23, 19–27.
6 Sangster, R.N. et al. (1986) J. Exp. Med. 163,1491–1508.
7 Siebert. R. et al. (1995) Brit. J. Haematol. 91, 350–354.
8 Kawamura, M. et al. (1999) Leukemia Res. 23, 115–126.
9 Ogawa, S. et al. (1994) Blood 84, 2431–2435.
10 Otsuki, T. et al. (1995) Cancer Res. 55, 1436–1440.
11 Hebert, J. et al. (1994) Blood 84, 4038–4044.
12 Borgonovo Brandter, L. et al. (1996) Eur. J. Haematol. 56, 313–318.
13 Nilsson, K. et al. (1977) Int. J. Cancer 19, 337–344.
14 Uckun, F.M. (1996) Blood 88, 1135–1146.
15 Micallef, M. et al. (1994) Int. J. Oncol. 4, 633–638.

MOLT 13

Culture characterization[1,2]

Sister cell lines (1) **MOLT 12** (at diagnosis – cell line is lost)
(2) **MOLT 14** (simultaneous sister cell line – from same sample – nearly identical features)

Culture medium 90% RPMI 1640 + 10% FBS

Authentication no

Primary reference Drexler and Minowada[1]

Availability of cell line MOLT 13: DSM ACC 436; MOLT 14: DSM ACC 437

Clinical characterization[1]

Patient 2-year-old female

Disease diagnosis T-ALL

Treatment status at relapse

Specimen site bone marrow

Year of establishment 1983

Immunophenotypic characterization[1–4]

T-/NK cell marker $CD1^-$, $CD2^-$, $CD3^+$, $cyCD3^+$, $CD4^-$, $CD5^+$, $CD7^+$, $CD8^-$, $CD28^+$, $CD57^{(+)}$, $TCR\alpha\beta^-$, $TCR\gamma\delta^+$, $cyTCR\alpha^-$, $cyTCR\beta^-$, $cyTCR\gamma^+$

B-cell marker $CD9^{(+)}$, $CD10^-$, $CD19^-$, $CD20^-$, $CD21^{(+)}$, $CD22^-$, $CD24^-$, sIg^-

Myelomonocytic marker $CD13^-$, $CD14^-$, $CD15^{(+)}$, $CD16^-$, $CD33^-$

Erythroid-megakaryocytic marker $CD41a^-$, $CD61^-$

Progenitor/activation/other marker $CD34^+$, $CD38^+$, $CD71^+$, $HLA\text{-}DR^-$, TdT^+

Adhesion marker $CD11b^-$, $CD29^+$

Cytokine receptor $CD25^{(+)}$

Genetic characterization[1,4–6]

Cytogenetic karyotype 46, XX, −17, +der(17)t(17;7)(p12;p15)

Receptor gene rearrangement *IGH* G, *TCRB* R, *TCRD* R (mRNA: IGH^-, $TCRA^-$, $TCRB^+$, $TCRD^+$)

Unique gene alteration *P15INK4B* deletion, *P16INK4A* deletion

Functional characterization[7]

Cytokine production MOLT 14: $mRNA^+$: GM-CSF, TGF-β1

Comments

- Immature $TCR\gamma\delta^+$ T-cell line (type T-IV) derived from relapsed childhood T-ALL.
- Available from cell line bank.

References

[1] Drexler, H.G. and Minowada, J. (1989) Hematol. Oncol. 7, 115–125.
[2] Drexler, H.G. et al. (1999) DSMZ Catalogue of Cell Lines, 7th edn, Braunschweig, Germany.
[3] Matsuo, Y. and Minowada, J. (1988) Human Cell 1, 263–274.
[4] Minowada, J. and Matsuo, Y. (1999) unpublished data.
[5] Furley, A.J. et al. (1996) Blood 68, 1101–1107.
[6] Borgonovo Brandter, L. et al. (1996) Eur. J. Haematol. 56, 313–318.
[7] Micallef, M. et al. (1994) Int. J. Oncol. 4, 633–638.

MOLT 16

Culture characterization[1,2]

Sister cell line	**MOLT 17** (simultaneous sister cell line derived from same specimen – nearly identical features)
Culture medium	90% RPMI 1640 + 10% FBS
Doubling time	30 h
Viral status	EBV^-, HBV^-, HCV^-, HHV-8^-, HIV^-, HTLV-I/II^-
Authentication	no
Primary reference	Drexler and Minowada[1]
Availability of cell line	MOLT 16: DSM ACC 29; MOLT 17: DSM ACC 36

Clinical characterization[1]

Patient	5-year-old female
Disease diagnosis	T-ALL (L2)
Treatment status	at relapse
Specimen site	peripheral blood
Year of establishment	1984

Immunophenotypic characterization[1–4]

T-/NK cell marker	$CD1^-$, $CD2^+$, $CD3^+$, $cyCD3^+$, $CD4^{(+)}$, $CD5^+$, $CD6^+$, $CD7^+$, $CD8^-$, $CD27^{(+)}$, $CD28^{(+)}$, $CD57^-$, $CD96^+$, $TCR\alpha\beta^+$, $TCR\gamma\delta^-$, $cyTCR\alpha^+$, $cyTCR\beta^+$, $cyTCR\gamma^+$, $cyTCR\delta^-$
B-cell marker	$CD9^{(+)}$, $CD10^{(+)}$, $CD19^-$, $CD20^-$, $CD21^-$, $CD22^-$, $CD24^-$, sIg^-
Myelomonocytic marker	$CD15^+$, $CD16^-$, $CD33^-$
Erythroid-megakaryocytic marker	$CD41a^-$, $CD61^-$
Progenitor/activation/other marker	$CD34^-$, $CD38^+$, $CD71^+$, HLA-DR^-, TdT^+
Adhesion marker	$CD29^+$
Cytokine receptor	$CD25^-$, $CD117^-$

Genetic characterization[1,2,4–11]

Cytogenetic karyotype	45, XX, −7, −9, −15, t(3;11)(q26;p14), t(8;14)(q24.1;q11.2), +der(7)t(7;7)(qter→p15::q11.2→qter), dup(9)(pter→p13::p24→ qter), der(15)t(15;9)(qter→p1?1::q1?1→qter)
Unique translocation/fusion gene	(1) t(8;14)(q24;q11) → *MYC-TCRA* genes altered (2) submicroscopic del(1)(p32) → *SIL-SCL* fusion gene
Receptor gene rearrangement	*IGH* G, *IGK* G, *TCRA* R, *TCRB* R, *TCRG* R, *TCRD* D (mRNA: IGH^-, $TCRA^+$, $TCRB^+$, $TCRG^+$, $TCRD^-$)
Unique gene alteration	*P15INK4B* deletion, *P16INK4A* deletion, *P53* mutation

Functional characterization[12]

Cytokine production mRNA$^+$: GM-CSF, TGF-β1

Comments

- Immature TCRαβ$^+$ T-cell line (type T-IV) derived from relapsed pediatric T-ALL.
- Carries t(8;14)(q24;q11) and del(1)(p32) causing alterations of the *MYC-TCRA* and *SIL-SCL* genes.
- Available from cell line bank.
- Cells grow well.

References

[1] Drexler, H.G. and Minowada, J. (1989) Hematol. Oncol. 7, 115–125.
[2] Drexler, H.G. et al. (1999) DSMZ Catalogue of Cell Lines, 7th edn, Braunschweig, Germany.
[3] Minowada, J. and Matsuo, Y. (1999) unpublished data.
[4] Burger, R. et al. (1999) Leukemia Res. 23, 19–27.
[5] McKeithan, T.W. et al. (1986) Proc. Natl. Acad. Sci. USA 83, 6636–6640.
[6] Brown, L. et al. (1990) EMBO J. 9, 3343–3351.
[7] Lange, B.J. et al. (1992) Leukemia 6, 613–618.
[8] Sangster, R.N. et al. (1986) J. Exp. Med. 163, 1491–1508.
[9] Ogawa, S. et al. (1994) Blood 84, 2431–2435.
[10] Borgonovo Brandter, L. et al. (1996) Eur. J. Haematol. 56, 313–318.
[11] Cheng, J. and Haas, M. (1990) Mol. Cell. Biol. 10, 5502–5509.
[12] Micallef, M. et al. (1994) Int. J. Oncol. 4, 633–638.

MT-ALL

Culture characterization[1]

Subclone GM-CSF/IL-3-dependent (**MN1.1**) and IL-2-dependent subclones with different immunophenotypic and functional features
Establishment initially grown with IL-3, then cytokine-independent
Culture medium 80% IMDM + 20% FBS
Authentication yes (by *TCRB, TCRG, TCRD* rearrangements)
Primary reference Griesinger et al.[1]
Availability of cell line not known

Clinical characterization[1]

Patient 25-year-old male
Disease diagnosis T-ALL
Treatment status at 2nd relapse
Specimen site peripheral blood
Year of establishment 1987

Immunophenotypic characterization[1]

T-/NK cell marker $CD2^+$, $CD3^+$, $CD4^-$, $CD5^+$, $CD7^+$, $CD8^-$, $CD56^-$, $TCR\alpha\beta^+$, $TCR\gamma\delta^-$
Myelomonocytic marker $CD13^+$, $CD14^{(+)}$, $CD33^+$
Progenitor/activation/other marker $CD34^-$, TdT^+
Cytokine receptor $CD25^-$

Genetic characterization[1]

Cytogenetic karyotype 47, XY, +19, del(6)(q15q25), t(1;10;12)(q25;p13;p13)
Receptor gene rearrangement *IGH* G, *TCRB* R, *TCRG* R, *TCRD* R

Functional characterization[1]

Cytochemistry $ANAE^-$, MPO^-, SBB^-, Toluidine $Blue^-$
Cytokine response growth stimulation: GM-CSF, IL-2, IL-3

Comments

- Immature $TCR\alpha\beta^+$ T-cell line derived from relapsed patient with T-ALL.
- Various cytokine-dependent subclones with different phenotypes were established.
- Immortalization and continuous culture of cell line and subclones are not documented.
- The IL-2-dependent subclone was described as showing cytotoxic (NK) activity and azurophilic granules.

Reference

[1] Griesinger, F. et al. (1989) J. Exp. Med. 169, 1101–1120.

P12/Ichikawa

Culture characterization[1–3]

Establishment	by serial passage in nude mice
Culture medium	90% RPMI 1640 + 10% FBS
Viral status	EBV⁻, HBV⁻, HCV⁻, HHV-8⁻, HIV⁻, HTLV-I/II⁻
Authentication	no
Primary reference	Kitahara et al.[1]
Availability of cell line	DSM ACC 34

Clinical characterization[1,2]

Patient	7-year-old male
Disease diagnosis	ALL
Specimen site	peripheral blood

Immunophenotypic characterization[2–5]

T-/NK cell marker	$CD1^+$, $CD2^+$, $CD3^{(+)}$, $cyCD3^+$, $CD4^+$, $CD5^+$, $CD6^+$, $CD7^+$, $CD8^+$, $CD27^-$, $CD28^+$, $CD57^{(+)}$, $TCR\alpha\beta^-$, $TCR\gamma\delta^-$, $cyTCR\alpha^-$, $cyTCR\beta^-$, $cyTCR\delta^-$
B-cell marker	$CD9^+$, $CD10^{(+)}$, $CD19^-$, $CD20^-$, $CD21^+$, $CD22^-$, $CD24^-$, sIg^-
Myelomonocytic marker	$CD15^+$, $CD16^-$, $CD33^-$
Erythroid-megakaryocytic marker	$CD41a^-$, $CD61^-$
Progenitor/activation/other marker	$CD34^{(+)}$, $CD38^+$, $CD71^+$, HLA-DR^-, TdT^+
Adhesion marker	$CD29^+$
Cytokine receptor	$CD25^-$

Genetic characterization[3–8]

Cytogenetic karyotype	hypotetraploid flat-moded, 1.6% polyploidy; 82–84<4n>XX, −Y, −Y, −2, +6, −9, −9, −10, −14, +20, −21, −21, −22, −22, +mar, der(2;9)(p10;q10), del(4)(q25), der(10;22)(q10;q10), add(19)(q13); sideline with N4, add(8)(p1?)
Receptor gene rearrangement	*IGH* G, *IGK* G, *TCRA* G, *TCRB* R, *TCRG* R, *TCRD* R
Unique gene alteration	*NRAS* mutation, *P15INK4B* rearrangement, *P16INK4A* deletion, *P53* mutation

Functional characterization[2]

Cytochemistry	ACP^+, GLC^+
Heterotransplantation	into nude mice

Comments

- Immature T-cell line (type T-III cortical) established from child with ALL.
- Available from cell line bank.

References

[1] Kitahara, T. et al. (1978) Acta Hematol Jpn. 41, 140.
[2] Watanabe, S. et al. (1978) Cancer Res. 38, 3494–3498.
[3] Drexler, H.G. et al. (1999) DSMZ Catalogue of Cell Lines, 7th edn, Braunschweig, Germany.
[4] Minowada, J. and Matsuo, Y. (1999) unpublished data.
[5] Burger, R. et al. (1999) Leukemia Res. 23, 19–27.
[6] Sangster, R.N. et al. (1986) J. Exp. Med. 163, 1491–1508.
[7] Borgonovo Brandter, L. et al. (1996) Eur. J. Haematol. 56, 313–318.
[8] Kawamura, M. et al. (1999) Leukemia Res. 23, 115–126.

Peer

Culture characterization[1,2]

Culture medium	90% RPMI 1640 + 10% FBS
Doubling time	50–60 h
Viral status	EBV⁻, HBV⁻, HCV⁻, HIV⁻, HTLV-I/II⁻
Authentication	no
Primary reference	Ravid et al.[1]
Availability of cell line	DSM ACC 115, JCRB 0830

Clinical characterization[1]

Patient	4-year-old female
Disease diagnosis	T-ALL
Treatment status	at 2nd relapse
Specimen site	peripheral blood
Year of establishment	1977

Immunophenotypic characterization[1-4]

T-/NK cell marker	$CD1^-$, $CD2^-$, $CD3^+$, $cyCD3^+$, $CD4^+$, $CD5^+$, $CD6^+$, $CD7^+$, $CD8^-$, $CD27^-$, $CD28^+$, $CD57^-$, $TCR\alpha\beta^-$, $TCR\gamma\delta^+$, $cyTCR\alpha^-$, $cyTCR\beta^+$, $cyTCR\gamma^+$
B-cell marker	$CD9^+$, $CD10^-$, $CD19^-$, $CD20^-$, $CD21^-$, $CD22^-$, $CD24^-$, $cyIg^-$, sIg^-
Myelomonocytic marker	$CD13^-$, $CD14^-$, $CD15^+$, $CD16^-$, $CD33^-$
Erythroid-megakaryocytic marker	$CD41a^-$, $CD61^-$
Progenitor/activation/other marker	$CD34^-$, $CD38^+$, $CD71^+$, HLA-DR^-, $TdT^{(+)}$
Adhesion marker	$CD11b^-$, $CD29^+$
Cytokine receptor	$CD25^-$

Genetic characterization[2-8]

Cytogenetic karyotype	94, XXXX, +2, −4, −4, +6, +7, +7, −8, −8, −9, +mar, del(5) (q13q15), del(5), del(6)(q1?2q2?1), del(6), del(9)(p22), del(9) (p22), del(9)(p22), +del(9)(q22), +del(9)(q22)
Receptor gene rearrangement	*IGH* G, *IGK* G, *TCRA* G, *TCRB* R, *TCRG* R, *TCRD* R (mRNA: $TCRA^-$, $TCRB^+$, $TCRG^+$, $TCRD^+$)
Unique gene alteration	*P15INK4B* rearrangement, *P16INK4A* deletion, *P53* mutation

Functional characterization[1,9]

Cytochemistry	ACP^+, $ANAE^+$, MPO^-, $Muraminidase^-$, Oil Red O^-, PAS^+, SBB^-
Cytokine production	$mRNA^+$: GM-CSF, TGF-β1

Comments

- Immature TCR$\gamma\delta^+$ T-cell line (type T-IV) derived from child with T-ALL (reference cell line).
- Available from cell line banks.

References

[1] Ravid, Z. et al. (1980) Int. J. Cancer 25, 705–710.
[2] Drexler, H.G. et al. (1999) DSMZ Catalogue of Cell Lines, 7th edn, Braunschweig, Germany.
[3] Minowada, J. and Matsuo, Y. (1999) unpublished data.
[4] Burger, R. et al. (1999) Leukemia Res. 23, 19–27.
[5] Heyman, M. et al. (1994) Leukemia 8, 425–434.
[6] Sangster, R.N. et al. (1986) J. Exp. Med. 163, 1491–1508.
[7] Borgonovo Brandter, L. et al. (1996) Eur. J. Haematol. 56, 313–318.
[8] Kawamura, M. et al. (1999) Leukemia Res. 23, 115–126.
[9] Micallef, M. et al. (1994) Int. J. Oncol. 4, 633–638.

PER-117

Culture characterization[1]

Culture medium	90% RPMI 1640 + 10% FBS or human serum
Doubling time	55–65 h
Authentication	yes (by cytogenetics)
Primary reference	Kees et al.[1]
Availability of cell line	not known

Clinical characterization[1]

Patient	2-year-old male
Disease diagnosis	ALL (L1/L2)
Treatment status	at 2nd relapse, post-BMT
Specimen site	bone marrow
Year of establishment	1984

Immunophenotypic characterization[1-3]

T-/NK cell marker	CD1⁻, CD2⁻, CD3⁻, CD4⁻, CD5⁺, CD6⁻, CD7⁺, CD8⁻, TCRαβ⁻
B-cell marker	CD10⁻, CD20⁻
Progenitor/activation/other marker	CD38⁺, CD71⁺, HLA-DR⁻, TdT⁻
Adhesion marker	CD11b⁻
Cytokine receptor	CD25⁻, CD117⁻

Genetic characterization[1,2,4]

Cytogenetic karyotype	46, XY, −16, +18, t(1;9;11)(p13;p22;p11)
Receptor gene rearrangement	*IGH* G, *TCRB* RR
Unique gene alteration	*P16INK4A* deletion

Functional characterization[1,2,5]

Colony formation	not clonable
Cytochemistry	ACP⁺, ANAE⁻, MPO⁻, PAS⁺
Cytokine production	mRNA⁺: IL-2, IL-4, IL-5
Inducibility of differentiation	TPA → growth arrest/activation

Comments

- Immature T-cell line (type T-II pre-T) derived from relapsed pediatric ALL.

References

[1] Kees, U.R. et al. (1987) Leukemia Res. 11, 489–498.
[2] Kees, U.R. (1988) Blood 72, 1524–1529.
[3] Kees, U.R. and Ashman, L.K. (1995) Leukemia 9, 1046–1050.
[4] Kees, U.R. et al. (1996) Oncogene 12, 2235–2239.
[5] Mordvinov, V.A. et al. (1999) J. Immunol. Methods 228, 163–168.

PER-255

Culture characterization[1]

Culture medium	90% RPMI 1640 + 10% FBS
Doubling time	84–95 h
Viral status	EBNA⁻
Authentication	yes (by *TCRB* rearrangement)
Primary reference	Kees et al.[1]
Availability of cell line	restricted

Clinical characterization[1]

Patient	5-year-old male
Disease diagnosis	T-ALL (L1)
Treatment status	at diagnosis
Specimen site	bone marrow
Year of establishment	1986

Immunophenotypic characterization[1,2]

T-/NK cell marker	$CD1^+$, $CD2^+$, $CD3^-$, $CD4^+$, $CD5^-$, $CD6^-$, $CD7^-$, $CD8^-$, $TCR\alpha\beta^-$
B-cell marker	$CD10^-$, $CD19^-$, $CD20^-$
Progenitor/activation/other marker	$CD38^+$, $CD71^-$, $HLA\text{-}DR^-$, TdT^+
Adhesion marker	$CD11b^-$
Cytokine receptor	$CD25^-$, $CD117^-$

Genetic characterization[1,3,4]

Cytogenetic karyotype	46, XY, t(7;10)(q32-34;q24), t(9;12)(p22;p12-13)
Unique translocation/fusion gene	t(7;10)(q35;q24) → *TCRB-HOX11* genes altered
Receptor gene rearrangement	*IGH* G, *TCRB* RR
Unique gene alteration	*P16INK4A* deletion

Functional characterization[1]

Colony formation	not clonable
Cytochemistry	ACP^+, $ANAE^-$, MPO^-, PAS^+, SBB^-

Comments

- Immature T-cell line (type T-III cortical) established from child with T-ALL.
- Carries t(7;10)(q35;q24) leading to involvement of the *TCRB-HOX11* genes.

References

1 Kees, U.R. et al. (1989) Blood 74, 369–373.
2 Kees, U.R. and Ashman, L.K. (1995) Leukemia 9, 1046–1050.
3 Kennedy, M.A. et al. (1991) Proc. Natl. Acad. Sci. USA 88, 8900–8904.
4 Kees, U.R. et al. (1996) Oncogene 12, 2235–2239.

PF-382

Culture characterization[1,2]

Culture medium	90% RPMI 1640 + 10% FBS
Doubling time	40 h
Viral status	EBV$^-$, HBV$^-$, HCV$^-$, HHV-8$^-$, HIV$^-$, HTLV-I/II$^-$
Authentication	no
Primary reference	Pegoraro et al.[1]
Availability of cell line	DSM ACC 38

Clinical characterization[1]

Patient	6-year-old female
Disease diagnosis	ALL
Treatment status	at 2nd relapse
Specimen site	pleural effusion
Year of establishment	1983

Immunophenotypic characterization[1,2]

T-/NK cell marker	CD1$^+$, CD2$^+$, CD3$^-$, CD4$^-$, CD5$^+$, CD6$^+$, CD7$^{(+)}$, CD8$^+$, CD57$^{(+)}$
B-cell marker	CD10$^-$, CD19$^-$
Myelomonocytic marker	CD13$^-$, CD16$^-$
Progenitor/activation/other marker	TdT$^-$
Adhesion marker	CD11b$^-$

Genetic characterization[1,2]

Cytogenetic karyotype	near-diploid, 9% near-tetraploid, 15% polyploidy; 45/46(43–47)<2n>X/XX, −10, +14; sideline with +add(?15)(p11), add(1) (p32)

Functional characterization[1]

Cytochemistry	ACP$^+$, ANAE$^-$, MPO$^-$, PAS$^-$

Comments

- Immature CD8$^+$ T-cell line (type T-III cortical) derived from relapsed child with ALL.
- Available from cell line bank.

References

[1] Pegoraro, L. et al. (1985) J. Natl. Cancer Inst. 75, 285–290.

[2] Drexler, H.G. et al. (1999) DSMZ Catalogue of Cell Lines, 7th edn, Braunschweig, Germany.

PFI-285

Culture characterization[1]

Culture medium	90% RPMI 1640 + 10% FBS
Doubling time	24 h
Viral status	HTLV-I$^-$
Authentication	no
Primary reference	Helgestad et al.[1]
Availability of cell line	original authors

Clinical characterization[1]

Patient	14-year-old male
Disease diagnosis	T-cell lymphoma
Treatment status	at 2nd relapse
Specimen site	peripheral blood
Year of establishment	1985

Immunophenotypic characterization[1]

T-/NK cell marker	CD1$^+$, CD2$^+$, CD3$^-$, CD4$^+$, CD5$^+$, CD7$^+$, CD8$^+$
B-cell marker	CD10$^-$, CD19$^-$
Myelomonocytic marker	CD15$^-$, CD16$^-$, CD36$^-$
Progenitor/activation/other marker	HLA-DR$^-$
Cytokine receptor	CD25$^-$

Genetic characterization[1]

Cytogenetic karyotype	46XY/92XXYY/87XXYY, −9, −13, −14, −18, −22

Functional characterization[1]

Heterotransplantation	into nude mice
Inducibility of differentiation	resistant to various biomodulators
Special features	sensitive to ascorbic acid (→ apoptosis); resistant to normal NK cells; weak NK activity

Comments

- Immature T-cell line (type T-III cortical) derived from child with T-cell lymphoma.

Reference

[1] Helgestad, J. et al. (1990) Eur. J. Haematol. 44, 9–17.

RPMI 8402

Culture characterization[1,2]

Sister cell line	**RPMI 8392** (= EBV+ B-LCL)
Culture medium	90% RPMI 1640 + 10% FBS
Doubling time	40–50 h
Viral status	EBV+, HBV−, HCV−, HHV-8−, HIV−, HTLV-I/II−
Authentication	no
Primary reference	Moore et al.[1]
Availability of cell line	ATCC CRL 1994, DSM ACC 290

Clinical characterization[1,3]

Patient	16-year-old female
Disease diagnosis	ALL
Specimen site	peripheral blood
Year of establishment	1972

Immunophenotypic characterization[2,3]

T-/NK cell marker	CD1−, CD2+, CD3−, cyCD3+, CD4+, CD5+, CD6+, CD7+, CD8−, CD27−, CD28+, CD57−, TCRαβ−, TCRγδ−, cyTCRα+, cyTCRβ+, cyTCRγ+, cyTCRδ−
B-cell marker	CD9+, CD10+, CD19−, CD20−, CD21−, CD22−, CD24−
Myelomonocytic marker	CD13−, CD14−, CD15+, CD16−, CD33−
Erythroid-megakaryocytic marker	CD41a−, CD61−
Progenitor/activation/other marker	CD34+, CD38+, CD71+, HLA-DR−, TdT+
Adhesion marker	CD11b−, CD29+
Cytokine receptor	CD25−

Genetic characterization[2-10]

Cytogenetic karyotype	hypotetraploid, 2% polyploidy; 90(79–91)<4n>XXX, −X, +3, +3, −10, −13, −14, +15, −18, −20, +2mar, dup(4)(q13q23)×2, del(6)(q14q22)×2, t(11;14)(p15;q11)×2, add(15)(p13); sideline with der(1)t(1;9)(p35/36;q11), add(13)(q34)
Unique translocation/fusion gene	(1) t(11;14)(p15;q11) → *TTG1-TCRD* genes altered (2) submicroscopic del(1)(p32) → *SIL-SCL* fusion gene
Receptor gene rearrangement	*IGH* G, *IGK* G, *TCRA* G, *TCRB* R, *TCRG* R, *TCRD* R (mRNA: TCRA−, TCRB+, TCRG+, TCRD+)
Unique gene alteration	*P15INK4B* deletion, *P16INK4A* deletion

Comments

- Immature T-cell line (type T-II pre-T) derived from young adult with ALL (reference cell line).
- Carries t(11;14)(p15;q11) causing alteration of the *TTG1-TCRD* genes.
- Carries submicroscopic del(1)(p32) leading to the *SIL-SCL* fusion gene.
- Available from cell line banks.
- Cells grow well.

References

[1] Moore, G.E. et al. (1973) In Vitro 8, 434.
[2] Drexler, H.G. et al. (1999) DSMZ Catalogue of Cell Lines, 7th edn, Braunschweig, Germany.
[3] Minowada, J. and Matsuo, Y. (1999) unpublished data.
[4] Huang, C.C. et al. (1974) J. Natl. Cancer Inst. 53, 655–660.
[5] Le Beau, M.M. et al. (1986) Proc. Natl. Acad. Sci. USA 83, 9744–9748.
[6] McGuire, E.A. et al. (1989) Mol. Cell. Biol. 9, 2124–2132.
[7] Aplan, P.D. et al. (1990) Science 250, 1426–1429.
[8] Brown, L. et al. (1990) EMBO J. 9, 3343–3351.
[9] Otsuki, T. et al. (1995) Cancer Res. 55, 1436–1440.
[10] Siebert, R. et al. (1995) Brit. J. Haematol. 91, 350–354.
[11] Sangster, R.N. et al. (1986) J. Exp. Med. 163, 1491–1508.
[12] Burger, R. et al. (1999) Leukemia Res. 23, 19–27.

ST-4

Culture characterization[1]

Establishment	initially serially transplanted into nude mice
Culture medium	90% RPMI 1640 + 10% FBS
Doubling time	36 h
Authentication	no
Primary reference	Arione et al.[1]
Availability of cell line	original authors

Clinical characterization[1]

Patient	12-year-old male
Disease diagnosis	T-NHL (convoluted type)
Treatment status	at diagnosis
Specimen site	axillary lymph node
Year of establishment	1984

Immunophenotypic characterization[1]

T-/NK cell marker	$CD1^+$, $CD2^-$, $CD3^-$, $CD4^-$, $CD7^+$, $CD8^+$, $CD57^+$
B-cell marker	$CD10^-$
Progenitor/activation/other marker	$CD38^+$, HLA-DR^-
Cytokine receptor	$CD25^-$

Genetic characterization[1]

Cytogenetic karyotype	47, X, +X, −Y, −15, +16, t(2;6)(q21;q23), +der(2)t(2;6) (q21;q23)
Receptor gene rearrangement	*TCRB* RR

Functional characterization[1]

Heterotransplantation	into nude mice

Comments

- Immature T-cell line (type T-III cortical) established from pediatric T-cell lymphoma.

Reference

[1] Arione, R. et al. (1988) Cancer Res. 48, 1312–1318.

SUP-T1

Culture characterization[1–4]

Establishment	initially on feeder layer (complete media, human serum, agar) in hypoxic environment
Culture medium	90% McCoy's 5A (or RPMI 1640) + 10% FBS
Doubling time	28–30 h
Viral status	EBV$^-$, HBV$^-$, HCV$^-$, HHV-8$^-$, HIV$^-$, HTLV-I/II$^-$
Authentication	no
Primary reference	Smith et al.[1]
Availability of cell line	ATCC CRL 1942, DSM ACC 140

Clinical characterization[1–3]

Patient	8-year-old male
Disease diagnosis	T-lymphoblastic lymphoma
Treatment status	at relapse
Specimen site	pleural effusion

Immunophenotypic characterization[1–5]

T-/NK cell marker	CD1$^+$, CD2$^{(+)}$, CD3$^+$, CD4$^+$, CD5$^+$, CD6$^+$, CD7$^+$, CD8$^+$, CD27$^-$, CD28$^+$, CD57$^+$, TCR$\alpha\beta^-$, TCRδ^-, cyTCRα^-, cyTCRβ^+, cyTCRδ^-
B-cell marker	CD9$^+$, CD10$^-$, CD19$^-$, sIg$^-$
Myelomonocytic marker	CD13$^-$, CD16$^-$
Progenitor/activation/other marker	CD38$^+$, CD71$^-$, HLA-DR$^-$, TdT$^+$
Adhesion marker	CD29$^+$
Cytokine receptor	CD25$^-$, CD117$^+$

Genetic characterization[1,3–11]

Cytogenetic karyotype	hypotetraploid, 1.8% polyploidy; 85<4n>XXX/XXX?Y, −8, −8, −9, −12, inv(2)(p22q11)×2, t(2;?20)(p13;?p11), del(4)(q31q35), del(6)(q25)×2, del(7)(q32), add(9)(q34)×2, inv(14)(q11q32)×2
Unique translocation/fusion gene	(1) t(7;9)(q34;q34.3) → *TCRB-TAL2* genes altered (2) inv(14)(q11q32) → *TCRA-IGH* genes altered
Receptor gene rearrangement	*IGH* G, *IGK* R, *TCRB* R, *TCRG* R, *TCRD* D
Unique gene alteration	*P15INK4B* deletion, *P16INK4A* deletion

Functional characterization[1]

Colony formation	in agar

Comments

- Immature T-cell line (type T-III cortical) established from relapsed child with lymphoblastic T-cell lymphoma (reference cell line).
- Carries t(7;9)(q34;q34) and inv(14)(q11q32) leading to alterations of the *TCRB-TAL2* and *TCRA-IGH* genes.
- Available from cell line banks.
- Cells grow very well.

References

[1] Smith, S.D. et al. (1986) Blood 67, 650–656.
[2] Smith, S.D. et al. (1984) Cancer Res. 44, 5657–5660.
[3] Smith, S.D. et al. (1988) Blood 71, 395–402.
[4] Drexler, H.G. et al. (1999) DSMZ Catalogue of Cell Lines, 7th edn, Braunschweig, Germany.
[5] Burger, R. et al. (1999) Leukemia Res. 23, 19–27.
[6] Baer, R. et al. (1985) Cell 43, 705–713.
[7] Denny, C.T. et al. (1986) Nature 320, 549–551.
[8] Denny, C.T. et al. (1986) Science 234, 197–200.
[9] Reynolds, T.C. et al. (1987) Cell 50, 107–117.
[10] Ellisen, L.W. et al. (1991) Cell 66, 649–661.
[11] Stranks, G. et al. (1995) Blood 85, 893–901.

SUP-T2

Culture characterization[1,2]

Establishment	initially on feeder layer (complete media, human serum, agar) in hypoxic environment
Culture medium	85% McCoy's 5A + 15% FBS
Doubling time	130 h
Viral status	EBV$^-$
Authentication	no
Primary reference	Smith et al.[1]
Availability of cell line	not known

Clinical characterization[1,2]

Patient	52-year-old female
Disease diagnosis	T-ALL
Treatment status	at diagnosis
Specimen site	bone marrow
Year of establishment	1984

Immunophenotypic characterization[1,2]

T-/NK cell marker	CD1$^-$, CD2$^+$, CD3$^-$, CD4$^-$, CD5$^+$, CD7$^+$, CD8$^+$
B-cell marker	CD10$^-$, sIg$^-$
Myelomonocytic marker	CD13$^-$, CD33$^-$
Progenitor/activation/other marker	CD38$^-$, CD71$^+$, HLA-DR$^-$, TdT$^+$

Genetic characterization[1]

Cytogenetic karyotype	46, XY, del(6)(q21q27)

Comments

- Immature T-cell line (type T-II pre-T) derived from adult T-ALL.

References

[1] Smith, S.D. et al. (1986) Blood 67, 650–656.
[2] Smith, S.D. et al. (1984) Cancer Res. 44, 5657–5660.

SUP-T3

Culture characterization[1,2]

Establishment	initially on feeder layer (complete media, human serum, agar) in hypoxic environment
Culture medium	85% McCoy's 5A + 15% FBS
Doubling time	61 h
Viral status	EBV$^-$
Authentication	no
Primary reference	Smith et al.[1]
Availability of cell line	not known

Clinical characterization[1–3]

Patient	12-year-old male
Disease diagnosis	T-ALL
Treatment status	at relapse
Specimen site	peripheral blood
Year of establishment	1984

Immunophenotypic characterization[1–3]

T-/NK cell marker	CD1$^+$, CD2$^+$, CD3$^-$, CD4$^+$, CD5$^+$, CD7$^+$, CD8$^+$
B-cell marker	CD10$^-$, sIg$^-$
Myelomonocytic marker	CD13$^-$, CD33$^-$
Progenitor/activation/other marker	CD38$^+$, CD71$^-$, HLA-DR$^-$, TdT$^+$
Cytokine receptor	CD25$^-$

Genetic characterization[1,3–8]

Cytogenetic karyotype	94, XXYY, +20, +20, del(4)(q31q35)×2, t(7;9)(q36;q34)×2
Unique translocation/fusion gene	t(7;9)(q34;q34.3) → *TCRB-TAL2/TCL3* genes altered
Receptor gene rearrangement	*TCRB* RD
Unique gene alteration	*P15INK4B* deletion, *P16INK4A* deletion

Functional characterization[1]

Colony formation	weakly in agar

Comments

- Immature T-cell line (type T-III cortical) derived from relapsed pediatric T-ALL.
- Carries t(7;9)(q34;q34) causing alterations of the *TCRB* and *TAL2/TCL3* genes.

References

[1] Smith, S.D. et al. (1986) Blood 67, 650–656.
[2] Smith, S.D. et al. (1984) Cancer Res. 44, 5657–5660.

[3] Smith, S.D. et al. (1988) Blood 71, 395–402.
[4] Reynolds, T.C. et al. (1987) Cell 50, 107–117.
[5] Westbrook, C.A. et al. (1987) Proc. Natl. Acad. Sci. USA 84, 251–255.
[6] Xia, Y. et al. (1991) Proc. Natl. Acad. Sci. USA 88, 11416–11420.
[7] Drexler, H.G. et al. (1998) Leukemia 12, 845–859.
[8] Drexler, H.G. et al. (1995) Leukemia 9, 480–500. (review on gene alterations in cell lines)

SUP-T4

Culture characterization[1,2]

Establishment	initially on feeder layer (complete media, human serum, agar) in hypoxic environment
Culture medium	85% McCoy's 5A + 15% FBS
Doubling time	33 h
Authentication	no
Primary reference	Smith et al.[1]
Availability of cell line	not known

Clinical characterization[1,2]

Patient	24-year-old male
Disease diagnosis	T-lymphoblastic lymphoma
Treatment status	at relapse
Specimen site	pleural effusion

Immunophenotypic characterization[1,2]

T-/NK cell marker	$CD1^+$, $CD2^+$, $CD3^+$, $CD4^+$, $CD5^+$, $CD7^+$, $CD8^+$
B-cell marker	$CD10^-$, sIg^-
Progenitor/activation/other marker	$CD38^+$, $CD71^+$, HLA-DR^-, TdT^+

Genetic characterization[1,3]

Cytogenetic karyotype	46, XY, −9, −18, −19, −20, +2mar, t(9;16)(q12;q13), t(10;14) (q23;q11.2), der(20)t(1;20)(p22;q13)
Unique translocation/fusion gene	t(10;14)(q23;q11) → *HOX11-TCRD* genes altered

Functional characterization[1]

Colony formation	in agar

Comments

- Immature T-cell line (type T-III cortical) established from relapsed T-lymphoblastic lymphoma.
- Carries t(10;14)(q23;q11) leading to alteration of the *HOX11-TCRD* genes.

References

[1] Smith, S.D. et al. (1986) Blood 67, 650–656.

[2] Smith, S.D. et al. (1984) Cancer Res. 44, 5657–5660.

[3] Drexler, H.G. et al. (1995) Leukemia 9, 480–500. (review on gene alterations in cell lines)

SUP-T6

Culture characterization[1]

Establishment	initially on feeder layer (complete media, human serum, agar) in hypoxic environment
Culture medium	85% McCoy's 5A + 15% FBS
Viral status	EBV$^-$
Authentication	yes (by cytogenetics, *TCRB*, *TCRG* rearrangements)
Primary reference	Smith et al.[1]
Availability of cell line	not known

Clinical characterization[1]

Patient	7-year-old male
Disease diagnosis	T-ALL
Treatment status	at diagnosis
Specimen site	bone marrow
Year of establishment	1986

Immunophenotypic characterization[1]

T-/NK cell marker	CD1$^-$, CD2$^+$, CD3$^-$, CD4$^-$, CD5$^+$, CD7$^+$, CD8$^+$
B-cell marker	CD10$^-$, sIg$^-$
Myelomonocytic marker	CD13$^-$, CD33$^-$
Progenitor/activation/other marker	HLA-DR$^-$, TdT$^+$
Cytokine receptor	CD25$^-$

Genetic characterization[1]

Cytogenetic karyotype	46,XY, t(7;9)(q34;q32), del(6)(q21)
Receptor gene rearrangement	*TCRB* R, *TCRG* R

Comments

- Immature T-cell (type T-II pre-T) derived from child with T-ALL.
- Carries a cytogenetic alteration at 7q34 which may involve the *TCRB* gene.

Reference

[1] Smith, S.D. et al. (1989) Blood 73, 2182–2187.

SUP-T7

Culture characterization[1]

Establishment	initially on feeder layer (complete media, human serum, agar) in hypoxic environment
Culture medium	85% McCoy's 5A + 15% FBS
Viral status	EBV⁻
Authentication	yes (by *TCRB* rearrangement)
Primary reference	Smith et al.[1]
Availability of cell line	not known

Clinical characterization[1]

Patient	19-year-old male
Disease diagnosis	T-ALL
Treatment status	at diagnosis
Specimen site	bone marrow
Year of establishment	1985

Immunophenotypic characterization[1,2]

T-/NK cell marker	$CD1^-$, $CD2^-$, $CD3^+$, $CD4^+$, $CD5^+$, $CD7^+$, $CD8^+$
B-cell marker	$CD9^+$, $CD10^-$, sIg^-
Myelomonocytic marker	$CD13^-$, $CD33^-$
Progenitor/activation/other marker	$CD38^+$, $CD71^+$, HLA-DR^-, TdT^+
Cytokine receptor	$CD25^-$

Genetic characterization[1-5]

Cytogenetic karyotype	46, XY, t(7;19)(q34;p13.1)
Unique translocation/fusion gene	t(7;19)(q34;p13) → *LYL1-TCRB* fusion gene
Receptor gene rearrangement	*TCRB* R

Comments

- Immature T-cell line (type T-IV) derived from T-ALL.
- Carries t(7;19)(q34;p13) leading to formation of the *LYL1-TCRB* fusion gene.

References

1 Smith, S.D. et al. (1989) Blood 73, 2182–2187.
2 Smith, S.D. et al. (1988) Blood 71, 395–402.
3 Cleary, M.L. et al. (1988) J. Exp. Med. 167, 682–687.
4 Mellentin, J.D. et al. (1989) Cell 58, 77–83.
5 Drexler, H.G. et al. (1995) Leukemia 9, 480–500. (review on gene alterations in cell lines)

SUP-T8

Culture characterization[1]

Establishment	initially on feeder layer (complete media, human serum, agar) in hypoxic environment
Culture medium	85% McCoy's 5A + 15% FBS
Viral status	EBV$^-$
Authentication	yes (by cytogenetics)
Primary reference	Smith et al.[1]
Availability of cell line	not known

Clinical characterization[1]

Patient	8-year-old female
Disease diagnosis	T-ALL
Treatment status	at relapse
Specimen site	bone marrow
Year of establishment	1987

Immunophenotypic characterization[1]

T-/NK cell marker	CD1$^-$, CD2$^-$, CD3$^-$, CD4$^+$, CD5$^+$, CD7$^+$, CD8$^-$
B-cell marker	CD10$^-$, sIg$^-$
Myelomonocytic marker	CD13$^-$, CD33$^+$
Progenitor/activation/other marker	HLA-DR$^-$, TdT$^-$
Cytokine receptor	CD25$^-$

Genetic characterization[1]

Cytogenetic karyotype	44, XX, −4, +8, −9, −12, −16, −17, −22, +mar, 1p+, +1p+, 2q+, 4q−, 7p+, 10p+, 19p+, del(1)(p32p35), t(4;19)(q21;p13), del(11)(q21q25), der(17)t(12;17)(p13;p13)
Receptor gene rearrangement	*TCRB* GG, *TCRG* GG

Functional characterization[1]

Cytochemistry	ANAE$^+$

Comments

- Either very immature T-cell line (type T-I pro-T) or myeloid cell line derived from relapsed pediatric leukemia: *TCRB/TCRG* in germline, TdT$^-$, co-expression of CD33/ANAE.

Reference

[1] Smith, S.D. et al. (1989) Blood 73, 2182–2187.

SUP-T9

Culture characterization[1]

Establishment	initially on feeder layer (complete media, human serum, agar) in hypoxic environment
Culture medium	85% McCoy's 5A + 15% FBS
Viral status	EBV⁻
Authentication	yes (by *TCRB*, *TCRG* rearrangements)
Primary reference	Smith et al.[1]
Availability of cell line	not known

Clinical characterization[1]

Patient	10-year-old female
Disease diagnosis	T-ALL
Treatment status	at relapse
Specimen site	bone marrow
Year of establishment	1987

Immunophenotypic characterization[1]

T-/NK cell marker	$CD1^+$, $CD2^+$, $CD3^-$, $CD4^+$, $CD5^+$, $CD7^+$, $CD8^+$
B-cell marker	$CD10^-$, sIg^-
Myelomonocytic marker	$CD13^-$, $CD33^-$
Progenitor/activation/other marker	$HLA\text{-}DR^-$, TdT^+
Cytokine receptor	$CD25^-$

Genetic characterization[1]

Cytogenetic karyotype	46, XX, t(6;14;21)(q23;q11.2;q22), del(11)(q23q25), t(15;21) (q15;q22)
Receptor gene rearrangement	*TCRB* R, *TCRG* R

Comments

- Immature T-cell line (type T-III cortical) derived from relapsed childhood T-ALL.

Reference

[1] Smith, S.D. et al. (1989) Blood 73, 2182–2187.

SUP-T10

Culture characterization[1]

Establishment	initially on feeder layer (complete media, human serum, agar) in hypoxic environment
Culture medium	85% McCoy's 5A + 15% FBS
Viral status	EBV$^-$
Authentication	no
Primary reference	Smith et al.[1]
Availability of cell line	not known

Clinical characterization[1]

Patient	8-year-old male
Disease diagnosis	T-ALL
Treatment status	at relapse
Specimen site	bone marrow
Year of establishment	1987

Immunophenotypic characterization[1]

T-/NK cell marker	CD1$^-$, CD2$^-$, CD3$^-$, CD4$^-$, CD5$^-$, CD7$^+$, CD8$^-$
B-cell marker	CD10$^-$, sIg$^-$
Myelomonocytic marker	CD13$^-$, CD33$^-$
Progenitor/activation/other marker	HLA-DR$^-$, TdT$^+$
Cytokine receptor	CD25$^-$

Genetic characterization[1]

Cytogenetic karyotype	47, XY, +mar, del(5)(q31), t(7;11)(p13;p13), t(8;12)(q13;p13), t(9;16)(p22;p13), t(17;18)(q11.2;q23)
Receptor gene rearrangement	*TCRB* GG, *TCRG* GG

Comments

- Very immature T-cell line (type T-I pro-T) (*TCR* genes in germline, most T-immunomarkers negative) established from relapsed pediatric T-ALL.

Reference

[1] Smith, S.D. et al. (1989) Blood 73, 2182–2187.

SUP-T11

Culture characterization[1]

Establishment	initially on feeder layer (complete media, human serum, agar) in hypoxic environment
Culture medium	85% McCoy's 5A + 15% FBS
Authentication	yes (by cytogenetics, *TCRB* rearrangement)
Primary reference	Smith et al.[1]
Availability of cell line	not known

Clinical characterization[1]

Patient	74-year-old male
Disease diagnosis	T-ALL
Treatment status	at diagnosis
Specimen site	bone marrow

Immunophenotypic characterization[1]

T-/NK cell marker	$CD1^-$, $CD2^+$, $CD3^+$, $CD4^-$, $CD5^+$, $CD7^+$, $CD8^-$
B-cell marker	$CD10^-$, sIg^-
Myelomonocytic marker	$CD13^-$, $CD33^-$
Progenitor/activation/other marker	$HLA\text{-}DR^-$, TdT^+
Cytokine receptor	$CD25^-$

Genetic characterization[1]

Cytogenetic karyotype	40, X, −Y, −1, −4, −9, −10, −13, −15, −16, −17, −21, −22, 1q−, 2p+, 6q+, 9q−, 12p+, 16q+, 19p+, 21q+, t(5;15;?)(q34;q22;?), t(5;9;?), der(5q), t(14;14)(q11.2;q32)
Receptor gene rearrangement	*TCRB* R, *TCRG* GG

Comments

- Immature T-cell line (type T-IV) from adult T-ALL.
- Carries unique t(14;14)(q11;q32).

Reference

[1] Smith, S.D. et al. (1989) Blood 73, 2182–2187.

SUP-T12

Culture characterization[1]

Establishment	initially on feeder layer (complete media, human serum, agar) in hypoxic environment
Culture medium	85% McCoy's 5A + 15% FBS
Viral status	EBV⁻
Authentication	yes (by cytogenetics, *TCRB, TCRG* rearrangements)
Primary reference	Smith et al.[1]
Availability of cell line	not known

Clinical characterization[1]

Patient	17-year-old male
Disease diagnosis	T-ALL
Treatment status	at diagnosis
Specimen site	peripheral blood
Year of establishment	1987

Immunophenotypic characterization[1]

T-/NK cell marker	$CD1^+$, $CD2^+$, $CD3^-$, $CD4^+$, $CD5^+$, $CD7^+$, $CD8^+$
B-cell marker	$CD10^-$, sIg^-
Myelomonocytic marker	$CD13^-$, $CD33^-$
Progenitor/activation/other marker	HLA-DR^-, TdT^+
Cytokine receptor	$CD25^-$

Genetic characterization[1-4]

Cytogenetic karyotype	60, XY, +Y, +4, +6, +7, +8, +8, +10, +13, +16, +17, +18, +19, +19, t(1;7)(p34;q34), +der(1)t(1;7)(p34;q34)
Unique translocation/fusion gene	t(1;7)(p34;q34) → *LCK-TCRB* genes altered
Receptor gene rearrangement	*TCRB* R, *TCRG* R

Comments

- Immature T-cell line (type T-III cortical) established from young adult with T-ALL.
- Carries t(1;7)(p34;q34) leading to alteration of the *LCK-TCRB* genes.

References

1 Smith, S.D. et al. (1989) Blood 73, 2182–2187.
2 Tycko, B. et al. (1991) J. Exp. Med. 174, 867–873.
3 Burnett, R.C. et al. (1994) Blood 84, 1232–1236.
4 Drexler, H.G. et al. (1995) Leukemia 9, 480–500. (review on gene alterations in cell lines)

Culture characterization[1]

Establishment	initially on feeder layer (complete media, human serum, agar) in hypoxic environment
Culture medium	85% McCoy's 5A + 15% FBS
Viral status	EBV⁻
Authentication	yes (by *TCRG* rearrangement)
Primary reference	Smith et al.[1]
Availability of cell line	not known

Clinical characterization[1]

Patient	2-year-old female
Disease diagnosis	T-ALL
Treatment status	at relapse
Specimen site	peripheral blood
Year of establishment	1987

Immunophenotypic characterization[1]

T-/NK cell marker	$CD1^-$, $CD2^+$, $CD3^+$, $CD4^+$, $CD5^+$, $CD7^+$, $CD8^+$
B-cell marker	$CD10^-$, sIg^-
Myelomonocytic marker	$CD13^-$, $CD33^-$
Progenitor/activation/other marker	$HLA\text{-}DR^-$, TdT^+
Cytokine receptor	$CD25^-$

Genetic characterization[1-3]

Cytogenetic karyotype	46, XY, t(1;5)(q41;p11), t(1;8)(q32;q24), del(9)(q24q34), t(11;19)(q24;p13)
Unique translocation/fusion gene	t(11;19)(q24;p13) → *MLL-ENL* fusion gene?
Receptor gene rearrangement	*TCRB* GG, *TCRG* R
Unique gene alteration	*P16INK4A* deletion

Comments

- Immature T-cell (type T-IV) derived from relapsed pediatric T-ALL.
- Carries t(11;19)(q24;p13) possibly leading to the *MLL-ENL* fusion gene.

References

[1] Smith, S.D. et al. (1989) Blood 73, 2182–2187.
[2] Jagasia, A.A. et al. (1996) Leukemia 10, 626–628.
[3] Drexler, H.G. et al. (1995) Leukemia 9, 480–500. (review on gene alterations in cell lines)

SUP-T14

Culture characterization[1]

Establishment	initially on feeder layer (complete media, human serum, agar) in hypoxic environment
Culture medium	85% McCoy's 5A + 15% FBS
Viral status	EBV$^-$
Authentication	yes (by *TCRG* rearrangement)
Primary reference	Smith et al.[1]
Availability of cell line	not known

Clinical characterization[1]

Patient	6-year-old male
Disease diagnosis	T-ALL
Treatment status	at diagnosis
Specimen site	bone marrow
Year of establishment	1987

Immunophenotypic characterization[1]

T-/NK cell marker	CD1$^-$, CD2$^+$, CD3$^+$, CD4$^+$, CD5$^+$, CD7$^+$, CD8$^-$
B-cell marker	CD10$^-$, sIg$^-$
Myelomonocytic marker	CD13$^-$, CD33$^-$
Progenitor/activation/other marker	HLA-DR$^-$, TdT$^+$
Cytokine receptor	CD25$^-$

Genetic characterization[1,2]

Cytogenetic karyotype	46, XY, del(6)(q15), del(9)(p22)
Receptor gene rearrangement	*TCRB* GG, *TCRG* R
Unique gene alteration	*P16INK4A* deletion

Comments

- Immature T-cell cell line (type T-IV) established from childhood T-ALL.

References

[1] Smith, S.D. et al. (1989) Blood 73, 2182–2187.
[2] Drexler, H.G. et al. (1998) Leukemia 12, 845–859.

TALL-1

Culture characterization[1]

Culture medium	85% RPMI 1640 + 15% FBS
Viral status	EBV⁻
Authentication	no
Primary reference	Miyoshi et al.[1]
Availability of cell line	JCRB 0086

Clinical characterization[1]

Patient	28-year-old male
Disease diagnosis	T-cell lymphosarcoma
Treatment status	leukemic phase, terminal
Specimen site	bone marrow
Year of establishment	1976

Immunophenotypic characterization[1–3]

T-/NK cell marker	$CD1^+$, $CD2^+$, $CD3^+$, $cyCD3^+$, $CD4^+$, $CD5^+$, $CD7^+$, $CD8^+$, $CD28^+$, $CD57^-$, $TCR\alpha\beta^+$, $TCR\gamma\delta^-$, $cyTCR\alpha^+$, $cyTCR\beta^+$, $cyTCR\gamma^+$
B-cell marker	$CD9^-$, $CD10^-$, $CD19^-$, $CD20^-$, $CD21^-$, $CD24^-$
Myelomonocytic marker	$CD13^-$, $CD14^-$, $CD15^+$, $CD16^-$, $CD33^-$
Erythroid-megakaryocytic marker	$CD41a^-$, $CD61^-$
Progenitor/activation/other marker	$CD34^-$, $CD38^+$, $CD71^+$, $HLA\text{-}DR^-$, TdT^+
Adhesion marker	$CD11b^-$, $CD29^+$
Cytokine receptor	$CD25^-$, $CD117^-$

Genetic characterization[2,4,5]

Cytogenetic karyotype	95, XXYY, −3, −9, +10, −12, −12, +13, +13, +14, +19, +20, −22, t(X;1)(q2?6;p3?2), del(2)(p2?1:), 5p+, +der(12)t(12;?) (q13;?), dup(21)(q11-q22), dup(21), +dup(21)
Unique gene alteration	*NRAS* mutation, *P15INK4B* deletion, *P16INK4A* deletion, *P53* mutation

Functional characterization[6]

Heterotransplantation	into newborn hamster or nude mice

Comments

- Immature $TCR\alpha\beta^+$ T-cell line (type T-III cortical) derived from T-lymphoblastic lymphoma.
- Available from cell line bank.
- Note that there is another immature T-cell line termed **TALL-1** which was derived independently from an unrelated patient in Philadelphia, PA, USA [reference: Lange, B. et al. (1987) Blood 70, 192–199].

References

[1] Miyoshi, I. et al. (1977) Nature 267, 843–844.
[2] Minowada, J. and Matsuo, Y. (1999) unpublished data.
[3] Inoue, K. et al. (1997) Blood 89, 1405–1412.
[4] Kawamura, M. et al. (1999) Leukemia Res. 23, 115–126.
[5] Borgonovo Brandter, L. et al. (1995) Eur. J. Haematol. 56, 313–318.
[6] Miyoshi, I. et al. (1981) Brit. J. Cancer 44, 124–126.

TALL-1

Culture characterization[1]

Culture medium	80% IMDM + 20% FBS
Viral status	EBNA$^-$, HTLV-II$^-$
Authentication	no
Primary reference	Lange et al.[1]
Availability of cell line	not known

Clinical characterization[1]

Patient	4-year-old male
Disease diagnosis	T-ALL (L1)
Treatment status	at relapse
Specimen site	bone marrow or peripheral blood

Immunophenotypic characterization[1]

T-/NK cell marker	CD2$^+$, CD3$^+$, CD5$^+$, CD7$^+$
B-cell marker	CD10$^-$
Progenitor/activation/other marker	HLA-DR$^-$

Genetic characterization[1]

Cytogenetic karyotype	46, XY, t(1;8)(q32;q24), del(6)(q23)
Receptor gene rearrangement	*IGH* G, *IGK* G, *IGL* G, *TCRB* RR (TCRB mRNA$^+$)

Functional characterization[1]

Cytochemistry	ANAE$^-$, CAE$^-$, MPO$^-$, PAS$^-$

Comments

- Immature T-cell line derived from child with relapsed T-ALL.
- Note that there is another immature T-cell line termed **TALL-1** which was derived independently from an unrelated patient in Okayama, Japan in 1976 [reference: Miyoshi, I. et al. (1977) Nature 267, 843–844].

Reference

[1] Lange, B. et al. (1987) Blood 70, 192–199.

TALL-101

Culture characterization[1-3]

Culture medium	80% IMDM + 20% FBS + 5 ng/ml GM-CSF (or 10% CM J-LB1)
Viral status	EBNA⁻, HTLV-I/II⁻
Authentication	yes (by cytogenetics)
Primary reference	Lange et al.[1]
Availability of cell line	ATCC CRL 9590

Clinical characterization[1-3]

Patient	3- or 9-year-old male
Disease diagnosis	T-ALL (L1)
Treatment status	at diagnosis or at relapse
Specimen site	bone marrow

Immunophenotypic characterization[1-5]

T-/NK cell marker	$CD1a^-$, $CD2^-$, $CD3^-$, $CD4^-$, $CD5^-$, $CD7^-$, $CD8^-$, $TCR\alpha\beta^-$, $TCR\gamma\delta^-$
B-cell marker	$CD10^-$
Myelomonocytic marker	$CD14^-$, $CD15^+$, $CD16^-$, $CD32^+$, $CD33^+$, $CD64^-$
Erythroid-megakaryocytic marker	$CD41^+$
Progenitor/activation/other marker	$CD38^+$, HLA-DR^-
Adhesion marker	$CD11b^+$
Cytokine receptor	$CD25^+$, $CD122^-$

Genetic characterization[1-5]

Cytogenetic karyotype	46, XY, t(8;14)(q24;q11), inv(9)
Unique translocation/fusion gene	t(8;14)(q24;q11) → *MYC-TCRA* genes altered?
Receptor gene rearrangement	*IGH* G, *IGK* G, *IGL* G, *TCRB* RG, *TCRG* R, *TCRD* R (TCRB $mRNA^-$)

Functional characterization[1-5]

Colony formation	in agar or methylcellulose
Cytochemistry	$ANAE^-$, CAE^-, MPO^-, PAS^-
Cytokine response	GM-CSF-dependent; growth stimulation: IL-3, IL-5

Comments

- Cell line derived from child with T-ALL.
- Differences in original authors' papers on age and disease status of patient.
- Controversial cell lineage assignment: (1) no expression of T-cell immunomarkers, but expression of myelomonocytic markers; (2) no proliferative response to 'lymphoid' cytokines, but response to 'myeloid' cytokines; (3) *TCRB* rearrangement, but non-productive; and (4) 'T-cell-associated' chromosome alteration.
- Cells are constitutively cytokine-dependent.
- Carries t(8;14)(q24;q11) presumably involving the *MYC* and *TCRA* genes.
- Available from cell line bank.

References
[1] Lange, B. et al. (1987) Blood 70, 192–199.
[2] Valtieri, M. et al. (1987) J. Immunol. 138, 4042–4050.
[3] Santoli, D. et al. (1987) J. Immunol. 139, 3348–3354.
[4] O'Connor, R. et al. (1991) Blood 77, 1534–1545.
[5] Lange, B. et al. (1992) Leukemia 6, 613–618.

TALL-103/2

Culture characterization[1–3]

Subclone	**TALL-103/2** is a subclone of **TALL-103/3** established by switching from IL-3 for TALL-103/3 to IL-2 for TALL-103/2 (clones are immunophenotypically, morphologically, and functionally totally different)
Culture medium	90% IMDM + 10% FBS + 5–10 U/ml IL-2
Viral status	HTLV-I/II$^-$
Authentication	yes (by cytogenetics, *TCRD* rearrangement)
Primary reference	Santoli et al.[1]
Availability of cell line	restricted

Clinical characterization[1–3]

Patient	6-year-old male
Disease diagnosis	T-lymphoblastic lymphoma
Treatment status	at relapse
Specimen site	bone marrow
Year of establishment	1988

Immunophenotypic characterization[1–4]

T-/NK cell marker	CD1a$^-$, CD2$^+$, CD3$^+$, CD4$^-$, CD7$^+$, CD8$^+$, CD56$^+$, TCR$\alpha\beta^-$, TCR$\gamma\delta^+$, TCRγ^-, TCRδ^+
Myelomonocytic marker	CD14$^-$, CD15$^-$, CD16$^-$, CD32$^-$, CD33$^+$, CD64$^-$
Progenitor/activation/other marker	CD34$^-$, CD38$^+$, CD71$^+$, HLA-DR$^-$
Adhesion marker	CD11b$^-$
Cytokine receptor	CD25$^+$, CD122$^+$

Genetic characterization[1–4]

Cytogenetic karyotype	47, XY, −9, +12, −13, +17, +mar, t(8;14)(q24;q11), der(12)t (12;13)(p11-13;q11-14)
Unique translocation/fusion gene	t(8;14)(q24;q11) → *MYC-TCRA* genes altered?
Receptor gene rearrangement	*TCRB* G, *TCRG* R, *TCRD* R (mRNA: TCRG$^+$, TCRD$^+$)

Functional characterization[1–3,5,6]

Cytochemistry	azurophilic granules
Cytokine production	mRNA$^+$: GM-CSF; inducible production: GM-CSF, IFN-γ, IL-3, TNF-α
Cytokine response	IL-2-dependent; growth inhibition: IFN-γ, IL-4

Inducibility of differentiation IFN-γ, IL-4, IL-6, IL-12 upregulate cytotoxic activity

Special features mRNA+: pore-forming protein, serine esterase; ADCC, cytotoxic activity

Comments

- Immature TCRγδ+ T-cell line (type T-IV) derived from relapsed child with T-lymphoblastic lymphoma.
- Carries t(8;14)(q24;q11) presumably involving the *MYC* and *TCRA* genes.
- Cells are constitutively cytokine-dependent.
- Displays cytotoxic activity and ADCC (cytotoxic T-lymphocytes).
- Subclone TALL-103/2 was established by switching from IL-3 to IL-2 leading to a conversion from a myeloid to a lymphoid phenotype – DNA fingerprinting is required to confirm this alleged derivation.

References

[1] Santoli, D. et al. (1990) J. Immunol. 144, 4703–4711.
[2] O'Connor, R. et al. (1990) J. Immunol 145, 3779–3787.
[3] O'Connor, R. et al. (1991) Blood 77, 1534–1545.
[4] Lange, B. et al. (1992) Leukemia 6, 613–618.
[5] Cesano, A. and Santoli, D. (1992) In Vitro Cell. Dev. Biol. 28A, 648–656.
[6] Cesano, A. and Santoli, D. (1992) In Vitro Cell. Dev. Biol. 28A, 657–662.

TALL-104

Culture characterization[1,2]

Culture medium	90% IMDM + 10% FBS + 5–10 U/ml IL-2
Viral status	HTLV-I/II$^-$
Authentication	yes (by cytogenetics)
Primary reference	O'Connor et al.[1]
Availability of cell line	ATCC CRL 11386

Clinical characterization[1,2]

Patient	2-year-old male
Disease diagnosis	T-lymphoblastic lymphoma
Treatment status	refractory
Specimen site	peripheral blood

Immunophenotypic characterization[1,2]

T-/NK cell marker	CD1a$^-$, CD2$^+$, CD3$^+$, CD4$^-$, CD7$^+$, CD8$^+$, CD56$^+$, TCR$\alpha\beta^+$, TCR$\gamma\delta^-$
Myelomonocytic marker	CD15$^-$, CD16$^-$, CD32$^-$, CD33$^-$, CD64$^-$
Progenitor/activation/other marker	CD34$^-$
Adhesion marker	CD11b$^-$, CD18$^+$, CD44$^+$, CD54$^+$
Cytokine receptor	CD25$^+$, CD122$^-$

Genetic characterization[1,2]

Cytogenetic karyotype	46, XY, t(11;14)(p13;q11)
Unique translocation/fusion gene	t(11;14)(p13;q11) → *TTG2-TCRD* genes altered?
Receptor gene rearrangement	*TCRB* R, *TCRD* R

Functional characterization[1-4]

Cytochemistry	azurophilic granules
Cytokine production	GM-CSF mRNA$^+$; inducible production: GM-CSF, IFN-γ, IL-3, TNF-α
Cytokine response	IL-2-dependent
Heterotransplantation	into SCID mice
Inducibility of differentiation	IFN-γ, IL-12 upregulate cytotoxic activity
Special features	mRNA$^+$: pore-forming protein, serine esterase; ADCC, cytotoxic activity

Comments

- Immature TCR$\alpha\beta^+$ T-cell line (type T-IV) derived from pediatric case with T-lymphoblastic lymphoma.
- Carries t(11;14)(p13;q11) which may alter the *TTG2-TCRD* genes.
- Cells are constitutively IL-2-dependent.
- Cytotoxic activity and ADCC (cytotoxic T-lymphocytes).
- Available from cell line bank.

References

[1] O'Connor, R. et al. (1991) Blood 77, 1534–1545.
[2] Cesano, A. et al. (1991) Blood 77, 2463–2474.
[3] Cesano, A. and Santoli, D. (1992) In Vitro Cell. Dev. Biol. 28A, 648–656.
[4] Cesano, A. and Santoli, D. (1992) In Vitro Cell. Dev. Biol. 28A, 657–662.

TALL-105

Culture characterization[1]

Culture medium	90% IMDM + 10% FBS
Viral status	HTLV-I/II$^-$
Authentication	yes (by cytogenetics)
Primary reference	O'Connor et al.[1]
Availability of cell line	not known

Clinical characterization[1]

Patient	9- or 13-year-old female
Disease diagnosis	T-ALL
Treatment status	at diagnosis
Specimen site	bone marrow

Immunophenotypic characterization[1,2]

T-/NK cell marker	CD1a$^+$, CD2$^+$, CD3$^+$, CD4$^-$, CD7$^+$, CD8$^+$, CD56$^-$, TCR$\alpha\beta^+$, TCR$\gamma\delta^-$
Myelomonocytic marker	CD15$^-$, CD16$^-$, CD32$^-$, CD33$^-$, CD64$^-$
Progenitor/activation/other marker	CD34$^+$
Adhesion marker	CD11b$^-$
Cytokine receptor	CD25$^-$, CD122$^-$

Genetic characterization[1,2]

Cytogenetic karyotype	46, XX, t(8;14)(q24;q11)
Unique translocation/fusion gene	t(8;14)(q24;q11) → *MYC-TCRA* genes altered?
Receptor gene rearrangement	*TCRB* R, *TCRD* D

Comments

- Immature TCR$\alpha\beta^+$ T-cell line (type T-III cortical) established from childhood T-ALL.
- Carries t(8;14)(q24;q11) possibly leading to alteration of the *MYC-TCRA* genes.
- Differences on age of patient in original authors' publications.

References

[1] O'Connor, R. et al. (1991) Blood 77, 1534–1545.
[2] Lange, B. et al. (1992) Leukemia 6, 613–618.

TALL-106

Culture characterization[1,2]

Culture medium	90% IMDM + 10% FBS
Viral status	HTLV-I/II⁻
Authentication	yes (by cytogenetics)
Primary reference	O'Connor et al.[1]
Availability of cell line	not known

Clinical characterization[1,2]

Patient	15-year-old male
Disease diagnosis	T-ALL
Treatment status	at diagnosis
Specimen site	bone marrow

Immunophenotypic characterization[1-3]

T-/NK cell marker	$CD1a^+$, $CD2^+$, $CD3^+$, $CD4^+$, $CD7^+$, $CD8^+$, $CD56^-$, $TCR\alpha\beta^+$, $TCR\gamma\delta^-$
Myelomonocytic marker	$CD15^-$, $CD16^-$, $CD32^-$, $CD33^-$, $CD64^-$
Progenitor/activation/other marker	$CD34^+$
Adhesion marker	$CD11b^-$, $CD18^+$, $CD44^+$, $CD54^-$
Cytokine receptor	$CD25^-$, $CD122^-$

Genetic characterization[1-3]

Cytogenetic karyotype	46, XY, t(8;14)(q24;q11), inv(14)(q11q32)
Unique translocation/fusion gene	t(8;14)(q24;q11) → *MYC-TCRA* genes altered?
Receptor gene rearrangement	*TCRB* R, *TCRD* D

Functional characterization[1,2]

Heterotransplantation	into SCID mice

Comments

- Immature $TCR\alpha\beta^+$ T-cell line (type T-III cortical) established from T-ALL.
- Carries t(8;14)(q24;q11) possibly leading to alteration of the *MYC-TCRA* genes.

References

[1] O'Connor, R. et al. (1991) Blood 77, 1534–1545.
[2] Cesano, A. et al. (1991) Blood 77, 2463–2474.
[3] Lange, B. et al. (1992) Leukemia 6, 613–618.

TALL-107

Culture characterization[1]

Establishment	primary material first heterotransplanted into SCID mice, then cultured *in vitro*
Culture medium	90% IMDM + 10% FBS + 100 U/ml IL-2
Authentication	yes (by *TCRB*, *TCRG*, *TCRD* rearrangements)
Primary reference	Cesano et al.[1]
Availability of cell line	not known

Clinical characterization[1]

Patient	8-year-old male
Disease diagnosis	T-ALL (L1)
Treatment status	at diagnosis
Specimen site	bone marrow

Immunophenotypic characterization[1]

T-/NK cell marker	CD2+, CD3+, CD4−, CD7+, CD8+, CD56+, TCRαβ+, TCRγδ−

Genetic characterization[1]

Cytogenetic karyotype	normal 46, XY
Receptor gene rearrangement	*TCRB* R, *TCRG* R, *TCRD* R

Functional characterization[1]

Cytochemistry	azurophilic granules
Cytokine production	inducible: IFN-γ, TNF-α
Cytokine response	IL-2-dependent; growth inhibition: IL-4
Special features	cytotoxic activity

Comments

- Immature TCRαβ+ T-cell line from pediatric T-ALL.
- Unusual normal karyotype.
- Constitutively IL-2-dependent.
- Displays cytotoxic activity and ADCC (cytotoxic T-lymphocytes).

Reference

[1] Cesano, A. et al. (1993) Blood 81, 2714–2722.

TK-6

Culture characterization[1]

Culture medium	90% RPMI 1640 + 10% FBS
Doubling time	60–72 h
Viral status	HTLV-I^{-}
Authentication	yes (by *BCR*, *TCRB* rearrangements)
Primary reference	Watanabe et al.[1]
Availability of cell line	original authors

Clinical characterization[1]

Patient	30-year-old male
Disease diagnosis	CML
Treatment status	at T-cell blast crisis (relapse post-BMT)
Specimen site	pleural effusion
Year of establishment	1992

Immunophenotypic characterization[1]

T-/NK cell marker	CD1$^{(+)}$, CD2^{-}, CD3$^{(+)}$, CD4^{+}, CD5^{+}, CD7^{+}, CD8^{+}, TCR$\alpha\beta^{-}$, TCR$\gamma\delta^{-}$
B-cell marker	CD10^{-}, CD19^{-}, CD20^{-}
Myelomonocytic marker	CD13^{-}, CD14^{-}, CD16^{+}, CD33^{-}
Erythroid-megakaryocytic marker	CD41b^{-}
Progenitor/activation/other marker	CD34^{-}, HLA-DR^{-}, TdT^{+}
Adhesion marker	CD11b^{-}
Cytokine receptor	CD25^{-}

Genetic characterization[1–3]

Cytogenetic karyotype	48, XY, +2mar, ins(1;?)(q21;?), del(1)(q21q32), del(6)(q21), dic(7)(:p13→cen→q32::q11.2→cen→pter), der(9)t(9;22)(q34;q11), der(22)t(9;22;?)(q34;q11;?)
Unique translocation/fusion gene	Ph^{+} t(9;22)(q34;q11) → *BCR-ABL* b3-a2 fusion gene
Receptor gene rearrangement	*IGH* GG, *TCRB* RD
Unique gene alteration	*MYB* mutation

Functional characterization[1]

Colony formation	in methylcellulose
Cytochemistry	ANAE^{-}, ANBE^{-}, MPO^{-}, PAS^{-}
Cytokine production	IL-1β, IFN-γ, TNF-β
Special features	production of parathyroid hormone-related protein

Comments

- Immature T-cell line (type T-III cortical) derived from CML in blast crisis.
- Carries Ph chromosome with *BCR-ABL* b3-a2 fusion gene.
- Rare CML-derived T-cell line.

References

[1] Watanabe, T. et al. (1995) Leukemia 9, 1926–1934.
[2] Tomita, A. et al. (1998) Leukemia 12, 1422–1429.
[3] Drexler, H.G. et al. (1999) Leukemia Res. 23, 207–215.

Part 2

Mature T-Cell Lines

Deglis

Culture characterization[1]

Culture medium	90% IMDM + 10% FBS
Viral status	EBNA$^+$
Authentication	yes (by EBV sequencing, *IGH* rearrangement)
Primary reference	Al Saati et al.[1]
Availability of cell line	original authors

Clinical characterization[1]

Patient	68-year-old male
Disease diagnosis	polymorphic centroblastic lymphoma
Treatment status	at diagnosis
Specimen site	extrapulmonary tumor mass
Year of establishment	1988

Immunophenotypic characterization[1]

T-/NK cell marker	CD1$^-$, CD2$^+$, cyCD3$^+$, CD4$^-$, CD5$^-$, CD7$^+$, CD8$^-$, CD57$^-$, TCR$\alpha\beta^-$
B-cell marker	CD10$^-$, CD19$^+$, CD20$^+$, CD21$^-$, CD22$^-$, CD23$^+$, CD37$^+$, cyIg$^-$, sIg$^-$
Myelomonocytic marker	CD14$^-$, CD15$^-$, CD16$^-$, CD68$^+$
Progenitor/activation/other marker	CD30$^+$, CD45$^+$, CD45RO$^-$, CD70$^+$, HLA-DR$^+$, TdT$^+$
Adhesion marker	CD11b$^-$, CD11c$^-$, CD43$^+$, CD54$^+$
Cytokine receptor	CD25$^-$

Genetic characterization[1]

Cytogenetic karyotype	46, XY, −2, −4, −8, +mar, +der(2)(?::2p23-2p37::?), +der(4)t(4;?)(q34;?), i(6p), del(6)(p22), +del(6)(p22), +der(8)t(8;?) (q24;?), t(13;15)(p11;p11), t(21;22)(p11;p11)
Receptor gene rearrangement	*IGH* RR, *IGK* RR, *TCRA* R, *TCRB* R, *TCRG* G, *TCRD* G (mRNA: IGH$^-$, IGK$^-$, TCRA$^+$, TCRB$^+$ truncated, TCRG$^-$, TCRD$^+$ truncated)

Functional characterization[1]

Special features	EBNA2$^+$, LMP$^+$

Comments

- Cell lineage assignment not completely determined, but rather T-cell than B-cell.
- Carries possible breaks at chromosome 2p23 (*ALK*?) and 8q24 (*MYC*?).

Reference

[1] Al Saati, T. et al. (1992) Blood 80, 209–216.

EBT-8

Culture characterization[1]

Culture medium	90% RPMI 1640 + 10% FBS + 40 U/ml IL-2
Doubling time	60–70 h
Viral status	EBV^{+}
Authentication	yes (by *TCRB*, *TCRG* rearrangements)
Primary reference	Asada et al.[1]
Availability of cell line	original authors

Clinical characterization[1,2]

Patient	36-year-old male
Disease diagnosis	LGL leukemia of T-cell type
Treatment status	at diagnosis
Specimen site	peripheral blood
Year of establishment	1988

Immunophenotypic characterization[1]

T-/NK cell marker	CD2^{+}, CD3^{+}, CD4^{-}, CD8^{+}, CD56^{-}, CD57^{-}, TCR$\alpha\beta^{+}$, TCR$\gamma\delta^{-}$
B-cell marker	CD19^{-}, CD20^{-}, CD21^{-}, CD23^{-}
Myelomonocytic marker	CD16^{-}
Progenitor/activation/other marker	HLA-DR^{+}
Cytokine receptor	CD25^{+}

Genetic characterization[1]

Cytogenetic karyotype	48, XY, +2, +17, der(3)t(3;3)(p25;q21), del(11)(q23), der(22)t (1;22)(q11;p11)
Receptor gene rearrangement	*TCRB* R, *TCRG* R

Functional characterization[1]

Cytochemistry	azurophilic granules
Special features	EBER1^{+}, EBNA1^{+}, EBNA2^{-}, LMP1^{+}; TPA → EBV production

Comments

- Mature EBV^{+}, CD4^{-}/CD8^{+}, TCR$\alpha\beta^{+}$ T-cell line established from patient with LGL T-cell type leukemia.
- Presumably cytotoxic T-cells.

References

[1] Asada, H. et al. (1994) Leukemia 8, 1415–1423.
[2] Asada, H. et al. (1994) J. Am. Acad. Dermatol. 31, 251–255. (case report)

HPB-MLp-W

Culture characterization[1]

Culture medium	80% RPMI 1640 + 20% FBS
Doubling time	72–96 h
Viral status	EBNA⁻, HTLV-I⁻
Authentication	no
Primary reference	Morikawa et al.[1]
Availability of cell line	not known

Clinical characterization[1]

Patient	50-year-old female
Disease diagnosis	T-NHL (diffuse mixed cell)
Treatment status	leukemic stage
Specimen site	lymph node
Year of establishment	1986

Immunophenotypic characterization[1]

T-/NK cell marker	CD1a$^-$, CD2$^+$, CD3$^-$, cyCD3$^+$, CD4$^+$, CD5$^-$, CD7$^-$, CD8$^-$, TCR$\alpha\beta^-$, TCR$\gamma\delta^-$
B-cell marker	CD9$^-$, CD10$^-$, sIg$^-$
Myelomonocytic marker	CD13$^-$, CD14$^-$, CD15$^-$, CD16$^-$
Progenitor/activation/other marker	CD30$^+$, CD38$^+$, CD45$^+$, CD71$^+$, HLA-DR$^+$
Cytokine receptor	CD25$^-$

Genetic characterization[1]

Cytogenetic karyotype	88, XX, +3q, +3q, −8p, iso(7q), +8mar
Receptor gene rearrangement	*IGH* G, *TCRB* RR, *TCRG* RR, *TCRD* DD

Functional characterization[1]

Cytochemistry	ACP$^{(+)}$, ALP$^-$, ANAE$^-$, CAE$^-$, GLC$^+$, MPO$^-$, PAS$^{(+)}$

Comments

- Mature T-cell line established from T-NHL.

Reference

[1] Morikawa, S. et al. (1991) Leukemia Res. 15, 381–389.

HUT 78

Culture characterization[1]

Subclone	H9 (permissive for HIV replication)
Culture medium	90% RPMI 1640 + 10% FBS
Doubling time	26 h
Viral status	EBNA$^-$
Authentication	no
Primary reference	Gazdar et al.[1]
Availability of cell line	HUT 78: ATCC TIB 161; H9: ATCC CRL 8543/HTB 176

Clinical characterization[1]

Patient	53-year-old male
Disease diagnosis	cutaneous T-cell lymphoma (Sézary syndrome)
Specimen site	peripheral blood

Immunophenotypic characterization[1,2]

T-/NK cell marker	CD1$^-$, CD2$^{(+)}$, CD3$^+$, cyCD3$^+$, CD4$^+$, CD5$^+$, CD7$^+$, CD8$^-$, CD28$^+$, CD57$^-$, TCR$\alpha\beta^+$, TCR$\gamma\delta^-$, cyTCRα^+, cyTCRβ^+, cyTCRγ^-
B-cell marker	CD9$^-$, CD10$^-$, CD19$^-$, CD20$^-$, CD21$^-$, CD24$^-$, sIg$^-$
Myelomonocytic marker	CD13$^-$, CD14$^-$, CD15$^+$, CD16$^-$, CD33$^-$
Erythroid-megakaryocytic marker	CD41a$^-$, CD61$^-$
Progenitor/activation/other marker	CD34$^-$, CD38$^+$, CD71$^+$, HLA-DR$^+$, TdT$^-$
Adhesion marker	CD11b$^-$, CD29$^+$
Cytokine receptor	CD25$^-$

Genetic characterization[1–8]

Cytogenetic karyotype	76, −X, dup(X)(p11.4→pter), dup(X), +1, −2, −3, −4, −4, −4, −5, − 5, −6, −6, −7, −7, −9, −10, −10, −10, −11, −11, −13, −13, −13, −14, −14, −16, −16, +17, −18, −18, −18, −20, −20, −20, −21, −22, +24mar, del(1)(p13p22), +der(2)t(2;?)(2pter→2q37::?), +der(3)t (3;?)(3pter→3q27::?), +der(4)t(4;?)(4pter→4q2?1::?)×2, +der(4)t (4;5)(4pter→4q1?1::5q1?1→5qter), +der(?)t(?;5)(?::5q1?1→5qter), +der(6)t(6;?)(6pter→6q1?1::?), +der(6), +der(7)t(7;?)(7qter→ 7p1?1::?)×2, +der(10?)t(10?;7)(10pter→10q1?1: :7q11.2→7q22:

	:7q11.2→7q22::7q22→7qter)×2, +del(12)(p1?2)×2, dup(19)(q?→ qter?), +der(?)t(?;14)(?::14q11.2→14qter)
Unique translocation/fusion gene	t(2;8)(q34;q24) → *TCL4-MYC* fusion gene
Unique gene alteration	*IFNA* deletion, *IFNB* deletion, *NFKB* rearrangement, *P15INK4B* deletion, *P16INK4A* deletion, *P53* mutation

Functional characterization[1,9,10]

Cytochemistry	ACP^+, ALP^-, $ANAE^+$, $ANBE^-$, GLC^+, $TRAP^-$
Cytokine production	$mRNA^+$: GM-CSF, IFN-γ, IL-10, IL-15, TGF-β1
Heterotransplantation	into nude mice

Comments

- Mature $TCR\alpha\beta^+$ T-cell line derived from Sézary syndrome (CTCL) (reference cell line).
- Carries t(2;8)(q34;q24) leading to *TCL4-MYC* fusion gene.
- Subclone H9 was used to first isolate HIV.
- Available from cell line bank.

References

1 Gazdar, A.F. et al. (1980) Blood 55, 409–417.
2 Minowada, J. and Matsuo, Y. (1999) unpublished data.
3 Chen, T.R. (1992) J. Natl. Cancer Inst. 84, 1922–1926.
4 Heyman, M. et al. (1994) Leukemia 8, 425–434.
5 Finger, L.R. et al. (1988) Proc. Natl. Acad. Sci. USA 85, 9158–9162.
6 Borgonovo Brandter, L. et al. (1996) Eur. J. Haematol. 56, 313–318.
7 Cheng, J. and Haas, M. (1990) Mol. Cell. Biol. 10, 5502–5509.
8 Zhang, J. et al. (1994) Oncogene 9, 1931–1937.
9 Micallef, M. et al. (1994) Int. J. Oncol. 4, 633–638.
10 Döbbeling, U. et al. (1998) Blood 91, 252–258.

HUT 102

Culture characterization[1-3]

Sister cell line — **CTCL-3** (serial sister cell line – at 2nd relapse – from peripheral blood – IL-2-dependent, HTLV-I^{+})
Culture medium — 80% RPMI 1640 + 20% FBS
Doubling time — 38 h
Viral status — EBNA^{-}, HTLV-I^{+}
Authentication — no
Primary reference — Gazdar et al.[1]
Availability of cell line — ATCC TIB 162

Clinical characterization[1]

Patient — 28-year-old male
Disease diagnosis — cutaneous T-cell lymphoma (mycosis fungoides)
Treatment status — at diagnosis
Specimen site — lymph node
Year of establishment — 1978

Immunophenotypic characterization[1,4,5]

T-/NK cell marker — CD1^{-}, CD2^{+}, CD3^{-}, cyCD3^{-}, CD4^{+}, CD5^{+}, CD7^{-}, CD8^{-}, CD28^{-}, CD57^{-}, TCR$\alpha\beta^{-}$, TCR$\gamma\delta^{-}$, cyTCRα^{+}, cyTCRβ^{+}, cyTCRγ^{+}
B-cell marker — CD9^{-}, CD10^{-}, CD19^{-}, CD20^{-}, CD21^{-}, CD22^{-}, CD24^{-}, sIg^{-}
Myelomonocytic marker — CD13^{-}, CD14^{-}, CD15^{+}, CD33^{-}
Erythroid-megakaryocytic marker — CD41a^{-}, CD61^{-}
Progenitor/activation/other marker — CD34^{-}, CD38^{-}, CD71^{+}, HLA-DR^{+}, TdT^{-}
Adhesion marker — CD11b^{-}, CD29^{+}
Cytokine receptor — CD25^{+}, CD122^{+}, CD132^{+}

Genetic characterization[1,4]

Cytogenetic karyotype — 91, XXXY, −13, −14, −14, +der(7)t(7;14)(7pter→7p1?1::7q11.2→7p1?1::14q11.2→14qter), +der(14?)t(14?;?)

Functional characterization[1,6]

Cytochemistry — ACP^{+}, ALP^{-}, ANAE^{+}, ANBE$^{(+)}$, GLC^{+}, TRAP^{-}
Cytokine production — mRNA^{+}: GM-CSF, IFN-γ, IL-12, TGF-β1

Comments

- Mature T-cell line derived from mycosis fungoides (CTCL) (reference cell line).
- Cells secrete HTLV-I.

- Serial sister cell lines established at presentation and at relapse.
- Available from cell line bank.

References

[1] Gazdar, A.F. et al. (1980) Blood 55, 409–417.
[2] Poiesz, B.J. et al. (1980) Proc. Natl. Acad. Sci. USA 77, 6815–6819.
[3] Poiesz, B.J. et al. (1980) Proc. Natl. Acad. Sci. USA 77, 7415–7419.
[4] Minowada, J. and Matsuo, Y. (1999) unpublished data.
[5] Schumann, R.R. et al. (1996) Blood 87, 2419–2427.
[6] Micallef, M. et al. (1994) Int. J. Oncol. 4, 633–638.

Karpas 384

Culture characterization[1]

Culture medium	80% RPMI 1640 + 20% FBS
Viral status	HTLV-I$^-$
Authentication	yes (by *TCRG, TCRD* rearrangements)
Primary reference	Dyer et al.[1]
Availability of cell line	original authors

Clinical characterization[1]

Patient	48-year-old male
Disease diagnosis	T-NHL (subcutaneous)
Treatment status	leukemic phase, refractory, terminal
Specimen site	peripheral blood
Year of establishment	1987

Immunophenotypic characterization[1]

T-/NK cell marker	CD1$^-$, CD2$^-$, CD3$^-$, cyCD3$^+$, CD4$^-$, CD5$^-$, CD7$^-$, cyCD7$^+$, CD8$^-$, TCR$\alpha\beta^-$
B-cell marker	CD10$^-$, CD19$^-$, CD20$^-$, CD37$^-$, sIg$^-$
Progenitor/activation/other marker	CD38$^+$, CD45$^+$, CD45RO$^-$, HLA-DR$^+$, TdT$^-$
Cytokine receptor	CD25$^-$

Genetic characterization[1,2]

Cytogenetic karyotype	47, XO, +20, +mar, t(1;2)(q11;q35), t(2;1;14)(q35;q11-q32.1; q22.1), t(7;14)(p13;q11.2), inv(7)(p13;q22.1), int del(12) (q24.1q24.3)
Receptor gene rearrangement	*TCRA* GG, *TCRB* GG, *TCRG* RR, *TCRD* RR

Comments

- Mature T-cell line derived from aggressive cutaneous T-cell lymphoma.

References

[1] Dyer, M.J.S. et al. (1993) Leukemia 7, 1047–1053.
[2] Stranks, G. et al. (1995) Blood 85, 893–901.

Kit 225

Culture characterization[1]

Culture medium	90% RPMI 1640 + 10% FBS + 1 U/ml IL-2
Viral status	HTLV-I/II$^-$
Authentication	yes (by *TCRB* rearrangement)
Primary reference	Hori et al.[1]
Availability of cell line	not known

Clinical characterization[1]

Patient	62-year-old male
Disease diagnosis	T-CLL
Specimen site	peripheral blood
Year of establishment	1983

Immunophenotypic characterization[1]

T-/NK cell marker	CD1$^-$, CD2$^+$, CD3$^+$, CD4$^+$, CD8$^-$
Progenitor/activation/other marker	CD71$^+$, HLA-DR$^+$
Cytokine receptor	CD25$^+$

Genetic characterization[1]

Cytogenetic karyotype	47, XY, −5, −6, −14, +19, ins inv(1)(pter→p36.3::p34.1→p31.2::p36.3→p34.1::p31.2→p22.1::q12→p22.1::q12→qter), inv(3)(pter→p26.2::q27.1→p26.2::q27.1→qter), +der(5)t(5;7)(7qter→7q22::5p15.3→cen→5qter), del(6)(pter→q21::q23.3→qter), +der(6)t(6;?) (q21;?), +der(14)t(6;14)(6q21→6q23.3::14p12→cen→14qter)
Receptor gene rearrangement	*TCRB* R

Functional characterization[1]

Cytokine response	IL-2-dependent

Comments

- Mature T-cell line derived from T-CLL.
- Constitutively IL-2-dependent.

Reference

[1] Hori, T. et al. (1987) Blood 70, 1069–1072.

My-La

Culture characterization[1]

Culture medium	80% RPMI 1640 + 20% FBS + IL-2
Doubling time	24–30 h
Viral status	HTLV-I/II$^-$
Authentication	yes (by *TCRB* sequencing)
Primary reference	Kaltoft et al.[1]
Availability of cell line	not known

Clinical characterization[1]

Patient	80-year-old male
Disease diagnosis	mycosis fungoides
Treatment status	at diagnosis, terminal, no therapy
Specimen site	skin biopsy
Year of establishment	1990

Immunophenotypic characterization[1]

T-/NK cell marker	CD1$^-$, CD2$^{(+)}$, CD3$^{(+)}$, CD4$^-$, CD8$^{(+)}$, CD56$^-$, TCR$\alpha\beta^{(+)}$
Progenitor/activation/other marker	CD71$^{(+)}$, HLA-DR$^-$
Cytokine receptor	CD25$^{(+)}$

Genetic characterization[1]

Cytogenetic karyotype	49, XYY, −10, −14, −15, −16, +17, +4mar, +del(7)(p15.2), t(1;?) (p36.2;?), del(1)(q32.2), t(2:?)(q37.2;?), del(4)(q31.2), del(5) (q21q23.1), del(6)(q23.2), del(7)(p15.2), t(10q13q)

Functional characterization[1,2]

Cytokine production	IL-15 mRNA$^+$
Cytokine response	IL-2-dependent

Comments

- Mature TCR$\alpha\beta^+$ T-cell line derived from mycosis fungoides (CTCL).
- Constitutively IL-2-dependent.

References

[1] Kaltoft, K. et al. (1992) In Vitro Cell. Dev. Biol. 28A, 161–167.
[2] Döbbeling, U. et al. (1998) Blood 92, 252–258.

Culture characterization[1]

Culture medium	90% RPMI 1640 + 10% FBS + IL-2
Doubling time	3 d
Viral status	EBNA⁻, HTLV-I⁻
Authentication	yes (by cytogenetics)
Primary reference	Kaltoft et al.[1]
Availability of cell line	not known

Clinical characterization[1]

Patient	66-year-old female
Disease diagnosis	severe exfoliative erythroderma → Sézary syndrome
Treatment status	refractory, progressive
Specimen site	peripheral blood
Year of establishment	1984/1985

Immunophenotypic characterization[1]

T-/NK cell marker	$CD1^-$, $CD2^+$, $CD3^-$, $cyCD3^+$, $CD4^-$, $CD5^-$, $CD8^-$
B-cell marker	$CD20^-$
Progenitor/activation/other marker	$CD38^+$, $CD71^+$, $HLA\text{-}DR^+$
Cytokine receptor	$CD25^+$

Genetic characterization[1]

Cytogenetic karyotype	85, XXX, −7, −8, −8, −8, −8, −9, −9, −10, −10, −12, −13, −15, +2mar1, +2mar2, +mar3, +mar4, i(1q), inv(1)(p36;q21)×2, del(2)(q21)×2, t(3;?)(q21;?)×2, t(4;?)(p16;?)×2, del(5)(p13)×2, del(6)(q21)×2, del(12)(p13), t(17;?)(p11;?)×2

Functional characterization[1,2]

Cytokine production	IL-15 $mRNA^+$
Cytokine response	IL-2–dependent; growth stimulation: IL-7, IL-15; growth inhibition: TNF-α

Comments

- Mature T-cell line derived from progressive Sézary syndrome (CTCL).
- Constitutively IL-2-dependent.

References

[1] Kaltoft, K. et al. (1987) Arch. Dermatol. Res. 279, 293–298.
[2] Döbbeling, U. et al. (1998) Blood 91, 252–258.

SKW-3

Culture characterization[1]

Culture medium	90% RPMI 1640 + 10% FBS
Doubling time	30–40 h
Viral status	EBV⁻, HBV⁻, HCV⁻, HHV-8⁻, HIV⁻, HTLV-I/II⁻
Authentication	no
Primary reference	unpublished
Availability of cell line	DSM ACC 53, RCB 1168

Clinical characterization[1]

Patient	61-year-old male
Disease diagnosis	CLL
Specimen site	peripheral blood
Year of establishment	1977

Immunophenotypic characterization[1-3]

T-/NK cell marker	$CD1a^-$, $CD2^+$, $CD3^-$, $cyCD3^+$, $CD4^+$, $CD5^+$, $CD6^+$, $CD7^+$, $CD8^+$, $CD27^{(+)}$, $CD28^+$, $CD57^-$, $TCR\alpha\beta^-$, $TCR\gamma\delta^-$, $cyTCR\alpha^-$, $cyTCR\beta^-$, $cyTCR\gamma^+$
B-cell marker	$CD9^+$, $CD10^-$, $CD19^-$, $CD20^-$, $CD21^-$, $CD22^-$, $CD24^-$
Myelomonocytic marker	$CD13^-$, $CD14^-$, $CD15^+$, $CD16^-$, $CD33^-$
Erythroid-megakaryocytic marker	$CD41a^-$, $CD61^-$
Progenitor/activation/other marker	$CD34^-$, $CD38^+$, $CD71^+$, HLA-DR^-, TdT^-
Adhesion marker	$CD11b^-$, $CD29^-$
Cytokine receptor	$CD25^-$

Genetic characterization[1-10]

Cytogenetic karyotype	near diploid, 4% polyploidy; 46(43–48)<2n>XY, +8, −14, t(3;3)(q11;q27), der(8)t(8;14)(q24;q11)×2, t(8;11)(p21;p12), der(12)t (12;?)(q24;?), der(14)t(8;14)(q24;q11)
Unique translocation/fusion gene	t(8;14)(q24;q11) → *MYC-TCRA* genes altered
Receptor gene rearrangement	*IGH* G, *IGK* G, *TCRB* R, *TCRG* R, *TCRD* D (mRNA: $TCRA^+$, $TCRB^+$)
Unique gene alteration	*P15INK4B* deletion, *P16INK4A* deletion

Comments

- Mature T-cell line derived from CLL (reference cell line).
- Carries t(8;14)(q24;q11) leading to alteration of the *MYC-TCRA* gene.
- Available from cell line banks.

References

[1] Drexler, H.G. et al. (1999) DSMZ Catalogue of Cell Lines, 7th edn, Braunschweig, Germany.
[2] Minowada, J. and Matsuo, Y. (1999) unpublished data.
[3] Burger, R. et al. (1999) Leukemia Res. 23, 19–27.
[4] Sangster, R.N. et al. (1986) J. Exp. Med. 163, 1491–1508.
[5] Heyman, M. et al. (1994) Leukemia 8, 425–434.
[6] Shima, E.A. et al. (1986) Proc. Natl. Acad. Sci. USA 83, 3439–3443.
[7] Erikson, J. et al. (1986) Science 232, 884–886.
[8] Ogawa, S. et al. (1994) Blood 84, 2431–2435.
[9] Stranks, G. et al. (1995) Blood 85, 893–901.
[10] Borgonovo Brandter, L. et al. (1996) Eur J. Haematol. 56, 313–318.

SMZ-1

Culture characterization[1]

Culture medium	85% RPMI 1640 + 15% FBS
Doubling time	34 h
Viral status	EBNA⁻, HTLV-I⁻
Authentication	yes (by cytogenetics)
Primary reference	Miyanishi and Ohno[1]
Availability of cell line	original authors

Clinical characterization[1]

Patient	46-year-old male
Disease diagnosis	systemic lupus erythematosus → T-NHL (diffuse large cell)
Treatment status	stage IV, at presentation, terminal
Specimen site	ascites

Immunophenotypic characterization[1]

T-/NK cell marker	$CD1b^-$, $CD2^+$, $CD3^+$, $CD4^+$, $CD5^{(+)}$, $CD7^+$, $CD8^-$, $CD28^+$
B-cell marker	$CD10^-$, $CD19^-$, $CD20^-$
Myelomonocytic marker	$CD13^-$, $CD14^-$, $CD16^-$
Progenitor/activation/other marker	$CD71^+$, HLA-DR^+
Adhesion marker	$CD11b^-$
Cytokine receptor	$CD25^+$

Genetic characterization[1]

Cytogenetic karyotype	47, XY, +8, −10, +21, t(6;14)(p21.1;q24), del(9)(p13q22), der(17)(?::p11-q23::?)
Receptor gene rearrangement	*IGH* G, *TCRB* R

Comments

• Mature T-cell line derived from T-NHL (DLCL).

Reference

[1] Miyanishi, S. and Ohno, H. (1992) Cancer Genet. Cytogenet. 59, 199–205.

Culture characterization[1]

Culture medium	80% RPMI 1640 + 20% FBS
Doubling time	36–48 h
Viral status	HTLV-I$^-$
Authentication	yes (by cytogenetics, *MYC* amplification, *TCRB* rearrangement)
Primary reference	Ohno et al.[1]
Availability of cell line	not known

Clinical characterization[1]

Patient	73-year-old male
Disease diagnosis	T-NHL (diffuse large cell)
Treatment status	at relapse
Specimen site	lymph node
Year of establishment	1985

Immunophenotypic characterization[1]

T-/NK cell marker	CD1$^-$, CD2$^+$, CD3$^+$, CD4$^+$, CD5$^+$, CD7$^-$, CD8$^-$, TCRαβ$^-$
B-cell marker	CD10$^-$, CD20$^-$, sIg$^-$
Progenitor/activation/other marker	HLA-DR$^-$, TdT$^-$
Cytokine receptor	CD25$^{(+)}$

Genetic characterization[1]

Cytogenetic karyotype	77, XXY, der(?)t(1;?)(1pter→1p12::?), del(2)(p21p23), i(3p)×2, del(11)(q21q23), der(12)t(12;?)(p11;?), der(13)t(13;?)(p12-13;?) ×3, del(14)(q24), der(15)t(15;?)(p12-13;?), der(15)t(15;?)(q24;?), der(16)t(16;?)(p11;?), der(19)t(19;?)(p13;?)×3
Receptor gene rearrangement	*TCRB* R
Unique gene alteration	*MYC* amplification

Functional characterization[1]

Proto-oncogene	MYC mRNA overexpression

Comments

- Mature T-cell line derived from T-NHL (DLCL).
- Carries rare *MYC* gene amplification.

Reference

[1] Ohno, H. et al. (1988) Cancer Res. 48, 4959–4963.

Section V

NATURAL KILLER CELL LEUKEMIA-LYMPHOMA CELL LINES

HANK1

Culture characterization[1]

Establishment	primary material first heterotransplanted into SCID mouse, then culture *in vitro*
Culture medium	Cosmedium 001 + 100 U/ml IL-2
Doubling time	3 d
Viral status	EBV$^+$
Authentication	yes (by EBV Southern blotting)
Primary reference	Kagami et al.[1]
Availability of cell line	original authors

Clinical characterization[1]

Patient	46-year-old female
Disease diagnosis	nasal-like NK/T-cell lymphoma (angiocentric)
Treatment status	at diagnosis
Specimen site	retroperitoneal lymph node

Immunophenotypic characterization[1]

T-/NK cell marker	CD1$^-$, CD2$^+$, CD3$^-$, CD3ϵ^+, CD4$^-$, CD5$^-$, CD7$^+$, CD8$^-$, CD56$^+$, CD57$^-$
B-cell marker	CD10$^-$
Myelomonocytic marker	CD16$^-$
Progenitor/activation/other marker	CD30$^+$, HLA-DR$^+$
Cytokine receptor	CD25$^+$

Genetic characterization[1]

Cytogenetic karyotype	48, XX, +2, +21, del(16q13)
Receptor gene rearrangement	*IGH* G, *TCRB* G, *TCRG* G

Functional characterization[1]

Cytochemistry	azurophilic granules
Cytokine response	IL-2-dependent
Heterotransplantation	into SCID mice
Special features	no NK activity; EBER$^+$, EBNA1$^+$, EBNA2$^-$, LMP1$^{(+)}$; granzyme B$^+$

Comments

- NK cell line positive for NK cell markers and negative for various T-cell markers.
- Expression of EBV type II latency pattern.
- No cytotoxic NK activity.
- Cells grow well.

Reference

[1] Kagami, Y. et al. (1998) Brit. J. Haematol. 103, 669–677.

KHYG-1

Culture characterization[1]

Culture medium	90% RPMI 1640 + 10% FBS + 100 U/ml IL-2
Doubling time	24–48 h
Viral status	EBV⁻
Authentication	yes (by cytogenetics, *P53* mutation)
Primary reference	Yagita et al.[1]
Availability of cell line	original authors

Clinical characterization[1]

Patient	45-year-old female
Disease diagnosis	aggressive NK leukemia
Treatment status	at diagnosis
Specimen site	peripheral blood
Year of establishment	1997

Immunophenotypic characterization[1]

T-/NK cell marker	CD2⁺, CD3⁻, cyCD3⁺, CD4⁻, CD5⁻, CD7⁺, CD8⁺, CD28⁻, CD56⁺, CD57⁻, CD94⁺, CD158a⁺, CD158b⁺, CD161⁻, TCRαβ⁻, TCRγδ⁻
B-cell marker	CD10⁻, CD19⁻, CD20⁻, CD21⁻, sIg⁻
Myelomonocytic marker	CD13⁻, CD16⁻, CD33⁺
Progenitor/activation/other marker	CD34⁻, CD38⁺, CD45⁺, CD45RA⁺, CD45RO⁻, CD95⁺, HLA-DR⁺, TdT⁻
Adhesion marker	CD11a⁺, CD18⁺, CD49d⁺, CD54⁺, CD58⁺
Cytokine receptor	CD25⁻, CD122⁺

Genetic characterization[1]

Cytogenetic karyotype	92–96<4n>, +X, +X, +X, +2, −5, −5, −7, −8, −8, −9, −9, −10, +12, +13, −14, −15, −16, −16, +17, −18, −18, +20, +20, −21, +22, +6-17mar, del(X)(q2?)×4 or i(X)(p10)×4, add(6)(q2?), add(6), ?del(6)(q?), ?del(6), add(7)(q11)×2, add(7)(q22), add(7), add(10)(p1?), add(10), add(11)(q23), add(11), add(15)(p11), add(15), add(20)(q13)×2, add(20)
Receptor gene rearrangement	*IGH* G, *TCRB* G, *TCRG* G, *TCRD* G
Unique gene alteration	*P53* mutation (same as patient)

Functional characterization[1]

Cytochemistry	MPO⁻; azurophilic granules
Cytokine production	IFN-γ⁺
Cytokine response	IL-2-dependent
Special features	strong cytotoxic activity

Comments

- NK cell line derived from patient with aggressive NK leukemia.
- Constitutively IL-2-dependent.
- Strong cytotoxic activity.

References

[1] Yagita, M. et al. (2000) Leukemia 14, 922–930.

NK-92

Culture characterization[1]

Culture medium	75% α-MEM + 12.5% FBS + 12.5% horse serum + 10–50 U/ml IL-2
Doubling time	24 h
Viral status	EBV⁻
Authentication	no
Primary reference	Gong et al.[1]
Availability of cell line	original authors

Clinical characterization[1]

Patient	50-year-old male
Disease diagnosis	LGL-NHL
Treatment status	at diagnosis
Specimen site	peripheral blood
Year of establishment	1992

Immunophenotypic characterization[1]

T-/NK cell marker	$CD1^-$, $CD2^+$, $CD3^-$, $CD4^-$, $CD5^-$, $CD7^+$, $CD8^-$, $CD28^+$, $CD56^+$
B-cell marker	$CD10^-$, $CD19^-$, $CD20^-$, $CD23^-$
Myelomonocytic marker	$CD14^-$, $CD16^-$
Progenitor/activation/other marker	$CD34^-$, $CD45^+$, HLA-DR^+
Adhesion marker	$CD11a^+$, $CD54^+$
Cytokine receptor	$CD25^+$, $CD122^+$

Genetic characterization[1]

Cytogenetic karyotype	near-tetraploid, multiple structural abnormalities
Receptor gene rearrangement	*TCRB* G, *TCRD* G

Functional characterization[1–4]

Cytochemistry	azurophilic granules
Cytokine response	IL-2-dependent; growth stimulation: IL-7
Special features	strong NK activity; transfection with IL-2 → IL-2-independent NK-92MI + NK-92CI subclones

Comments

- NK cell line expressing NK cell markers, but not specific T-cell markers (reference cell line).
- Strong cytotoxic NK activity.
- Cells are difficult to culture.

References

[1] Gong, J.H. et al. (1994) Leukemia 8, 652–658.
[2] Klingemann, H.G. et al. (1996) Biol. Blood Marrow Transplant 2, 68–75.
[3] Komatsu, F. and Kajiwara, M. (1999) Oncol. Res. 10, 483–489.
[4] Tam, Y.K. et al. (1999) Human Gene Therapy 10, 1359–1373.

NKL

Culture characterization[1]

Culture medium	85% RPMI 1640 + 15% FBS + 100 pM IL-2
Doubling time	24–48 h
Authentication	no
Primary reference	Robertson et al.[1]
Availability of cell line	original authors

Clinical characterization[1]

Patient	63-year-old male
Disease diagnosis	NK-LGL leukemia
Treatment status	terminal, refractory
Specimen site	peripheral blood

Immunophenotypic characterization[1]

T-/NK cell marker	CD1$^-$, CD2$^+$, CD3$^-$, CD4$^-$, CD5$^-$, CD6$^+$, CD7$^-$, CD8$^-$, CD27$^+$, CD28$^-$, CD31$^+$, CD56$^{(+)}$, CD57$^-$, CD94$^+$, TCR$\alpha\beta^-$, TCR$\gamma\delta^-$
B-cell marker	CD19$^-$, CD20$^-$, CD81$^+$
Myelomonocytic marker	CD14$^-$, CD16$^-$
Progenitor/activation/other marker	CD26$^+$, CD38$^+$, CD45$^+$, CD69$^-$, CD71$^+$, CD95$^+$, HLA-DR$^+$
Adhesion marker	CD11a$^+$, CD11b$^-$, CD11c$^-$, CD18$^+$, CD29$^+$, CD43$^+$, CD44$^+$, CD58$^+$
Cytokine receptor	CD25$^+$, CD122$^-$, CD132$^-$

Genetic characterization[1]

Cytogenetic karyotype	47(41–50)XY, +6, add(1)(q42), del(6)(q15q23), del(17)(p11)
Receptor gene rearrangement	*TCRB* G, *TCRG* R

Functional characterization[1]

Cytokine response	IL-2-dependent
Special features	ADCC, NK activity

Comments

- NK cell line expressing NK cell markers.
- Displays cytotoxic NK activity and ADCC.
- Cells grow well.

Reference

[1] Robertson, M.J. et al. (1996) Exp. Hematol. 24, 406–415.

NK-YS

Culture characterization[1]

Establishment	first cultured on mouse spleen stroma cell line
Culture medium	90% IMDM + 10% FBS + 100 U/ml IL-2
Doubling time	48 h
Viral status	EBV$^+$
Authentication	yes (by cytogenetics, EBV Southern blotting)
Primary reference	Tsuchiyama et al.[1]
Availability of cell line	original authors

Clinical characterization[1]

Patient	19-year-old female
Disease diagnosis	NK cell lymphoma, nasal angiocentric, leukemic state with systemic skin infiltration
Treatment status	terminal, refractory
Specimen site	peripheral blood
Year of establishment	1996

Immunophenotypic characterization[1]

T-/NK cell marker	CD1a$^-$, CD2$^+$, CD3$^-$, CD4$^-$, CD5$^+$, CD6$^-$, CD7$^+$, CD8$^-$, CD56$^+$, CD57$^-$, TCR$\alpha\beta^-$, TCR$\gamma\delta^-$
B-cell marker	CD10$^-$, CD19$^-$, CD20$^-$, CD21$^-$, sIg$^-$
Myelomonocytic marker	CD14$^-$, CD15$^-$, CD16$^-$, CD33$^-$
Progenitor/activation/other marker	CD34$^-$, CD38$^+$, CD45$^+$, CD95$^+$
Adhesion marker	CD11a$^+$, CD11b$^-$, CD11c$^+$, CD58$^+$
Cytokine receptor	CD25$^+$

Genetic characterization[1]

Cytogenetic karyotype	46, XX, add(3)(q26.2), der(4)t(1;4)(q12;p16)
Receptor gene rearrangement	*TCRB* G, *TCRG* G

Functional characterization[1]

Cytochemistry	azurophilic granules, MPO$^-$
Cytokine production	IFN-γ
Cytokine response	IL-2-dependent
Special features	NK activity; EBNA2$^-$, LMP1$^+$, ZEBRA$^-$

Comments

- NK cell line corresponding to the NK immunoprofile and functional definition.
- Cytotoxic NK activity.

- Expression of EBV type II latency pattern.
- Constitutively IL-2-dependent.
- Cells are difficult to culture.

Reference

[1] Tsuchiyama, J. et al. (1998) Blood 92, 1374–1383.

NOI-90

Culture characterization[1]

Culture medium	90% RPMI 1640 + 10% FBS
Authentication	yes (by cytogenetics)
Primary reference	Sahraoui et al.[1]
Availability of cell line	original authors

Clinical characterization[1]

Patient	45-year-old male
Disease diagnosis	NK-NHL in leukemic phase
Treatment status	at diagnosis
Specimen site	peripheral blood
Year of establishment	1989

Immunophenotypic characterization[1]

T-/NK cell marker	$CD1^+$, $CD2^-$, $CD3^+$, $CD4^{(+)}$, $CD5^-$, $CD7^+$, $CD8^-$, $CD56^+$, $CD57^+$
B-cell marker	$CD10^-$
Myelomonocytic marker	$CD13^-$, $CD14^-$, $CD16^-$
Progenitor/activation/other marker	$CD34^-$
Cytokine receptor	$CD25^{(+)}$, $CD122^{(+)}$

Genetic characterization[1]

Cytogenetic karyotype	47, XY, +8, 17p−
Receptor gene rearrangement	*IGH* G, *TCRB* G, *TCRG* G

Functional characterization[1]

Cytochemistry	ACP^+, $ANAE^-$, MPO^-
Cytokine production	IL-2
Cytokine receptor	IL-2Rγ $mRNA^+$
Special features	no NK activity

Comments

- NK cell line expressing NK cell and T-cell immunomarkers.
- No cytotoxic NK activity.
- A NOI-90 culture which was received from the original investigators was found to be cross-contaminated (real identity of this culture: precursor B-cell line Reh) [2]. It is not known whether the original, correct NOI-90 still exists.

References

[1] Sahraoui, Y. et al. (1997) Leukemia 11, 245–252.
[2] Drexler, H.G. and Dirks, W.G. (1999) unpublished data.

YT

Culture characterization[1–3]

Culture medium	90% RPMI 1640 + 10% FBS
Doubling time	40–50 h
Viral status	EBV^{+}, HBV^{-}, HCV^{-}, HHV-8^{-}, HIV^{-}, HTLV-I/II^{-}
Authentication	no
Primary reference	Yodoi et al.[1]
Availability of cell line	DSM ACC 434

Clinical characterization[1,2]

Patient	15-year-old male
Disease diagnosis	acute lymphoblastic lymphoma (with thymoma)
Treatment status	at relapse
Specimen site	pericardial fluid
Year of establishment	1983

Immunophenotypic characterization[1–4]

T-/NK cell marker	CD1^{-}, CD2^{-}, CD3^{-}, CD4^{-}, CD5^{-}, CD6^{-}, CD7^{+}, CD8^{-}, CD28^{+}, CD56^{+}, CD57^{-}, TCR$\alpha\beta^{-}$, TCR$\gamma\delta^{-}$
B-cell marker	CD10^{-}, CD19^{-}, CD20^{-}, CD21^{-}, CD23^{-}, CD24^{-}
Myelomonocytic marker	CD13^{-}, CD15^{-}, CD16^{-}, CD33^{-}
Progenitor/activation/other marker	CD30^{+}, CD34^{-}, CD38^{-}, CD45RA^{-}, CD45RO^{+}, CD71^{+}, HLA-DR^{+}
Adhesion marker	CD11a^{+}, CD11b^{-}
Cytokine receptor	CD25^{+}, CD122^{+}, CD132^{+}

Genetic characterization[1–3]

Cytogenetic karyotype	flat-moded, near-tetraploid, 12% polyploidy; 84–98<4n>XXYY, −6, −9, −15, −19, +6-8mar, der(X)t(X;7)(q25;q21), der(1)dup(1)(q12.2q2?2)t(1;17)(q3?1;q1?2), der(4)t(1;4)(q32.2;q35)×2, der(5)t(X;5)(p1?3;q21)×2, del(6)(p11), der(10;17)(q10;q10), dup(14)(q1?q2?), add(16)(q24), add(17)(q11)
Receptor gene rearrangement	*IGH* G, *TCRB* G, *TCRG* G, *TCRD* G

Functional characterization[1,2,4]

Cytochemistry	azurophilic granules, MPO^{-}, PAS^{+}
Special features	ADCC, weak NK activity; RT-PCR: EBNA1^{+}, EBNA2^{-}, LMP1^{+}, LMP2A^{+}, LMP2B^{-}

Comments

- First NK cell line expressing various NK cell-associated features (reference cell line).
- Cytotoxic NK activity and ADCC.
- Expression of EBV type II latency.
- Available from cell line bank.
- Cells grow well.

References

[1] Yodoi, J. et al. (1985) J. Immunol. 134, 1623–1630.
[2] Yoneda, N. et al. (1992) Leukemia 6, 136–141.
[3] Drexler, H.G. et al. (1999) DSMZ Catalogue of Cell Lines, 7th edn, Braunschweig, Germany.
[4] Kanegane, H. et al. (1998) Leukemia Lymphoma 29, 491–498.

Section VI

HODGKIN'S DISEASE CELL LINES

DEV

Culture characterization[1]

Culture medium 90% RPMI 1640 + 10% FBS
Doubling time 3 d
Viral status EBNA$^-$
Authentication no
Primary reference Poppema et al.[1]
Availability of cell line restricted

Clinical characterization[1,2]

Patient 51-year-old male
Disease diagnosis Hodgkin's disease (nodular sclerosis)
Treatment status stage II, at relapse
Specimen site pleural effusion

Immunophenotypic characterization[1–4]

T-/NK cell marker CD1a$^-$, CD2$^-$, CD3$^-$, CD4$^-$, CD5$^-$, CD7$^-$, CD8$^-$, CD57$^-$
B-cell marker CD10$^-$, CD20$^+$, CD21$^-$, CD22$^+$, CD37$^-$, cyIgA$^+$, sIg$^-$
Myelomonocytic marker CD15$^+$
Progenitor/activation/other marker CD30$^+$, CD45$^+$, CD71$^+$, HLA-DR$^-$
Adhesion marker CD11b$^-$, CD11c$^-$, CD43$^-$
Cytokine receptor CD25$^+$

Genetic characterization[1–4]

Cytogenetic karyotype 48, XXY, −2, +12, +mar, t(3;14)(3;22), t(3;7), del(3)
Receptor gene rearrangement *IGH* R, *IGK* R, *TCRA* G, *TCRB* G

Functional characterization[1–4]

Cytochemistry ACP$^{(+)}$, ALP$^-$, Aminopeptidase$^-$, ANAE$^-$, CAE$^+$, MPO$^-$
Cytokine production IL-8
Heterotransplantation into nude mice
Proto-oncogene mRNA$^+$: JUN, MYB, MYC, NRAS, PIM, RAF

Comments

- Hodgkin cell line of B-cell type.

References

[1] Poppema, S. et al. (1985) Cancer 55, 683–690.
[2] Poppema, S. et al. (1989) Recent Results Cancer Res. 117, 67–74.
[3] Drexler, H.G. (1993) Leukemia Lymphoma 9, 1–25. (review on Hodgkin cell lines)
[4] Diehl, V. et al. (1990) Semin. Oncol. 17, 660–672. (review on Hodgkin cell lines)

HD-70

Culture characterization[1,2]

Culture medium 70% RPMI 1640 + 20% FBS + 10% human cord blood serum
Doubling time 28 h
Viral status EBNA$^-$
Authentication no
Primary reference Kanzaki et al.[1]
Availability of cell line restricted

Clinical characterization[1,2]

Patient 69-year-old male
Disease diagnosis Hodgkin's disease (nodular sclerosis)
Treatment status Hodgkin's cell leukemia – refractory/terminal
Specimen site peripheral blood
Year of establishment 1989

Immunophenotypic characterization[1,3]

T-/NK cell marker CD1a$^-$, CD2$^-$, CD3$^-$, CD4$^-$, CD5$^-$, CD8$^-$, CD57$^-$, TCRαβ$^-$
B-cell marker CD19$^-$, CD20$^-$, CD21$^-$, CD24$^-$, cyIgAκ$^+$, sIg$^-$
Myelomonocytic marker CD13$^-$, CD14$^-$, CD15$^+$, CD33$^-$, CD36$^-$
Progenitor/activation/other marker CD30$^+$, CD38$^-$, CD45$^-$, CD45RO$^-$, CD71$^+$, HLA-DR$^-$
Cytokine receptor CD25$^-$

Genetic characterization[1,3]

Cytogenetic karyotype 73, Y, −X, +1, +1, +2, +3, +5, +7, +9, +10, −13, −13, +16, +19, +20, +21, +5mar, +ins(6;?)(p21;?), +der(7)t(7;?)(q22;?), +der(8)t(8;14)(q24?;q32)×2, +der(11)t(11;1)(p13;q13)×2, +der(11)t(11;?)(p11;?), +der(12)t(12;?)(p13;?)×2, +i(13)(qcen→qter)×2, +der(17)t(17;?)(p13;?)
Receptor gene rearrangement *IGH* R, *IGK* R, *TCRB* G, *TCRG* G, *TCRD* G

Functional characterization[1,3]

Cytochemistry ACP$^+$, ALP$^-$, ANAE$^{(+)}$, CAE$^-$, MPO$^-$, PAS$^{(+)}$, SBB$^-$
Heterotransplantation into newborn hamster

Comments

- Hodgkin cell line of B-cell type.

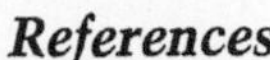

References

[1] Kanzaki, T. et al. (1992) Cancer 69, 1034–1041.

[2] Kubonishi, I. et al. (1990) Brit. J. Haematol. 75, 286–287.

[3] Drexler, H.G. (1993) Leukemia Lymphoma 9, 1–25. (review on Hodgkin cell lines)

HDLM-2

Culture characterization[1–3]

Sister cell lines	**HDLM-1, HDLM-3** (simultaneous sister cell lines – similar or identical features)
Culture medium	80% RPMI 1640 + 20% FBS
Doubling time	70–75 h
Viral status	EBV$^-$, HBV$^-$, HCV$^-$, HHV-8$^-$, HIV$^-$, HTLV-I/II$^-$
Authentication	no
Primary reference	Drexler et al.[1]
Availability of cell line	DSM ACC 17

Clinical characterization[1–5]

Patient	74-year-old male
Disease diagnosis	Hodgkin's disease (nodular sclerosis)
Treatment status	stage IV
Specimen site	pleural effusion
Year of establishment	1982

Immunophenotypic characterization[1–6]

T-/NK cell marker	CD1a$^-$, CD2$^+$, CD3$^-$, cyCD3$^-$, CD4$^{(+)}$, CD5$^-$, CD6$^-$, CD7$^-$, CD8$^-$, CD56$^-$, CD57$^-$, TCR$\alpha\beta^-$, TCR$\gamma\delta^-$
B-cell marker	CD9$^-$, CD10$^-$, CD19$^-$, CD20$^-$, CD21$^-$, CD22$^-$, CD24$^-$, sIg$^-$
Myelomonocytic marker	CD13$^-$, CD14$^-$, CD15$^+$, CD16$^-$, CD33$^-$, CD68$^-$
Erythroid-megakaryocytic marker	CD41a$^-$, CD61$^-$, GlyA$^-$
Progenitor/activation/other marker	CD30$^+$, CD34$^-$, CD38$^-$, CD71$^+$, HLA-DR$^+$, TdT$^-$
Adhesion marker	CD11a$^-$, CD11b$^-$, CD11c$^-$, CD18$^-$, CD54$^+$, CD58$^+$
Cytokine receptor	CD25$^+$, CD115$^+$, CD120$^+$, CD126$^+$, CD129$^+$

Genetic characterization[3,4,5,7]

Cytogenetic karyotype	hypodiploid, large number of rearrangements of all chromosomes except ch10; 36(35–38)<2n>X, der(Y), breakpoints at: 1p21, 1q11, 2p14, 2q37, 3q21, 3p14, 3q11, 3q25.2, 3q27, 4q25, 5q11, 5q35, 6q15/16, 7p14, 8q21.1, 8q23, 8q24.2, 9p11, 11q13, 12p12, 12q24.3, 13p11, 13q14, 14q21, 14q23/24.2, 15p11, 15q13, 17p11, 17q25, 18p11, 19p13, 19q13, 21q21, 22q33, Ycen, Yq12

Receptor gene rearrangement	*IGH* G, *IGK* G, *IGL* G, *TCRA* R, *TCRB* R, *TCRG* R, *TCRD* D
Unique gene alteration	mutations: *HRAS, KRAS, NRAS*

Functional characterization[1,2,4,5,8]

Cytochemistry	ACP+, ALP−, ANAE+, CAE−, MPO−, TRAP+
Cytokine production	IL-1α, IL-4, IL-6, IL-9, IL-13, M-CSF, TNF-α, TNF-β; post-TPA: GM-CSF, IL-3, IL-8; mRNA+: IL-5, LIF, TGF-β
Heterotransplantation	into nude mice
Inducibility of differentiation	responsive to TPA (various effects)
Proto-oncogene	mRNA+: FOS, JUN
Transcription factor	AP1+, NFAT1+, NFκB+

Comments

- Hodgkin cell line of T-cell type (reference cell line).
- Available from cell line bank.
- Cells grow very well.
- Large number of giant multinucleated cells.

References

[1] Drexler, H.G. et al. (1986) Leukemia Res. 10, 487–500.

[2] Drexler, H.G. et al. (1989) Recent Results Cancer Res. 117, 75–82.

[3] Drexler, H.G. et al. (1999) DSMZ Catalogue of Cell Lines, 7th edn, Braunschweig, Germany.

[4] Drexler, H.G. (1993) Leukemia Lymphoma 9, 1–25. (review on Hodgkin cell lines)

[5] Diehl, V. et al. (1990) Semin. Oncol. 17, 660–672. (review on Hodgkin cell lines)

[6] Drexler, H.G. et al. (1989) Int. J. Cancer 43, 1083–1090.

[7] Drexler, H.G. et al. (1988) Leukemia 2, 371–376.

[8] Kapp, U. et al. (1999) J. Exp. Med. 189, 1939–1945.

HD-MyZ

Culture characterization[1,2]

Culture medium 90% RPMI 1640 + 10% FBS
Doubling time 35 h
Viral status EBV⁻, HBV⁻, HCV⁻, HHV-8⁻, HIV⁻, HTLV-I/II⁻
Authentication no
Primary reference Bargou et al.[1]
Availability of cell line DSM ACC 346

Clinical characterization[1]

Patient 29-year-old patient
Disease diagnosis Hodgkin's disease (nodular sclerosis)
Treatment status stage III → IV, refractory, terminal
Specimen site pleural effusion
Year of establishment 1991

Immunophenotypic characterization[1,2]

T-/NK cell marker $CD1a^-$, $CD2^-$, $CD3^-$, $CD4^-$, $CD5^-$, $CD7^-$, $CD8^-$, $CD56^-$
B-cell marker $CD10^+$, $CD19^-$, $CD20^-$, $CD21^-$, $CD22^-$, $CD23^-$, $CD37^-$, $CD39^-$, $CD40^-$, $CD76^-$, $CD77^-$, sIg^-
Myelomonocytic marker $CD13^+$, $CD14^+$, $CD15^+$, $CD33^-$, $cyCD68^+$
Progenitor/activation/other marker $CD30^-$, $cyCD30^{(+)}$, $CD34^-$, $CD38^-$, $CD71^+$, $CD95^-$, $HLA\text{-}DR^+$
Adhesion marker $CD11b^-$, $CD29^+$
Cytokine receptor $CD25^-$

Genetic characterization[1,2]

Cytogenetic karyotype highly rearranged, hyperdiploid, 24% polyploidy; 52–55<2n>X /Xdel(X)(p11), +1, +2, +5, +6, +8, +9, −17, −17, −18, +19, +20, +21, +22, +2mar, der(1)t(1;3)(q32;p21), der(2)t(2;?10) (p25;q22), add(3)(q27), del(4)(q22), i(5p), i(5q), der(5)t(5;8) (p14.1;p12), del(6)(q15), add(7)(p15), add(7)(p11), del(8)(q11), der(9)add(9)(p24)add(9)(q21), der(10)add(10)(p11)add(10)(q22), add(11)(q24), i(13q), der(13)t(7;13)(q11;p13), add(18)(p13), add(19)(q13.1); additional rearrangements present in subclones include der(1)t(1;2)(q11;p25), del(3)(q21), add(16)(q24)
Receptor gene rearrangement *IGH* G, *TCRB* G

Functional characterization[1]

Cytokine production	RT-PCR+: IL-1α, IL-5, IL-7, IL-10; protein+: IL-1β, IL-6, IL-8
Cytokine receptor	RT-PCR+: IL-1R, IL-6R, IL-8R (after TPA)
Heterotransplantation	into SCID mice
Inducibility of differentiation	TPA → mono differentiation
Proto-oncogene	RT-PCR+: BCL1, FMS (after TPA), MYB, MYC, NRAS
Special features	Restin mRNA+

Comments

- Hodgkin cell line with myelomonocytic features.
- Available from cell line bank.
- Cells grow very easily.

References

[1] Bargou, R.C. et al. (1993) J. Exp. Med. 177, 1257–1268.

[2] Drexler, H.G. et al. (1999) DSMZ Catalogue of Cell Lines, 7th edn, Braunschweig, Germany.

HKB-1

Culture characterization[1]

Culture medium	90% RPMI 1640 + 10% FBS
Viral status	EBV$^-$
Authentication	no
Primary reference	Wagner et al.[1]
Availability of cell line	original authors

Clinical characterization[1]

Patient	14-year-old female
Disease diagnosis	Hodgkin's disease (nodular sclerosis) + simultaneous ALCL
Treatment status	at relapse
Specimen site	intrapulmonary tumor nodules
Year of establishment	1994

Immunophenotypic characterization[1]

T-/NK cell marker	CD3$^-$, CD4$^-$, CD8$^-$, CD56$^-$
B-cell marker	CD19$^+$, CD20$^+$, CD21$^-$, CD23$^+$
Myelomonocytic marker	CD15$^+$, CD16$^-$
Progenitor/activation/other marker	CD30$^+$, CD45$^+$, HLA-DR$^-$
Cytokine receptor	CD25$^+$

Genetic characterization[1]

Cytogenetic karyotype	46, X, der(X)add(X)(p11)add(q21), t(1;13)(q2?5;q21), add(2) (p23), t(6;14)(p21;q22-23), t(8;14)(q24;q32); sideline with idem, der(5)t(5;6)(q23;q16), t(6;14)(p21;q22-23), der(6)t(6;14)(p21;q22-23)t(5;6)(q23;q16), der(14)t(6;14)(p21;q22-23)
Receptor gene rearrangement	*IGH* R, *TCR* G

Functional characterization[1]

Cytokine production	IL-6

Comments

- Hodgkin cell line of B-cell type.
- Derivation from Hodgkin/Reed–Sternberg cells not confirmed and questionable.
- Cells grow very well.
- Cells have FAB L3 morphology (Burkitt´s lymphoma-associated).

Reference

[1] Wagner, H.J. et al. (1998) Med. Ped. Oncol. 31, 138–143.

KM-H2

Culture characterization[1,2]

Culture medium	90% RPMI 1640 + 10% FBS
Doubling time	48–60 h
Viral status	EBV⁻, HBV⁻, HCV⁻, HHV-8⁻, HIV⁻, HTLV-I/II⁻
Authentication	yes (by cytogenetics)
Primary reference	Kamesaki et al.[1]
Availability of cell line	DSM ACC 8

Clinical characterization[1]

Patient	37-year-old male
Disease diagnosis	Hodgkin's disease (mixed cellularity → lymphocyte depletion)
Treatment status	stage IV, at relapse
Specimen site	pleural effusion
Year of establishment	1974

Immunophenotypic characterization[1-4]

T-/NK cell marker	CD1a$^-$, CD2$^-$, CD3$^-$, CD4$^-$, CD5$^-$, CD6$^-$, CD7$^-$, CD8$^-$, CD56$^-$, CD57$^-$, TCR$\alpha\beta^-$, TCRδ^-
B-cell marker	CD9$^+$, CD10$^-$, CD19$^-$, CD20$^-$, CD21$^+$, CD22$^+$, CD23$^-$, CD24$^{(+)}$, cyIg$^-$, sIg$^-$
Myelomonocytic marker	CD13$^-$, CD14$^-$, CD15$^+$, CD16$^-$, CD33$^-$, CD68$^-$
Erythroid-megakaryocytic marker	GlyA$^-$
Progenitor/activation/other marker	CD30$^+$, CD34$^-$, CD45$^-$, CD70$^+$, CD71$^+$, HLA-DR$^+$, TdT$^-$
Adhesion marker	CD11a$^-$, CD11b$^-$, CD11c$^-$, CD18$^{(+)}$, CD54$^+$, CD58$^+$
Cytokine receptor	CD25$^+$, CD115$^+$, CD120$^+$, CD126$^+$, CD129$^+$

Genetic characterization[1-4]

Cytogenetic karyotype	hypotriploid, 3.5% polyploidy; 59(56–60)<3n>der(X)der(Y), −X, −1, −1, −1, −2, −2, −3, −4, −5, −5, −6, −8, −9, −9, −10, −10, −10, −11, −12, −13, −14, −15, −15, −16, −16, −18, −19, −20, −21, −22, +25mar, add(1)(q32), add(3)(q26), add(3)(q26), del(4) (q26q31), i(4q), add(5)(p11), add(6)(q14), add(7)(p14), add(7) (q34), del(7)(q31), add(8)(q24), del(8)(p12), del(11)(q23), i(13q), der(13)t(11;13) (p13;q13), add(14)(p12)×2, del(17)(p12), i(18q)
Receptor gene rearrangement	*IGH* R, *IGK* R, *IGL* G, *TCRB* G, *TCRG* G, *TCRD* G

Functional characterization[1,3–5]

Cytochemistry	ACP+, ALP−, ANAE+, CAE−, MPO−
Cytokine production	GM-CSF, IL-5, IL-6, IL-9, IL-13, M-CSF, TNF-α, TNF-β; post-TPA: IL-1α, IL-8; mRNA+: IL-1β, IL-3, LIF, TGF-β
Cytokine response	growth stimulation: IL-9
Inducibility of differentiation	responsive to ECM, retinoic acid, TPA (various effects)
Proto-oncogene	mRNA+: BCL2, BCL6, FOS, JUN
Transcription factor	AP1+, NFAT1+, NFκB+
Special features	production of prostaglandin E2

Comments

- Hodgkin cell line of B-cell type (reference cell line).
- Available from cell line bank.
- Cells grow very well.

References

[1] Kamesaki, H. et al. (1986) Blood 68, 285–292.

[2] Drexler, H.G. et al. (1999) DSMZ Catalogue of Cell Lines, 7th edn, Braunschweig, Germany.

[3] Drexler, H.G. (1993) Leukemia Lymphoma 9, 1–25. (review on Hodgkin cell lines)

[4] Diehl, V. et al. (1990) Semin. Oncol. 17, 660–672. (review on Hodgkin cell lines)

[5] Kapp, U. et al. (1999) J. Exp. Med. 189, 1939–1945.

Culture characterization[1–3]

Subclones	**L 428 KS** (adapted to calf serum), **L 428 KSA** (post-TPA, adherent)
Culture medium	90% RPMI 1640 + 10% FBS
Doubling time	35–46 h
Viral status	EBV⁻, HBV⁻, HCV⁻, HHV-8⁻, HIV⁻, HTLV-I/II⁻
Authentication	no
Primary reference	Schaadt et al.[1]
Availability of cell line	DSM ACC 197

Clinical characterization[1,2,4]

Patient	37-year-old female
Disease diagnosis	Hodgkin's disease (nodular sclerosis)
Treatment status	stage IVB, refractory, terminal
Specimen site	pleural effusion
Year of establishment	1978

Immunophenotypic characterization[3–6]

T-/NK cell marker	CD1a⁻, CD1b⁻, CD2⁻, CD3⁻, CD4⁻, CD5⁻, CD6⁻, CD7⁻, CD8⁻, CD28⁻, CD56⁻, CD57⁻, TCRαβ⁻
B-cell marker	CD9⁻, CD10⁻, CD19⁻, CD20⁻, CD21⁽⁺⁾, CD22⁻, CD23⁺, CD24⁺, CD75⁻, sIg⁻
Myelomonocytic marker	CD13⁻, CD14⁻, CD15⁺, CD16⁻, CD33⁻
Erythroid-megakaryocytic marker	CD41a⁻, CD42b⁻, CD61⁻, GlyA⁻
Progenitor/activation/other marker	CD30⁺, CD34⁻, CD38⁻, CD45⁻, CD70⁺, CD71⁺, HLA-DR⁺, TdT⁻
Adhesion marker	CD11a⁻, CD11b⁻, CD11c⁺, CD18⁻, CD54⁺, CD58⁺
Cytokine receptor	CD25⁽⁺⁾, CD115⁺, CD124⁺

Genetic characterization[3,5,6]

Cytogenetic karyotype	flat-moded hypertetraploid, 10% polyploidy; 94(75–99)<4n> XXXX, +2, −5, +6, −9, −9, +12, −13, +17, +4mar, +2r/dmin, del(1)(q13), der(2)t(2;4)(p14;q21)×2–3, add(5)(p11), add(6) (q24/25)×2, add(7)(q35)×2, del(7)(p14), add(9)(p22), del(11) (q22.3), del(12)(q21)×1–2, add(13)(p11)×2, add(14)(q32)×2, der(16)t(16;?)(q24;?)t(1;?)(q11;?), del(17)(p12), del(21)(q21.2)×2
Receptor gene rearrangement	*IGH* R, *IGK* R?, *IGL* R?, *TCRA G*, *TCRB* G, *TCRG* G, *TCRD* G (mRNA: IGH⁺, TCRA⁺?)

Functional characterization[5–7]

Cytochemistry	ACP+, ALP−, ANAE+, CAE−, MPO−, PAS−
Cytokine production	GM-CSF, IL-4, IL-5, IL-6, IL-13, M-CSF, TGF-β, TNF-α, TNF-β
Cytokine receptor	IL-6R mRNA+
Heterotransplantation	into nude mice
Inducibility of differentiation	responsive to TPA (various effects)
Proto-oncogene	mRNA+: FES, JUN, MET, MYB, MYC, NRAS, P53, PIM, RAF, SYN
Special features	antigen presentation; stimulation of MLR; accessory cell function

Comments

- Hodgkin cell line of B-cell type (reference cell line).
- Available from cell line bank.
- Cells grow very well.

References

[1] Schaadt, M. et al. (1979) Blut 38, 185–190.
[2] Schaadt, M. et al. (1980) Int. J. Cancer 26, 723–731.
[3] Drexler, H.G. et al. (1999) DSMZ Catalogue of Cell Lines, 7th edn, Braunschweig, Germany.
[4] Schaadt, M. et al. (1989) Recent Results Cancer Res. 117, 53–61.
[5] Diehl, V. et al. (1990) Semin. Oncol. 17, 660–672.
[6] Drexler, H.G. (1993) Leukemia Lymphoma 9, 1–25.
[7] Kapp, U. et al. (1999) J. Exp. Med. 189, 1939–1945.

L 540

Culture characterization[1,2]

Sister cell line	**L 538** (simultaneous sister cell line – from peripheral blood – identical features)
Culture medium	90% RPMI 1640 + 10% FBS
Doubling time	70 h
Viral status	EBV⁻, HBV⁻, HCV⁻, HHV-8⁻, HIV⁻, HTLV-I/II⁻
Authentication	no
Primary reference	Diehl et al.[1]
Availability of cell line	DSM ACC 72

Clinical characterization[1]

Patient	20-year-old female
Disease diagnosis	Hodgkin's disease (nodular sclerosis)
Treatment status	stage IVB, prefinal
Specimen site	bone marrow

Immunophenotypic characterization[1–5]

T-/NK cell marker	$CD2^+$, $CD3^-$, $CD4^+$, $CD5^-$, $CD7^-$, $CD8^-$, $TCR\alpha\beta^-$
B-cell marker	$CD10^-$, $CD19^-$, $CD20^-$, $CD21^-$, $CD22^-$, $CD23^+$, sIg^-
Myelomonocytic marker	$CD13^-$, $CD14^-$, $CD15^+$
Progenitor/activation/other marker	$CD30^+$, $CD34^-$, $CD70^+$, $HLA\text{-}DR^+$, TdT^-
Adhesion marker	$CD11b^+$, $CD11c^-$
Cytokine receptor	$CD25^+$, $CD126^+$

Genetic characterization[1–5]

Cytogenetic karyotype	flat-moded hypotriploid, 16% polyploidy; 59–64<3n>XX, −X, −1, +2, −6, +7, −8, −9, −10, −11, −13, −14, −16, −17, −18, +22, +5-6mar, der(1;5)(q10;p10)×1–2, der(2;12)(q10;q10), der(2)t(2;?8)(q3?7;q2?1)×1–2, ins(5;?)(p11;?), add(7)(p1?), add(8)(q24)×1–2, add(9)(p1?), add(11)(q24), der(15)t(1;15)(p21;p13)×2, add(19)(p11), add(21)(p13)
Receptor gene rearrangement	*IGH* G, *IGK* G, *IGL* G, *TCRA* R, *TCRB* R, *TCRG* R, *TCRD* D (TCRA $mRNA^+$)

Functional characterization[1,3–5]

Cytochemistry	ACP^+, ALP^-, $ANAE^-$, CAE^-, MPO^-

Cytokine production IL-3, TNF-α, TNF-β; mRNA+: GM-CSF

Heterotransplantation into nude mice

Proto-oncogene mRNA+: JUN, MET, MYB, MYC, NRAS, P53, PIM, RAF

Comments

- Hodgkin cell line of T-cell type.
- Available from cell line bank.
- Cells grow very well.

References

[1] Diehl, V. et al. (1981) J. Cancer Res. Clin. Oncol. 101, 111–124.

[2] Drexler, H.G. et al. (1999) DSMZ Catalogue of Cell Lines, 7th edn, Braunschweig, Germany.

[3] Diehl, V. et al. (1982) Cancer Treat. Rep. 66, 615–632.

[4] Diehl, V. et al. (1990) Semin. Oncol. 17, 660–672. (review on Hodgkin cell lines)

[5] Drexler, H.G. (1993) Leukemia Lymphoma 9, 1–25. (review on Hodgkin cell lines)

L1236

Culture characterization[1,2]

Culture medium 90% RPMI 1640 + 10% FBS
Viral status EBV⁻
Authentication yes (by *IGH* rearrangement)
Primary reference Wolf et al.[1]
Availability of cell line original authors

Clinical characterization[1]

Patient 34-year-old male
Disease diagnosis Hodgkin's disease (mixed cellularity)
Treatment status stage IV, refractory, terminal (3rd relapse)
Specimen site peripheral blood
Year of establishment 1994

Immunophenotypic characterization[1]

T-/NK cell marker $CD3^-$, $CD4^-$, $CD5^-$, $CD8^-$, $TCR\gamma\delta^-$
B-cell marker $CD10^-$, $CD19^-$, $CD20^-$, $CD23^+$, $CD80^+$, $CD86^+$, sIg^-
Myelomonocytic marker $CD14^-$, $CD15^+$, $CD16^-$, $CD33^-$
Progenitor/activation/other marker $CD30^+$, $CD34^-$, $CD38^-$, $CD45^-$, $CD45RA^-$, $CD45RO^-$, $CD71^+$, $HLA\text{-}DR^+$
Adhesion marker $CD54^+$, $CD58^+$
Cytokine receptor $CD25^-$

Genetic characterization[1]

Cytogenetic karyotype near-triploid, XY, der(1)t(1;14)(p34;?), der(1)t(1;8)(p22;?), der(1) dup(1)(q21q44)add(1)(p31-32), dup(2)(p15p23), der(3)t(3;16) (p25;?), der(4)t(4;8)(q31;?), der(6)dup(6)(p11p25), der(6)t(1;6) (p34;q15), del(7)(q11), der(7)t(7;17)(p22;?), der(7)t(7;7) (p22;q22), der(7), der(8), del(10)(q11), add(11)(p13), del(11) (q13), der(12), del(12)(q22), del(12)(q15), der(14)t(1;14)(p34-35;q22), del(14)(q22-24), der(15)t(15;20)(q22;q11), der(16), del(17)(q23)×2, der(19)t(7;19)(?;p13), add(20)t(15;20)(q22;q11)×2, 2mar
Receptor gene rearrangement *IGH* R, *IGK* R, *IGL* G, *TCRB* G

Functional characterization[1]

Cytokine production GM-CSF, IFN-γ, IL-6, IL-8, IL-10, TGF-β, TNF-α
Heterotransplantation into SCID mice (producing ALCL)

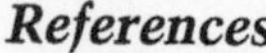

Comments

- Hodgkin cell line of B-cell type.
- Only Hodgkin cell line shown formally to be derived from Hodgkin/Reed–Sternberg cells (verification).
- Cells grow very well.

References

[1] Wolf, J. et al. (1996) Blood 87, 3418–3428.
[2] Kanzler, H. et al. (1996) Blood 87, 3429–3436.

SBH-1

Culture characterization[1]

Culture medium	90% α-MEM + 10% FBS
Doubling time	2–3 d
Viral status	EBV$^-$
Authentication	no
Primary reference	DeCoteau et al.[1]
Availability of cell line	original authors

Clinical characterization[1]

Patient	78-year-old female
Disease diagnosis	Hodgkin's disease
Treatment status	at diagnosis
Specimen site	pleural effusion

Immunophenotypic characterization[1]

T-/NK cell marker	CD2$^-$, CD3$^-$, CD4$^-$, CD5$^-$, CD7$^-$, CD8$^-$
B-cell marker	CD10$^-$, CD19$^+$, CD20$^+$, CD22$^+$, sIgκ$^+$, sIgλ$^+$
Myelomonocytic marker	CD13$^-$, CD14$^-$, CD15$^+$, CD33$^-$
Progenitor/activation/other marker	cyCD30$^+$, CD34$^+$, CD45$^+$, CD71$^+$, EMA$^-$, HLA-DR$^-$
Cytokine receptor	CD25$^+$

Genetic characterization[1]

Cytogenetic karyotype	+7, +10, del(3)(p11p25), del(4)(p12p15), del(4)(q21q28), del(6)(q21), dup(8)(q13q22), dup(9)(p13p22), del(11)(q23), add(12) (p13), t(14;18)(q32;q21), i(15)(q10)
Unique translocation/fusion gene	no *IGH-BCL2* fusion gene
Receptor gene rearrangement	*IGH* R, *IGK* R, *IGL* R, *TCRB* G

Functional characterization[1]

Colony formation	in methylcellulose
Cytochemistry	ACP$^+$, ANAE$^{(+)}$, PAS$^{(+)}$, POX$^-$, SBB$^-$
Cytokine production	RT-PCR$^+$: IL-1β, TGF-β, TNF-α
Cytokine receptor	RT-PCR$^+$: IL-4R, IL-6R, IL-7R, TNF-αRI
Heterotransplantation	into SCID mice

Comments

- Hodgkin cell line of B-cell type.
- Original diagnosis of patient not unequivocally established.
- Cells grow well.

Reference

[1] DeCoteau, J.F. et al. (1995) Blood 85, 2829–2838.

SUP-HD1

Culture characterization[1]

Establishment	initially on feeder layer (complete media, human serum, agar, IGF-I) in hypoxic environment
Culture medium	90% McCoy's 5A + 10% FBS + 10 ng/ml IGF-I (hypoxic conditions: 5% O_2, 6% CO_2, 89% N_2)
Doubling time	3–4 d
Viral status	EBV⁻
Authentication	yes (by HLA Southern blot profiling)
Primary reference	Naumovski et al.[1]
Availability of cell line	original authors

Clinical characterization[1]

Patient	37-year-old male
Disease diagnosis	Hodgkin's disease (nodular sclerosis → lymphocyte depletion)
Treatment status	stage IIISA → IV, refractory, terminal
Specimen site	pleural effusion
Year of establishment	1987

Immunophenotypic characterization[1-3]

T-/NK cell marker	$CD2^-$, $CD3^-$, $CD4^-$, $CD5^-$, $CD7^-$, $CD8^-$, $CD56^-$, $CD57^-$, $TCR\alpha\beta^-$
B-cell marker	$CD19^-$, $CD20^-$, $CD21^-$, $CD22^-$, $CD37^-$, $CD74^+$, $sIg\kappa^+$
Myelomonocytic marker	$CD14^-$, $CD15^+$, $CD16^-$, $CD68^-$
Progenitor/activation/other marker	$CD30^-$, $CD45^-$, $CD71^+$, $HLA\text{-}DR^+$, TdT^-
Adhesion marker	$CD11c^-$
Cytokine receptor	$CD25^+$, $CD120^+$, $CD126^+$

Genetic characterization[1-3]

Cytogenetic karyotype	44, X, −Y, −1, −2, −4, −7, −9, −13, +2mar, dup(1)(p13q32), +der(1)t(1;1;6)(q44→q25::p34→q32::p25), del(2)(p23p25), +der(2)t(2;7)(p25;p15), del(5)(p13p15.3), +der(7), dic(4;7)(q31;p15), del(8)(p21p23), t(8;22)(q22;q13), del(11)(q23q25), t(11;?;11;?)(p15;?;q23;?), t(14;?)(p11.2-q11.2;?), del(21)(q21q22.3)
Receptor gene rearrangement	*IGH* R, *IGK* R, *IGL* G, *TCRB* R (IGK $mRNA^+$)

Functional characterization[1–3]

Colony formation in agar
Cytochemistry ACP+, ANAE(+), MPO−, PAS(+), SBB−
Cytokine production IFN-γ, IL-6, TNF-α
Inducibility of differentiation responsive to TPA
Transcription factor NFκB+

Comments

- Hodgkin cell line of B-cell type.
- Cells are difficult to culture under standard conditions (5–10% CO_2 in air).

References

[1] Naumovski, L. et al. (1989) Blood 74, 2733–2742.
[2] Drexler, H.G. (1993) Leukemia Lymphoma 9, 1–25. (review on Hodgkin cell lines)
[3] Diehl, V. et al. (1990) Semin. Oncol. 17, 660–672. (review on Hodgkin cell lines)

SU/RH-HD-1

Culture characterization[1-5]

Culture medium	85% RPMI 1640 + 15% FBS
Doubling time	70–100 h
Viral status	EBNA⁻
Authentication	no
Primary reference	Olsson et al.[1]
Availability of cell line	not known

Clinical characterization[1]

Patient	12-year-old male
Disease diagnosis	Hodgkin's disease (nodular sclerosis)
Treatment status	stage III, at diagnosis
Specimen site	spleen
Year of establishment	1980

Immunophenotypic characterization[1-5]

T-/NK cell marker	$CD2^-$, $CD4^-$, $CD5^-$, $CD8^-$, $CD57^-$
B-cell marker	sIg^-
Myelomonocytic marker	$CD14^+$, $CD15^-$
Progenitor/activation/other marker	$CD30^-$, $HLA\text{-}DR^+$

Genetic characterization[1-5]

Cytogenetic karyotype	hyperdiploid, 44–47

Functional characterization[1-5]

Colony formation	in methylcellulose, clonable
Cytochemistry	$ANAE^+$, MPO^-
Heterotransplantation	into nude mice (low efficiency)
Special features	phagocytosis$^{(+)}$; antigen presentation

Comments

- Hodgkin cell line with monocyte-macrophage features.
- Further characterization required.

References

[1] Olsson, L. et al. (1984) J. Natl. Cancer Inst. 73, 809–830.
[2] Olsson, L. and Behnke, O. (1985) Cancer Surveys 4, 421–438.
[3] Olsson, L. and Behnke, O. (1988) Hematol. Oncol. 6, 213–222.
[4] Olsson, L. (1985) Int. J. Rad. Oncol. Biol. Phys. 11, 37–48.
[5] Drexler, H.G. (1993) Leukemia Lymphoma 9, 1–25. (review on Hodgkin cell lines)

Section VII

ANAPLASTIC LARGE CELL LYMPHOMA CELL LINES

Culture characterization[1–3]

Establishment	by serial passage in nude mice
Culture medium	90% RPMI 1640 + 10% FBS
Doubling time	30 h
Viral status	EBV⁻, HBV⁻, HCV⁻, HHV-8⁻, HIV⁻, HTLV-I/II⁻
Authentication	no
Primary reference	Barbey et al.[1]
Availability of cell line	DSM ACC 338

Clinical characterization[1,2]

Patient	12-year-old male
Disease diagnosis	malignant histiocytosis
Treatment status	at diagnosis
Specimen site	pleural effusion
Year of establishment	1987

Immunophenotypic characterization[1,3–5]

T-/NK cell marker	CD1a⁻, CD1⁻, CD2⁻, CD3⁻, CD4⁻, CD5⁽⁺⁾, CD7⁻, CD8⁻
B-cell marker	CD10⁻, CD19⁻, CD20⁻, CD21⁻, CD22⁻, CD24⁻, CD37⁻, sIg⁻
Myelomonocytic marker	CD13⁺, CD14⁻, CD15⁺, CD33⁺, CD68⁽⁺⁾
Progenitor/activation/other marker	CD30⁺, CD34⁻, CD38⁻, CD45⁺, CD45RO⁻, CD71⁺, EMA⁺, HLA-DR⁺
Adhesion marker	CD11a⁻, CD11b⁻, CD11c⁻
Cytokine receptor	CD25⁺

Genetic characterization[1,3–8]

Cytogenetic karyotype	biclonal, hyperdiploid/hypertriploid, 6.7% polyploidy; 74(74–77)<3n>XXY, +1, +3, +5, +5, +6, +7, −10, +13, −18, +20, t(5;6)(q35;p21)×2, add(10)(q24)×1–2, der(13) t(1;13)(q32;p11)t(1;13)(q21;q34)×2, add(16)(q23), add(19)(p13)
Unique translocation/fusion gene	t(2;5;6)(p23;q35;p21) → *NPM-ALK* fusion gene
Receptor gene rearrangement	*IGH* R, *IGK* G, *TCRB* G

Functional characterization[1,9–11]

Cytochemistry	ACP⁺, ALP⁻, ANAE⁽⁺⁾, CAE⁻, Lysozyme⁺, MPO⁻, NBT⁺, Oil Red O⁺, PAS⁽⁺⁾
Heterotransplantation	into nude mice
Inducibility of differentiation	TPA → macro differentiation
Proto-oncogene	mRNA⁺: FGR, FMS, KRAS, MYB, MYC, PIM
Special features	α1-antitrypsin⁺, α1-antichymotrypsin⁺

Comments

- Well-characterized ALCL cell line derived from patient with malignant histiocytosis (reference cell line)[12,13].
- ALCL cell line of apparently null-cell type.
- Carries cytogenetically not detectable t(2;5)(p23;q35) leading to *NPM-ALK* fusion gene.
- Available from cell line bank.
- Cells grow well.

References

[1] Barbey, S. et al. (1990) Int. J. Cancer 45, 546–553.
[2] Rousseau-Merck, M.F. et al. (1982) Virchows Arch. Pathol. Anat. 397, 171–181. (case report)
[3] Drexler, H.G. et al. (1999) DSMZ Catalogue of Cell Lines, 7th edn, Braunschweig, Germany.
[4] Dirks, W.G. et al. (1996) Leukemia 10, 142–149.
[5] Mason, D.Y. et al. (1990) Brit. J. Haematol. 74, 161–168.
[6] Soulie, J. et al. (1986) Virchows Arch. Cell Pathol. 50, 339–344.
[7] Bullrich, F. et al. (1994) Cancer Res. 54, 2873–2877.
[8] Gogusev, J. et al. (1990) Int. J. Cancer 46, 106–112.
[9] Gogusev, J. et al. (1991) Cancer Res. 51, 5712–5715.
[10] Gogusev, J. et al. (1996) Anticancer Res. 16, 455–460.
[11] Gogusev, J. et al. (1993) Anticancer Res. 13, 1043–1048.
[12] Nezelof, C. et al. (1992) Semin. Diagn. Pathol. 9, 75–89. (review on ALCL cell lines)
[13] Gogusev, J. and Nezelof, C. (1998) Hematol. Oncol. Clin. North Am. 12, 445–463. (review on ALCL cell lines)

DL-40

Culture characterization[1]

Culture medium	80% RPMI 1640 + 20% FBS
Doubling time	72 h
Viral status	EBNA⁻, HTLV-I⁻
Authentication	no
Primary reference	Kubonishi et al.[1]
Availability of cell line	restricted

Clinical characterization[1]

Patient	64-year-old female
Disease diagnosis	Ki-1 large cell lymphoma
Treatment status	leukemic conversion
Specimen site	peripheral blood
Year of establishment	1989

Immunophenotypic characterization[1,2]

T-/NK cell marker	$CD1^-$, $CD2^+$, $CD3^-$, $CD4^+$, $CD5^-$, $CD7^-$, $CD8^-$, $CD57^-$
B-cell marker	$CD10^-$, $CD19^-$, $CD20^-$, sIg^-
Myelomonocytic marker	$CD13^-$, $CD14^-$, $CD15^-$, $CD16^-$, $CD33^-$
Progenitor/activation/other marker	$CD30^+$, $CD34^-$, $CD38^{(+)}$, $CD45^+$, HLA-DR^+, TdT^-
Cytokine receptor	$CD25^-$, $CD117^-$

Genetic characterization[1]

Cytogenetic karyotype	74(63–78) XX, +X, +1, +2, +3, +4, +5, +5, +7, +7, +9, +14, +15, +15, −17, +19, +20, −21, +7mar, +ins(3)(q26;qm), +del(6)(q15), +der(8)t(8;10)(cen;cen), +i(11q), +der(1)t(1;13)(cen;cen), +der(1)t(1;13)(cen;cen)
Unique translocation/fusion gene	no t(2;5)(p23;q35)
Receptor gene rearrangement	*IGH* G, *TCRB* R

Functional characterization[1]

Cytochemistry	ACP^+, ALP^-, $ANBE^-$, CAE^-, MPO^-, PAS^-, SBB^-
Heterotransplantation	into newborn hamster or nude mice

Comments

- ALCL cell line of T-cell type (not formally confirmed as ALCL cell line).
- Does not carry the characteristic t(2;5)(p23;q35).

References

[1] Kubonishi, I. et al. (1990) Cancer Res. 50, 7682–7685.
[2] Inoue, K. et al. (1997) Blood 89, 1405–1412.

Culture characterization[1]

Culture medium	90% RPMI 1640 + 10% FBS
Doubling time	56 h
Viral status	EBV⁻, HTLV-I/II⁻
Authentication	yes (by *TCRB* rearrangement)
Primary reference	Del Mistro et al.[1]
Availability of cell line	original authors

Clinical characterization[1]

Patient	46-year-old female
Disease diagnosis	initially diagnosed as Hodgkin's disease (nodular sclerosis; stage IIIAE) – later revised to Hodgkin-like ALCL
Treatment status	at relapse in leukemic phase
Specimen site	peripheral blood
Year of establishment	1991

Immunophenotypic characterization[1,2]

T-/NK cell marker	$CD2^+$, $CD3^-$, $CD4^+$, $CD5^-$, $CD7^-$, $CD8^-$
B-cell marker	$CD10^-$, $CD19^-$, $CD20^-$
Myelomonocytic marker	$CD13^-$, $CD14^-$, $CD15^-$, $CD33^-$
Progenitor/activation/other marker	$CD30^+$, $CD34^-$, $CD45RO^+$, $CD71^+$, EMA^+, $HLA\text{-}DR^+$
Adhesion marker	$CD11b^-$
Cytokine receptor	$CD25^+$

Genetic characterization[1,2]

Cytogenetic karyotype	48, XX, −4, −13, +14, −17, +mar, t(1;8)(8qter→8q22.3::1p22.3→1q22::1q32→1q12::1q12→1qter;8pter→8q22.3::1p22.3→1pter), +der(4)t(4;?)(q21.3;?), 9p+, del(9)(pter→q32:), +der(13)t(2;13)(q21;q22), +der(17)t(17;?)(p13;?)
Unique translocation/fusion gene	no t(2;5)(p23;q35) – no *NPM-ALK* fusion gene
Receptor gene rearrangement	*IGH* G, *TCRB* R

Functional characterization[1]

Special features	secretion of sCD30

Comments

- ALCL cell line of T-cell type.
- Does not carry the characteristic t(2;5)(p23;q35) leading to *NPM-ALK* fusion gene.

References

[1] Del Mistro, A. et al. (1994) Leukemia 8, 1214–1219.
[2] Dirks, W.G. et al. (1996) Leukemia 10, 142–149.

JB6

Culture characterization[1]

Other name of cell line	JB
Doubling time	60 h
Culture medium	95% RPMI 1640 or IMDM + 5% FBS
Authentication	no
Primary reference	Kadin et al.[1]
Availability of cell line	original authors

Clinical characterization[1]

Patient	12-year-old male
Disease diagnosis	ALCL
Treatment status	advanced, resistant
Specimen site	peripheral blood

Immunophenotypic characterization[1–3]

T-/NK cell marker	$CD2^+$, $CD3^-$, $CD5^-$, $CD7^+$, $TCR\beta^+$
B-cell marker	$CD10^-$, $CD19^-$, $CD20^-$, $CD70^+$
Myelomonocytic marker	$CD13^+$, $CD14^-$, $CD15^+$, $CD33^-$, $CD68^+$
Progenitor/activation/other marker	$CD30^+$, $CD34^-$, $CD45^+$, $CD71^+$, HLA-DR^+
Adhesion marker	$CD11b^-$
Cytokine receptor	$CD25^+$

Genetic characterization[1–4]

Cytogenetic karyotype	49, XY, +1, +8, +13, der(2)t(2;5)(p23;q35)t(2;21)(q14;q11), der(5)t(2;5)(p23;q35), add(15)(q26), der(21)t(2;21)(q32;q11), i(22) (q10)
Unique translocation/fusion gene	t(2;5)(p23;q35) → *NPM-ALK* fusion gene
Receptor gene rearrangement	*IGH* G, *TCRB* R

Functional characterization[1,3,5]

Colony formation	in methylcellulose
Cytokine production	RT-PCR+: IL-1β, IL-7, TGF-β, TNF-α
Cytokine receptor	RT-PCR+: IL-1R, IL-6R, TNF-αR
Heterotransplantation	into SCID mice

Comments

- ALCL cell line of T-cell type.
- Carries t(2;5)(p23;q35) leading to *NPM-ALK* fusion gene.
- Cells grow very well.

References

1 Kadin, M.E. et al. (1990) Blood 76, 354a.
2 Dirks, W.G. et al. (1996) Leukemia 10, 142–149.
3 Pasqualucci, L. et al. (1995) Blood 85, 2139–2146.
4 Ott, G. et al. (1998) Genes Chromosomes Cancer 22, 114–121.
5 Al Hashmi, I. et al. (1993) Blood 82, 132a.

Karpas 299

Culture characterization[1,2]

Other name of cell line	**K299**
Culture medium	90% RPMI 1640 + 10% FBS
Doubling time	15–30 h
Viral status	EBV^{-}, HBV^{-}, HCV^{-}, HHV-8^{-}, HIV^{-}, HTLV-I/II^{-}
Authentication	yes (by cytogenetics)
Primary reference	Fischer et al.[1]
Availability of cell line	DSM ACC 31

Clinical characterization[1]

Patient	25-year-old male
Disease diagnosis	histiocytic high-grade lymphoma
Treatment status	refractory, terminal stage
Specimen site	peripheral blood
Year of establishment	1986

Immunophenotypic characterization[1–6]

T-/NK cell marker	CD2^{-}, CD3^{-}, CD4^{+}, CD5^{+}, CD6^{-}, CD7^{-}, CD8^{-}
B-cell marker	CD10^{-}, CD19^{-}, CD20^{+}, CD37^{-}, sIg^{-}
Myelomonocytic marker	CD13^{+}, CD14^{-}, CD15^{-}, CD33^{-}, CD68^{-}
Progenitor/activation/other marker	CD30^{+}, CD34^{-}, CD45^{+}, CD71^{+}, EMA^{+}, HLA-DR^{+}
Adhesion marker	CD11b^{-}, CD11c^{+}
Cytokine receptor	CD25^{+}

Genetic characterization[1–3,7,8]

Cytogenetic karyotype	hypodiploid, 14% polyploidy; 44(42–45)<2n>XY, −10, −22, t(1;17)(q22;p11), t(2;5)(p23;q35), del(6)(q23), der(13)t (13;?) (p12;?), der(14)t(14;22)(p12;q11), der(19)t(19;?)(q13;?), del(22) (q12)
Unique translocation/fusion gene	t(2;5)(p23;q35) → *NPM-ALK* fusion gene
Receptor gene rearrangement	*TCRB* R (TCRB mRNA^{+})

Functional characterization[1]

Cytochemistry	ACP^{+}, ANAE^{+}, ANBE^{+}, CAE^{-}, MPO^{-}, PAS^{+}, SBB^{-}
Special features	α1-antitrypsin^{+}

Comments

- ALCL cell line of T-cell type (reference cell line).
- Carries t(2;5)(p23;q35) leading to *NPM-ALK* fusion gene.
- Cells grow well.
- Available from cell line bank.

References

[1] Fischer, P. et al. (1988) Blood 72, 234–240.

[2] Drexler, H.G. et al. (1999) DSMZ Catalogue of Cell Lines, 7th edn, Braunschweig, Germany.

[3] Dirks, W.G. et al. (1996) Leukemia 10, 142–149.

[4] Mason, D.Y. et al. (1990) Brit. J. Haematol. 74, 161–168.

[5] Nezelof, C. et al. (1992) Semin. Diagn. Pathol. 9, 75–89. (review on ALCL cell lines)

[6] Gogusev, J. and Nezelof, C. (1998) Hematol. Oncol. Clin. North Am. 12, 445–463. (review on ALCL cell lines)

[7] Pulford, K. et al. (1997) Blood 89, 1394–1404.

[8] Pébusque, M.J. et al. (1993) Genes Chromosomes Cancer 8, 119–126.

Ki-JK

Culture characterization[1]

Culture medium	90% RPMI 1640 + 10% FBS
Doubling time	72 h
Viral status	EBV+
Authentication	yes (by cytogenetics)
Primary reference	Shimakage et al.[1]
Availability of cell line	original authors

Clinical characterization[1]

Patient	15-year-old male
Disease diagnosis	Ki-1 lymphoma
Treatment status	at diagnosis
Specimen site	pleural effusion
Year of establishment	1989

Immunophenotypic characterization[1,2]

T-/NK cell marker	CD2−, CD3−, CD4−, CD5+, CD7+, CD8−
B-cell marker	CD10−, CD19−, CD20−, CD21−, CD22−, CD23−, sIg−
Myelomonocytic marker	CD13+, CD14−, CD15+, CD33+
Progenitor/activation/other marker	CD30+, CD34−, CD45+, CD45RO+, CD71+, EMA+, HLA-DR+
Adhesion marker	CD11b+
Cytokine receptor	CD25+

Genetic characterization[1,2]

Cytogenetic karyotype	pseudotetraploid, 82
Unique translocation/fusion gene	*NPM-ALK* fusion gene
Receptor gene rearrangement	*IGH* G, *TCRB* G

Functional characterization[1]

Cytochemistry	ANBE+, Lysozyme+
Heterotransplantation	into nude mice
Special features	EBNA2+; LMP mRNA+

Comments

- ALCL cell line of null-cell type.
- Carries *NPM-ALK* fusion gene.

References

[1] Shimakage, M. et al. (1993) Intervirology 36, 215–224.
[2] Dirks, W.G. et al. (1996) Leukemia 10, 142–149.

Mac-2A

Culture characterization[1–3]

Other names of cell line	**McG-2A, PB-2A**
Sister cell lines	(1) **Mac-1** (serial sister cell line – at indolent stage in 1985 – different immunological, functional features) (2) **Mac-2B** (simultaneous sister cell line – similar features) (3) **Mac-LCL** (= EBV$^+$ B-LCL)
Culture medium	85% RPMI 1640 + 15% FBS
Viral status	HTLV-I$^-$
Authentication	yes (by *TCR* rearrangement, cytogenetics)
Primary reference	Davis et al.[1]
Availability of cell line	original authors

Clinical characterization[1]

Patient	47-year-old male
Disease diagnosis	lymphomatoid papulosis (1971) → Hodgkin's disease (mixed cellularity, stage IIA, 1975) → cutaneous T-cell lymphoma (clinically indolent, 1985) → ALCL (clinically aggressive, 1987)
Treatment status	terminal stage
Specimen site	skin tumor nodules
Year of establishment	1987

Immunophenotypic characterization[1–5]

T-/NK cell marker	CD2$^-$, CD3$^-$, CD4$^-$, CD8$^-$
Myelomonocytic marker	CD15$^+$
Progenitor/activation/other marker	CD30$^+$, CD45$^{(+)}$, HLA-DR$^+$
Cytokine receptor	CD25$^+$, CD122$^+$, CD124$^-$, TGF-βRI/II$^-$

Genetic characterization[1]

Cytogenetic karyotype	t(8;9)(p22;p24)
Unique translocation/fusion gene	no t(2;5)(p23;q35) – no *NPM-ALK* fusion gene
Receptor gene rearrangement	*TCRA* R, *TCRB* G (TCRA mRNA$^+$)

Functional characterization[3,5–7]

Colony formation	in methylcellulose
Cytokine production	mRNA$^+$: IL-2, IL-4, PDGF, TGF-β; protein$^+$: GM-CSF, IL-9
Cytokine receptor	mRNA$^+$: IL-4Rα, common γ
Special features	10–20% multinucleated giant cells (Reed–Sternberg cell-like)

Comments

- ALCL cell line of T-cell type.
- Clinically well-characterized case.
- Serial and simultaneous sister cell lines from various disease stages available.

References

[1] Davis, T.H. et al. (1992) New Engl. J. Med. 326, 1115–1122. (case report)
[2] Su, I.J. and Kadin, M.E. (1989) Am. J. Pathol. 135, 439–445.
[3] Wasik, M.A. et al. (1996) Leukemia Lymphoma 23, 125–136.
[4] Kadin, M.E. et al. (1994) Proc. Natl. Acad. Sci. USA 91, 6002–6006.
[5] Su, I.J. et al. (1988) Am. J. Pathol. 132, 192–198.
[6] Kadin, M.E. et al. (1990) Blood 76, 354a.
[7] Marti, R.M. et al. (1996) Cytokine 8, 323–329.

Culture characterization[1]

Culture medium	80% RPMI 1640 + 20% FBS
Authentication	no
Primary reference	Rimokh et al.[1]
Availability of cell line	restricted

Clinical characterization[1]

Patient	74-year-old male
Disease diagnosis	poorly differentiated lymphoma
Treatment status	at diagnosis
Specimen site	pleural effusion

Immunophenotypic characterization[1-3]

T-/NK cell marker	CD1⁻, CD2⁻, CD3⁻, CD4⁻, CD5⁻, CD8⁻
B-cell marker	CD10⁻, CD19⁻, CD20⁻, CD21⁻, CD23⁻, CD24⁻, CD37⁺, sIg⁻
Myelomonocytic marker	CD13⁻, CD14⁻, CD15⁻, CD33⁻, CD35⁻
Progenitor/activation/other marker	CD30⁺, CD38⁺, CD45⁺, CD71⁺, EMA⁽⁺⁾, HLA-DR⁺
Adhesion marker	CD11b⁻

Genetic characterization[1,4]

Cytogenetic karyotype	50, XY, −4, −6, +7, +8, −15, +19, t(2;8)(q22;q22), t(3;5) (q12;q35), t(3;17)(p11;q12), +der(4)t(4;?)(q33;?), del(6)(q23-q25), +der(6)t(6;?)(p25;?), ins(7;16)(7pter→q11::16pter→q11::7q11→7qter::16q11→qter), del(12)(q24), dir dup(12)(p11p13), +der(12)(q24?dup12q), +der(15)t(15;15)(p11;q22)
Receptor gene rearrangement	*IGH* DR, *IGK* DD, *IGL* DR, *TCRB* G (mRNA: IGH⁺, IGL⁺)

Functional characterization[1]

Cytochemistry	Cytokeratin⁻

Comments

- ALCL cell line of B-cell type.
- Carries t(3;5)(q12;q35) with 5q3b breakpoint typical for ALCL.

References

[1] Rimokh, R. et al. (1989) Brit. J. Haematol. 71, 31–36.
[2] Mason, D.Y. et al. (1990) Brit. J. Haematol. 74, 161–168.
[3] Nezelof, C. et al. (1992) Semin. Diagn. Pathol. 9, 75–89. (review on ALCL cell lines)
[4] Pébusque, M.J. et al. (1993) Genes Chromosomes Cancer 8, 119–126.

SR-786

Culture characterization[1–4]

Other name of cell line	**SR**
Culture medium	85% RPMI 1640 + 15% FBS
Doubling time	30 h
Viral status	EBV⁻, HBV⁻, HCV⁻, HHV-8⁻, HIV⁻, HTLV-I/II⁻
Authentication	no
Primary reference	Su et al.[1]
Availability of cell line	ATCC CRL 2262, DSM ACC 369

Clinical characterization[1,3]

Patient	11-year-old male
Disease diagnosis	Ki-1 lymphoma (large T-cell lymphoma)
Specimen site	pleural effusion
Year of establishment	1983

Immunophenotypic characterization[1–6]

T-/NK cell marker	$CD2^-$, $CD3^-$, $CD4^+$, $CD5^-$, $CD7^-$, $CD8^-$, $TCR\alpha\beta^-$, $TCR\gamma\delta^-$
B-cell marker	$CD10^-$, $CD19^-$, $CD20^+$, $CD21^-$, $CD22^-$, sIg^-
Myelomonocytic marker	$CD13^+$, $CD14^-$, $CD15^+$, $CD33^-$, $CD68^+$
Progenitor/activation/other marker	$CD30^+$, $CD34^-$, $CD45^+$, $CD71^+$, HLA-DR^+
Adhesion marker	$CD11b^-$
Cytokine receptor	$CD25^+$

Genetic characterization[1,3–5]

Cytogenetic karyotype	hypertriploid, flat-moded; 70–84<3n>XX, +1, +2, −4, +5, +6, +7, +8, −13, −13, +14, −18, +19, +22, +6-9mar, add(1)(q11), del(1)(p11)/der(?)t(1;?)(q11;?), der(2)t(2;5)(p23;q35)inv(2) (p23q14)×2, del(4)(q22), der(5)t(2;5)(p23;q35)×2, del(7)(q21), der(9)t(1;9)(q11;p24)×2, der(12)t(12;13)(q24.32;q11)×1–2, del(13) (q13q31), add(14)(p11)/der(?)t(14;?)(q11;?)×2, del(21)(q22); carries masked t(2;5)
Unique translocation/fusion gene	t(2;5)(p23;q35) → *NPM-ALK* fusion gene
Receptor gene rearrangement	*IGH* G, *TCRB* R (TCRA $mRNA^+$)

Functional characterization[1,2,6]

Cytokine production	$mRNA^+$: IL-2, TGF-β1
Proto-oncogene	$mRNA^+$: HRAS, MYC
Special features	retinoic acid → apoptosis; 5% multinucleated giant cells

Comments

- ALCL cell line of T-cell type (or null-cell type).
- Carries t(2;5)(p23;q35) leading to *NPM-ALK* fusion gene.
- Available from cell line banks.

References

[1] Su, I.J. et al. (1988) Am. J. Pathol. 132, 192–198.
[2] Su, I.J. and Kadin, M.E. (1989) Am. J. Pathol. 135, 439–445.
[3] Beckwith, M. et al. (1990) J. Natl. Cancer Inst. 82, 501–509.
[4] Drexler, H.G. et al. (1999) DSMZ Catalogue of Cell Lines, 7th edn, Braunschweig, Germany.
[5] Dirks, W.G. et al. (1996) Leukemia 10, 142–149.
[6] Su, I.J. et al. (1994) Int. J. Oncol. 4, 1089–1095.

SU-DHL-1

Culture characterization[1,2]

Culture medium 80% RPMI 1640 + 20% FBS
Doubling time 40–50 h
Viral status EBV⁻, HBV⁻, HCV⁻, HHV-8⁻, HIV⁻, HTLV-I/II⁻
Authentication no
Primary reference Epstein et al.[1]
Availability of cell line DSM ACC 356

Clinical characterization[1,3,4]

Patient 10-year-old male
Disease diagnosis initially diffuse histiocytic lymphoma, later changed to diffuse large cell lymphoma (non-cleaved type) and malignant histiocytosis
Treatment status at diagnosis
Specimen site pleural effusion
Year of establishment 1973

Immunophenotypic characterization[1–10]

T-/NK cell marker $CD1^-$, $CD2^-$, $CD3^-$, $CD4^-$, $CD5^+$, $CD7^-$, $CD8^-$
B-cell marker $CD9^-$, $CD10^-$, $CD19^-$, $CD20^-$, $CD21^-$, $CD22^-$, $CD24^+$, $CD37^-$, $CD75^-$, $cyIg^-$, sIg^-
Myelomonocytic marker $CD13^-$, $CD14^-$, $CD15^-$, $CD33^-$, $CD68^+$
Erythroid-megakaryocytic marker $CD41a^-$, $CD61^-$
Progenitor/activation/other marker $CD30^+$, $CD34^-$, $CD71^+$, EMA^-, $HLA\text{-}DR^-$, TdT^-
Adhesion marker $CD11b^-$, $CD11c^-$
Cytokine receptor $CD25^+$

Genetic characterization[1,2,4,10–14]

Cytogenetic karyotype flat-moded hypertriploid, 16% polyploidy; 74(67–75)<3n>XX, −Y, +1, +2, +3, +5, −7, +12, −16, −18, +19, −20, +21, +3mar, del(1)(p21), t(2;5)(p23;q35)×2, del(6)(q23)×1–2, add(8) (p12), add(9)(p21), del(10)(p14), add(12)(q24), add(14)(p12), add(16)(q24), dup(19)(q13.1qter); sideline with der(12)t(12;19) (q13;q13)
Unique translocation/fusion gene t(2;5)(p23;q35) → *NPM-ALK* fusion gene
Receptor gene rearrangement *IGH* G, *TCRB* RG

Functional characterization[1,3,5,15,16]

Cytochemistry	ACP⁺, ALP⁻, ANAE⁺, ANBE⁺, CAE⁻, GLC⁺, Lysozyme⁻, MGP⁺, MPO⁻, Oil Red⁺, PAS⁺, SBB⁻, TRAP⁻
Cytokine production	RT-PCR⁺: IL-1β, IL-7, TNF-α
Cytokine receptor	RT-PCR⁺: IL-1R, IL-6R, TNF-αR
Heterotransplantation	into nude mice
Proto-oncogene	FMS mRNA⁺
Special features	phagocytosis

Comments

- First ALCL cell line of null-cell type (reference cell line).
- Carries t(2;5)(p23;q35) leading to *NPM-ALK* fusion gene.
- Available from cell line bank.
- Cells grow well.

References

[1] Epstein, A.L. et al. (1974) Cancer 34, 1851–1872.
[2] Drexler, H.G. et al. (1999) DSMZ Catalogue of Cell Lines, 7th edn, Braunschweig, Germany.
[3] Winter, J.N. et al. (1984) Blood 63, 140–146.
[4] Morgan, R. et al. (1989) Blood 73, 2155–2164.
[5] Epstein, A.L. et al. (1978) Cancer 42, 2379–2391.
[6] Mason, D.Y. et al. (1990) Brit. J. Haematol. 74, 161–168.
[7] Nezelof, C. et al. (1992) Semin. Diagn. Pathol. 9, 75–89. (review on ALCL cell lines)
[8] Gogusev, J. and Nezelof, C. (1998) Hematol. Oncol. Clin. North Am. 12, 445–463. (review on ALCL cell lines)
[9] Minowada, J. and Matsuo, Y. (1999) unpublished data.
[10] Dirks, W.G. et al. (1996) Leukemia 10, 142–149.
[11] Kaiser-McCaw Hecht, B. et al. (1985) Cancer Genet. Cytogenet. 14, 205–218.
[12] Pulford, K. et al. (1997) Blood 89, 1394–1404.
[13] Morris, S.W. et al. (1994) Science 263, 1281–1284.
[14] Pébusque, M.J. et al. (1993) Genes Chromosomes Cancer 8, 119–126.
[15] Al Hashmi, I. et al. (1993) Blood 82, 132a.
[16] Epstein, A.L. et al. (1976) Cancer 37, 2158–2176.

SUP-M2

Culture characterization[1]

Culture medium	90% RPMI 1640 + 10% FBS
Authentication	yes (by *TCR* rearrangement, cytogenetics)
Primary reference	Morgan et al.[1]
Availability of cell line	first author [5]

Clinical characterization[1]

Patient	5-year-old female
Disease diagnosis	malignant histiocytosis
Treatment status	refractory
Specimen site	cerebrospinal fluid

Immunophenotypic characterization[1–4]

T-/NK cell marker	$CD1^-$, $CD2^+$, $CD3^-$, $CD4^+$, $CD5^-$, $CD7^-$, $CD8^-$
B-cell marker	$CD19^-$, $CD20^-$, $CD21^-$, $CD22^-$, $CD37^-$
Myelomonocytic marker	$CD14^-$, $CD15^-$, $CD68^+$
Progenitor/activation/other marker	$CD30^+$, $CD45RO^+$, $CD71^+$, EMA^+, HLA-DR^+
Adhesion marker	$CD11c^-$
Cytokine receptor	$CD25^+$

Genetic characterization[1,5]

Cytogenetic karyotype	47, XX, +X, −9, +der(1)t(1;?)(q44;?), del(1)(p34), t(2;5) (p23;q35)
Unique translocation/fusion gene	t(2;5)(p23;q35) → *NPM-ALK* fusion gene
Receptor gene rearrangement	*IGH* G, *TCRB* RR (TCRB $mRNA^+$)

Functional characterization[1,6]

Cytokine production	RT-PCR^+: IL-1β, IL-7, IL-9, SCF, TNF-α
Cytokine receptor	RT-PCR^+: IL-1R, IL-6R, TNF-αR
Proto-oncogene	FMS $mRNA^+$

Comments

- ALCL cell line of T-cell type.
- Carries t(2;5)(p23;q35) leading to *NPM-ALK* fusion gene.
- Cells grow well.

References

[1] Morgan, R. et al. (1989) Blood 73, 2155–2164.
[2] Mason, D.Y. et al. (1990) Brit. J. Haematol. 74, 161–168.
[3] Nezelof, C. et al. (1992) Semin. Diagn. Pathol. 9, 75–89. (review on ALCL cell lines)
[4] Gogusev, J. and Nezelof, C. (1998) Hematol. Oncol. Clin. North Am. 12, 445–463. (review on ALCL cell lines)
[5] Morris, S.W. et al. (1994) Science 263, 1281–1284.
[6] Al Hashmi, I. et al. (1993) Blood 82, 132a.

USP-91

Culture characterization[1]

Establishment	on feeder murine stromal cell line
Culture medium	90% RPMI 1640 + 10% FBS
Doubling time	26 h
Authentication	yes (by *TCR* rearrangement)
Primary reference	Umiel et al.[1]
Availability of cell line	original authors

Clinical characterization[1]

Patient	14-year-old male
Disease diagnosis	NHL (large, histiocytic)
Treatment status	refractory, terminal
Specimen site	pleural effusion

Immunophenotypic characterization[1]

T-/NK cell marker	$CD2^-$, $CD3^+$, $CD4^+$, $CD5^-$, $CD7^+$, $CD8^-$, $CD56^+$, $TCR\alpha\beta^+$, $TCR\gamma\delta^-$
B-cell marker	$CD10^+$, $CD19^-$
Myelomonocytic marker	$CD13^+$, $CD14^{(+)}$, $CD16^+$, $CD33^+$
Erythroid-megakaryocytic marker	$CD41^+$, $CD61^+$, PPO^-
Progenitor/activation/other marker	$CD30^-$, $CD34^+$, EMA^-, HLA-DR^+
Adhesion marker	$CD11b^+$
Cytokine receptor	$CD25^+$, $CD115^+$, $CD122^+$

Genetic characterization[1]

Cytogenetic karyotype	45–47, X, −Y, −1, −2, −5, −7, −8, −10, −16, −16, −17, −21, +11mar, +der(2)t(2p?;5q), +der(5)t(5q;2p?), +der(16)t(16q;?), +der(16)t(16q;?), +22q+
Unique translocation/fusion gene	t(2;5)(p23;q35) not shown, but likely present
Receptor gene rearrangement	*IGH* G, *IGK* G, *TCRB* R

Functional characterization[1]

Cytochemistry	$ANAE^+$, $Lysozyme^+$, MPO^-, PAS^+, SBB^-
Inducibility of differentiation	TPA → macro differentiation (phagocytosis)
Special features	α1-$antitrypsin^+$, α1-$antichymotrypsin^+$

Comments

- (CD30-negative, but likely) ALCL cell line of T-cell type.
- t(2;5)(p23;q35) and *NPM-ALK* fusion gene not formally shown, but probably present.

Reference

[1] Umiel, T. et al. (1993) Blood 82, 1829–1837.

Section VIII

MYELOID LEUKEMIA CELL LINES

Part 1
Myelocytic Cell Lines

AML14

Culture characterization[1]

Subclone	**AML14.3D10** (spontaneous differentiation to eosinophils)
Culture medium	90% RPMI 1640 + 10% FBS
Doubling time	24–36 h
Authentication	no
Primary reference	Paul et al.[1]
Availability of cell line	restricted

Clinical characterization[1]

Patient	68-year-old male
Disease diagnosis	AML M2
Treatment status	refractory
Specimen site	peripheral blood
Year of establishment	1991

Immunophenotypic characterization[1]

T-/NK cell marker	$CD2^-$
B-cell marker	$CD20^-$
Myelomonocytic marker	$CD13^+$, $CD14^{(+)}$, $CD33^+$
Progenitor/activation/other marker	$CD34^-$
Adhesion marker	$CD11b^+$

Genetic characterization[1,3]

Cytogenetic karyotype	48–50, X, −Y, +13, −16, +18, −20, +3-5mar, +add(1)(p31), del(5)(q23q34), dup(5)(q13q31), add(8)(q24.3), del(9)(q22), add(14)(p13), add(17)(p13), add(22)(q13)

Functional characterization[1–4]

Cytokine receptor	$mRNA^+$: GM-CSFRα, IL-3Rα, IL-5Rα, commonRβ
Cytokine response	growth stimulation: GM-CSF, IL-3, IL-5
Inducibility of differentiation	GM-CSF+IL-3+IL-5 → eosino differentiation; retinoic acid → neutro differentiation; TPA, Vit. D3 → mono/macro differentiation
Special features	spontaneous eosino differentiation; spontaneous/induced expression of specific eosino granula proteins: Charcot-Leyden crystal protein, eosino-cationic protein, eosino-derived neurotoxin, eosino-lysophospholipase, eosino-peroxidase, major basic protein

Comments

- Cell line with myelocytic characteristics.
- Spontaneous and induced differentiation to eosinophils.
- Weak constitutive expression of eosinophil features which can be up-regulated by differentiation-inducing reagents.

References

[1] Paul, C.C. et al. (1993) Blood 81, 1193–1199.
[2] Paul, C.C. et al. (1995) Blood 86, 3737–3744.
[3] Baumann, M.A. and Paul, C.C. (1998) Stem Cells 16, 16–24. (review on AML14)
[4] Paul, C.C. et al. (1994) J. Leuk. Biol. 56, 74–79.

AR230

Culture characterization[1]

Culture medium	90% RPMI 1640 + 10% FBS
Authentication	yes (by *BCR* rearrangement)
Primary reference	Wada et al.[1]
Availability of cell line	original authors

Clinical characterization[1]

Patient	52-year-old female
Disease diagnosis	CML
Treatment status	at myeloid blast crisis
Specimen site	peripheral blood
Year of establishment	1991

Immunophenotypic characterization[1]

T-/NK cell marker	$CD2^-$, $CD3^-$, $CD4^+$, $CD5^-$, $CD7^-$, $CD8^-$
B-cell marker	$CD10^-$, $CD19^-$, $CD20^-$, $CD22^-$, $CD24^-$
Myelomonocytic marker	$CD13^+$, $CD33^+$, $CD35^+$
Progenitor/activation/other marker	$CD34^-$, $CD45^+$, HLA-DR^-
Adhesion marker	$CD11b^{(+)}$

Genetic characterization[1,2]

Cytogenetic karyotype	44, XX, −11, −14, −17, −17, +mar, inv(1)(p31-32p36) or add(1) (p31-32), −2 or add(2)(p13), del(9)(q22) or der(9)t(9;?)(q11-13;?), der(9)t(9;?)(p13;?)t(9;22)(q34;q11), add(15)(p11), add(18)(q21), der(11;21)(q10;p10), der(22)t(9;22)(q34;q11), +der(22)t(9;22)(q34;q11)
Unique translocation/fusion gene	Ph^+ t(9;22)(q34;q11) → *BCR-ABL* c3/e19-a2 fusion gene

Functional characterization[1]

Special features	p230 BCR-ABL fusion protein

Comments

- Cell line with myelocytic characteristics.
- Carries Ph chromosome leading to *BCR-ABL* e19-a2 fusion gene.
- Only cell line with the Ph chromosome involving the μ-bcr leading to the p230 BCR-ABL fusion protein.

References

[1] Wada, H. et al. (1995) Cancer Res. 55, 3192–3196.

[2] Drexler, H.G. et al. (1999) Leukemia Res. 23, 207–215.

CML-C-1

Culture characterization[1]

Subclone	**CML-C-1** (*in vitro* variant of **CML-N-1**, established and serially transplanted in nude mice)
Culture medium	90% RPMI 1640 + 10% FBS on mouse bone marrow stromal feeder layer
Authentication	yes (by cytogenetics)
Primary reference	Kubonishi et al.[1]
Availability of cell line	restricted

Clinical characterization[1]

Patient	42-year-old female
Disease diagnosis	CML
Treatment status	at blast crisis
Specimen site	peripheral blood
Year of establishment	1990

Immunophenotypic characterization[1]

T-/NK cell marker	CD3⁻
B-cell marker	CD19⁻
Myelomonocytic marker	CD13⁺, CD33⁻
Erythroid-megakaryocytic marker	CD41a⁻
Progenitor/activation/other marker	CD34⁺, CD38⁻, HLA-DR⁺
Adhesion marker	CD11b⁻

Genetic characterization[1]

Cytogenetic karyotype	49, XX, +8, +8, +21, del(3)(q21q23), t(9;22)(q34;q11)
Unique translocation/fusion gene	Ph⁺ t(9;22)(q34;q11) → *BCR-ABL* fusion gene

Functional characterization[1]

Cytochemistry	ACP⁺, Alcian Blue⁻, ANBE⁻, CAE⁻, MPO⁻, PAS⁺, SBB⁻, Toluidine Blue⁻
Heterotransplantation	into nude mice

Comments

- Cell line with myelocytic features.
- Carries Ph chromosome leading to *BCR-ABL* fusion gene.

Reference

[1] Kubonishi, I. et al. (1995) Jpn. J. Cancer Res. 86, 451–459.

Culture characterization[1]

Culture medium	90% RPMI 1640 + 10% FBS
Doubling time	72 h
Authentication	no
Primary reference	Kakuda et al.[1]
Availability of cell line	original authors

Clinical characterization[1]

Patient	13-year-old female
Disease diagnosis	AML M1
Treatment status	at relapse (post-BMT)
Specimen site	peripheral blood
Year of establishment	1992

Immunophenotypic characterization[1]

T-/NK cell marker	$CD1^-$, $CD2^-$, $CD3^-$, $CD4^-$, $CD5^-$, $CD7^+$, $CD8^-$
B-cell marker	$CD10^-$, $CD19^-$, $CD20^-$
Myelomonocytic marker	$CD13^+$, $CD14^-$, $CD33^+$
Erythroid-megakaryocytic marker	$CD41a^-$, $GlyA^-$
Progenitor/activation/other marker	$CD34^+$, HLA-DR^+
Cytokine receptor	$CD25^-$

Genetic characterization[1]

Cytogenetic karyotype	46, XX, −17, −22, +2mar, t(6;11)(q27;q23)
Unique translocation/fusion gene	t(6;11)(q27;q23) → *MLL-AF6* fusion gene
Receptor gene rearrangement	*IGH* R, *IGK* R, *IGL* G, *TCRB* G, *TCRG* G, *TCRD* D

Functional characterization[1]

Cytochemistry	ACP^-, ALP^-, $ANAE^-$, $ANBE^-$, CAE^-, MPO^-, PAS^-; electron microscopy: MPO^+
Inducibility of differentiation	resistant to differentiation induction with DMSO, hemin, retinoic acid, TPA

Comments

- Cell line with myelocytic characteristics.
- Carries t(6;11)(q27;q11) leading to *MLL-AF6* fusion gene.

Reference

[1] Kakuda, H. et al. (1996) Brit. J. Haematol. 95, 306–318.

EM-2

Culture characterization[1–3]

Sister cell line — **EM-3** (serial sister cell line – similar features)

Culture medium — 90% RPMI 1640 + 10% FBS

Doubling time — 24–48 h

Viral status — EBV$^-$, HBV$^-$, HCV$^-$, HHV-8$^-$, HIV$^-$, HTLV-I/II$^-$

Authentication — yes (by cytogenetics)

Primary reference — Keating et al.[1]

Availability of cell line — EM-2: DSM ACC 135; EM-3: DSM ACC 134

Clinical characterization[1,2]

Patient — 5-year-old female

Disease diagnosis — CML

Treatment status — at myeloid blast crisis (2nd relapse post-BMT)

Specimen site — bone marrow

Year of establishment — 1982

Immunophenotypic characterization[1–4]

T-/NK cell marker — CD3$^-$, CD4$^+$

B-cell marker — CD10$^-$, CD19$^-$

Myelomonocytic marker — CD13$^+$, CD14$^-$, CD15$^+$, CD33$^+$, CD68$^{(+)}$

Erythroid-megakaryocytic marker — CD41$^-$, CD42b$^-$, vWF$^-$

Progenitor/activation/other marker — CD34$^-$, HLA-DR$^-$, TdT$^-$

Cytokine receptor — CD117$^-$

Genetic characterization[1–3,5–8]

Cytogenetic karyotype — hypertriploid, hypotetraploid sideline; 74(70–86)<3n>X, −X, −X, +3, +4, +6, +6, +6, +8, −9, +11, −14, −14, +15, +17, −19, +21, +22, +mar, der(5)t(5;?)(q13-15;?), der(9)t(9;22)(q34;q11), i(17q)×2

Unique translocation/fusion gene — Ph$^+$ t(9;22)(q34;q11) → *BCR-ABL* b3-a2 fusion gene

Receptor gene rearrangement — *IGH* G

Unique gene alteration — *P53* mutation

Functional characterization[1,2,9]

Colony formation — yes

Heterotransplantation — into SCID mice

Inducibility of differentiation — DMSO, retinoic acid → neutro differentiation; TPA → mono/macro differentiation

Comments

- Prototypical myelocytic cell line (reference cell line).
- Carries Ph chromosome leading to *BCR-ABL* b3-a2 fusion gene.
- Responsive to various types of induction of differentiation.
- Available from cell line bank.
- Cells grow very well.

References

[1] Keating, A. et al. (1983) in Normal and Neoplastic Hematopoiesis (Golde, D.W. and Marks, P.A. eds.) Alan R. Liss, New York, pp. 513-520.
[2] Keating, A. (1987) Baillière's Clin. Haematol. 1, 1021–1029. (review)
[3] Drexler, H.G. et al. (1999) DSMZ Catalogue of Cell Lines, 7th edn, Braunschweig, Germany.
[4] Papayannopoulou, T. et al. (1987) J. Clin. Invest. 79, 859–866.
[5] Raskind, W.A. et al. (1987) Cancer Genet. Cytogenet. 25, 271–284.
[6] Feinstein, E. et al. (1991) Proc. Natl. Acad. Sci. USA 88, 6293–6297.
[7] Bi, S. et al. (1993) Leukemia 7, 1840–1845.
[8] Drexler, H.G. et al. (1999) Leukemia Res. 23, 207–215.
[9] Uckun, F.M. (1996) Blood 88, 1135–1146.

EoL-1

Culture characterization[1,2]

Sister cell line	**Eo-B** (= EBV^+ B-LCL)
Subclones	**EoL-2, EoL-3** (similar immunophenotypic and functional features)
Culture medium	90% RPMI 1640 + 10% FBS
Doubling time	48–60 h
Viral status	EBV^-, HBV^-, HCV^-, $HHV\text{-}8^-$, HIV^-, $HTLV\text{-}I/II^-$
Authentication	yes (by cytogenetics)
Primary reference	Saito et al.[1]
Availability of cell line	DSM ACC 386, RCB 0641

Clinical characterization[1]

Patient	33-year-old male
Disease diagnosis	hypereosinophilic syndrome → eosinophilic leukemia
Treatment status	at diagnosis
Specimen site	peripheral blood
Year of establishment	1984

Immunophenotypic characterization[1–4]

T-/NK cell marker	$CD2^-$, $CD3^-$, $CD4^+$, $CD5^-$, $CD7^-$, $CD8^-$, $CD28^-$, $CD57^-$
B-cell marker	$CD9^-$, $CD10^-$, $CD19^-$, $CD20^-$, $CD21^-$, $CD23^-$, $cyIg^-$, sIg^-
Myelomonocytic marker	$CD13^+$, $CD14^-$, $CD15^+$, $CD16^-$, $CD32^+$, $CD33^+$, $CD64^-$
Progenitor/activation/other marker	$CD34^-$, $CD71^-$, $CD95^+$, $HLA\text{-}DR^+$, TdT^-
Adhesion marker	$CD11a^+$, $CD11b^+$, $CD49d^+$, $CD54^{(+)}$
Cytokine receptor	$CD25^+$, $CD116^+$, $CD117^-$, $CD123^+$, $CD125^-$

Genetic characterization[1,2]

Cytogenetic karyotype	hyperdiploid, 7.5% polyploidy; 50(48–51)<2n> XY, +4, +6, +8, +19, del(9)(q22)

Functional characterization[1,3–6]

Colony formation	in agarose
Cytochemistry	ACP^+, $ANAE^+$, $ANBE^{(+)}$, CAE^-, $Luxol^{(+)}$, MPO^+, MSE^+, $PAS^{(+)}$, SBB^+, Toluidine $Blue^-$
Inducibility of differentiation	cAMP, DMSO, G-CSF, TNF-α → eosino differentiation
Transcription factor	EoL-3: GATA-2 $mRNA^+$

Comments

- First human leukemia cell line with eosinophilic features (reference cell line).
- Cells grow well.
- Available from cell line banks.

References

[1] Saito, H. et al. (1985) Blood 66, 1233–1240.
[2] Drexler, H.G. et al. (1999) DSMZ Catalogue of Cell Lines, 7th edn, Braunschweig, Germany.
[3] Mayumi, M. (1992) Leukemia Lymphoma 7, 243–250. (review on EoL-1)
[4] Wong, C.K. et al. (1999) Immunol. Letters 68, 317–323.
[5] Shimamoto, T. et al. (1995) Blood 86, 3173–3180.
[6] Uphoff, C.C. et al. (1994) Leukemia 8, 1510–1526.

FKH-1

Culture characterization[1]

Culture medium	90% RPMI 1640 + 10% FBS + 10 ng/ml G-CSF
Doubling time	54 h
Viral status	EBV$^-$
Authentication	yes (by rearrangement of *DEK-CAN*)
Primary reference	Hamaguchi et al.[1]
Availability of cell line	not known

Clinical characterization[1]

Patient	61-year-old male
Disease diagnosis	Ph-negative CML with trilineage myelodysplasia → AML M4
Treatment status	refractory
Specimen site	peripheral blood
Year of establishment	1993

Immunophenotypic characterization[1]

Myelomonocytic marker	CD13$^+$, CD33$^+$
Progenitor/activation/other marker	CD34$^-$, HLA-DR$^+$
Adhesion marker	CD11a$^+$, CD11b$^+$, CD11c$^+$

Genetic characterization[1]

Cytogenetic karyotype	46, XY, −7, t(6;9)(p23;q34)
Unique translocation/fusion gene	t(6;9)(p23;q34) → *DEK-CAN* fusion gene

Functional characterization[1]

Cytochemistry	ANBE$^+$, CAE$^-$, MPO$^-$, Toluidine Blue$^-$
Cytokine response	G-CSF/GM-CSF-dependent; growth stimulation: IL-3, SCF
Special features	granules positive for sulfate glycoconjugates

Comments

- Cell line with some myelocytic features.
- Constitutively cytokine-dependent.
- Only cell line carrying the t(6;9)(p23;q34) leading to *DEK-CAN* fusion gene.

Reference

[1] Hamaguchi, H. et al. (1998) Brit. J. Haematol. 102, 1249–1256.

GDM-1

Culture characterization[1,2]

Culture medium	80% RPMI 1640 + 20% FBS
Doubling time	96–120 h
Viral status	EBV$^-$, HBV$^-$, HCV$^-$, HHV-8$^-$, HIV$^-$, HTLV-I/II$^-$
Authentication	yes (by cytogenetics)
Primary reference	Ben-Bassat et al.[1]
Availability of cell line	DSM ACC 87

Clinical characterization[1]

Patient	66-year-old female
Disease diagnosis	MPD (CML-like) → AML M4
Treatment status	refractory, terminal
Specimen site	peripheral blood
Year of establishment	1979

Immunophenotypic characterization[1-3]

T-/NK cell marker	CD1b$^-$, CD2$^-$, CD3$^-$, cyCD3$^-$, CD4$^+$, CD5$^-$, CD7$^-$, CD8$^-$, CD28$^-$, TCR$\alpha\beta^-$
B-cell marker	CD10$^-$, CD19$^-$, CD20$^-$, CD21$^-$, cyIg$^-$, sIg$^-$
Myelomonocytic marker	CD13$^-$, CD14$^-$, CD15$^+$, CD33$^+$
Erythroid-megakaryocytic marker	CD9$^+$, CD41a$^-$, CD61$^-$
Progenitor/activation/other marker	CD34$^+$, CD71$^+$, HLA-DR$^+$, TdT$^-$
Adhesion marker	CD11b$^+$
Cytokine receptor	CD25$^-$

Genetic characterization[1,2]

Cytogenetic karyotype	48(46–48)<2n>XX, +8, +13, −16, +mar, t(2;11)(q36;q13), del(6)(q21), t(7;?)(q35;?), del(12)(p13)

Functional characterization[1]

Cytochemistry	ACP$^+$, ALP$^+$, ANBE$^+$ (NaF inhibitable), CAE$^-$, GLC$^+$, Lysozyme$^+$, MPO$^{(+)}$, MSE$^-$, PAS$^-$, SBB$^+$
Inducibility of differentiation	TPA → macro differentiation
Special features	phagocytosis

Comments

- Cell line with some myelocytic characteristics.
- Available from cell line bank.
- Slowly growing difficult cell line.

References

[1] Ben-Bassat, H. et al. (1982) Leukemia Res. 6, 743–752.

[2] Drexler, H.G. et al. (1999) DSMZ Catalogue of Cell Lines, 7th edn, Braunschweig, Germany.

[3] Minowada, J. and Matsuo, Y. (1999) unpublished data.

GF-D8

Culture characterization[1]

Culture medium	80% RPMI 1640 + 20% FBS (or serum-free) + 50 ng/ml GM-CSF
Doubling time	48–72 h
Viral status	EBV$^-$
Authentication	no
Primary reference	Rambaldi et al.[1]
Availability of cell line	original authors

Clinical characterization[1]

Patient	82-year-old male
Disease diagnosis	AML M1
Treatment status	at diagnosis
Specimen site	peripheral blood
Year of establishment	1989

Immunophenotypic characterization[1,2]

T-/NK cell marker	CD2$^-$, CD3$^-$, CD4$^-$, CD5$^-$, CD7$^-$, CD8$^-$
B-cell marker	CD10$^-$, CD19$^-$, CD20$^-$, CD21$^-$, CD22$^-$
Myelomonocytic marker	CD13$^+$, CD14$^-$, CD15$^-$, CD33$^+$
Erythroid-megakaryocytic marker	CD41$^-$
Progenitor/activation/other marker	CD34$^-$, HLA-DR$^-$, TdT$^-$
Adhesion marker	CD11a$^+$, CD11b$^+$, CD11c$^-$, CD49d$^-$, CD54$^-$
Cytokine receptor	CD25$^-$, CD132$^-$

Genetic characterization[1,3]

Cytogenetic karyotype	44, XY, −5, −15, −17, +mar, del(7)(q32qter), inv(7)(q31.2q36), add(8q), add(11q), del(12p)
Receptor gene rearrangement	*IGH* G, *TCRD* G
Unique gene alteration	*MYC* amplification

Functional characterization[1,2,4]

Colony formation	in agar
Cytochemistry	ACP$^+$, ALP$^-$, ANAE$^+$ (NaF inhibitable), ANBE$^-$, MPO$^{(+)}$, SBB$^+$
Cytokine production	IL-1β
Cytokine receptor	mRNA$^+$: G-CSFR, GM-CSFRα, GM-CSFRβ, IL-2Rα, IL-2Rβ, IL-2Rγ, IL-3Rα, IL-4Rα, IL-7Rα, IL-9Rα
Cytokine response	GM-CSF/IL-3-dependent; growth stimulation: IFN-γ, PIXY-321, SCF; growth inhibition: TGF-β1
Inducibility of differentiation	TPA → mono/macro differentiation

Comments

- Typical myelocytic cell line.
- Constitutively cytokine-dependent and responsive to cytokine stimulation.
- Cells are difficult to grow.

References

[1] Rambaldi, A. et al. (1993) Blood 81, 1376–1383.
[2] Schumann, R.R. et al. (1996) Blood 87, 2419–2427.
[3] Tosi, S. et al. (1999) Genes Chromosomes Cancer 24, 213–221.
[4] Drexler, H.G. et al. (1997) Leukemia 11, 701–708.

Culture characterization[1]

Culture medium 90% RPMI 1640 + 10% FBS + 10 ng/ml GM-CSF
Authentication yes (by cytogenetics)
Primary reference Oez et al.[1]
Availability of cell line original authors

Clinical characterization[1]

Patient 50-year-old female
Disease diagnosis CML
Treatment status at myeloid blast crisis
Specimen site bone marrow
Year of establishment 1988

Immunophenotypic characterization[1]

T-/NK cell marker CD3$^-$, CD7$^-$
B-cell marker CD10$^-$, CD20$^-$
Myelomonocytic marker CD13$^+$, CD14$^-$, CD16$^-$
Erythroid-megakaryocytic marker CD36$^+$, CD42b$^-$
Progenitor/activation/other marker CD34$^+$, HLA-DR$^+$
Adhesion marker CD11b$^-$

Genetic characterization[1,2]

Cytogenetic karyotype 45, XX, −9, −17, −19, −22, 7p−, 9q+, +3mar, der(9)t(9;22) (q34;q11), der(13q)
Unique translocation/fusion gene Ph$^+$ t(9;22)(q34;q11) → *BCR-ABL* b3-a2 fusion gene

Functional characterization[1,3]

Cytochemistry ANAE$^-$, MPO$^-$, PAS$^-$
Cytokine response GM-CSF-dependent; growth stimulation: IFN-γ, IL-1α, IL-4, IL-13, PIXY-321, SCF; growth inhibition: TGF-β1
Inducibility of differentiation TPA → mono/macro differentiation

Comments

- Cell line with myelocytic markers.
- Constitutively cytokine-dependent.
- Carries Ph chromosome leading to *BCR-ABL* b3-a2 fusion gene.

References

[1] Oez, S. et al. (1990) Blood 76, 578–582.
[2] Drexler, H.G. et al. (1999) Leukemia Res. 23, 207–215.
[3] Drexler, H.G. et al. (1997) Leukemia 11, 701–708.

Culture characterization[1,2]

Subclones	multiple subclones with different unique features including eosinophilic subclones **HL-60-C15**[3] and **YY-1**[4]
Establishment	initially with CM from human embryonic fibroblast cultures
Culture medium	90% RPMI 1640 + 10% FBS
Doubling time	25–40 h
Viral status	EBV⁻, HBV⁻, HCV⁻, HHV-8⁻, HIV⁻, HTLV-I/II⁻
Authentication	yes (by cytogenetics)
Primary reference	Collins et al.[1]
Availability of cell line	ATCC CCL 240, DSM ACC 3, IFO 50022, JCRB 0085, RCB 0041; HL-60-C15: ATCC CRL 1964

Clinical characterization[1,5–7]

Patient	35-year-old female
Disease diagnosis	initially AML M3, later changed to AML M2
Treatment status	at diagnosis
Specimen site	peripheral blood
Year of establishment	1976

Immunophenotypic characterization[1,2,8–12]

T-/NK cell marker	CD1⁻, CD2⁻, CD3⁻, CD4⁺, CD5⁻, CD6⁻, CD7⁻, CD8⁻, CD28⁻, CD57⁻, TCRαβ⁻, TCRγδ⁻
B-cell marker	CD10⁻, CD19⁻, CD20⁻, CD22⁻, CD24⁺, CD80⁻, CD86⁻, sIg⁻
Myelomonocytic marker	CD13⁺, CD14⁻, CD15⁺, CD16⁻, CD33⁺
Erythroid-megakaryocytic marker	CD9⁺, CD41a⁻, CD42b⁻, CD61⁻, GlyA⁻, vWF⁻
Progenitor/activation/other marker	CD34⁻, CD38⁺, CD45⁺, CD71⁺, HLA-DR⁻, TdT⁻
Adhesion marker	CD11b⁻, CD44⁺, CD54⁻, CD62L⁻
Cytokine receptor	CD25⁻, CD117⁻, CD122⁺, CD132⁺

Genetic characterization[1,2,12–15]

Cytogenetic karyotype	flat-moded hypotetraploid, hypodiploid sideline, 1.5% polyploidy; 82(78–88)<4n>XX, −X, −X, −8, −8, −16, −17, −17, +18, +22, +2mar, ins(1;8)(p?31;q24hsr)×2, der(5)t(5;17)(q11;q11)×2, add(6) (q27)×2, der(9)del(9)(p13)t(9;14)(q?22;q?22)×2, der(14)t(9;14) (q?22;q?22)×2,

	der(16)t(16;17)(q22;q22)×1–2, add(18)(q21); sideline with −2, −5, −15, del(11)(q23.1q23.2)
Unique translocation/fusion gene	does not carry t(15;17)(q22;q11) nor *PML-RARA* fusion gene
Receptor gene rearrangement	*IGH* G, *TCRA* G, *TCRB* G (mRNA: IGH$^{(+)}$, TCRA$^-$, TCRB$^-$)
Unique gene alteration	amplification of *MYC*, *NEU*, *NRAS*, *P15INK4B* GG/DG, *P16INK4A* GG/DG, *P53* alteration

Functional characterization[1,5,6,11,16–20]

Colony formation	in agar or in methylcellulose
Cytochemistry	ACP$^{(+)}$, Alcian Blue$^-$, ALP$^-$, ANAE$^-$, ANBE$^{(+)}$, CAE$^+$, MGP$^-$, MPO$^+$, MSE$^-$, Muramidase$^{(+)}$, NBT$^-$, Oil Red O$^+$, PAS$^+$, Pyronin$^+$, SBB$^+$
Cytokine receptor	mRNA$^+$: IGF-1R, IGF-2R, IL-2Rα, IL-4Rα, IL-7Rα, IL-9Rα
Heterotransplantation	into nude or SCID mice
Inducibility of differentiation	inducible to neutro, mono, macro, eosino, baso differentiation with various reagents
Transcription factor	mRNA$^+$: GATA-1, GATA-2
Special features	mRNA$^+$: azurocidin, cathepsin G, Charcot Leyden crystal protein, defensin, major basic protein, myeloblastin, *N*-elastase; production of lysozyme

Comments

- Widely used and distributed cell line (reference cell line).
- First human myelocytic leukemia cell line.
- Probably most often studied and published human leukemia cell line (more than 5000 entries in Medline, 1984–1999).
- Model for granulocytic and monocytic/macrophage differentiation induction[20].
- Does not carry the typical t(15;17), but valuable model for promyelocytic leukemia[21].
- Available from cell line banks.
- Cells grow quickly and easily.

References

[1] Collins, S.J. et al. (1977) Nature 270, 347–349.

[2] Drexler, H.G. et al. (1999) DSMZ Catalogue of Cell Lines, 7th edn, Braunschweig, Germany.

[3] Fishkoff, S.A. (1988) Leukemia Res. 12, 679–686.

[4] Uneo, M. et al. (1994) Int. Arch. Allergy Immunol. 104 (Suppl. 1), 60–62.

[5] Gallagher, R. et al. (1979) Blood 54, 713–733.

[6] Collins, S.J. (1987) Blood 70, 1233–1244. (review on HL-60)
[7] Dalton, W.T. et al. (1988) Blood 71, 242–247.
[8] Minowada, J. and Matsuo, Y. (1999) unpublished data.
[9] Inoue, K. et al. (1997) Blood 89, 1405–1412.
[10] Kojika, S. et al. (1996) Leukemia 10, 994–999.
[11] Schumann, R.R. et al. (1996) Blood 86, 3173–3180.
[12] Furley, A.J. et al. (1986) Blood 68, 1101–1107.
[13] Aguiar, R.C.T. et al. (1997) Leukemia 11, 233–238.
[14] Nakamaki, T. et al. (1995) Brit. J. Haematol. 91, 139–149.
[15] Drexler, H.G. (1998) Leukemia 11, 845–859.
[16] Potter, G.K. et al. (1984) Am. J. Pathol. 114, 360–366.
[17] Machado, E.A. et al. (1984) Blood 63, 1015–1022.
[18] Uckun, F.M. (1996) Blood 88, 1135–1146.
[19] Shimamoto, T. et al. (1995) Blood 86, 3173–3180.
[20] Koeffler, H.P. (1983) Blood 62, 709–721. (review on differentiation of HL-60)
[21] Drexler, H.G. et al. (1995) Leukemia Res. 19, 681–691. (review on AML M3 cell lines)

HMC-1

Culture characterization[1,2]

Culture medium	80–90% IMDM + 10–20% FBS
Doubling time	80 h
Viral status	EBV⁻, HBV⁻, HCV⁻, HHV-8⁻, HIV⁻, HTLV-I/II⁻
Authentication	yes (by cytogenetics)
Primary reference	Butterfield et al.[1]
Availability of cell line	DSM ACC 283

Clinical characterization[1]

Patient	52-year-old female
Disease diagnosis	mast cell leukemia
Treatment status	refractory
Specimen site	peripheral blood

Immunophenotypic characterization[1–4]

T-/NK cell marker	$CD1^-$, $CD2^+$, $CD3^-$, $CD4^-$, $CD5^-$, $CD7^-$, $CD8^-$, $CD56^-$, $CD57^-$
B-cell marker	$CD10^-$, $CD19^-$, $CD20^-$, $CD21^-$, $CD22^-$, $CD23^-$, $CD24^-$, $CD37^+$, $CD40^+$, $CD74^-$
Myelomonocytic marker	$CD13^+$, $CD14^-$, $CD15^-$, $CD16^-$, $CD32^+$, $CD33^-$, $CD35^-$, $CD65^-$, $CD68^-$, $CD88^+$
Erythroid-megakaryocytic marker	$CD9^+$, $CD31^-$, $CD41^-$, $CD61^+$, $CD63^+$
Progenitor/activation/other marker	$CD34^-$, $CD38^-$, $CD45^+$, $CD69^-$, $CD71^-$, HLA-DR^-, TdT^-
Adhesion marker	$CD11a^+$, $CD11b^-$, $CD11c^{(+)}$, $CD18^+$, $CD43^+$, $CD44^+$, $CD54^+$
Cytokine receptor	$CD25^{(+)}$, $CD116^+$, $CD117^+$, $CD121b^{(+)}$, $CD122^{(+)}$, $CD123^-$, $CD126^-$, $CD127^-$, $CD131^+$
Comment	additional data on immunoprofile[2]

Genetic characterization[1,2,5]

Cytogenetic karyotype	pseudodiploid, 8% polypoidy; 46(42–46)<2n>XX, ins(10;16)(q24;q21q24), add(13)(q32); sideline with add(20)(p11/12)
Unique gene alteration	*KIT* mutation

Functional characterization[1,4,6]

Colony formation	in methylcellulose
Cytochemistry	Alcian Blue⁺, CAE⁺, Luxol⁻, MPO⁻, Toluidine Blue⁺
Cytokine production	MCP-1
Heterotransplantation	into nude mice
Inducibility of differentiation	IL-4, TPA → differentiation

Special features mRNA⁺: IgE-Rγ, β-tryptase, carboxypeptidase A, Charcot Leyden crystal protein, core protein, lysozyme, mast cell tryptase; protein⁺: aminocaproate esterase, cathepsin G, eosino major basic protein, eosino peroxidase, tryptase; no histamine production; no IgE binding

Comments

- Only available mast cell leukemia-derived cell line (reference cell line).
- Carries various mast cell-associated features.
- Extensive immunological and functional characterization.
- Available from cell line bank.
- Cells are difficult to grow.

References

[1] Butterfield, J.H. et al. (1988) Leukemia Res. 12, 345–355.
[2] Drexler, H.G. et al. (1999) DSMZ Catalogue of Cell Lines, 7th edn, Braunschweig, Germany.
[3] Nilsson, G. et al. (1994) Scand. J. Immunol. 39, 489–498.
[4] Agis, H. et al. (1996) Leukemia Lymphoma 22, 187–204. (review, partly on HMC-1)
[5] Furitsu, T. et al. (1993) J. Clin. Invest. 92, 1736–1744.
[6] Sillaber, C. et al. (1991) J. Immunol. 147, 4224–4228.

HNT-34

Culture characterization[1]

Culture medium	90% RPMI 1640 + 10% FBS
Doubling time	26–27 h
Authentication	yes (by cytogenetics)
Primary reference	Hamaguchi et al.[1]
Availability of cell line	RCB 1296

Clinical characterization[1]

Patient	47-year-old female
Disease diagnosis	MDS (CMML) → AML M4
Treatment status	refractory, terminal stage
Specimen site	peripheral blood
Year of establishment	1994

Immunophenotypic characterization[1]

T-/NK cell marker	CD1$^-$, CD2$^-$, CD3$^-$, CD4$^+$, CD5$^-$, CD7$^-$, CD8$^-$
B-cell marker	CD10$^-$, CD19$^-$, CD20$^-$
Myelomonocytic marker	CD13$^+$, CD14$^-$, CD33$^+$
Erythroid-megakaryocytic marker	CD41a$^-$, CD42b$^-$, GlyA$^-$
Progenitor/activation/other marker	CD34$^+$, HLA-DR$^-$

Genetic characterization[1]

Cytogenetic karyotype	46, XX, t(3;3)(q21;q26), t(9;22)(q34;q11), 20q−
Unique translocation/fusion gene	(1) Ph$^+$ t(9;22)(q34;q11) → *BCR-ABL* fusion genes (M-bcr/m-bcr) (2) t(3;3)(q21;q26) → EVI1 overexpression

Functional characterization[1]

Cytochemistry	ALP$^-$, ANBE$^{(+)}$, CAE$^-$, MPO$^-$, PAS$^-$
Inducibility of differentiation	resistant to DMSO, TPA

Comments

- Myelocytic cell line with two unique chromosomal translocations.
- Carries Ph chromosome with *BCR-ABL* M-bcr and m-bcr fusion genes.
- Carries t(3;3)(q21;q26) leading to EVI1 overexpression.
- Available from cell line bank.

Reference

[1] Hamaguchi, H. et al. (1997) Brit. J. Haematol. 98, 399–407.

Culture characterization[1]

Culture medium	90% RPMI 1640 + 10% FBS
Doubling time	48 h
Authentication	no
Primary reference	Kishi et al.[1]
Availability of cell line	restricted

Clinical characterization[1]

Patient	66-year-old male
Disease diagnosis	AML M3
Treatment status	at relapse
Specimen site	peripheral blood
Year of establishment	1993

Immunophenotypic characterization[1,2]

T-/NK cell marker	CD1⁻, CD2⁻, CD3⁻, CD4⁻, CD7⁻, CD8⁻, CD56⁺
B-cell marker	CD10⁻, CD19⁻, CD20⁻, CD21⁻, CD22⁻, CD23⁺
Myelomonocytic marker	CD13⁻, CD14⁻, CD15⁽⁺⁾, CD33⁺
Erythroid-megakaryocytic marker	CD36⁻, CD41⁻, GlyA⁻
Progenitor/activation/other marker	CD34⁺, CD38⁻, HLA-DR⁻
Adhesion marker	CD11b⁻
Cytokine receptor	CD25⁻, CD122⁻

Genetic characterization[1,3]

Cytogenetic karyotype	46, XY, t(1;12)(q25;p13), 2q+, t(4;6)(q12;q13), t(15;17) (q22;q11)
Unique translocation/fusion gene	(1) t(15;17)(q22;q11) → *PML-RARA* bcr3 fusion gene (2) t(1;12)(q25;p13) → *ETV6-ARG* fusion gene

Functional characterization[1,4]

Colony formation	in methylcellulose
Cytochemistry	ALP⁻, ANAE⁻, MPO⁽⁺⁾
Cytokine response	growth stimulation: G-CSF, GM-CSF
Inducibility of differentiation	ATRA → neutro/eosino differentiation; ATRA+G-CSF → neutro differentiation; ATRA+GM-CSF, IL-3, IL-5 → eosino/baso differentiation
Special features	induction of eosino peroxidase, major basic protein

Comments

- Myelocytic cell line with two unique chromosomal translocations.
- Carries t(15;17)(q22;q11) leading to *PML-RARA* fusion gene.

- Carries t(1;12)(q25;p13) leading to *ETV6-ARG* fusion gene.
- Differentiation potential with acquirement of neutrophil, eosinophil and basophil features.

References

[1] Kishi, K. et al. (1998) Exp. Hematol. 26, 135–142.
[2] Toba, K. et al. (1996) Exp. Hematol. 24, 894–901.
[3] Sato, Y. et al. (1998) Blood 92, 592a.
[4] Kishi, K. et al. (1995) Blood 86, 669a.

K051

Culture characterization[1]

Sister cell line	**K052** (serial sister cell line – at relapse – different immunophenotypic, cytogenetic, oncogene features)
Culture medium	90% α-MEM + 10% FBS
Doubling time	48 h
Viral status	EBNA⁻
Authentication	yes (by *P53* point mutation)
Primary reference	Abo et al.[1]
Availability of cell line	K051: JCRB 0122; K052: JCRB 0123

Clinical characterization[1]

Patient	46-year-old male
Disease diagnosis	AML M2
Treatment status	at diagnosis
Specimen site	bone marrow
Year of establishment	1991

Immunophenotypic characterization[1]

T-/NK cell marker	$CD2^-$, $CD3^-$, $CD5^-$, $CD7^-$
B-cell marker	$CD10^-$, $CD19^-$, $CD20^-$
Myelomonocytic marker	$CD13^+$, $CD14^-$, $CD33^+$
Erythroid-megakaryocytic marker	$CD41^+$, $GlyA^+$
Progenitor/activation/other marker	$CD34^-$, HLA-DR^-

Genetic characterization[1]

Cytogenetic karyotype	48, XY, +1, +2, 7q−, 17p−, 21q−
Unique gene alteration	*P53* mutation; K052: *NRAS* mutation

Functional characterization[1]

Cytochemistry	$ANAE^-$, MPO^+, MSE^-, PAS^-, SBB^+
Inducibility of differentiation	retinoic acid → ery differentiation
Special features	MDR1 $mRNA^+$

Comments

- Cell line with myelocytic characteristics.
- Cell lines established at diagnosis and at relapse.
- Available from cell line bank.

Reference

[1] Abo, J. et al. (1993) Blood 82, 2829–2836.

Kasumi-1

Culture characterization[1,2]

Culture medium	80% RPMI 1640 + 20% FBS
Doubling time	40–72 h
Viral status	EBV$^-$, HBV$^-$, HCV$^-$, HHV-8$^-$, HIV$^-$, HTLV-I/II$^-$
Authentication	yes (by cytogenetics)
Primary reference	Asou et al.[1]
Availability of cell line	DSM ACC 220

Clinical characterization[1]

Patient	7-year-old male
Disease diagnosis	AML M2
Treatment status	at 2nd relapse (post-BMT)
Specimen site	peripheral blood
Year of establishment	1989

Immunophenotypic characterization[1,2]

T-/NK cell marker	CD2$^-$, CD3$^-$, CD4$^+$, CD5$^-$, CD7$^-$, CD8$^-$
B-cell marker	CD10$^-$, CD19$^-$, CD20$^-$, CD21$^-$
Myelomonocytic marker	CD13$^+$, CD14$^-$, CD15$^+$, CD33$^+$, CD68$^-$
Progenitor/activation/other marker	CD34$^+$, CD38$^+$, CD71$^+$, HLA-DR$^+$
Adhesion marker	CD11b$^-$, CD11c$^-$
Cytokine receptor	CD117$^+$

Genetic characterization[1,2]

Cytogenetic karyotype	45<2n>X, −Y, −9, −13, −16, +3mar, t(8;21)(q22;q22), der(9)t (9;?)(p22;?), der(15)t(?9;15)(?q11;?p11)
Unique translocation/fusion gene	t(8;21)(q22;q22) → *AML1-ETO* fusion gene
Receptor gene rearrangement	*IGH* G, *TCRB* G

Functional characterization[1,3]

Cytochemistry	ALP$^-$, ANBE$^-$, CAE$^+$, MPO$^+$
Cytokine response	growth stimulation: G-CSF, GM-CSF, IL-3, IL-6
Inducibility of differentiation	TPA → macro differentiation
Transcription factor	GATA-2 mRNA$^+$

Comments

- Cell line carrying myelocytic features.
- First cell line with t(8;21)(q22;q22) leading to *AML1-ETO* fusion gene (reference cell line).
- Available from cell line bank.
- Cells are difficult to culture.

References

[1] Asou, H. et al. (1991) Blood 77, 2031–2036.

[2] Drexler, H.G. et al. (1999) DSMZ Catalogue of Cell Lines, 7th edn, Braunschweig, Germany.

[3] Shimamoto, T. et al. (1995) Blood 86, 3173–3180.

Kasumi-3

Culture characterization[1]

Culture medium	80% RPMI 1640 + 20% FBS (at 7.5% CO_2)
Doubling time	55–60 h
Authentication	yes (by cytogenetics)
Primary reference	Asou et al.[1]
Availability of cell line	original authors

Clinical characterization[1]

Patient	57-year-old male
Disease diagnosis	AML M0
Treatment status	at diagnosis
Specimen site	bone marrow
Year of establishment	1990

Immunophenotypic characterization[1]

T-/NK cell marker	$CD2^-$, $CD3^-$, $cyCD3^-$, $CD4^+$, $CD5^-$, $CD7^+$, $CD8^-$
B-cell marker	$CD10^-$, $CD19^-$, $CD20^-$
Myelomonocytic marker	$CD13^+$, $CD14^-$, $CD15^-$, $CD33^+$
Erythroid-megakaryocytic marker	$CD36^-$, $CD41^-$, $CD42^-$
Progenitor/activation/other marker	$CD34^+$, $HLA\text{-}DR^+$
Adhesion marker	$CD11a^+$, $CD11b^{(+)}$, $CD11c^{(+)}$, $CD54^+$
Cytokine receptor	$CD25^+$, $CD117^+$

Genetic characterization[1,2]

Cytogenetic karyotype	46, XY, −8, +mar, t(2;5)(p13;q33), t(3;7)(q27;q22), del(5)(q15), del(9)(q32), add(12)(p11), add(16)(q13); FISH: der(3)t(3;7;8) (3pter→3q26::7q35→7q22::8q22→8qter)
Unique translocation/fusion gene	t(3;7)(q27;q22) → *EVI1-TCRB* genes altered (→ EVI1 overexpression)
Receptor gene rearrangement	*IGH* G, *TCRB* R, *TCRG* G, *TCRD* G, (TCRB $mRNA^+$)

Functional characterization[1]

Cytochemistry	ACP^+, ALP^-, $ANBE^-$, CAE^-, MPO^-, PAS^-
Cytokine response	growth stimulation: GM-CSF, IL-2, IL-3, IL-4, SCF
Inducibility of differentiation	TPA → mono differentiation

Comments

- Myelocytic cell line.
- Carries t(3;7)(q27;q22) leading to alteration of the *EVI1-TCRB* genes and to EVI1 overexpression.
- Cells are difficult to culture.

References

[1] Asou, H. et al. (1996) Jpn. J. Cancer Res. 87, 269–274.

[2] Suzuku, K. et al. (1999) Leukemia 13, 1359–1366.

Kasumi-4

Culture characterization[1]

Culture medium	80% RPMI 1640 + 20% FBS (at 7.5–10% CO_2)
Doubling time	55 h
Authentication	yes (by cytogenetics)
Primary reference	Asou et al.[1]
Availability of cell line	original authors

Clinical characterization[1]

Patient	6-year-old female
Disease diagnosis	CML
Treatment status	at myeloid blast crisis
Specimen site	peripheral blood
Year of establishment	1993

Immunophenotypic characterization[1]

T-/NK cell marker	CD2⁻, CD3⁻, CD4⁻, CD7⁻, CD8⁻
B-cell marker	CD10⁻, CD19⁻, CD20⁻
Myelomonocytic marker	CD13⁺, CD14⁻, CD33⁺
Erythroid-megakaryocytic marker	CD36⁻, CD41a⁻, CD42b⁻, GlyA⁻, PPO⁻
Progenitor/activation/other marker	CD34⁺, HLA-DR⁺
Adhesion marker	CD11b⁻

Genetic characterization[1,2]

Cytogenetic karyotype	46, XX, inv(3)(q21q26), t(9;22;11)(q34;q11;q13)
Unique translocation/fusion gene	(1) Ph⁺ t(9;22)(q34;q11) → *BCR-ABL* b2-a2 fusion gene (2) inv(3)(q21q26) → EVI1 overexpression

Functional characterization[1]

Cytochemistry	ACP⁺, ALP⁻, ANBE⁻, CAE⁻, MPO⁻, PAS⁻
Cytokine response	growth stimulation: GM-CSF, IL-3, IL-6, SCF
Inducibility of differentiation	resistant to differentiation induction by cytokines, DMSO, retinoic acid, TPA

Comments

- Cell line with various myelocytic markers.
- Carries Ph chromosome leading to *BCR-ABL* b2-a2 fusion gene.
- Carries inv(3)(q21q26) leading to EVI1 overexpression.
- Cells are very difficult to culture.

References

[1] Asou, H. et al. (1996) Brit. J. Haematol. 93, 68–74.
[2] Drexler, H.G. et al. (1999) Leukemia Res. 23, 207–215.

KBM-7

Culture characterization[1–3]

Subclones	**KBM-7/B5** and **KBM-7 P1-55** (near-haploid – similar immunological and functional features)
Establishment	initially in hypoxic environment
Culture medium	90% IMDM + 10% FBS
Doubling time	22–24 h
Viral status	EBNA$^-$
Authentication	yes (by isoenzyme analysis)
Primary reference	Andersson et al.[1]
Availability of cell line	original authors

Clinical characterization[1,2]

Patient	39-year-old male
Disease diagnosis	CML
Treatment status	at 2nd myeloid blast crisis (post-BMT)
Specimen site	bone marrow
Year of establishment	1984

Immunophenotypic characterization[1]

T-/NK cell marker	CD4$^+$
B-cell marker	CD20$^{(+)}$
Myelomonocytic marker	CD13$^+$, CD14$^{(+)}$, CD33$^+$
Progenitor/activation/other marker	CD34$^+$, HLA-DR$^{(+)}$, TdT$^-$
Adhesion marker	CD11b$^{(+)}$

Genetic characterization[1–6]

Cytogenetic karyotype	48, X, −Y, +8, +8, −9, +15, −22, +22q−, del(6)t(6;?)(p23;?), t(9;22)(q34;q11), +t(9;22)(q34;q11), del(12p)
Unique translocation/fusion gene	Ph$^+$ t(9;22)(q34;q11) → *BCR-ABL* b2-a2 fusion gene
Unique gene alteration	*P53* mutation

Functional characterization[1,2]

Colony formation	in agar
Cytochemistry	ACP$^+$, ANAE$^+$, ANBE$^-$, CAE$^+$, MPO$^-$, MSE$^-$, PAS$^+$
Heterotransplantation	into nude mice (→ granulocytic sarcoma)

Comments

- Cell line with various myelocytic characteristics.
- Carries multiple Ph chromosomes leading to *BCR-ABL* b2-a2 fusion gene.
- Cells grow well.

References

[1] Andersson, B.S. et al. (1995) Leukemia 9, 2100–2108.
[2] Andersson, B.S. et al. (1987) Cancer Genet. Cytogenet. 24, 335–343.
[3] Kotecki, M. et al. (1999) Exp. Cell Res. 252, 273–280.
[4] Drexler, H.G. et al. (1999) Leukemia Res. 23, 207–215.
[5] Sen, S. et al. (1995) Cancer Genet. Cytogenet. 82, 35–40.
[6] Sen, S. et al. (1993) Leukemia Res. 17, 639–647.

KCL-22

Culture characterization[1,2]

Culture medium	80% RPMI 1640 + 20% FBS
Doubling time	24 h
Viral status	EBNA$^-$
Authentication	no
Primary reference	Kubonishi and Miyoshi[1]
Availability of cell line	restricted

Clinical characterization[1,2]

Patient	32-year-old female
Disease diagnosis	CML
Treatment status	at blast crisis
Specimen site	pleural effusion
Year of establishment	1981

Immunophenotypic characterization[1–3]

T-/NK cell marker	CD1b$^-$, CD2$^-$, CD4$^{(+)}$, CD28$^-$
B-cell marker	CD20$^-$, CD80$^-$, CD86$^-$, cyIg$^-$, sIg$^-$
Myelomonocytic marker	CD13$^+$, CD14$^-$, CD15$^+$, CD33$^+$, CD65$^+$
Erythroid-megakaryocytic marker	CD9$^+$, CD41a$^+$, CD61$^+$, GlyA$^-$
Progenitor/activation/other marker	CD34$^+$, HLA-DR$^-$, TdT$^-$
Adhesion marker	CD11b$^-$, CD54$^+$
Cytokine receptor	CD25$^+$

Genetic characterization[1,4–7]

Cytogenetic karyotype	52, XX, +1p−, +6, +8, +8, +8, t(9;22), +22q−
Unique translocation/fusion gene	Ph$^+$ t(9;22)(q34;q11) → *BCR-ABL* b2-a2 fusion gene
Unique gene alteration	*P53* mutation

Functional characterization[1]

Cytochemistry	ACP$^+$, ALP$^-$, ANBE$^-$, Benzidine$^-$, CAE$^-$, MPO$^-$, MSE$^-$, PAS$^+$, SBB$^-$
Heterotransplantation	into newborn hamsters

Comments

- Cell line with some myelocytic parameters.
- Carries Ph chromosome leading to *BCR-ABL* b2-a2 fusion gene.

References

[1] Kubonishi, I. and Miyoshi, I. (1983) Int. J. Cell Cloning 1, 105–117.
[2] Kubonishi, I. et al. (1983) Gann 74, 319–322.
[3] Minowada, J. and Matsuo, Y. (1999) unpublished data.
[4] Nakamaki, T. et al. (1995) Brit. J. Haematol. 91, 139–149.
[5] Bi, S. et al. (1992) Leukemia 6, 839–842.
[6] Bi, S. et al. (1993) Leukemia 7, 1840–1845.
[7] Drexler, H.G. et al. (1999) Leukemia Res. 23, 207–215.

KG-1

Culture characterization[1–3]

Subclone	**KG-1a** (established at passages 15–35 of KG-1 – morphologically, cytochemically, immunologically and functionally less mature)
Culture medium	90% α-MEM or RPMI 1640 + 10% FBS
Doubling time	40–50 h
Viral status	EBV⁻, HBV⁻, HCV⁻, HHV-8⁻, HIV⁻, HTLV-I/II⁻
Authentication	yes (by cytogenetics)
Primary reference	Koeffler and Golde[1]
Availability of cell line	KG-1: ATCC CCL 246/CRL 8031, DSM ACC 14, JCRB 0065/9051, RCB 1166; KG-1a: ATCC CCL 246.1, DSM ACC 421

Clinical characterization[1,3]

Patient	59-year-old male
Disease diagnosis	erythroleukemia → AML
Treatment status	at relapse
Specimen site	bone marrow
Year of establishment	1977

Immunophenotypic characterization[1,2,4–6]

T-/NK cell marker	CD1⁻, CD2⁻, CD3⁻, cyCD3⁻, CD4⁻, CD5⁻, CD7⁻, CD8⁻, CD28⁻, CD56⁻, CD57⁻, TCRαβ⁻, TCRγδ⁻
B-cell marker	CD10⁻, CD19⁻, CD20⁻, CD21⁻, CD22⁻, CD24⁺, CD80⁺, CD86⁻, sIg⁻
Myelomonocytic marker	CD13⁺, CD14⁻, CD15⁺, CD33⁺, CD65⁺
Erythroid-megakaryocytic marker	CD9⁻, CD36⁻, CD41⁻, CD42b⁻, CD61⁺, GlyA⁻, vWF⁻
Progenitor/activation/other marker	CD34⁺, CD38⁺, CD71⁺, HLA-DR⁺, TdT⁻
Adhesion marker	CD11b⁺, CD44⁺, CD54⁺
Cytokine receptor	CD25⁺, CD117⁻, CD122⁺, CD132⁺

Genetic characterization[1,2,7–10]

Cytogenetic karyotype	hypodiploid, 4.5% polyploidy; 45(42–47)<2n> X/XY, −4, +8, +8, −12, −17, −20, +2mar, der(5;17)(q10;q10) del(5)(q?11q?13), dup(7)(q12q33), del(7)(q22q35), i(8q)×2, der(8)t(6;8)(p11;q22), der(8)t(8;12)(p11;q13), der(11)t(1;11)(q13-21;p11-p13), der(16)t(?12;16)(?p13;q13/21)

Unique gene alteration *NRAS* mutation, *P15INK4B* methylation, *P16INK4A* methylation, *P53* mutation, *RB1* rearrangement

Functional characterization[1,11-15]

Colony formation	in agar
Cytochemistry	ACP+, CAE+, MPO+, MSE−, PAS+
Cytokine receptor	IL-9Rα mRNA+
Heterotransplantation	into nude or SCID mice
Inducibility of differentiation	TPA→ mono/macro differentiation
Transcription factor	mRNA+: GATA-1, GATA-2

Comments

- Prototypical myelocytic cell line (reference cell line).
- Second-oldest human myelocytic leukemia cell line.
- Slightly less mature subclone (KG-1a) available.
- Model cell line for induction of differentiation.
- Cells grow very well.
- Available from cell line banks.

References

1 Koeffler, H.P. and Golde, D.W. (1978) Science 200, 1153–1154.
2 Drexler, H.G. et al. (1999) DSMZ Catalogue of Cell Lines, 7th edn, Braunschweig, Germany.
3 Koeffler, H.P. and Golde, D.W. (1980) Blood 56, 344–350. (review on early AML cell lines)
4 Minowada, J. and Matsuo, Y. (1999) unpublished data.
5 Toba, K. et al. (1996) Exp. Hematol. 24, 894–901.
6 Furley, A.J. et al. (1986) Blood 68, 1101–1107.
7 Drexler, H.G. (1998) Leukemia 12, 845–859.
8 Sheng, X.M. et al. (1997) Leukemia Res. 21, 697–701.
9 Sugimoto, K. et al. (1992) Blood 79, 2378–2383.
10 Nakamaki, T. et al. (1995) Brit. J. Haematol. 91, 139–149.
11 Schumann, R.R. et al. (1996) Blood 87, 2419–2427.
12 Shimamoto, T. et al. (1995) Blood 86, 3173–3180.
13 Machado, E.A. et al. (1984) Blood 63, 1015–1022.
14 Uckun, F.M. (1996) Blood 88, 1135–1146.
15 Koeffler, H.P. (1983) Blood 62, 709–721. (review on induced differentiation of AML cell lines)

KOPM-28

Culture characterization[1]

Culture medium	90% RPMI 1640 + 10% FBS
Doubling time	22–24 h
Authentication	no
Primary reference	Mori et al.[1]
Availability of cell line	original authors

Clinical characterization[1]

Patient	64-year-old female
Disease diagnosis	CML
Treatment status	at myeloid blast crisis
Specimen site	peripheral blood
Year of establishment	1982

Immunophenotypic characterization[1–3]

T-/NK cell marker	$CD1a^-$, $CD2^-$, $CD3^-$, $CD4^+$, $CD8^-$
B-cell marker	$CD10^-$, $CD19^-$, $CD20^-$, $CD22^-$, $cyIg^-$, sIg^-
Myelomonocytic marker	$CD13^+$, $CD15^-$, $CD33^+$
Erythroid-megakaryocytic marker	$CD41^+$, PPO^-
Progenitor/activation/other marker	$CD38^+$, $CD71^+$, HLA-DR^+, TdT^-
Adhesion marker	$CD11b^+$
Cytokine receptor	$CD25^+$

Genetic characterization[1,4]

Cytogenetic karyotype	44, XX, −5, −17, −19, t(9;22)(q34;q11), +der(19)t(19;?)(p13;?)
Unique translocation/fusion gene	Ph^+ t(9;22)(q34;q11) → *BCR-ABL* b3-a2 fusion gene

Functional characterization[1–3]

Colony formation	in agar
Cytochemistry	ACP^+, ALP^-, $ANBE^+$, CAE^+, MPO^-, MSE^-, PAS^-
Inducibility of differentiation	TPA → meg differentiation

Comments

- Cell line displays myelocytic characteristics.
- Carries Ph chromosome leading to *BCR-ABL* b3-a2 fusion gene.

References

1 Mori, T. et al. (1987) Leukemia Res. 11, 241–249.
2 Kojika, S. et al. (1996) Leukemia 10, 994–999.
3 Tsuda, H. et al. (1988) Int. J. Cell Cloning 6, 209–220.
4 Drexler, H.G. et al. (1999) Leukemia Res. 23, 207–215.

KPB-M15

Culture characterization[1]

Culture medium	originally 80% α-medium + 20% horse serum, later 90% RPMI 1640 + 10% FBS
Doubling time	24 h
Viral status	EBNA$^-$
Authentication	no
Primary reference	Kamamoto et al.[1]
Availability of cell line	first author[2]

Clinical characterization[1]

Patient	38-year-old male
Disease diagnosis	CML
Treatment status	at blast crisis
Specimen site	peripheral blood
Year of establishment	1981

Immunophenotypic characterization[1,2]

T-/NK cell marker	CD2$^-$, CD5$^-$
B-cell marker	CD10$^-$, CD20$^-$, cyIg$^-$, sIg$^-$
Myelomonocytic marker	CD13$^+$, CD14$^+$, CD15$^-$
Erythroid-megakaryocytic marker	GlyA$^-$
Progenitor/activation marker	HLA-DR$^-$, TdT$^-$
Adhesion marker	CD11b$^+$

Genetic characterization[1]

Cytogenetic karyotype	86(72–88), −X, +Xp+, +Xp+, −Y, +1, +1, +2, +2, +3p–, +3p–, +4, +4, +5, +5, +5, +6, +6, −7, −7, +i(7q), +i(7q), +8, +8, +8q+, −9, −9, +10, +10, +11, +11, +13, +13, −15, +16, +17, +18, +18, +19, +19, +21, +21, +22q−(Ph), +22q−(Ph), +9mar
Unique translocation/fusion gene	Ph$^+$ t(9;22)(q34;q11) → *BCR-ABL* fusion gene

Functional characterization[1,2]

Cytochemistry	ACP$^+$, ANAE$^{(+)}$, ANBE$^-$, CAE$^-$, MPO$^-$, PAS$^-$
Cytokine production	IL-6, M-CSF
Inducibility of differentiation	resistant to DMSO, sodium butyrate, TPA, Vit. D3

Comments

- Myelocytic cell line derived from CML in blast crisis.
- Carries Philadelphia chromosome leading to *BCR-ABL* fusion gene.

References

[1] Kamamoto, T. et al. (1986) Jpn. J. Clin. Oncol. 16, 107–115.
[2] Hiraoka, A. et al. (1998) Acta Haematol. 100, 174–180.

KT-1

Culture characterization[1]

Culture medium	90% RPMI 1640 + 10% FBS
Doubling time	18–24 h
Viral status	EBNA$^-$
Authentication	yes (by cytogenetics)
Primary reference	Yanagisawa et al.[1]
Availability of cell line	original authors

Clinical characterization[1]

Patient	32-year-old male
Disease diagnosis	CML
Treatment status	at blast crisis (refractory)
Specimen site	peripheral blood
Year of establishment	1991

Immunophenotypic characterization[1]

T-/NK cell marker	CD2$^-$, CD3$^-$, CD4$^+$, CD5$^-$, CD7$^-$, CD8$^-$
B-cell marker	CD10$^-$, CD19$^-$, CD20$^-$
Myelomonocytic marker	CD13$^-$, CD14$^-$, CD33$^+$
Erythroid-megakaryocytic marker	CD41a$^-$
Progenitor/activation/other marker	CD34$^-$, HLA-DR$^-$, TdT$^-$
Adhesion marker	CD11b$^-$

Genetic characterization[1,2]

Cytogenetic karyotype	51, XXYY, +X, +Y, −5, −6, +8, +8, +19, +19, +mar, t(9;22)(q34;q11)×2
Unique translocation/fusion gene	Ph$^+$ t(9;22)(q34;q11) → *BCR-ABL* b2-a2 fusion gene
Receptor gene rearrangement	*IGH* G, *TCR* G

Functional characterization[1]

Colony formation	in methylcellulose
Cytochemistry	ANBE$^-$, CAE$^-$, MPO$^+$, PAS$^{(+)}$
Inducibility of differentiation	IFN-α, IFN-γ → growth arrest (no differentiation)

Comments

- Cell line with few myelocytic markers.
- Carries two Ph chromosomes leading to *BCR-ABL* b2-a2 fusion gene.

References

[1] Yanagisawa, K. et al. (1998) Blood 91, 641–648.
[2] Drexler, H.G. et al. (1999) Leukemia Res. 23, 207–215.

KU812

Culture characterization[1–4]

Subclones	**KU812E, KU812F** (isolated by semi-solid culture – similar properties)
Culture medium	80–90% McCoy's 5A or RPMI 1640 + 10–20% FBS
Doubling time	80–100 h
Viral status	EBV⁻, HBV⁻, HCV⁻, HIV⁻, HTLV-I/II⁻
Authentication	no
Primary reference	Kishi[1]
Availability of cell line	KU812: DSM ACC 378, IFO 50363, JCRB 0104, RCB 0495; KU812E: JCRB 104.1, RCB 0496; KU812F: JCRB 104.2, RCB 0497

Clinical characterization[1]

Patient	38-year-old male
Disease diagnosis	CML
Treatment status	at blast crisis
Specimen site	peripheral blood
Year of establishment	1981

Immunophenotypic characterization[1,2,5–9]

T-/NK cell marker	$CD2^-$, $CD3^-$, $CD5^-$, $CD7^-$
B-cell marker	$CD10^-$, $CD19^-$, $CD20^-$, $CD21^-$, $CD22^-$, $CD23^-$, $CD24^+$, $CD37^-$, $CD39^-$, $CD40^+$, $CD80^-$, $CD86^+$, $cyIg^-$, sIg^-
Myelomonocytic marker	$CD13^+$, $CD14^-$, $CD15^-$, $CD16^-$, $CD17^+$, $CD32^+$, $CD33^+$, $CD35^+$, $CD65^-$, $CD68^+$, $CD88^+$
Erythroid-megakaryocytic marker	$CD9^+$, $CD31^+$, $CD41^+$, $CD42^+$, $CD61^+$, $CD63^+$, $GlyA^{(+)}$
Progenitor/activation/other marker	$CD34^-$, $CD38^-$, $CD45^+$, $CD71^+$, HLA-DR^-, TdT^-
Adhesion marker	$CD11a^-$, $CD11b^{(+)}$, $CD11c^-$, $CD18^{(+)}$, $CD43^+$, $CD44^+$, $CD54^+$, $CD69^-$
Cytokine receptor	$CD25^-$, $CD116^+$, $CD117^-$, $CD121b^+$, $CD122^-$, $CD123^+$, $CD126^-$, $CD127^-$, $CD131^+$, EPO-R^+
Comment	additional data on immunoprofile[5]

Genetic characterization[1,2,10–14]

Cytogenetic karyotype	hypotriploid, 8% polyploidy; 61(58–62)<3n> XYY, −2, −3, −5, +6, −7, +8, −10, −12, −16, −17, −18, +19, −20, t(9;22) (q34;q11)×2, i(11q), i(17q)
Unique translocation/fusion gene	Ph^+ t(9;22)(q34;q11) → *BCR-ABL* b3-a2 fusion gene
Unique gene alteration	*P53* mutation

Functional characterization[1,5,6,8,9,15–17]

Colony formation	in agar
Cytochemistry	ACP$^+$, Alcian Blue$^{(+)}$, ALP$^-$, ANAE$^+$, ANBE$^{(+)}$, Astra Blue$^+$, CAE$^-$, MPO$^-$, PAS$^{(+)}$, SBB$^-$, Toluidine Blue$^+$
Inducibility of differentiation	spontaneous ery differentiation; serum-free culture → baso differentiation; TPA → macro differentiation
Transcription factor	mRNA$^+$: GATA-1, GATA-2, PU1, SCL
Special features	mRNA$^+$: α-/β-/γ-/δ-globin, δ-aminolevulin synthase; synthesis of HbA, HbF, Hb Bart's inducible; mRNA$^+$: β-tryptase, carboxy-peptidase A, Charcot Leyden crystal protein, heparin core protein, lysozyme, mast cell tryptase, IgE-Rα/β/γ; production of histamine; no IgE binding

Comments

- Well-characterized model cell line for basophilic leukemia (reference cell line).
- Cells display morphology, cytochemistry and functional aspects of basophils.
- Pluripotential cell line with spontaneous and inducible expression of myelocytic, monocytic/macrophage, basophil, erythrocytic and megakaryocytic features.
- Carries Ph chromosome leading to *BCR-ABL* b3-a2 fusion gene and i(17q).
- Parental cell line and subclones available from cell line banks.

References

1 Kishi, K. (1985) Leukemia Res. 9, 381–390.
2 Drexler, H.G. et al. (1999) DSMZ Catalogue of Cell Lines, 7th edn, Braunschweig, Germany.
3 Fukuda, T. et al. (1987) Blood 70, 612–619.
4 Toba, K. et al. (1996) Exp. Hematol. 24, 894–901.
5 Nilsson, G. et al. (1994) Scand. J. Immunol. 39, 489–498.
6 Nakazawa, M. et al. (1989) Blood 73, 2003–2013.
7 Inoue, K. et al. (1997) Blood 89, 1405–1412.
8 Hirata, J. et al. (1990) Leukemia 4, 365–372.
9 Agis, H. et al. (1996) Leukemia Lymphoma 22, 187–204.
10 Nakamaki, T. et al. (1995) Brit. J. Haematol. 91, 139–149.
11 Bi, S. et al. (1993) Leukemia 7, 1840–1845.
12 Bi, S. et al. (1992) Leukemia 6, 839–842.
13 Feinstein, E. et al. (1991) Proc. Natl. Acad. Sci. USA 88, 6293–6297.
14 Drexler, H.G. et al. (1999) Leukemia Res. 23, 207–215.
15 Shimamoto, T. et al. (1995) Blood 86, 3173–3180.
16 Endo, K. et al. (1993) Brit. J. Haematol. 85, 653–662.
17 Blom, T. et al. (1992) Eur. J. Immunol. 22, 2025–2032.

KY821

Culture characterization[1]

Culture medium	90% RPMI 1640 + 10% FBS
Authentication	no
Primary reference	Saito et al.[1]
Availability of cell line	JCRB 0105

Clinical characterization[1]

Patient	28-year-old male
Disease diagnosis	AML M2
Treatment status	at relapse
Specimen site	meninges
Year of establishment	1982

Immunophenotypic characterization[1]

T-/NK cell marker	$CD2^-$, $CD3^-$
B-cell marker	$CD19^-$
Myelomonocytic marker	$CD13^+$, $CD14^-$, $CD15^+$, $CD33^+$
Progenitor/activation/other marker	$CD34^-$, HLA-DR^-

Genetic characterization[1,2]

Cytogenetic karyotype	86, X, X, −Y, +1, +1, +3, +3, +4, +4, +5, +5, +6, +6, +7, +7, +8, +8, +10, +10, +11, +11, +12, +13, +13, +13, +13, +14, +14, +15, +15, +16, +16, +18, +19, +19, +20, +20, +21, +22, +22, +mar, +der(2)t(2;?)(p25;?), +der(2)t(2;?)(p25;?), del(9)(p22p24), del(9)(p22p24)
Unique gene alteration	*P53* mutation

Functional characterization[1]

Colony formation	in methylcellulose
Cytochemistry	$ANBE^{(+)}$, CAE^+, MPO^+, MSE^-
Special features	resistant to methotrexate

Comments

- Cell line with few myelocytic features.
- Further characterization required.
- Available from cell line bank.

References

[1] Saito, H. et al. (1992) Leukemia Res. 16, 217–226.
[2] Sugimoto, K. et al. (1992) Blood 79, 2378–2383.

KYO-1

Culture characterization[1,2]

Sister cell line	**KPB-M8** (simultaneous sister cell line – independently from the same primary sample using horse serum – cytogenetic differences)
Culture medium	80% RPMI 1640 + 20% FBS
Doubling time	22 h
Viral status	EBNA⁻
Authentication	no
Primary reference	Ohkubo et al.[1]
Availability of cell line	original authors

Clinical characterization[1]

Patient	22-year-old male
Disease diagnosis	CML
Treatment status	at myeloid blast crisis
Specimen site	peripheral blood
Year of establishment	1981

Immunophenotypic characterization[1]

T-/NK cell marker	$CD2^-$
B-cell marker	sIg^-
Erythroid-megakaryocytic marker	$GlyA^-$
Progenitor/activation marker	$HLA\text{-}DR^-$, TdT^-
Adhesion marker	$CD11b^+$

Genetic characterization[1,3–5]

Cytogenetic karyotype	47, XY, +15, −16, −18, 6q−, 8q+, 12p+, 15p+, 17p+, 22p+, 22q−(Ph), +22q−(Ph)
Unique translocation/fusion gene	Ph^+ t(9;22)(q34;q11) → *BCR-ABL* b2-a2 fusion gene
Unique gene alteration	*P53* mutation

Functional characterization[1]

Colony formation	in agar
Cytochemistry	$ANBE^-$, CAE^-, MPO^-, MSE^-

Comments

- Incompletely characterized myelocytic cell line.
- Carries Ph chromosome leading to *BCR-ABL* b2-a2 fusion gene.

References

[1] Ohkubo, T. et al. (1985) Leukemia Res. 9, 921–926.
[2] Kamamoto, T. et al. (1986) Jpn. J. Clin. Oncol. 16, 107–115.
[3] Bi, S. et al. (1992) Leukemia 6, 839–842.
[4] Bi, S. et al. (1993) Leukemia 7, 1840–1845.
[5] Drexler, H.G. et al. (1999) Leukemia Res. 23, 207–215.

Culture characterization[1]

Culture medium	90% RPMI 1640 + 10% FBS
Authentication	no
Primary reference	Oez et al.[1]
Availability of cell line	not known

Clinical characterization[1]

Patient	60-year-old female
Disease diagnosis	AML M4 → AML M2
Treatment status	at relapse
Specimen site	bone marrow

Immunophenotypic characterization[1]

T-/NK cell marker	CD2$^-$, CD3$^{(+)}$, CD4$^+$, CD5$^+$, CD7$^+$, CD8$^-$, CD56$^-$
B-cell marker	CD10$^-$
Myelomonocytic marker	CD15$^+$
Erythroid-megakaryocytic marker	CD41$^-$, CD42$^-$
Progenitor/activation/other marker	CD34$^{(+)}$, CD38$^{(+)}$, CD45$^+$, CD71$^+$, HLA-DR$^-$, TdT$^-$
Cytokine receptor	CD25$^+$, CD117$^-$, CD120a$^+$, CD120b$^+$, CD122$^-$

Genetic characterization[1]

Cytogenetic karyotype	45, X, −X, der(9)inv(9)(p12q13), del(9)(p22?)

Functional characterization[1]

Colony formation	not clonable
Cytochemistry	ACP$^{(+)}$, ANAE$^-$, PAS$^{(+)}$, SBB$^-$

Comments

- Apparently myelocytic cell line.
- Further characterization required.

Reference

[1] Oez, S. et al. (1996) Ann. Hematol. 72, 307–316.

Marimo

Culture characterization[1]

Culture medium	90% RPMI 1640 + 10% FBS
Doubling time	25–28 h
Authentication	no
Primary reference	Yoshida et al.[1]
Availability of cell line	original authors

Clinical characterization[1]

Patient	68-year-old female
Disease diagnosis	essential thrombocythemia → t-AML M2
Treatment status	at terminal stage (no treatment)
Specimen site	bone marrow
Year of establishment	1993

Immunophenotypic characterization[1]

T-/NK cell marker	T-markers negative
B-cell marker	B-markers negative
Myelomonocytic marker	CD13$^+$, CD14$^-$, CD15$^+$, CD33$^-$
Erythroid-megakaryocytic marker	ery-meg markers negative
Progenitor/activation/other marker	CD34$^-$, HLA-DR$^-$

Genetic characterization[1]

Cytogenetic karyotype	44, X, −X, −5, +18, +18, ins(1;?)(q21;?), del(8)(q22), t(10;14;11)(q22;q32;q13), der(14)t(10;14;11)(q22;q32;q13), add(17)(p11), psu dic(18;9)(q23;p21)×2
Unique gene alteration	*MYC* amplification

Functional characterization[1]

Cytochemistry	ANAE$^-$, MPO$^-$
Cytokine response	growth stimulation: G-CSF, GM-CSF, IL-3
Inducibility of differentiation	DMSO, retinoic acid → granulo differentiation

Comments

- Interesting, but incompletely characterized myelocytic cell line.

Reference

[1] Yoshida, H. et al. (1998) Cancer Genet. Cytogenet. 100, 21–24.

MDS92

Culture characterization[1]

Culture medium	90% RPMI 1640 + 10% FBS + 10 ng/ml IL-3
Doubling time	80–90 h
Authentication	no
Primary reference	Tohyama et al.[1]
Availability of cell line	original authors

Clinical characterization[1]

Patient	52-year-old male
Disease diagnosis	MDS (RARS/RAEB) (prior to leukemic transformation)
Treatment status	at diagnosis
Specimen site	bone marrow
Year of establishment	1991

Immunophenotypic characterization[1,2]

T-/NK cell marker	CD2$^-$, CD3$^-$, CD7$^-$
B-cell marker	CD19$^-$, CD20$^{(+)}$
Myelomonocytic marker	CD13$^+$, CD14$^{(+)}$, CD33$^+$
Erythroid-megakaryocytic marker	CD41$^{(+)}$, CD61$^-$, GlyA$^-$
Progenitor/activation/other marker	CD34$^+$, HLA-DR$^+$
Adhesion marker	CD11b$^+$

Genetic characterization[1]

Cytogenetic karyotype	44, XY, −7, −12, −13, +mar, del(5)(q13q35), add(14)(p11), add(22)(q13)
Unique gene alteration	*NRAS* mutation

Functional characterization[1,2]

Colony formation	in agar
Cytochemistry	MPO$^+$, MSE$^-$
Cytokine response	IL-3-dependent; growth stimulation: GM-CSF, SCF, TPO
Inducibility of differentiation	G-CSF, GM-CSF, IL-3, SCF → neutro differentiation

Comments

- MDS-derived myelocytic cell line.
- Constitutively cytokine-dependent.
- Cells are very difficult to culture.

References

[1] Tohyama, K. et al. (1994) Brit. J. Haematol. 87, 235–242.

[2] Tohyama, K. et al. (1995) Brit. J. Haematol. 91, 795–799.

MTO-94

Culture characterization[1]

Culture medium	85% RPMI 1640 + 15% FBS
Viral status	EBNA$^-$
Authentication	yes (by cytogenetics)
Primary reference	Mizobuchi et al.[1]
Availability of cell line	not known

Clinical characterization[1]

Patient	74-year-old male
Disease diagnosis	MDS (RAEB) → AML
Treatment status	refractory, terminal
Specimen site	bone marrow
Year of establishment	1994

Immunophenotypic characterization[1]

T-/NK cell marker	CD7$^+$
Myelomonocytic marker	CD13$^+$, CD33$^+$
Progenitor/activation marker	CD34$^-$, HLA-DR$^+$

Genetic characterization[1]

Cytogenetic karyotype	46, XY, i(17q)

Functional characterization[1]

Cytochemistry	ACP$^+$, ALP$^+$, CAE$^+$, MPO$^+$, NBT$^+$, PAS$^+$, SBB$^+$
Cytokine response	growth stimulation: G-CSF, GM-CSF, IL-3, SCF
Special features	phagocytosis$^{(+)}$

Comments

- AML (post-MDS)-derived myelocytic cell line.
- Possible deletion of *P53* gene.

Reference

[1] Mizobuchi, N. et al. (1997) Acta Medica Okayama 51, 227–232.

MUTZ-2

Culture characterization[1,2]

Culture medium	80% α-MEM + 20% FBS + 50 ng/ml SCF
Doubling time	48 h
Viral status	EBV$^-$, HBV$^-$, HCV$^-$, HHV-8$^-$, HIV$^-$, HTLV-I/II$^-$
Authentication	yes (by DNA fingerprinting)
Primary reference	Hu et al.[1]
Availability of cell line	DSM ACC 271

Clinical characterization[1]

Patient	62-year-old male
Disease diagnosis	AML M2
Treatment status	at diagnosis
Specimen site	peripheral blood
Year of establishment	1993

Immunophenotypic characterization[1,2]

T-/NK cell marker	CD1$^-$, CD3$^-$, CD4$^+$, CD5$^-$, CD7$^-$, CD8$^-$, CD56$^-$
B-cell marker	CD10$^-$, CD19$^-$, CD20$^-$
Myelomonocytic marker	CD13$^+$, CD14$^-$, CD15$^+$, CD16$^-$, CD33$^+$, CD65$^+$, CD68$^+$
Erythroid-megakaryocytic marker	CD41a$^-$, CD42b$^-$, CD61$^-$, GlyA$^-$
Progenitor/activation/other marker	CD30$^+$, CD34$^+$, CD38$^+$, CD71$^+$, HLA-DR$^+$, TdT$^-$
Adhesion marker	CD11b$^-$
Cytokine receptor	CD25$^+$, CD115$^+$, CD116$^+$, CD117$^+$, CD123$^+$

Genetic characterization[1,2]

Cytogenetic karyotype	hyperdiploid, 1.8% polyploidy; 48(46–50)<2n> XY, +8, +10
Unique gene alteration	*P53* mutation

Functional characterization[1,3]

Colony formation	in methylcellulose
Cytochemistry	MSE$^-$, TRAP$^-$
Cytokine response	SCF-dependent; growth stimulation: bFGF, FL, G-CSF, IFN-β, IFN-γ, IGF-I, IL-6, M-CSF, TNF-α
Inducibility of differentiation	TPA → mono differentiation
Proto-oncogene	BCL2 mRNA$^+$

Comments

- Cell line with myelocytic characteristics.
- Constitutively cytokine-dependent and -responsive.
- Cells are difficult to culture.
- Available from cell line bank.

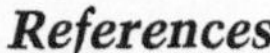

References

[1] Hu, Z.B. et al. (1996) Leukemia 10, 1025–1040.
[2] Drexler, H.G. et al. (1999) DSMZ Catalogue of Cell Lines, 7th edn, Braunschweig, Germany.
[3] Drexler, H.G. et al. (1997) Leukemia 11, 701–708.

NB4

Culture characterization[1–4]

Subclones	various retinoic acid-resistant subclones
Culture medium	90% RPMI 1640 + 10% FBS
Doubling time	36–48 h
Viral status	EBV⁻, HBV⁻, HCV⁻, HHV-8⁻, HIV⁻, HTLV-I/II⁻
Authentication	yes (by cytogenetics)
Primary reference	Lanotte et al.[1]
Availability of cell line	DSM ACC 207

Clinical characterization[1]

Patient	23-year-old female
Disease diagnosis	AML M3
Treatment status	at 2nd relapse
Specimen site	bone marrow
Year of establishment	1989

Immunophenotypic characterization[1,4,5]

T-/NK cell marker	$CD2^+$, $CD3^-$, $CD4^+$, $CD5^-$, $CD6^-$, $CD7^-$, $CD8^-$, $TCR\alpha\beta^-$, $TCR\gamma\delta^-$
B-cell marker	$CD9^+$, $CD10^-$, $CD19^-$, $CD20^-$, $CD23^-$
Myelomonocytic marker	$CD13^+$, $CD14^-$, $CD15^+$, $CD33^+$, $CD68^-$
Erythroid-megakaryocytic marker	$CD36^-$, $CD41^-$, $CD42^-$, $GlyA^-$
Progenitor/activation/other marker	$CD34^-$, $CD38^+$, HLA-DR^-
Adhesion marker	$CD11b^+$, $CD11c^-$
Cytokine receptor	$CD25^-$

Genetic characterization[1,4,6–8]

Cytogenetic karyotype	hypertriploid, 3% polyploidy; 78(71–81)<3n>XX, −X, +2, +6, +7, +7, +11, +12, +13, +14, +17, −19, +20, +4mar, der(8)t (8;?)(q24;?), der(11)t(11;?)(?→::11p15→11q22.1::11q 13→22.1:), der(12)t(12;?)(p11;?), 14p+, t(15;17)(q22;q11-12.1), der(19)t (10;19)(q21.1;p13.3)×2
Unique translocation/fusion gene	t(15;17)(q22;q11) → *PML-RARA* fusion gene (bcr1-2)
Unique gene alteration	*P15INK4B* deletion, *P16INK4A* deletion

Functional characterization[1]

Cytochemistry	ALP^-, $ANBE^-$, CAE^-, MPO^+, MSE^+
Inducibility of differentiation	retinoic acid → neutro differentiation

Comments

- First immortalized promyelocytic leukemia cell line (reference cell line)[9].
- Carries t(15;17)(q22;q11) leading to *PML-RARA* fusion gene.
- Responsive to differentiation induction with retinoic acids.
- Cells grow well.
- Available from cell line bank.

References

[1] Lanotte, M. et al. (1991) Blood 77, 1080–1086.
[2] Dermine, S. et al. (1993) Blood 82, 1573–1577.
[3] Duprez, E. et al. (1992) Leukemia 6, 1281–1287.
[4] Drexler, H.G. et al. (1999) DSMZ Catalogue of Cell Lines, 7th edn, Braunschweig, Germany.
[5] Kizaki, M. et al. (1996) Blood 88, 1824–1833.
[6] de Thé, H. et al. (1990) Nature 347, 558–561.
[7] Lo Coco, F. et al. (1998) Leukemia 12, 1866–1880.
[8] Nakamaki, T. et al. (1995) Brit. J. Haematol. 91, 139–149.
[9] Drexler, H.G. et al. (1995) Leukemia Res. 19, 681–691. (review on AML M3 cell lines)

NKM-1

Culture characterization[1]

Culture medium	90% RPMI 1640 + 10% FBS
Doubling time	36–48 h
Viral status	EBV⁻
Authentication	no
Primary reference	Kataoka et al.[1]
Availability of cell line	restricted

Clinical characterization[1]

Patient	33-year-old male
Disease diagnosis	AML M2
Treatment status	at diagnosis
Specimen site	peripheral blood
Year of establishment	1981

Immunophenotypic characterization[1]

T-/NK cell marker	$CD2^-$, $CD3^-$, $CD5^-$, $CD7^-$
B-cell marker	$CD10^-$, $CD19^-$, $CD20^-$
Myelomonocytic marker	$CD13^-$, $CD14^-$, $CD15^+$, $CD16^-$, $CD32^+$, $CD33^-$
Erythroid-megakaryocytic marker	$CD41^-$, $GlyA^-$
Progenitor/activation/other marker	HLA-DR^+
Adhesion marker	$CD11b^-$, $CD18^+$
Cytokine receptor	$CD114^+$, $CD115^+$

Genetic characterization[1]

Cytogenetic karyotype	47, XY, −6, +8, +der(2)t(2;?)(q37;?), +der(6)t(6;?)(p23;?)

Functional characterization[1]

Colony formation	in agar
Cytochemistry	$ANBE^-$, CAE^+, MPO^+, NBT^-
Cytokine response	growth stimulation: G-CSF, M-CSF
Inducibility of differentiation	resistant to induction with DMSO, Nabutyrate, retinoic acid, TPA

Comments

- Cell line expresses various myelocytic features.

Reference

[1] Kataoka, T. et al. (1990) Cancer Res. 50, 7703–7709.

OCI/AML1

Culture characterization[1-4]

Subclone	**OCI/AML1a** (similar features)
Culture medium	90% α-MEM + 10% FBS + 5637 CM or 10 ng/ml G-CSF
Doubling time	32 h
Authentication	no
Primary reference	Wang et al.[1]
Availability of cell line	original authors

Clinical characterization[1]

Patient	73-year-old female
Disease diagnosis	AML M4
Treatment status	at diagnosis
Specimen site	peripheral blood
Year of establishment	1987

Immunophenotypic characterization[2,3,5]

T-/NK cell marker	$CD7^-$, $CD8^-$
B-cell marker	$CD19^-$, $CD20^-$
Myelomonocytic marker	$CD13^+$, $CD14^{(+)}$, $CD33^+$
Erythroid-megakaryocytic marker	$CD36^+$, $CD41a^-$, $GlyA^-$
Progenitor/activation/other marker	$CD34^+$, $CD38^+$, $CD45^+$, $CD71^+$, $HLA\text{-}DR^+$
Adhesion marker	$CD11b^+$

Genetic characterization[3]

Cytogenetic karyotype	OCI/AML1a: 46, XX, −6, +der(6)t(6;8)(p25;q22)

Functional characterization[1-6]

Colony formation	in methylcellulose
Cytochemistry	OCI/AML1a: $ANBE^{(+)}$, CAE^+, MPO^+, MSE^-, NBT^-
Cytokine production	GM-CSF $mRNA^+$
Cytokine response	G-CSF-dependent; growth stimulation: GM-CSF, IFN-β, IGF-1, IL-3, IL-4, IL-6, M-CSF, PIXY-321, SCF; growth inhibition: TGF-β1, TNF-α, TNF-β

Comments

- Cell line committed to myelocytic differentiation.
- Constitutively cytokine-dependent and -responsive cell line.

References

[1] Wang, C. et al. (1989) Leukemia 3, 264–269.
[2] Wang, C. et al. (1991) Leukemia 5, 493–499.
[3] Nara, N. et al. (1990) Jpn. J. Cancer Res. 81, 625–631.
[4] Nara, N. (1992) Leukemia Lymphoma 7, 331–335. (review on OCI/AML1a)
[5] Taetle, R. et al. (1993) Cancer Res. 53, 3386–3393.
[6] Drexler, H.G. et al. (1997) Leukemia 11, 701–708.

OCI/AML5

Culture characterization[1,2]

Culture medium	90% α-MEM + 20% FBS + 5–10 ng/ml GM-CSF or 5637 CM
Doubling time	30–50 h
Viral status	EBV$^-$, HBV$^-$, HCV$^-$, HHV-8$^-$, HIV$^-$, HTLV-I/II$^-$
Authentication	no
Primary reference	Wang et al.[1]
Availability of cell line	DSM ACC 247

Clinical characterization[1]

Patient	77-year-old male
Disease diagnosis	AML M4
Treatment status	at relapse
Specimen site	peripheral blood
Year of establishment	1990

Immunophenotypic characterization[1,2]

T-/NK cell marker	CD3$^-$, CD7$^-$, CD8$^+$
B-cell marker	CD19$^-$
Myelomonocytic marker	CD13$^+$, CD14$^-$, CD15$^+$, CD33$^+$, CD68$^+$
Progenitor/activation/other marker	CD34$^+$, HLA-DR$^+$, TdT$^+$

Genetic characterization[2]

Cytogenetic karyotype	48(44–48)<2n>XY, +1, +8, der(1)t(1;19)(p13;p13)

Functional characterization[1,3]

Colony formation	in methylcellulose
Cytokine response	GM-CSF/IL-3-dependent; growth stimulation: FL, G-CSF, IFN-β, IFN-γ, IL-6, M-CSF, PIXY-321; growth inhibition: TGF-β1, TNF-α

Comments

- Myelocytic cell line available from cell line bank.
- Constitutively cytokine-dependent cell line but cells become quickly independent.
- Cell line grows very well.

References

[1] Wang, C. et al. (1991) Leukemia 5, 493–499.
[2] Drexler, H.G. et al. (1999) DSMZ Catalogue of Cell Lines, 7th edn, Braunschweig, Germany.
[3] Drexler, H.G. et al. (1998) Leukemia 11, 701–708.

OHN-GM

Culture characterization[1]

Culture medium	80% RPMI 1640 + 20% FBS + 10 ng/ml GM-CSF
Doubling time	60 h
Viral status	EBV$^-$
Authentication	yes (by cytogenetics)
Primary reference	Nagai et al.[1]
Availability of cell line	original authors

Clinical characterization[1]

Patient	60-year-old male
Disease diagnosis	Hodgkin (mixed cellularity) → Hodgkin (relapse) → t-MDS (RA) → t-MDS (RAEB) → t-AML
Treatment status	refractory
Specimen site	bone marrow
Year of establishment	1995

Immunophenotypic characterization[1]

T-/NK cell marker	CD3$^-$, CD4$^-$, CD5$^-$, CD7$^+$, CD8$^-$
B-cell marker	CD10$^-$, CD19$^-$, CD20$^-$
Myelomonocytic marker	CD13$^+$, CD33$^+$
Erythroid-megakaryocytic marker	CD41$^-$, GlyA$^-$
Progenitor/activation/other marker	CD34$^+$, CD38$^+$, CD71$^+$, HLA-DR$^+$

Genetic characterization[1,2]

Cytogenetic karyotype	48, XY, −7, +8, +22, del(5)(q11.2q31), t(10;13)(q24;q14), +der(13)t(10;13)(q24;q14); SKY: der(12)t(10;12)(q?26;p13), der(13)t(5;12;13), +der(13)t(5;12;13)
Unique gene alteration	*P53* mutation, *RB1* deletion

Functional characterization[1]

Cytochemistry	ANAE$^{(+)}$, MPO$^+$
Cytokine response	GM-CSF-dependent

Comments

- Cytokine-dependent myelocytic cell line.
- Expanded characterization required.

References

[1] Nagai, M. et al. (1997) Brit. J. Haematol. 98, 392–398.
[2] Kakazu, N. et al. (1999) Genes Chromosomes Cancer 26, 336–345.

OIH-1

Culture characterization[1]

Culture medium	90% RPMI 1640 + 10% FBS + 10 ng/ml G-CSF or GM-CSF
Doubling time	48 h
Viral status	EBV$^-$
Authentication	yes (by cytogenetics)
Primary reference	Hamaguchi et al.[1]
Availability of cell line	not known

Clinical characterization[1]

Patient	72-year-old male
Disease diagnosis	MDS (RAEB) → AML
Treatment status	at diagnosis of AML
Specimen site	peripheral blood
Year of establishment	1993

Immunophenotypic characterization[1]

T-/NK cell marker	CD1$^-$, CD2$^-$, CD3$^-$, CD4$^+$, CD5$^-$, CD7$^+$, CD8$^-$
B-cell marker	CD10$^-$, CD19$^{(+)}$, CD20$^-$
Myelomonocytic marker	CD13$^+$, CD14$^-$, CD33$^+$
Erythroid-megakaryocytic marker	CD41a$^-$, CD42b$^-$, GlyA$^-$
Progenitor/activation/other marker	CD34$^+$, HLA-DR$^+$

Genetic characterization[1]

Cytogenetic karyotype	49, X, −Y, −5, −11, +18, +20, +21, add(7)(q11), add(7)(p21), +del(7)(q22), add(8)(q22), +add(8)(q22), inv(9), add(15)(p11)
Unique gene alteration	*DCC* alteration

Functional characterization[1]

Cytochemistry	ANBE$^+$, CAE$^-$, MPO$^-$, PAS$^{(+)}$
Cytokine response	G-CSF/GM-CSF-dependent; growth stimulation: IL-3, SCF

Comments

- Cell line with some myelocytic markers.
- Constitutively cytokine-dependent cell line.

Reference

[1] Hamaguchi, H. et al. (1998) Int. J. Hematol. 67, 153–164.

PL-21

Culture characterization[1,2]

Culture medium	85% RPMI 1640 + 15% FBS
Doubling time	48–64 h
Viral status	EBNA⁻
Authentication	no
Primary reference	Kubonishi et al.[1]
Availability of cell line	restricted

Clinical characterization[1,2]

Patient	24-year-old male
Disease diagnosis	(mediastinal) granulocytic sarcoma → AML M3
Treatment status	refractory
Specimen site	peripheral blood
Year of establishment	1981

Immunophenotypic characterization[2,3]

T-/NK cell marker	$CD2^-$, $CD4^+$
B-cell marker	$CD10^-$, $CD20^-$, sIg^-
Myelomonocytic marker	$CD13^+$, $CD14^+$, $CD15^+$, $CD33^+$
Erythroid-megakaryocytic marker	$CD41a^+$, $CD61^+$
Progenitor/activation/other marker	$HLA\text{-}DR^+$, TdT^-
Cytokine receptor	$CD25^-$

Genetic characterization[2,3]

Cytogenetic karyotype	46, XY, −11, +der(11)t(11;?)(11pter-11q23::?)
Unique translocation/fusion gene	no t(15;17)(q22;q11) *PML-RARA*

Functional characterization[2]

Cytochemistry	ACP^+, ALP^-, $ANBE^{(+)}$, CAE^+, $Lysozyme^+$, MPO^+, MSE^+, $PAS^{(+)}$, SBB^+
Inducibility of differentiation	DMSO → macro differentiation
Special features	phagocytosis

Comments

- Cell line with myelocytic and monocytic features.
- Allegedly derived from AML M3, but cells do not carry typical t(15;17) *PML-RARA*[4].

References

[1] Kubonishi, I. et al. (1983) Gann 74, 319–322.
[2] Kubonishi, I. et al. (1984) Blood 63, 254–259.
[3] Minowada, J. and Matsuo, Y. (1999) unpublished data.
[4] Drexler, H.G. et al. (1995) Leukemia Res. 19, 681–691. (review on AML M3 cell lines)

SKNO-1

Culture characterization[1]

Culture medium	90% RPMI 1640 + 10% FBS + 10 ng/ml GM-CSF
Doubling time	48–72 h
Viral status	EBNA⁻
Authentication	yes (by cytogenetics)
Primary reference	Matozaki et al.[1]
Availability of cell line	original authors

Clinical characterization[1]

Patient	22-year-old male
Disease diagnosis	AML M2
Treatment status	at 2nd relapse
Specimen site	bone marrow
Year of establishment	1990

Immunophenotypic characterization[1]

T-/NK cell marker	$CD4^+$
B-cell marker	$CD19^+$
Myelomonocytic marker	$CD13^+$, $CD33^+$
Progenitor/activation/other marker	$CD34^+$

Genetic characterization[1]

Cytogenetic karyotype	44, X, −Y, 6q+, 11p+, −16, −17.2q, 19p+, +der(16q17q), t(8;21)(q22;q22)
Unique translocation/fusion gene	t(8;21)(q22;q22) → *AML1-ETO* fusion gene
Unique gene alteration	*P53* mutation

Functional characterization[1-3]

Cytochemistry	$ANAE^-$, CAE^+, MPO^+
Cytokine response	GM-CSF-dependent; growth stimulation: G-CSF, IFN-α, IFN-β, IFN-γ, IL-3, IL-5, IL-6, IL-13, M-CSF, PIXY-321, SCF; growth inhibition: TNF-α
Inducibility of differentiation	TPA → mono/macro differentiation

Comments

- Cell line with myelocytic characteristics.
- Second cell line to carry the t(8;21)(q22;q22) *AML1-ETO* fusion gene.

References

[1] Matozaki, S. et al. (1995) Brit. J. Haematol. 89, 805–811.
[2] Kawaguchi, R. et al. (1993) Cancer Genet. Cytogenet. 67, 157.
[3] Drexler, H.G. et al. (1997) Leukemia 11, 701–708.

TSU1621MT

Culture characterization[1]

Culture medium	90% RPMI 1640 + 10% FBS + 10 ng/ml G-CSF
Authentication	no
Primary reference	Shikami et al.[1]
Availability of cell line	original authors

Clinical characterization[1]

Patient	81-year-old female
Disease diagnosis	AML M4
Treatment status	progressive disease, terminal stage
Specimen site	peripheral blood or bone marrow
Year of establishment	1997

Immunophenotypic characterization[1]

T-/NK cell marker	CD2$^-$, CD4$^-$, CD7$^+$, CD56$^+$
B-cell marker	CD19$^-$, CD20$^-$
Myelomonocytic marker	CD13$^+$, CD14$^-$, CD15$^-$, CD33$^+$
Progenitor/activation/other marker	CD34$^+$, HLA-DR$^+$
Adhesion marker	CD11b$^+$, CD11c$^-$
Cytokine receptor	CD25$^+$, CD114$^{(+)}$, CD115$^+$, CD116$^+$, CD117$^+$, CD122$^+$, CD123$^+$, CD124$^-$, CD125$^-$, CD127$^-$, CD131$^+$, CD132$^+$, MPL$^-$

Genetic characterization[1]

Cytogenetic karyotype	48, XX, +4, +4, del(9)(q13q22), add(12)(q24), add(14)(p10), t(16;21)(p11;q22)
Unique translocation/fusion gene	t(16;21)(p11;q22) → *TLS/FUS-ERG* fusion gene

Functional characterization[1]

Cytochemistry	ANBE$^-$, CAE$^-$, MPO$^{(+)}$
Cytokine response	G-CSF-dependent; growth stimulation: GM-CSF, IL-3, IL-4, SCF; growth inhibition: IL-5

Comments

- Cell line expressing various myelocytic characteristics.
- Constitutively cytokine-dependent cell line.
- Carries t(16;21)(p11;q22) leading to *TLS/FUS-ERG* fusion gene.
- Cells grow very well.

Reference

[1] Shikami, M. et al. (1999) Brit. J. Haematol. 105, 711–719.

UCSD/AML1

Culture characterization[1]

Culture medium	90% RPMI 1640 + 10% FBS + 3 ng/ml GM-CSF
Doubling time	48–72 h
Authentication	yes (by cytogenetics)
Primary reference	Oval et al.[1]
Availability of cell line	original authors

Clinical characterization[1]

Patient	73-year-old female
Disease diagnosis	AML
Treatment status	at relapse
Specimen site	bone marrow
Year of establishment	1989

Immunophenotypic characterization[1,2]

T-/NK cell marker	$CD2^-$, $CD4^-$, $CD5^-$, $CD7^+$, $CD8^-$
B-cell marker	$CD19^-$, $CD20^-$
Myelomonocytic marker	$CD13^+$, $CD14^-$, $CD15^+$, $CD33^+$
Erythroid-megakaryocytic marker	$CD36^+$, $CD41a^-$, $CD42b^+$, $GlyA^-$
Progenitor/activation/other marker	$CD34^+$, $CD38^-$, $CD45^+$, $CD71^+$, $HLA\text{-}DR^+$, TdT^+
Adhesion marker	$CD49b^+$

Genetic characterization[1,3,4]

Cytogenetic karyotype	45, XX, −7, t(3;3)(q21;q26), t(12;22)(p13;q12)
Unique translocation/fusion gene	(1) t(3;3)(q21;q26) → EVI1 overexpression (2) t(12;22)(p13;q12) → *ETV6/TEL-MN1* fusion gene
Receptor gene rearrangement	*IGH* G, *TCRB* G

Functional characterization[1,2,5,6]

Cytochemistry	ACP^+, $ANAE^+$, CAE^-, MSE^-
Cytokine response	GM-CSF-dependent; growth stimulation: CNTF, IFN-β, IL-3, IL-4, IL-6, IL-12, IL-13, IL-15, LIF, M-CSF, OSM, PIXY-321, SCF; growth inhibition: TGF-β1
Inducibility of differentiation	TPA → macro differentiation; IL-6 → meg differentiation

Comments

- Myelocytic cell line with unique cytogenetic and cytokine-related features.
- Constitutively cytokine-dependent and responsive to multiple cytokines.
- Carries t(3;3)(q21;q26) leading to EVI1 overexpression.
- Carries t(12;22)(p13;q12) leading to *ETV6/TEL-MN1* fusion gene.
- Cells are difficult to culture.

References

[1] Oval, J. et al. (1990) Blood 76, 1369–1374.

[2] Taetle, R. et al. (1993) Cancer Res. 53, 3386–3393.

[3] Hu, Z.B. et al. (1996) Leukemia 10, 1025–1040.

[4] Oval, J. et al. (1992) Leukemia 6, 446–451.

[5] Drexler, H.G. et al. (1997) Leukemia 11, 701–708.

[6] Oval, J. and Taetle, R. (1990) Blood Reviews 4, 270–279. (review on cytokine-dependent AML cell lines)

UF-1

Culture characterization[1]

Culture medium	90% RPMI 1640 + 10% FBS
Doubling time	72 h
Authentication	no
Primary reference	Kizaki et al.[1]
Availability of cell line	original authors

Clinical characterization[1]

Patient	33-year-old female
Disease diagnosis	AML M3
Treatment status	at 2nd relapse
Specimen site	peripheral blood
Year of establishment	1994

Immunophenotypic characterization[1]

T-/NK cell marker	$CD3^-$, $CD4^-$, $CD5^-$, $CD7^+$, $CD8^-$
B-cell marker	$CD10^-$, $CD19^-$, $CD20^-$
Myelomonocytic marker	$CD13^+$, $CD14^+$, $CD33^+$
Erythroid-megakaryocytic marker	$CD41^-$
Progenitor/activation/other marker	$CD34^-$, $CD38^+$, HLA-DR^-
Adhesion marker	$CD11b^-$

Genetic characterization[1]

Cytogenetic karyotype	46, XX, add(1)(q44), add(6)(q12), add(7)(q36), t(15;17)(q21;q21)
Unique translocation/fusion gene	t(15;17)(q21;q21) → *PML-RARA* fusion gene

Functional characterization[1-3]

Cytochemistry	MPO^+, NBT^-
Cytokine response	growth stimulation: GM-CSF, IL-3, SCF; growth inhibition TGF-β1
Heterotransplantation	into SCID mice
Inducibility of differentiation	resistant to retinoic acids (ATRA, 9-cis-RA)

Comments

- Cell line with myelocytic features.
- Carries t(15;17)(q21;q21) leading to *PML-RARA* fusion gene.
- Constitutively resistant to differentiation induction with retinoic acids.

References

[1] Kizaki, M. et al. (1996) Blood 88, 1824–1833.
[2] Kizaki, M. et al. (1998) J. Natl. Cancer Inst. 90, 1906–1907.
[3] Kizaki, M. et al. (1999) Int. J. Mol. Med. 4, 359–364. (review on UF-1)

YNH-1

Culture characterization[1]

Culture medium	90% RPMI 1640 + 10% FBS + 10 ng/ml GM-CSF
Doubling time	82 h
Viral status	EBV−
Authentication	yes (by cytogenetics)
Primary reference	Yamamoto et al.[1]
Availability of cell line	RCB 1291

Clinical characterization[1]

Patient	46-year-old male
Disease diagnosis	AML M1
Treatment status	at diagnosis
Specimen site	peripheral blood
Year of establishment	1994

Immunophenotypic characterization[1,2]

T-/NK cell marker	CD1$^-$, CD2$^-$, CD3$^-$, CD4$^-$, CD5$^-$, CD7$^-$, CD8$^-$
B-cell marker	CD10$^+$, CD19$^-$, CD20$^-$
Myelomonocytic marker	CD13$^+$, CD14$^-$, CD15$^-$, CD33$^+$
Erythroid-megakaryocytic marker	CD41a$^+$, CD42b$^-$
Progenitor/activation/other marker	CD34$^+$, HLA-DR$^-$
Adhesion marker	CD11a$^+$, CD11b$^+$, CD11c$^+$, CD56$^+$
Cytokine receptor	CD25$^+$, CD114$^{(+)}$, CD115$^+$, CD116$^+$, CD117$^+$, CD122$^-$, CD123$^+$, CD124$^+$, CD125$^-$, CD127$^-$, CD131$^+$, CD132$^+$, MPL$^-$

Genetic characterization[1,2]

Cytogenetic karyotype	46, XY, der(16)t(16;21)(p11;q22)t(1;16)(q12;q13), der(21)t (16;21)(p11;q22), der(6)t(6;12)(q13;q13), der(12)t(6;12)(q21;q13)
Unique translocation/fusion gene	t(16;21)(p11;q22) → *TLS/FUS-ERG* fusion gene
Unique gene alteration	*P53* mutation

Functional characterization[1,2]

Cytochemistry	ANBE$^-$, CAE$^-$, MPO$^+$, PAS$^-$
Cytokine response	G-CSF/GM-CSF/IL-3-dependent; growth stimulation: EPO, IL-4, SCF, TPO

Comments

- Cell line expressing various myelocytic characteristics.
- Constitutively cytokine-dependent cell line.

- Carries t(16;21)(p11;q22) leading to *TLS/FUS-ERG* fusion gene.
- Available from cell line bank.

References

[1] Yamamoto, K. et al. (1997) Leukemia 11, 599–608.
[2] Shikami, M. et al. (1999) Brit. J. Haematol. 105, 711–719.

YOS-M

Culture characterization[1]

Sister cell line	**YOS-B** (= EBV$^+$ B-LCL, Ph$^+$)
Culture medium	90% RPMI 1640 + 10% FBS
Doubling time	5–6 d
Viral status	EBNA$^-$
Authentication	yes (by cytogenetics)
Primary reference	Yasukawa et al.[1]
Availability of cell line	original authors

Clinical characterization[1]

Patient	77-year-old male
Disease diagnosis	CML
Treatment status	at myeloid blast crisis
Specimen site	peripheral blood
Year of establishment	1990

Immunophenotypic characterization[1]

T-/NK cell marker	CD2$^-$, CD3$^-$, CD4$^+$, CD5$^-$, CD7$^-$, CD8$^-$
B-cell marker	CD9$^-$, CD10$^-$, CD19$^-$, CD20$^-$, CD21$^-$
Myelomonocytic marker	CD13$^{(+)}$, CD14$^+$, CD33$^+$
Progenitor/activation/other marker	CD34$^+$, HLA-DR$^+$, TdT$^-$
Cytokine receptor	CD25$^+$, CD122$^-$

Genetic characterization[1,2]

Cytogenetic karyotype	46, XY, −20, +mar, del(7)(q22q32), t(9;22)(q34;q11)
Unique translocation/fusion gene	Ph$^+$ t(9;22)(q34;q11) → *BCR-ABL* b2-a2 fusion gene
Receptor gene rearrangement	*IGH* G

Functional characterization[1]

Cytochemistry	ANBE$^-$, CAE$^{(+)}$, MPO$^+$, MSE$^-$

Comments

- Cell line displaying various myelocytic parameters.
- Carries Ph chromosome leading to *BCR-ABL* b2-a2 fusion gene.

References

[1] Yasukawa, M. et al. (1992) Brit. J. Haematol. 82, 515–521.
[2] Drexler, H.G. et al. (1999) Leukemia Res. 23, 207–215.

Part 2
Monocytic Cell Lines

AML-1

Culture characterization[1]

Culture medium	90% IMDM + 10% FBS
Viral status	EBNA$^-$, HTLV-II$^-$
Authentication	yes (by cytogenetics)
Primary reference	Lange et al.[1]
Availability of cell line	not known

Clinical characterization[1]

Patient	12-year-old male
Disease diagnosis	AML M4
Treatment status	at relapse

Immunophenotypic characterization[1]

T-/NK cell marker	CD2$^-$, CD3$^-$, CD5$^-$, CD7$^+$
B-cell marker	CD10$^-$
Myelomonocytic marker	CD13$^+$, CD14$^+$, CD15$^+$, CD33$^+$
Progenitor/activation/other marker	HLA-DR$^+$

Genetic characterization[1]

Cytogenetic karyotype	51, XY, +1p−, 2p−, +6, +8, +8, 12p+, +19
Receptor gene rearrangement	*IGH* G, *IGK* G, *IGL* G, *TCRB* G

Functional characterization[1]

Cytochemistry	ANAE$^+$, CAE$^+$, MPO$^-$, PAS$^-$

Comments

- Cell line expresses monocytic surface markers.

Reference

[1] Lange, B. et al. (1987) Blood 70, 192–199.

AML-193

Culture characterization[1]

Culture medium	IMDM + 2 ng/ml GM-CSF or 3 U/ml IL-3
Viral status	EBNA⁻, HTLV-II⁻
Authentication	yes (by cytogenetics)
Primary reference	Lange et al.[1]
Availability of cell line	ATCC CRL 9589/HTB 188

Clinical characterization[1]

Patient	13-year-old female
Disease diagnosis	AML M5
Treatment status	at relapse

Immunophenotypic characterization[1–3]

T-/NK cell marker	$CD2^-$, $CD3^-$, $CD4^-$, $CD5^-$, $CD7^-$, $CD8^-$
B-cell marker	$CD10^-$, $CD19^-$
Myelomonocytic marker	$CD13^+$, $CD14^-$, $CD15^+$, $CD16^-$, $CD33^+$
Erythroid-megakaryocytic marker	$CD41a^-$, $GlyA^-$
Progenitor/activation/other marker	$CD34^-$, $CD38^-$, $CD45^+$, $CD71^+$, $HLA\text{-}DR^+$
Cytokine receptor	$CD25^-$

Genetic characterization[1,4]

Cytogenetic karyotype	49, X, +3, +6, +8, −17, +der(17)t(17;17)(p13.1;q21.3)
Receptor gene rearrangement	*IGH* G, *IGK* G, *IGL* G, *TCRB* G

Functional characterization[1,2,5]

Cytochemistry	$ANAE^{(+)}$, CAE^-, MPO^-, $PAS^{(+)}$
Cytokine response	GM-CSF/IL-3-dependent; growth stimulation: G-CSF, IGF-I, PIXY-321; growth inhibition: TGF-β1

Comments

- Cell line with monocytic features.
- Constitutively cytokine-dependent.
- Available from cell line bank.
- Cells are rather difficult to grow.

References

[1] Lange, B. et al. (1987) Blood 70, 192–199.
[2] Santoli, D. et al. (1987) J. Immunol. 139, 3348–3354.
[3] Taetle, R. et al. (1993) Cancer Res. 53, 3386–3393.
[4] Hay, R. et al. (1994) ATCC Cell Lines and Hybridomas, 8th edn, Rockville, MD, USA.
[5] Drexler, H.G. et al. (1997) Leukemia 11, 701–708.

Culture characterization[1,2]

Culture medium	80–90% RPMI 1640 + 10–20% FBS
Doubling time	36–40 h
Viral status	EBV^-, HBV^-, HCV^-, HHV-8^-, HIV^-, HTLV-I/II^-
Authentication	no
Primary reference	Chen et al.[1]
Availability of cell line	DSM ACC 40

Clinical characterization[1]

Patient	40-year-old female
Disease diagnosis	AML M5
Treatment status	at relapse
Specimen site	peripheral blood
Year of establishment	1982

Immunophenotypic characterization[1,2]

T-/NK cell marker	$CD2^-$, $CD3^-$, $CD4^-$, $CD5^-$, $CD6^+$, $CD7^+$, $CD8^-$, $CD28^-$, $CD57^+$, $TCR\alpha\beta^-$, $TCR\gamma\delta^-$
B-cell marker	$CD10^-$, $CD19^-$, $CD20^-$, $CD21^-$, $cyIg^-$, sIg^-
Myelomonocytic marker	$CD13^-$, $CD14^{(+)}$, $CD15^+$, $CD33^-$, $CD68^-$
Erythroid-megakaryocytic marker	$CD41a^-$, $CD61^-$
Progenitor/activation/other marker	$CD34^-$, HLA-DR^-, TdT^-
Adhesion marker	$CD11b^-$
Cytokine receptor	$CD25^-$

Genetic characterization[1,2]

Cytogenetic karyotype	near tetraploid, 9% polyploidy; 92(89–94)<4n> XX/XXY/ XXYY, −1, +6, +6, −17, t(1;7)(p34.2;q34)×2, del(3)(p21), i(6q)×2, t(12;16)(q24.32;q11)×2
Unique gene alteration	*P53* mutation

Functional characterization[1,3]

Cytochemistry	ACP^-, $ANAE^+$ (NaF inhibitable), $ANBE^+$, MPO^-, MSE^-, PAS^+, SBB^-, $TRAP^-$
Inducibility of differentiation	resistant to various inducing agents

Comments

- Only weak expression of monocytic markers.
- Cells grow well.
- Available from cell line bank.

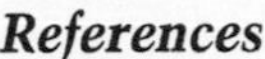

References

[1] Chen, P.M. et al. (1984) Gann 75, 660–664.

[2] Drexler, H.G. et al. (1999) DSMZ Catalogue of Cell Lines, 7th edn, Braunschweig, Germany.

[3] Uphoff, C.C. et al. (1994) Leukemia 8, 1510–1526.

DOP-M1

Culture characterization[1]

Culture medium	90% RPMI 1640 + 10% FBS
Doubling time	48 h
Authentication	yes (by cytogenetics)
Primary reference	Sugita et al.[1]
Availability of cell line	not known

Clinical characterization[1]

Patient	1-year-old female
Disease diagnosis	AML M5a
Treatment status	during therapy (refractory)
Specimen site	cerebrospinal fluid
Year of establishment	1989

Immunophenotypic characterization[1]

T-/NK cell marker	$CD2^-$, $CD3^-$, $CD4^-$, $CD7^-$, $CD8^-$
B-cell marker	$CD10^-$, $CD19^-$
Myelomonocytic marker	$CD13^-$, $CD14^-$, $CD15^+$, $CD33^+$, $CD65^+$
Erythroid-megakaryocytic marker	$CD41^-$
Progenitor/activation/other marker	$HLA\text{-}DR^-$
Adhesion marker	$CD11b^-$

Genetic characterization[1]

Cytogenetic karyotype	47, X, −X, −13, +19, +20, +mar

Functional characterization[1]

Cytochemistry	$ANAE^+$ (NaF inhibitable), CAE^+, MPO^+

Comments

- Described as monocytic cell line, but expressing few monocytic markers.

Reference

[1] Sugita, K. et al. (1992) Int. J. Hematol. 55, 42–51.

FLG 29.1

Culture characterization[1]

Culture medium	90% RPMI 1640 + 10% FBS
Viral status	EBV⁻
Authentication	no
Primary reference	Gattei et al.[1]
Availability of cell line	restricted

Clinical characterization[1]

Patient	38-year-old female
Disease diagnosis	AML M5a
Treatment status	at diagnosis
Specimen site	bone marrow
Year of establishment	1987

Immunophenotypic characterization[1,2]

T-/NK cell marker	$CD1a^-$, $CD2^-$, $CD3^-$, $CD4^-$, $CD5^-$, $CD7^-$, $CD8^-$
B-cell marker	$CD19^-$, $CD20^-$, $CD21^-$, $CD22^-$, $CD23^-$, $CD24^-$
Myelomonocytic marker	$CD13^+$, $CD14^{(+)}$, $CD15^{(+)}$, $CD32^+$ $CD33^{(+)}$, $CD35^{(+)}$, $CD68^+$
Erythroid-megakaryocytic marker	$CD9^+$, $CD31^-$, $CD36^-$, $CD41a^-$, $CD42b^{(+)}$, $CD61^+$
Progenitor/activation/other marker	$CD34^+$, $CD45^+$, $CD71^+$, HLA-DR^-
Adhesion marker	$CD11a^-$, $CD11b^-$, $CD18^-$, $CD44^+$, $CD51^{(+)}$, $CD54^+$, $CD55^+$
Cytokine receptor	$CD25^-$, $CD114^-$, $CD116^+$, $CD117^+$, $CD119^+$, $CD120a^+$, $CD120b^+$, $CD121a^+$, $CD123^+$, $CD124^-$, $CD126^-$

Genetic characterization[1]

Cytogenetic karyotype	45–69, 3p+
Receptor gene rearrangement	*IGH* G, *TCRB* G, *TCRG* G

Functional characterization[1,2]

Cytochemistry	ACP^+, $ANAE^+$ (NaF inhibitable), CAE^-, MSE^+, NBT^+, PAS^-, SBB^-, $TRAP^+$
Cytokine receptor	$mRNA^+$: IGF-RI, IGF-RII
Inducibility of differentiation	TPA → osteoclast differentiation
Special features	responsive to calcitonin; expression of calcitonin receptor and estrogen receptor

Comments

- Expression of monocytic features.
- Induced differentiation to an osteoclast-like phenotype.
- Described as 'pre-osteoclast cell line'.

References

[1] Gattei, V. et al. (1992) J. Cell Biol. 116, 437–447.
[2] Gattei, V. et al. (1997) Brit. J. Haematol. 97, 934–936.

IMS-M1

Culture characterization[1]

Culture medium	90% IMDM + 10% FBS
Doubling time	36–48 h
Authentication	no
Primary reference	Nagamura et al.[1]
Availability of cell line	restricted

Clinical characterization[1,2]

Patient	33-year-old male
Disease diagnosis	AML M5a
Specimen site	bone marrow
Year of establishment	1988

Immunophenotypic characterization[1,3]

T-/NK cell marker	$CD4^+$
Myelomonocytic marker	$CD14^+$, $CD33^+$
Progenitor/activation/other marker	$CD34^+$, $CD38^+$, HLA-DR^+
Cytokine receptor	$CD117^-$

Genetic characterization[1,2,4]

Cytogenetic karyotype	46, XY, −7, +8, −10, +12, 1p−, t(9;11)(p22;q23)
Unique translocation/fusion gene	t(9;11)(p22;q23) → *MLL-AF9* fusion gene
Unique gene alteration	*P53* mutation

Functional characterization[1]

Cytochemistry	$ANBE^+$, CAE^-, MPO^-, MSE^+
Cytokine response	growth stimulation: G-CSF, GM-CSF, M-CSF
Inducibility of differentiation	Vit. D3 → mono/macro differentiation
Special features	retinoic acids induce apoptosis

Comments

- Mature monocytic cell line.
- Carries t(9;11)(p22;q23) leading to *MLL-AF9* fusion gene.
- Slow cell growth.

References

[1] Nagamura, F. et al. (1995) Blood 86, 150a.
[2] Yamamoto, K. et al. (1994) Blood 83, 2912–2921.
[3] Inoue, K. et al. (1997) Blood 89, 1405–1412.
[4] Iida, S. et al. (1993) Oncogene 8, 3085–3092.

KBM-3

Culture characterization[1]

Establishment	initially in hypoxic environment
Culture medium	90% IMDM + 10% FBS
Doubling time	23 h
Authentication	yes (by cytogenetics)
Primary reference	Andersson et al.[1]
Availability of cell line	original authors

Clinical characterization[1]

Patient	32-year-old female
Disease diagnosis	AML M4
Treatment status	refractory
Specimen site	peripheral blood

Immunophenotypic characterization[1]

T-/NK cell marker	$CD2^+$, $CD3^-$, $CD4^+$, $CD5^-$, $CD8^-$
B-cell marker	$CD10^-$, $CD19^-$
Myelomonocytic marker	$CD14^+$, $CD33^+$
Progenitor/activation/other marker	$CD34^-$, HLA-DR^+, TdT^-

Genetic characterization[1-3]

Cytogenetic karyotype	48(41–51), −4, −12, +13, del(12)ins(12q+?), del(5)(p12-ter), del(9)(p11-ter), del(16)t(16q+?), del(7)ins(7p+HSR)
Unique gene alteration	*P53* mutation

Functional characterization[1,4]

Colony formation	in agar
Cytochemistry	$ANBE^+$, CAE^+, MPO^+, MSE^+, PAS^-
Cytokine production	GM-CSF $mRNA^+$
Cytokine receptor	M-CSFR $mRNA^+$
Heterotransplantation	into nude or SCID mice

Comments

- Monocytic cell line expressing monocytic immuno- and enzyme markers.
- Large cells growing well.

References

[1] Andersson, B.S. et al. (1992) Exp. Hematol. 20, 361–367.
[2] Sen, S. et al. (1993) Leukemia Res. 17, 639–647.
[3] Sen, S. et al. (1995) Cancer Genet. Cytogenet. 82, 35–40.
[4] Uphoff, C.C. et al. (1994) Leukemia 8, 1510–1526.

KBM-5

Culture characterization[1]

Establishment	initially in hypoxic environment
Culture medium	90% IMDM + 10% FBS
Doubling time	24–60 h
Viral status	EBNA⁻
Authentication	yes (by isoenzyme analysis)
Primary reference	Beran et al.[1]
Availability of cell line	original authors

Clinical characterization[1]

Patient	67-year-old female
Disease diagnosis	CML
Treatment status	at monocytic blast crisis
Specimen site	peripheral blood

Immunophenotypic characterization[1]

T-/NK cell marker	$CD2^-$, $CD4^+$
B-cell marker	sIg^-
Myelomonocytic marker	$CD13^+$, $CD14^-$, $CD33^+$
Progenitor/activation/other marker	$CD34^-$, HLA-DR^-, TdT^-
Adhesion marker	$CD11b^-$

Genetic characterization[1–5]

Cytogenetic karyotype	69–87, XX, +6, +7, +8, +8, +8, −9, −9, 9q+, −11, −11, −13, 14q−, +15, 17p+, +20, +22q−, +22q−, multiple Ph, +frag, +DMs
Unique translocation/fusion gene	Ph^+ t(9;22)(q34;q11) → *BCR-ABL* b3-a2 fusion gene
Unique gene alteration	*ABL* amplification, *BCR* amplification, *P53* mutation

Functional characterization[1,6]

Colony formation	in agar or methylcellulose
Cytochemistry	ACP^+, $ANBE^+$, CAE^+, $MPO^{(+)}$, MSE^+, PAS^-
Cytokine production	RT-PCR^+: TNF-α
Cytokine response	growth inhibition: TNF-α
Heterotransplantation	into SCID mice
Inducibility of differentiation	various differentiation inducers without effect
Special features	resistant to NK activity, IFN-α, IFN-γ

Comments

- Well-characterized monocytic cell line.
- Resistant to induction of differentiation and treatment with various agents.
- Carries Ph chromosome leading to *BCR-ABL* b3-a2 fusion gene.

References

[1] Beran, M. et al. (1993) Cancer Res. 53, 3603–3610.
[2] Blick, M.B. et al. (1986) Leukemia Res. 10, 1401–1409.
[3] Sen, S. et al. (1993) Leukemia Res. 17, 639–647.
[4] Sen, S. et al. (1995) Cancer Genet. Cytogenet. 82, 35–40.
[5] Drexler, H.G. et al. (1999) Leukemia Res. 23, 207–215.
[6] Uphoff, C.C. et al. (1994) Leukemia 8, 1510–1526.

KP-1

Culture characterization[1]

Culture medium	90% RPMI 1640 + 10% FBS
Doubling time	96 h
Authentication	no
Primary reference	Adachi et al.[1]
Availability of cell line	restricted

Clinical characterization[1]

Patient	2-year-old female
Disease diagnosis	AML M5
Treatment status	at diagnosis
Specimen site	peripheral blood
Year of establishment	1986

Immunophenotypic characterization[1]

T-/NK cell marker	CD2⁻, CD3⁻, CD4⁺, CD8⁻
B-cell marker	CD19⁻, CD20⁻
Myelomonocytic marker	CD13⁺, CD14⁺, CD33⁺
Progenitor/activation/other marker	HLA-DR⁺
Adhesion marker	CD11a⁺, CD11c⁺, CD18⁺

Genetic characterization[1]

Cytogenetic karyotype	47(46–48)XX, 1p−, 10p+, 16q+, 19p+, +mar, t(11q+;19p−)
Unique translocation/fusion gene	t(11;19)(q23;p13)? → *MLL-MLLT1* fusion gene?

Functional characterization[1]

Cytochemistry	ANBE⁺ (NaF inhibitable), CAE⁻, MPO⁻, PAS⁻
Inducibility of differentiation	TPA → macro differentiation
Special features	phagocytosis; production of lysozyme; scavenger receptor expression

Comments

- Monocytic cell line with typical monocytic features.
- Responsive to induction of differentiation.

Reference

[1] Adachi, N. et al. (1993) Pediat. Res. 34, 258–264.

KP-MO-TS

Culture characterization[1]

Culture medium	90% RPMI 1640 + 10% FBS
Doubling time	48 h
Viral status	EBNA$^-$
Authentication	yes (by cytogenetics)
Primary reference	Ikushima et al.[1]
Availability of cell line	original authors

Clinical characterization[1]

Patient	12-year-old male
Disease diagnosis	AML M5b
Treatment status	at diagnosis
Specimen site	peripheral blood
Year of establishment	1987

Immunophenotypic characterization[1]

T-/NK cell marker	CD1a$^-$, CD2$^-$, CD3$^-$, CD4$^+$, CD5$^-$, CD8$^-$
B-cell marker	CD10$^-$, CD19$^-$, CD20$^-$, sIg$^-$
Myelomonocytic marker	CD13$^-$, CD14$^-$, CD15$^+$, CD33$^+$, CD35$^-$
Progenitor/activation/other marker	HLA-DR$^+$, TdT$^+$
Adhesion marker	CD11b$^+$

Genetic characterization[1,2]

Cytogenetic karyotype	46, XY, −17, +mar, t(10;11)(p13;q21)
Unique translocation/fusion gene	t(10;11)(p13;q14) → *CALM-AF10* fusion gene
Receptor gene rearrangement	*IGH* G, *IGK* R, *TCRB* G, *TCRG* G

Functional characterization[1]

Cytochemistry	ACP$^+$, ANBE$^+$, CAE$^{(+)}$, MPO$^+$, PAS$^-$
Inducibility of differentiation	retinoic acid, TPA → mono/macro differentiation
Special features	α1-antitrypsin$^+$; production of lysozyme; phagocytosis

Comments

- Cell line carries monocytic features.
- Carries t(10;11)(p13;q14) leading to *CALM-AF10* fusion gene.
- Amenable to induced monocytic/macrophage differentiation.

References

[1] Ikushima, S. et al. (1990) Acta Haematol. Jpn. 53, 678–687.
[2] Narita, M. et al. (1999) Brit. J. Haematol. 105, 928–937.

ME-1

Culture characterization[1,2]

Subclones	**ME-2/ME-3, ME-F1/ME-F2/ME-F3** (different morphological, cytochemical, cytogenetic features)
Culture medium	90% RPMI 1640 + 10% FBS
Doubling time	4–5 d
Viral status	EBNA⁻
Authentication	yes (by cytogenetics)
Primary reference	Yanagisawa et al.[1]
Availability of cell line	original authors

Clinical characterization[1]

Patient	40-year-old male
Disease diagnosis	AML M4eo
Treatment status	at 2nd relapse
Specimen site	peripheral blood
Year of establishment	1988

Immunophenotypic characterization[1,2]

T-/NK cell marker	$CD2^-$, $CD7^-$
B-cell marker	$CD10^-$, $CD20^-$
Myelomonocytic marker	$CD13^+$, $CD14^+$, $CD33^+$
Progenitor/activation/other marker	$CD34^+$, $HLA\text{-}DR^+$
Adhesion marker	$CD11b^+$

Genetic characterization[1-3]

Cytogenetic karyotype	47, XY, +8, inv(16)(p13q22), del(17)(p12p13)
Unique translocation/fusion gene	inv(16)(p13q22) → *CBFB-MYH11* fusion gene

Functional characterization[1,4]

Colony formation	in methylcellulose
Cytochemistry	$ANBE^{(+)}$, CAE^+, $Luxol^{(+)}$, MPO^+, MSE^+, Toluidine $Blue^{(+)}$
Cytokine response	growth stimulation: GM-CSF, IL-3
Inducibility of differentiation	GM-CSF, IL-3, IL-4, PHA-LCM → macro differentiation; serum-free culture → neutro differentiation; PHA-LCM → eosino differentiation
Special features	production of lysozyme; phagocytosis

Comments

- Cell line with monocytic characteristics.
- Responsive to various differentiation inducers.
- Weak constitutive expression of eosinophil features which can be upregulated by differentiation-inducing agents.
- Carries inv(16) with *CBFB-MYH11* fusion gene (reference cell line).

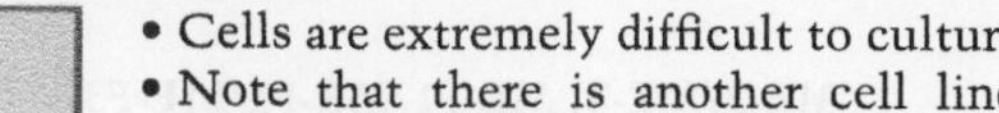

- Cells are extremely difficult to culture.
- Note that there is another cell line termed **ME-1** which was derived independently from a patient with multiple myeloma in New York, USA [reference: Hjyek, E. et al. (1998) Blood 92, 96a.].

References

[1] Yanagisawa, K. et al. (1991) Blood 78, 451–457.
[2] Yanagisawa, K. et al. (1994) Blood 84, 83–92.
[3] Liu, P. et al. (1993) Blood 82, 716–721.
[4] Yanagisawa, K. et al. (1992) Brit. J. Haematol. 80, 293–297.

ML-2

Culture characterization[1-4]

Sister cell lines	**ML-1, ML-3** (simultaneous sister cell lines – similar features)
Culture medium	90% RPMI 1640 + 10% FBS
Doubling time	60 h
Viral status	EBV^{-}, HBV^{-}, HCV^{-}, HHV-8^{-}, HIV^{-}, HTLV-I/II^{-}
Authentication	yes (by cytogenetics)
Primary reference	Ohyashiki et al.[3]
Availability of cell line	DSM ACC 15

Clinical characterization[1]

Patient	26-year-old male
Disease diagnosis	T-NHL → T-ALL → AML M4
Treatment status	at diagnosis of AML
Specimen site	peripheral blood
Year of establishment	1978

Immunophenotypic characterization[1-4]

T-/NK cell marker	CD1^{-}, CD2^{-}, CD3^{-}, cyCD3^{-}, CD4^{+}, CD5^{-}, CD7^{+}, CD8^{-}, CD28^{-}, CD57^{-}, TCR$\alpha\beta^{-}$, TCRδ^{-}
B-cell marker	CD10$^{(+)}$, CD19^{-}, CD20^{-}, CD21^{-}
Myelomonocytic marker	CD13^{+}, CD14$^{(+)}$, CD15^{+}, CD33^{+}
Erythroid-megakaryocytic marker	CD9^{-}, CD41a^{-}, CD61$^{(+)}$
Progenitor/activation/other marker	CD34^{-}, CD38^{+}, CD71^{+}, HLA-DR^{+}, TdT^{-}
Adhesion marker	CD11b^{+}
Cytokine receptor	CD25^{+}

Genetic characterization[2-7]

Cytogenetic karyotype	92(84–94)<4n>XX, −Y, −Y, −7, −9, −10, −10, +11, +12, +12, +13, +13, −15, −16, −17, −17, +18, +18, −20, −20, +4mar, der(1)t(1;?)(p21;?)×2, del(6)(q23)×2, der(6)t(6;11)(q27;?q23)×2, ?der(11)t(6;11)(q27;?q23)/del(11)(q23)×2, der(11)t(11;?) (?→11p11→11q23:)×2, der(11)t(11;?)(q11-13;?), dup(13)(q32→qter)×2, der(18)t(15;?;18)(q21;?;q11)×2
Unique translocation/fusion gene	t(6;11)(q27;q23)→ *MLL-AF6* fusion gene
Receptor gene rearrangement	*IGH* GR, *TCRA* GG, *TCRB* GR, *TCRG* GR
Unique gene alteration	ML-1: *P16INK4A* deletion, *P53* alteration, *RB1* rearrangement

Functional characterization[1,8-10]

Cytochemistry	MSE+
Inducibility of differentiation	AraC, DMSO, TPA → mono/macro differentiation
Transcription factor	ML-1: GATA-2 mRNA+

Comments

- Widely distributed and used monocytic cell line (reference cell line).
- Cytogenetically well-characterized cell line.
- Carries t(6;11)(q27;q23) leading to *MLL-AF6* fusion gene.
- Available from cell line bank.
- Cells are growing well.

References

1 Herrmann, R. et al. (1980) Cancer 46, 1383–1388. (case report)

2 Drexler, H.G. et al. (1999) DSMZ Catalogue of Cell Lines, 7th edn, Braunschweig, Germany.

3 Ohyashiki, K. et al. (1986) Cancer Res. 46, 3642–3647.

4 Ohyashiki, J.H. et al. (1989) Cancer Genet. Cytogenet. 37, 193–200.

5 Tanabe, S. et al. (1996) Genes Chromosomes Cancer 15, 206–216.

6 Nakamaki, T. et al. (1995) Brit. J. Haematol. 91, 139–149.

7 Sugimoto, K. et al. (1992) Blood 79, 2378–2383.

8 Uphoff, C.C. et al. (1994) Leukemia 8, 1510–1526.

9 Shimamoto, T. et al. (1995) Blood 86, 3173–3180.

10 Takeda, K. et al. (1982) Cancer Res. 42, 5152–5158.

MOLM-13

Culture characterization[1]

Sister cell line	**MOLM-14** (simultaneous sister cell line – some immunological and minor cytogenetic differences)
Culture medium	90% RPMI 1640 + 10% FBS
Doubling time	3–4 days
Authentication	yes (by cytogenetics)
Primary reference	Matsuo et al.[1]
Availability of cell line	original authors

Clinical characterization[1]

Patient	20-year-old male
Disease diagnosis	MDS (RAEB) → AML M5a
Treatment status	at relapse
Specimen site	peripheral blood
Year of establishment	1995

Immunophenotypic characterization[1]

T-/NK cell marker	CD1$^-$, CD2$^-$, CD3$^-$, CD4$^{(+)}$, CD5$^-$, CD7$^-$, CD8$^-$, CD57$^-$
B-cell marker	CD10$^-$, CD19$^-$, CD20$^-$, CD21$^-$, CD22$^-$, CD23$^-$, CD40$^-$
Myelomonocytic marker	CD13$^-$, CD14$^-$, CD15$^+$, CD32$^+$, CD33$^+$, CD64$^-$, CD65$^-$, CD68$^-$, CD87$^{(+)}$, CD91$^-$, CD92$^{(+)}$, CD93$^+$, CD155$^-$
Erythroid-megakaryocytic marker	CD9$^-$, CD41a$^-$, CD42b$^-$, CD61$^-$, CD62P$^-$
Progenitor/activation/other marker	CD34$^+$, HLA-DR$^-$, TdT$^-$
Adhesion marker	CD11a$^+$
Cytokine receptor	CD115$^-$, CD116$^+$, CD119$^+$, CD120a$^-$, CD120b$^-$

Genetic characterization[1]

Cytogenetic karyotype	49<2n>XY, +6, +8, +13, ins(11;9)(q23;p22p23), del(14)(q23.3q31.3)
Unique translocation/fusion gene	ins(11;9)(q23;p22p23) → *MLL-AF9* fusion gene
Unique gene alteration	*P15INK4B* deletion, *P16INK4A* deletion

Functional characterization[1]

Cytochemistry	MPO$^-$, MSE$^+$
Inducibility of differentiation	IFN-γ, IFN-γ+TNF-α → macro differentiation

Comments

- Cell line with some monocytic markers.
- Carries ins(11;9) leading to *MLL-AF9* fusion gene.
- Large cells growing slowly.

Reference

[1] Matsuo, Y. et al. (1997) Leukemia 11, 1469–1477.

Mono Mac 6

Culture characterization[1-3]

Sister cell line Mono Mac 1 (simultaneous sister cell line – diploid karyotype, otherwise similar)

Culture medium 90% RPMI 1640 + 10% FBS

Doubling time 50–60 h

Viral status EBV⁻, HBV⁻, HCV⁻, HHV-8⁻, HIV⁻, HTLV-I/II⁻

Authentication no

Primary reference Ziegler-Heitbrock et al.[1]

Availability of cell line Mono Mac 6: DSM ACC 124; Mono Mac 1: DSM ACC 252

Clinical characterization[1]

Patient 64-year-old male

Disease diagnosis myeloid metaplasia → AML M5

Treatment status at relapse

Specimen site peripheral blood

Year of establishment 1985

Immunophenotypic characterization[1-5]

T-/NK cell marker $CD2^-$, $CD3^-$, $CD4^{(+)}$

B-cell marker $CD19^-$, $CD20^-$, $CD23^-$

Myelomonocytic marker $CD13^+$, $CD14^+$, $CD15^+$, $CD33^+$, $CD68^+$

Erythroid-megakaryocytic marker $GlyA^-$

Progenitor/activation/other marker $CD34^-$, $HLA\text{-}DR^+$, TdT^-

Adhesion marker $CD11b^+$

Cytokine receptor $CD132^-$

Genetic characterization[1-3,6]

Cytogenetic karyotype hypotetraploid, flat-moded, near-diploid (8%), polyploid (17%) sidelines; 84–90<4n>XX/XXX, −Y, +6, +7, −12, −13, −13, −16, −16, +2mar, t(9;11)(p22;q23)×2, add(10)(p11)×2, add(12)(q21), del(13)(q13q14)der(13)t(13;14)(p11;q12) ×2, der(17)t(13;17) (q21;p11)×2

Unique translocation/fusion gene t(9;11)(p22;q23) → *MLL-AF9* fusion gene

Functional characterization[1,3,5,7]

Cytochemistry ACP^-, $ANAE^+$ (NaF inhibitable), CAE^-, MPO^-, MSE^+, PAS^-

Cytokine production IL-1

Cytokine receptor $mRNA^+$: IL-2Rα, IL-2Rβ, IL-2Rγ, IL-4Rα, IL-7Rα, IL-9Rα

Inducibility of differentiation IFN-γ, TPA → mono differentiation

Special features production of lysozyme; phagocytosis; mRNA+: azurocidin, C3 complement, cathepsin G, myeloblastin, *N*-elastase, tryptase

Comments

- Prototypical monocytic cell line (reference cell line).
- Available from cell line bank.
- Carries t(9;11)(p22;q23) leading to *MLL-AF9* fusion gene.
- Additional information on website <*www.med.uni-muenchen.de/immuno/ziegler*>.

References

[1] Ziegler-Heitbrock, H.W.L. et al. (1988) Int. J. Cancer 41, 456–461.

[2] Drexler, H.G. et al. (1999) DSMZ Catalogue of Cell Lines, 7th edn, Braunschweig, Germany.

[3] Steube, K.G. et al. (1997) Leukemia Res. 21, 327–335.

[4] Schumann, R.R. et al. (1996) Blood 87, 2419–2427.

[5] Abrink, M. et al. (1994) Leukemia 8, 1579–1584.

[6] Super, H.G. et al. (1995) Blood 85, 855–856.

[7] Uphoff, C.C. et al. (1994) Leukemia 8, 1510–1526.

MUTZ-3

Culture characterization[1,2]

Culture medium	80% α-MEM + 20% FBS + 5–10 ng/ml GM-CSF, IL-3 or PIXY-321
Doubling time	90–110 h
Viral status	EBV⁻, HBV⁻, HCV⁻, HHV-8⁻, HIV⁻, HTLV-I/II⁻
Authentication	yes (by DNA fingerprinting)
Primary reference	Hu et al.[1]
Availability of cell line	DSM ACC 295

Clinical characterization[1]

Patient	29-year-old male
Disease diagnosis	AML M4
Treatment status	at diagnosis
Specimen site	peripheral blood
Year of establishment	1993

Immunophenotypic characterization[1,2]

T-/NK cell marker	$CD1^-$, $CD3^-$, $CD4^+$, $CD5^-$, $CD7^-$, $CD8^-$, $CD56^-$
B-cell marker	$CD10^-$, $CD19^-$, $CD20^-$, $CD23^-$
Myelomonocytic marker	$CD13^+$, $CD14^+$, $CD15^+$, $CD16^-$, $CD32^+$, $CD33^+$, $CD64^{(+)}$, $CD65^+$, $CD68^+$
Erythroid-megakaryocytic marker	$CD41a^-$, $CD42b^-$, $CD61^-$, $GlyA^-$
Progenitor/activation/other marker	$CD30^+$, $CD34^{(+)}$, $CD38^+$, $CD71^+$, $HLA\text{-}DR^+$, TdT^-
Adhesion marker	$CD11b^+$
Cytokine receptor	$CD25^-$, $CD115^+$, $CD116^+$, $CD117^+$, $CD123^+$

Genetic characterization[1,3,4]

Cytogenetic karyotype	near-diploid, 6% polyploidy; 46(44–48)<2n>XY, t(1;3)(q43;q13) −inv(3)(q21q26), t(2;7)(q36;q36)inv(7)(p15q36), t(12;22)(p13;q12)
Unique translocation/fusion gene	(1) t(12;22)(p13;q12), but no *TEL-MN1* or *MN1-TEL* fusion (2) inv(3)(q21q26)
Unique gene alteration	*P53* mutation

Functional characterization[1,5]

Colony formation	in methylcellulose
Cytochemistry	MPO^+, MSE^+, $TRAP^+$

Cytokine response GM-CSF/IL-3-dependent; growth stimulation: G-CSF, IFN-β, IFN-γ, IGF-I, IL-4, IL-6, IL-7, insulin, M-CSF, MIP-1α, PIXY-321, SCF, TPO

Inducibility of differentiation retinoic acids, TPA → mono/macro differentiation

Comments

- Cell line with several monocytic features.
- Constitutively cytokine-dependent and responsive to various cytokines.
- Responsive to induction of monocytic differentiation.
- Carries t(12;22)(p13;q12) and inv(3)(q21q26).
- Available from cell line bank.

References

[1] Hu, Z.B. et al. (1996) Leukemia 10, 1025–1040.
[2] Drexler, H.G. et al. (1999) DSMZ Catalogue of Cell Lines, 7th edn, Braunschweig, Germany.
[3] MacLeod, R.A.F. et al. (1996) Genes Chromosomes Cancer 16, 144–148.
[4] Poirel, H. et al. (1998) Oncogene 16, 2895–2905.
[5] Drexler, H.G. et al. (1997) Leukemia 11, 701–708.

MV4-11

Culture characterization[1,2]

Culture medium	90% IMDM or RPMI 1640 + 10% FBS or IMDM + 5 ng/ml GM-CSF
Doubling time	50 h
Viral status	EBV⁻, HBV⁻, HCV⁻, HHV-8⁻, HIV⁻, HTLV-I/II⁻
Authentication	yes (by cytogenetics)
Primary reference	Lange et al.[1]
Availability of cell line	ATCC CRL 9591/HTB 189, DSM ACC 102

Clinical characterization[1]

Patient	10-year-old male
Disease diagnosis	AML M5
Treatment status	at diagnosis

Immunophenotypic characterization[1-3]

T-/NK cell marker	$CD2^-$, $CD3^-$, $CD4^+$, $CD5^-$, $CD7^-$, $CD8^-$
B-cell marker	$CD10^-$, $CD19^-$, $CD21^-$, $CD37^-$, $CD138^+$
Myelomonocytic marker	$CD13^+$, $CD14^{(+)}$, $CD15^+$, $CD16^-$, $CD33^+$
Progenitor/activation/other marker	$CD34^-$, $CD38^-$, HLA-DR^+
Cytokine receptor	$CD25^-$

Genetic characterization[1,2]

Cytogenetic karyotype	48(46–48)<2n>XY, +8, +18, +19, −21, t(4;11)(q21;q23)
Unique translocation/fusion gene	t(4;11)(q21;q23) → *MLL-AF4* fusion gene
Receptor gene rearrangement	*IGH* G, *IGK* RG, *IGL* G, *TCRB* G

Functional characterization[1,3,4]

Colony formation	in agar
Cytochemistry	$ANAE^+$, $CAE^{(+)}$, MPO^-, MSE^+, PAS^-
Cytokine response	growth stimulation: G-CSF, GM-CSF, IL-3

Comments

- Cell line with some monocytic features.
- Carries t(4;11)(q21;q23) leading to *MLL-AF4* fusion gene.
- Available from cell line banks.

References

[1] Lange, B. et al. (1987) Blood 70, 192–199.

[2] Drexler, H.G. et al. (1999) DSMZ Catalogue of Cell Lines, 7th edn, Braunschweig, Germany.

[3] Santoli, D. et al. (1987) J. Immunol. 139, 3348–3354.

[4] Uphoff, C.C. et al. (1994) Leukemia 8, 1510–1526.

NOMO-1

Culture characterization[1]

Culture medium	90% RPMI 1640 + 10% FBS or serum-free
Doubling time	24 h
Viral status	EBNA$^-$
Authentication	no
Primary reference	Kato et al.[1]
Availability of cell line	original authors

Clinical characterization[1]

Patient	31-year-old female
Disease diagnosis	AML M5a
Treatment status	at 2nd relapse
Specimen site	bone marrow
Year of establishment	1986

Immunophenotypic characterization[1]

T-/NK cell marker	CD2$^-$
B-cell marker	cyIg$^-$, sIg$^-$
Progenitor/activation/other marker	HLA-DR$^+$

Genetic characterization[1]

Cytogenetic karyotype	47, XX, 1p+, 7q+, 13p+

Functional characterization[1-3]

Cytochemistry	ACP$^+$, ANBE$^+$, CAE$^-$, MPO$^+$, MSE$^+$, NBT$^+$
Inducibility of differentiation	TPA → macro differentiation
Special features	production of lysozyme; phagocytosis

Comments

- Monocytic cell line requiring further immunological and cytogenetic characterization.
- Responsive to induction of differentiation.

References

[1] Kato, Y. et al. (1986) Acta Haematol. Jpn. 49, 277.
[2] Hamaguchi, M. et al. (1991) Blood 77, 94–100.
[3] Uphoff, C.C. et al. (1994) Leukemia 8, 1510–1526.

OCI/AML2

Culture characterization[1,2]

Culture medium	80–90% α-MEM + 10–20% FBS
Viral status	EBV⁻, HBV⁻, HCV⁻, HIV⁻, HTLV-I/II⁻
Authentication	no
Primary reference	Wang et al.[1]
Availability of cell line	DSM ACC 99

Clinical characterization[1]

Patient	65-year-old male
Disease diagnosis	AML M4
Treatment status	at diagnosis
Specimen site	peripheral blood
Year of establishment	1986

Immunophenotypic characterization[2]

T-/NK cell marker	$CD3^-$
B-cell marker	$CD19^-$
Myelomonocytic marker	$CD13^+$, $CD14^-$, $CD15^+$, $CD33^+$

Genetic characterization[2]

Cytogenetic karyotype	hyperdiploid, 3.8% polyploidy; 48(43–49)<2n>XY, +6, +8, der(1)inv(1)(p36q31)t(1;6)(q13;p12), der(2)t(2;17)(p23; q24.1) del(2)(q14.2q36), der(3)t(1;3)(p36;p25), ins(3;2)(q21;q14.2q36), t(5;8)(q11.2;q24), der(6)t(1;6)(q31;p12)t(3;6)(q26;q24), inv(12) (p13.3q13.2), t(13;14)(q32/33;q24.2), der(17)t(2;17)(p23;q24.1); sideline with +der(5)

Functional characterization[1,3]

Colony formation	in methylcellulose
Cytochemistry	MSE^+, NBT^-
Cytokine response	growth stimulation: G-CSF, GM-CSF, M-CSF

Comments

- Cell line with some monocytic features.
- More extensive immunoprofiling required.
- Available from cell line bank.

References

[1] Wang, C. et al. (1989) Leukemia 3, 264–269.

[2] Drexler, H.G. et al. (1999) DSMZ Catalogue of Cell Lines, 7th edn, Braunschweig, Germany.

[3] Uphoff, C.C. et al. (1994) Leukemia 8, 1510–1526.

OMA-AML-1

Culture characterization[1]

Culture medium	80% RPMI 1640 + 10% FBS + 10% horse serum
Authentication	no
Primary reference	Pirruccello et al.[1]
Availability of cell line	original authors

Clinical characterization[1,2]

Patient	male
Disease diagnosis	AML M4
Treatment status	at 5th relapse
Specimen site	peripheral blood
Year of establishment	1989

Immunophenotypic characterization[1,2]

Myelomonocytic marker	$CD13^+$, $CD14^{(+)}$, $CD15^+$, $CD33^+$
Progenitor/activation/other marker	$CD34^+$, HLA-DR^-
Adhesion marker	$CD11c^+$

Genetic characterization[1,2]

Cytogenetic karyotype	46, XY, t(2;7;11)(p21;p12;p15), +mar

Functional characterization[1,2]

Colony formation	clonable in suspension
Cytochemistry	$ANAE^+$ (NaF inhibitable), NBT^-
Cytokine response	growth stimulation: G-CSF, GM-CSF, IL-3, IL-6, LIF
Heterotransplantation	into nude or SCID mice
Inducibility of differentiation	GM-CSF, IL-3 → eosino differentiation
Special features	phagocytosis; formation of cobblestone areas

Comments

- Cell line with few monocytic features.
- Weak constitutive expression of eosinophil features which can be upregulated by differentiation-inducing agents.
- Further characterization required.

References

[1] Pirruccello, S.J. et al. (1992) Blood 80, 1026–1032.

[2] Pirruccello, S.J. et al. (1994) Leukemia Lymphoma 13, 169–178. (review on OMA-AML-1)

P31/Fujioka

Culture characterization[1]

Culture medium	90% RPMI 1640 + 10% FBS
Doubling time	80 h
Viral status	EBNA$^-$
Authentication	no
Primary reference	Hirose et al.[1]
Availability of cell line	JCRB 0091

Clinical characterization[1]

Patient	7-year-old male
Disease diagnosis	AML M5
Treatment status	at relapse
Specimen site	peripheral blood
Year of establishment	1980

Immunophenotypic characterization[1]

T-/NK cell marker	CD1$^-$, CD2$^-$, CD3$^-$, CD4$^-$, CD5$^-$, CD8$^-$
B-cell marker	CD10$^-$, CD20$^-$, CD80$^-$, CD86$^+$, cyIg$^-$, sIg$^-$
Myelomonocytic marker	CD14$^+$, CD15$^+$
Progenitor/activation/other marker	HLA-DR$^+$, TdT$^-$
Adhesion marker	CD11b$^+$, CD54$^+$

Genetic characterization[1-3]

Cytogenetic karyotype	47, XY, −10, +15, +19, t(7;11)(7qter→7p13::11q22→11qter; 11pter→11q22:), inv(9)(pter→p11::q13→q11:q13→qter), del(17) (qter→p11:); FISH: t(10;11)(p13;q14)
Unique translocation/fusion gene	t(10;11)(p13;q14) → *CALM-AF10* fusion gene
Unique gene alteration	*NRAS* mutation

Functional characterization[1]

Cytochemistry	ACP$^+$, ALP$^-$, ANBE$^+$ (NaF inhibitable), CAE$^+$, MPO$^-$, PAS$^+$
Special features	phagocytosis

Comments

- Cell line with typical monocytic characteristics.
- Carries t(10;11)(p13;q14) leading to *CALM-AF10* fusion gene.
- Available from cell line bank.

References

[1] Hirose, M. et al. (1982) Gann 73, 735–741.
[2] Sheng, X.M. et al. (1997) Leukemia Res. 21, 697–701.
[3] Narita, M. et al. (1999) Brit. J. Haematol. 105, 928–937.

P39/Tsugane

Culture characterization[1]

Culture medium	90% RPMI 1640 + 10% FBS
Authentication	no
Primary reference	Nagai et al.[1]
Availability of cell line	JCRB 0092

Clinical characterization[1]

Patient	69-year-old male
Disease diagnosis	MDS (CMML) → AML M2
Treatment status	at diagnosis
Specimen site	peripheral blood
Year of establishment	1983

Immunophenotypic characterization[1]

T-/NK cell marker	CD1⁻, CD2⁻, CD3⁻, CD4⁺, CD5⁻, CD6⁻, CD8⁻
B-cell marker	CD10⁻, CD20⁻, CD80⁻, CD86⁻, cyIg⁻, sIg⁻
Myelomonocytic marker	CD13⁺, CD14⁺, CD15⁺
Progenitor/activation/other marker	HLA-DR⁻, TdT⁻
Adhesion marker	CD11b⁻, CD54⁺

Genetic characterization[1,2]

Cytogenetic karyotype	45(39–46)XY, −7, −8, −16, −17, +22, +del(6)(pter→q15:), t(9;?)(q34;?), t(14;16)(14pter→14q24::16q21→16qter; 16pter→16q21::14q24→14qter)
Unique gene alteration	*NRAS* mutation

Functional characterization[1]

Cytochemistry	ACP⁺, ANBE⁺ (NaF inhibitable), CAE⁺, MPO⁻, PAS⁺
Heterotransplantation	into nude mice
Special features	phagocytosis

Comments

- Cell line with various monocytic features.
- Available from cell line bank.

References

[1] Nagai, M. et al. (1984) Gann 75, 1100–1107.
[2] Sheng, X.M. et al. (1997) Leukemia Res. 21, 697–701.

Culture characterization[1]

Sister cell line	**CESS-B** (= EBV^+ B-LCL)
Subclone	RC-2A is subclone of parental **RC-2** (CSF-dependent)
Culture medium	90% RPMI 1640 + 10% FBS
Doubling time	48 h
Authentication	no
Primary reference	Bradley et al.[1]
Availability of cell line	not known

Clinical characterization[1]

Patient	adult male
Disease diagnosis	AML M4
Treatment status	refractory
Specimen site	peripheral blood

Immunophenotypic characterization[2]

Myelomonocytic marker	$CD14^-$
Progenitor/activation/other marker	$HLA\text{-}DR^+$
Adhesion marker	$CD11b^+$

Genetic characterization[1]

Cytogenetic karyotype	47, XY, +11, t(11;12)

Functional characterization[1,2]

Colony formation	in agar
Cytochemistry	$ACP^{(+)}$, $ANAE^+$, $ANBE^+$, MSE^-
Inducibility of differentiation	PHA-LCM → mono/macro differentiation
Special features	antigen presentation

Comments

- Cell line with features suggestive of monocytic commitment.
- Further characterization required.
- A RC-2A culture which is in widespread circulation was shown to be cross-contaminated (real identity of this culture: immature T-cell line CCRF-CEM)[3]. It is not known whether the original, correct RC-2A still exists.

References

[1] Bradley, T.R. et al. (1982) Brit. J. Haematol. 51, 595–604.

[2] Lyons, A.B. and Ashman, L.K. (1987) Leukemia Res. 11, 797–805.

[3] Drexler, H.G. et al. (1999) Leukemia 13, 1601–1607.

RWLeu-4

Culture characterization[1,2]

Culture medium	90% RPMI 1640 + 10% FBS
Doubling time	25–26 h
Viral status	EBNA$^-$
Authentication	no
Primary reference	Wiemann et al.[1]
Availability of cell line	original authors

Clinical characterization[1]

Disease diagnosis	CML
Treatment status	at blast crisis
Specimen site	peripheral blood

Immunophenotypic characterization[1]

T-/NK cell marker	CD8$^-$
B-cell marker	CD21$^+$, sIg$^-$
Myelomonocytic marker	CD14$^-$, CD15$^+$
Progenitor/activation/other marker	HLA-DR$^{(+)}$
Adhesion marker	CD11b$^+$

Genetic characterization[1,3]

Cytogenetic karyotype	56, 7p+, 12p+, 17p+, Ph
Unique translocation/fusion gene	Ph$^+$ t(9;22)(q34;q11) → *BCR-ABL* b2-a2 fusion gene

Functional characterization[1,2,4,5]

Cytochemistry	CAE$^+$, MPO$^+$, MSE$^+$, NBT$^-$, PAS$^+$
Heterotransplantation	into nude mice
Inducibility of differentiation	DMSO, TPA, Vit. D3 → mono/macro differentiation
Proto-oncogene	MYC$^+$

Comments

- Cell line expressing some monocytic parameters.
- Carries Ph chromosome with *BCR-ABL* b2-a2 fusion gene.
- A more detailed characterization would be useful.

References

1 Wiemann, M. et al. (1985) Clin. Res. 33, 460A.
2 Lasky, S.R. et al. (1990) Cancer Res. 50, 3087–3094.
3 Drexler, H.G. et al. (1999) Leukemia Res. 23, 207–215.
4 Uphoff, C.C. et al. (1994) Leukemia 8, 1510–1526.
5 Clark, J.W. et al. (1992) J. Cancer Res. Clin. Oncol. 118, 190–194.

SCC-3

Culture characterization[1]

Culture medium	90% RPMI 1640 + 10% FBS
Doubling time	48 h
Viral status	EBNA$^-$, HTLV-I$^-$
Authentication	no
Primary reference	Kimura et al.[1]
Availability of cell line	JCRB 0115

Clinical characterization[1]

Patient	20-year-old female
Disease diagnosis	NHL (DLCL – stage IVA)
Treatment status	at relapse
Specimen site	pleural effusion
Year of establishment	1985

Immunophenotypic characterization[1]

T-/NK cell marker	CD1, CD2$^-$, CD3$^-$, CD4$^-$, CD5$^-$, CD8$^-$, CD57$^-$
B-cell marker	CD10$^+$, CD19$^-$, CD20$^-$, CD21$^-$, CD24$^-$, PCA-1$^+$
Myelomonocytic marker	CD13$^+$, CD14$^+$, CD15$^-$, CD16$^-$, CD33$^-$
Erythroid-megakaryocytic marker	CD36$^-$
Progenitor/activation/other marker	CD38$^+$, CD71$^+$, HLA-DR$^-$
Adhesion marker	CD11b$^-$
Cytokine receptor	CD25$^+$

Genetic characterization[1]

Cytogenetic karyotype	50, XX, −2, −4, −7, −15, +16, −20, +1p−, 1q−, 3q−, 6q−, 7p+, 8q−, 10q+, +11q+, +12q+, 12p−, 13p+, 17p+, + 5mar

Functional characterization[1]

Cytochemistry	ANBE$^+$ (NaF inhibitable), CAE$^-$, MPO$^-$, MSE$^-$, PAS$^+$
Inducibility of differentiation	TPA → macro differentiation
Special features	phagocytosis

Comments

- Cell line with some monocytic markers.
- Available from cell line bank.

Reference

[1] Kimura, Y. et al. (1986) Jpn. J. Cancer Res. 77, 862–865.

SKM-1

Culture characterization[1]

Culture medium	80% RPMI 1640 + 20% FBS (+ 1 ng/ml GM-CSF)
Doubling time	48 h
Viral status	EBNA⁻
Authentication	yes (by cytogenetics)
Primary reference	Nakagawa et al.[1]
Availability of cell line	JCRB 0118

Clinical characterization[1,2]

Patient	76-year-old male
Disease diagnosis	MDS (RAEBT) → AML M5
Treatment status	refractory
Specimen site	peripheral blood
Year of establishment	1989

Immunophenotypic characterization[1]

T-/NK cell marker	$CD2^-$, $CD3^-$, $CD4^+$, $CD5^-$, $CD7^-$, $CD8^-$
Myelomonocytic marker	$CD13^+$, $CD14^-$, $CD33^+$
Progenitor/activation/other marker	$HLA\text{-}DR^+$
Adhesion marker	$CD11b^+$

Genetic characterization[1]

Cytogenetic karyotype	46, XY, del(9)(q13q22), der(17)t(17;?)(p13;?)
Unique gene alteration	*P53* mutation

Functional characterization[1,3,4]

Colony formation	in agar
Cytochemistry	$ANBE^+$, MPO^+, MSE^+
Cytokine response	growth stimulation: GM-CSF
Special features	secretion of MPO; p53 overexpression

Comments

- Cytokine-responsive cell line with some monocytic markers.
- Available from cell line bank.

References

[1] Nakagawa, T. et al. (1993) Brit. J. Haematol. 85, 469–476.
[2] Nakagawa, T. et al. (1991) Brit. J. Haematol. 77, 250–252. (case report)
[3] Kawaguchi, R. et al. (1992) Leukemia 6, 1296–1301.
[4] Uphoff, C.C. et al. (1994) Leukemia 8, 1510–1526.
[5] Nakagawa, T. and Matozaki, S. (1995) Leukemia Lymphoma 17, 335–339. (review on SKM-1)

THP-1

Culture characterization[1,2]

Culture medium	90% RPMI 1640 + 10% FBS
Doubling time	35–40 h
Viral status	EBV$^-$, HBV$^-$, HCV$^-$, HHV-8$^-$, HIV$^-$, HTLV-I/II$^-$
Authentication	no
Primary reference	Tsuchiya et al.[1]
Availability of cell line	ATCC TIB 202, DSM ACC 16, JCRB 012, RCB 1189

Clinical characterization[1]

Patient	1-year-old male
Disease diagnosis	AML M5
Treatment status	at relapse
Specimen site	peripheral blood
Year of establishment	1978

Immunophenotypic characterization[1–4]

T-/NK cell marker	CD1$^-$, CD2$^-$, CD3$^-$, cyCD3$^-$, CD4$^+$, CD7$^+$, CD28$^-$, CD57$^-$, TCR$\alpha\beta^-$
B-cell marker	CD19$^-$, CD20$^-$, CD23$^{(+)}$, CD24$^+$, CD80$^-$, CD86$^+$, cyIg$^-$, sIg$^-$
Myelomonocytic marker	CD13$^+$, CD14$^{(+)}$, CD15$^+$, CD33$^+$, CD68$^-$
Erythroid-megakaryocytic marker	CD9$^-$, CD41a$^+$, CD61$^+$
Progenitor/activation/other marker	CD34$^-$, CD38$^+$, CD71$^+$, HLA-DR$^+$
Adhesion marker	CD11b$^+$, CD54$^+$
Cytokine receptor	CD25$^+$, CD117$^-$

Genetic characterization[1,2,5–9]

Cytogenetic karyotype	94(88–96)<4n>XY/XXY, −Y, +1, +3, +6, +6, −8, −13, −19, −22, −22, +2mar, add(1)(p11), del(1)(q42.2), i(2q), del(6)(p21)×2–4, i(7p), der(9)t(9;11)(p22;q23)i(9)(p10)×2, der(11)t(9;11)(p22;q23)×2, add(12)(q24)×1–2, der(13)t(8;13)(p11;p12), add(?18) (q21)
Unique translocation/fusion gene	t(9;11)(p21;q23) → *MLL-AF9* fusion gene
Receptor gene rearrangement	*TCRA* G, *TCRB* G
Unique gene alteration	*NRAS* mutation, *P15INK4B* deletion, *P16INK4A* deletion, *P53* mutation, *RB1* rearrangement

Functional characterization[1,10–12]

Cytochemistry	ALP$^-$, ANBE$^+$ (NaF inhibitable), CAE$^-$, MPO$^-$, MSE$^+$, PAS$^+$, SBB$^-$

Inducibility of differentiation	TPA → macro differentiation
Special features	antigen presentation; phagocytosis; production of lysozyme; mRNA+: azurocidin, C3 complement, *N*-elastase, myeloblastin

Comments

- Widely distributed and used monocytic cell line (reference cell line).
- Second-oldest monocytic leukemia cell line.
- Cell line displays various monocyte-specific features.
- Carries t(9;11)(p21;q23) leading to *MLL-AF9* fusion gene.
- Available from cell line banks.
- Cells grow very well.

References

1 Tsuchiya, S. et al. (1980) Int. J. Cancer 26, 171–176.
2 Drexler, H.G. et al. (1999) DSMZ Catalogue of Cell Lines, 7th edn, Braunschweig, Germany.
3 Inoue, K. et al. (1997) Blood 89, 1405–1412.
4 Minowada, J. and Matsuo, Y. (1999) unpublished data.
5 Nakamaki, T. et al. (1995) Brit. J. Haematol. 91, 139–149.
6 Ogawa, S. et al. (1994) Blood 84, 2431–2435.
7 Sugimoto, K. et al. (1992) Blood 79, 2378–2383.
8 Sheng, X.M. et al. (1997) Leukemia Res. 21, 697–701.
9 Sangster, R.N. et al. (1986) J. Exp. Med. 163, 1491–1508.
10 Tsuchiya, S. et al. (1982) Cancer Res. 42, 1530–1536.
11 Abrink, M. et al. (1994) Leukemia 8, 1579–1584.
12 Uphoff, C.C. et al. (1994) Leukemia 8, 1510–1526.

Culture characterization[1,2]

Subclones — **TK-1B** (pseudodiploid with complex chromosomal aberrations), **TK-1D** (diploid normal)
Culture medium — 90% RPMI 1640 + 10% FBS
Viral status — EBNA⁻
Doubling time — 36–48 h
Authentication — no
Primary reference — Ohno et al.[1]
Availability of cell line — original authors

Clinical characterization[1]

Patient — 22-year-old male
Disease diagnosis — T-lymphoblastic lymphoma → biclonal T-ALL L2 + AML M4
Treatment status — at diagnosis of AML M4
Specimen site — peripheral blood
Year of establishment — 1984

Immunophenotypic characterization[1]

T-/NK cell marker — $CD1^-$, $CD2^-$, $CD3^-$, $CD4^-$, $CD5^-$, $CD7^-$, $CD8^-$, $TCR\alpha\beta^-$
B-cell marker — $CD10^-$, $CD20^-$ $cyIg^-$, sIg^-
Myelomonocytic marker — $CD13^-$, $CD14^-$, $CD15^+$
Progenitor/activation/other marker — $HLA\text{-}DR^+$, TdT^-
Adhesion marker — $CD11b^-$

Genetic characterization[1,2]

Cytogenetic karyotype — 46, XY, −14, −17, +der(14)t(14;17)(14pter→14q22::17q23 →17qter), +der(17)t(11;14;17)(17pter→17q23::14q 22→14qter: :11q13→11qter)
Receptor gene rearrangement — *IGH* G, *TCRB* G, *TCRG* R

Functional characterization[1,3]

Colony formation — clonable
Cytochemistry — $ANAE^+$ (NaF inhibitable), CAE^+, MPO^+, MSE^+, NBT^-
Inducibility of differentiation — TPA, Vit. D3 → mono/macro differentiation

Comments

- Cell line with monocytic features.
- Responsive to induction of differentiation.

References

1 Ohno, H. et al. (1986) Int. J. Cancer 37, 761–767.
2 Nosaka, T. et al. (1988) J. Clin. Invest. 81, 1824–1828.
3 Uphoff, C.C. et al. (1994) Leukemia 8, 1510–1526.

Culture characterization[1,2]

Subclones	numerous drug-resistant subclones
Establishment	initially as grid organ culture, then on feeder layer (adult allogeneic skin fibroblasts or glia cells)
Culture medium	originally: 90% Ham's F10 + 10% newborn calf serum; now: 90% RPMI 1640 + 10% FBS
Doubling time	30–40 h
Viral status	EBV⁻, HBV⁻, HCV⁻, HHV-8⁻, HIV⁻, HTLV-I/II⁻
Authentication	no
Primary reference	Sundström and Nilsson[1]
Availability of cell line	ATCC CRL 1593, DSM ACC 5, IFO 50038, JCRB 9021, RCB 0435

Clinical characterization[1]

Patient	37-year-old male
Disease diagnosis	generalized, diffuse histiocytic lymphoma
Treatment status	refractory
Specimen site	pleural effusion
Year of establishment	1974

Immunophenotypic characterization[1–7]

T-/NK cell marker	$CD1^-$, $CD2^-$, $CD3^-$, $cyCD3^-$, $CD4^+$, $CD7^+$, $CD8^-$, $CD28^-$, $CD57^-$, $TCR\alpha\beta^-$, $TCR\gamma\delta^-$
B-cell marker	$CD10^-$, $CD19^-$, $CD20^-$, $CD21^-$, $CD23^+$, $CD24^{(+)}$, $CD75^-$, $CD80^-$, $CD86^-$, $cyIg^-$, sIg^-
Myelomonocytic marker	$CD13^+$, $CD14^{(+)}$, $CD15^+$, $CD17^-$, $CD33^+$, $CD35^+$, $CD68^+$, $CD88^{(+)}$
Erythroid-megakaryocytic marker	$CD9^-$, $CD41a^-$, $CD42b^-$, vWF^-
Progenitor/activation/other marker	$CD34^-$, $CD38^{(+)}$, $CD45^+$, $CD71^+$, $HLA\text{-}DR^-$, TdT^-
Adhesion marker	$CD11b^+$, $CD44^+$, $CD54^+$
Cytokine receptor	$CD25^-$, $CD116^+$, $CD117^-$, $CD121b^{(+)}$, $CD122^+$, $CD123^{(+)}$, $CD126^+$, $CD127^-$, $CD131^+$, $CD132^+$

Genetic characterization[1,2,8–10]

Cytogenetic karyotype	flat-moded, hypotriploid; 63(58–69)<3n>XXY, −2, −4, −6, +7, −9, −20, −21, +3mar, t(1;12)(q21;p13), der(5)t(1;5)(p22;q35), add(9)(p22), t(10;11)(p14;q23), i(11q), i(12p), add(16)(q22), add(19)(q13)
Unique translocation/fusion gene	t(10;11)(p13;q14) → *CALM-AF10* fusion gene
Unique gene alteration	*P53* mutation

Functional characterization[1,3,5,11,12]

Cytochemistry	ACP⁻, Alcian Blue(+), ALP⁻, ANAE⁺ (NaF inhibitable), ANBE⁺, CAE⁺, GLC⁺, MPO⁻, MSE⁺, Oil Red O⁻, PAS(+), SBB⁺, Toluidine Blue⁻, TRAP⁻
Cytokine receptor	mRNA⁺: IL-2Rα, IL-4Rα, IL-7Rα, IL-9Rα
Inducibility of differentiation	IFN-γ, retinoic acid, TPA, Vit. D3 → mono/macro differentiation
Special features	ADCC; phagocytosis; production of lysozyme; mRNA⁺: α1-antitrypsin, C3 complement, MPO, tryptase

Comments

- First human continuous monocytic cell line.
- Widely distributed and used monocytic cell line (reference cell line).
- Displays various monocytic characteristics.
- Carries t(10;11)(p13;q14) leading to *CALM-AF10* fusion gene.
- Available from cell line banks.
- Cells grow well.

References

1 Sundström, C. and Nilsson, K. (1976) Int. J. Cancer 17, 565–577.
2 Drexler, H.G. et al. (1999) DSMZ Catalogue of Cell Lines, 7th edn, Braunschweig, Germany.
3 Schumann, R.R. et al. (1996) Blood 87, 2419–2427.
4 Papayannopoulou, T. et al. (1987) J. Clin. Invest. 79, 859–866.
5 Winter, J.N. et al. (1984) Blood 63, 140–146.
6 Inoue, K. et al. (1997) Blood 89, 1405–1412.
7 Agis, H. et al. (1996) Leukemia Lymphoma 22, 187–204.
8 Dreyling, M.H. et al. (1996) Proc. Natl. Acad. Sci. USA 93, 4804–4809.
9 Sugimoto, K. et al. (1992) Blood 79, 2378–2383.
10 Narita, M. et al. (1999) Brit. J. Haematol. 105, 928–937.
11 Uphoff, C.C. et al. (1994) Leukemia 8, 1510–1526.
12 Abrink, M. et al. (1994) Leukemia 8, 1579–1584.

Culture characterization[1]

Culture medium	95% IMDM + 5% FBS + 5 ng/ml GM-CSF or 5 ng/ml IL-3
Doubling time	60–70 h
Authentication	no
Primary reference	Ikeda et al.[1]
Availability of cell line	original authors

Clinical characterization[1]

Patient	26-year-old female
Disease diagnosis	AML M5
Treatment status	during therapy
Specimen site	peripheral blood
Year of establishment	1997

Immunophenotypic characterization[1]

Myelomonocytic marker	$CD13^+$, $CD14^{(+)}$, $CD16^-$, $CD33^+$, $CD64^+$, $CD68^+$
Erythroid-megakaryocytic marker	$CD36^+$
Progenitor/activation/other marker	$CD71^+$, HLA-DR^+
Adhesion marker	$CD11b^+$, $CD11c^-$, $CD54^+$
Cytokine receptor	$CD114^+$

Genetic characterization[1]

Cytogenetic karyotype	46, XX, −7, +8, t(9;11)(p22;q23)
Unique translocation/fusion gene	t(9;11)(p22;q23) → *MLL-AF9* fusion gene

Functional characterization[1,2]

Colony formation	in methylcellulose (with cytokines)
Cytochemistry	ALP^-, $ANBE^+$, CAE^+, MPO^-
Cytokine production	IL-3 cultured cells: IL-6, IL-8, MCP-1, TNF-α; GM-CSF cultured cells: G-CSF, IL-8, MCP-1; M-CSF cultured cells: G-CSF, IL-6, IL-8, MCP-1, M-CSF, TNF-α
Cytokine receptor	RT-PCR^+: FLT3, G-CSFR, GM-CSFRα, IL-1R, IL-3Rα, IL-6Rα, IL-6R gp130, IL-7Rα, kit, M-CSFR
Cytokine response	GM-CSF/IL-3-dependent; growth stimulation: G-CSF, M-CSF
Inducibility of differentiation	G-CSF → neutro differentiation; GM-CSF, M-CSF → macro differentiation; M-CSF + IL-4/IL-13 → osteoclast differentiation

Comments

- Constitutively cytokine-dependent cell line with monocytic characteristics.
- Production of various cytokines.
- Responds proliferatively and with differentiation to several cytokines.
- Carries t(9;11)(p22;q23) leading to *MLL-AF9* fusion gene.

References

[1] Ikeda, T. et al. (1998) Blood 91, 4543–4553.
[2] Ikeda, T. et al. (1998) Biochem. Biophys. Res. Commun. 253, 265–272.

YK-M2

Culture characterization[1]

Culture medium	90% RPMI 1640 + 10% FBS
Doubling time	60 h
Viral status	EBNA$^-$
Authentication	yes (by cytogenetics)
Primary reference	Ohno et al.[1]
Availability of cell line	original authors

Clinical characterization[1]

Patient	31-year-old male
Disease diagnosis	AML M5
Treatment status	during therapy
Specimen site	peripheral blood
Year of establishment	1985

Immunophenotypic characterization[1]

T-/NK cell marker	CD1a$^-$, CD2$^-$, CD3$^-$, CD4$^-$, CD5$^-$, CD8$^-$, TCR$\alpha\beta^-$
B-cell marker	CD10$^-$, CD20$^-$, CD21$^-$, sIg$^-$
Myelomonocytic marker	CD13$^+$, CD14$^+$, CD15$^+$
Progenitor/activation/other marker	HLA-DR$^+$, TdT$^-$
Adhesion marker	CD11b$^-$

Genetic characterization[1]

Cytogenetic karyotype	68<3n>, X, −Y, +3, −4, −4, −5, +6, +7, −10, −11, −16, +4mar, del(17)(p11)

Functional characterization[1]

Cytochemistry	ANBE$^+$ (NaF inhibitable), CAE$^{(+)}$, MPO$^+$, NBT$^-$
Inducibility of differentiation	Vit. D3 → mono/macro differentiation
Special features	phagocytosis

Comments

- Cell line expressing a few monocytic markers.

Reference

[1] Ohno, H. et al. (1986) Cancer Res. 46, 6400–6405.

Part 3

Erythrocytic-Megakaryocytic Cell Lines

Culture characterization[1]

Culture medium	90% RPMI 1640 + 10% FBS
Doubling time	24–30 h
Authentication	yes (by cytogenetics)
Primary reference	Mossuz et al.[1]
Availability of cell line	original authors

Clinical characterization[1]

Patient	42-year-old male
Disease diagnosis	CML
Treatment status	at blast crisis
Specimen site	peripheral blood
Year of establishment	1992

Immunophenotypic characterization[1]

T-/NK cell marker	$CD2^-$, $CD3^-$
B-cell marker	$CD19^-$, $CD22^-$
Myelomonocytic marker	$CD15^+$, $CD33^+$
Erythroid-megakaryocytic marker	$CD36^+$, $CD41a^+$, $GlyA^{(+)}$
Progenitor/activation/other marker	$CD34^-$, $HLA\text{-}DR^+$
Adhesion marker	$CD11a^+$, $CD11b^+$
Cytokine receptor	MPL^+

Genetic characterization[1,2]

Cytogenetic karyotype	87, XXYY, −7, −9, der(5)t(5;?)(q31orq32;?), t(9;22)(q34;q11), del(17)(p11), der(19)t(11;19)(q13;q13), der(21)t(21;?)(p13;?)
Unique translocation/fusion gene	Ph^+ t(9;22)(q34;q11) → *BCR-ABL* b3-a2 fusion gene

Functional characterization[1]

Colony formation	in methylcellulose, clonable
Cytokine receptor	$RT\text{-}PCR^+$: EPO-R
Inducibility of differentiation	hemin, retinoic acid → ery differentiation; TPA → meg differentiation
Proto-oncogene	MYC $mRNA^+$
Transcription factor	GATA-1 $mRNA^+$
Special features	β-/γ-globin $mRNA^+$

Comments

- Model for erythrocytic-megakaryocytic precursor cell.
- Carries Ph chromosome with *BCR-ABL* b3-a2 fusion gene.
- Model for induction towards erythroid or megakaryocytic differentiation.
- Cells grow well.

References

[1] Mossuz, P. et al. (1997) Leukemia Res. 21, 529–537.
[2] Drexler, H.G. et al. (1999) Leukemia Res. 23, 207–215.

Culture characterization[1]

Culture medium 80% IMDM + 20% FBS + 2 U/ml EPO
Doubling time 49 h
Viral status EBV^-
Authentication yes (by DNA fingerprinting)
Primary reference Miyazaki et al.[1]
Availability of cell line original authors

Clinical characterization[1]

Patient 62-year-old male
Disease diagnosis AML M6
Treatment status at relapse
Specimen site bone marrow
Year of establishment 1993

Immunophenotypic characterization[1]

T-/NK cell marker $CD2^-$, $CD3^-$
B-cell marker $CD10^-$, $CD19^-$
Myelomonocytic marker $CD13^-$, $CD14^-$, $CD33^-$
Erythroid-megakaryocytic marker $CD36^+$, $CD41^-$, $GlyA^+$
Progenitor/activation/other marker $CD34^-$, $CD38^+$, $CD71^+$, HLA-DR^-
Adhesion marker $CD11b^{(+)}$
Cytokine receptor $CD25^-$, EPO-R^+

Genetic characterization[1]

Cytogenetic karyotype 67–82, XXYY, −2, −2, −4, −5, −10, −10, −13, −13, +14, −15, −15, −16, −16, −17, −18, −19, −19, −22, −22, +mar, del(1)(p32p36), del(3) (p21), del(3)(q21), der(4)t(1;4)(q21;p16)×2, add(8)(p23), add(8) (q24)×2, add(9)(q34), +del(9)(q11q22), add(10)(p14), add(11) (q23), i(11)(q10), add(15)(p12)

Functional characterization[1]

Colony formation in methylcellulose
Cytochemistry $ANBE^-$, $Benzidine^+$, CAE^-, MPO^-, $PAS^{(+)}$
Cytokine response EPO-dependent; no proliferative response to any other cytokine
Transcription factor $mRNA^+$: GATA-1, GATA-2
Special features γ-globin $mRNA^+$

Comments

- Model for erythrocytic precursor cell.
- Constitutively EPO-dependent, not responsive to other cytokines.
- Large cells growing slowly.

Reference

[1] Miyazaki, Y. et al. (1997) Leukemia 11, 1941–1949.

B1647

Culture characterization[1]

Culture medium	90% IMDM + 10% FBS
Authentication	no
Primary reference	Bonsi et al.[1]
Availability of cell line	not known

Clinical characterization[1]

Patient	14-year-old male
Disease diagnosis	AML M2
Specimen site	bone marrow

Immunophenotypic characterization[1]

T-/NK cell marker	$CD3^-$, $CD4^-$, $CD8^-$, $CD56^-$, $CD57^-$
B-cell marker	$CD19^-$, $CD20^-$, $CD23^-$
Myelomonocytic marker	$CD14^-$, $CD16^-$, $CD33^+$
Erythroid-megakaryocytic marker	$CD41^+$, $CD42b^-$, $GlyA^+$, vWF^+
Progenitor/activation/other marker	$CD34^-$, $CD38^+$, $HLA\text{-}DR^+$
Adhesion marker	$CD11c^-$, $CD54^+$
Cytokine receptor	$CD25^-$, $CD117^{(+)}$, $CD122^-$, $EPO\text{-}R^+$, MPL^+

Genetic characterization[1]

Cytogenetic karyotype	53, XY, +2, +5, +8, +13, −14, +19, +21, t(10;11)(p11;q21), +der(14)t(14;?)(p11;?)

Functional characterization[1]

Colony formation	in agar
Cytokine response	growth stimulation: TPO
Inducibility of differentiation	hemin, hydroxyurea, TPO → ery differentiation
Special features	γ-globin$^+$, α-tubulin$^+$; RT-PCR$^+$: α-/β-globin

Comments

- Cells express predominantly erythroid markers.
- TPO induces erythroid (not megakaryocytic) differentiation.

Reference

[1] Bonsi, L. et al. (1997) Brit. J. Haematol. 98, 549–559.

CHRF-288-11

Culture characterization[1,2]

Establishment established from the solid tumor line **CHRF-288** that since 1985 was continuously passaged by heterotransplantation into nude mice

Culture medium 75% Fischer + 25% horse serum (or FBS)

Authentication yes (by cytogenetics)

Primary reference Fugman et al.[1]

Availability of cell line original authors

Clinical characterization[1]

Patient 2-year-old male

Disease diagnosis AML M7 + myelofibrosis

Treatment status during therapy

Specimen site metastatic tumor

Year of establishment 1988

Immunophenotypic characterization[1]

T-/NK cell marker $CD1^-$, $CD2^-$, $CD3^-$, $CD4^-$, $CD5^-$, $CD7^-$, $CD8^-$, $CD56^-$

B-cell marker $CD10^-$, $CD19^-$, $CD20^-$

Myelomonocytic marker $CD13^+$, $CD14^-$, $CD33^+$

Erythroid-megakaryocytic marker $CD36^+$, $CD41a^+$, $GlyA^-$, PPO^+, vWF^+

Progenitor/activation/other marker $HLA\text{-}DR^+$

Genetic characterization[1]

Cytogenetic karyotype 50, XY, 1p+, 6p−, +6q−, +8, −10, 12p+, −15, +17, +19p+

Functional characterization[1,3]

Cytochemistry MPO^-

Cytokine production $RT\text{-}PCR^+$: bFGF, GM-CSF, IFN-α, IL-1β, IL-3, IL-7, IL-8, IL-11, SCF, TGF-β, TNF-α; $ELISA^+$: bFGF, GM-CSF, IFN-α, SCF, TGF-β, TNF-α

Cytokine receptor $RT\text{-}PCR^+$: FGF-R1, FGF-R2, GM-CSFRα, IFN-αR, IFN-βR, IL-1R, IL-6R, Kit, M-CSFR, MPL, TNF-αR

Cytokine response no proliferative response to various cytokines

Heterotransplantation into nude mice

Inducibility of differentiation TPA → meg differentiation; TPA enhances secretion: GM-CSF, IFN-α, IL-6R, TNF-α

Special features $\alpha\text{-granules}^+$

Comments

- Well-characterized regarding expression of cytokines and their receptors and response to cytokines.
- Model for megakaryocytic precursor cell.
- Responsive to megakaryocytic differentiation induction.

References

[1] Fugman, D.A. et al. (1990) Blood 75, 1252–1261.
[2] Witte, D.P. et al. (1986) Cancer 58, 238–244. (case report)
[3] Sandrock, B. et al. (1996) In Vitro Cell. Dev. Biol. 32, 225–233.

CMK

Culture characterization[1-4]

Subclones	**CMK6** (poorly differentiated), **CMK11-5** (well differentiated)
Culture medium	90% RPMI 1640 + 10% FBS
Doubling time	40–50 h
Viral status	EBV^{-}, HBV^{-}, HCV^{-}, HHV-8^{-}, HIV^{-}, HTLV-I/II^{-}
Authentication	yes (by cytogenetics)
Primary reference	Sato et al.[1]
Availability of cell line	DSM ACC 392, IFO 50428/30

Clinical characterization[1,2,5]

Patient	1-year-old male
Disease diagnosis	AML M7 + Down's syndrome
Treatment status	at relapse
Specimen site	peripheral blood
Year of establishment	1985

Immunophenotypic characterization[1,3,4]

T-/NK cell marker	CD1^{-}, CD2^{-}, CD3^{-}, CD4$^{(+)}$, CD7$^{(+)}$, CD8^{-}, CD56^{-}
B-cell marker	CD10^{-}, CD19^{-}, CD20^{-}, CD21^{-}, CD22^{-}
Myelomonocytic marker	CD13^{+}, CD14$^{(+)}$, CD15^{-}, CD33^{+}
Erythroid-megakaryocytic marker	CD36^{+}, CD41a^{+}, CD42b$^{(+)}$, GlyA^{+}, PPO^{+}, vWF^{+}
Progenitor/activation/other marker	CD34^{+}, CD38^{-}, CD71^{+}, HLA-DR^{+}
Cytokine receptor	CD25^{-}, CD122^{-}, MPL$^{(+)}$

Genetic characterization[1,3,6,7]

Cytogenetic karyotype	flat-moded hypotetraploid, 8% polyploidy; 85–90<4n>XY, −X, −Y, −2, −3, +5, −6, −6, −8, +11, −15, −15, +16, −17, −19, +21, +22, +7-11mar, add(1)(p36), add(1)(q31), add(3)(q11), del(3) (p14)×2–3, add(5)(q11), add(5)(q13), dup(8)(q11q21), add(8)(q13-21), del(8)(q11), del(9)(p21)×2, add(9)(q11)×2, del(10)(q22q24), der(11;17)(q10;q10), der(11)dup(11)(p13p15)t(5;11)(q11;p15) ×1–2, del(11)(q23), add(12)(p13)×2, add(17)(p1?), add(18)(q23)×2–3, add(19)(p13), der(20)t(1;20)(q2?5;q1?2)×2, add(22)(q13)
Unique gene alteration	*P53* mutation

Functional characterization[1,2,4]

Colony formation	in methylcellulose
Cytochemistry	ACP+, ALP−, ANAE+, MPO−, PAS+
Cytokine production	GM-CSF, IL-1α, IL-1β, IL-6, TGF-β, TNF-α
Cytokine response	growth stimulation: GM-CSF, IL-3, TPO
Inducibility of differentiation	TPA → meg differentiation
Proto-oncogene	SIS mRNA+
Transcription factor	mRNA+: GATA-1, GATA-2
Special features	α-granules+, demarcation system+

Comments

- Well-characterized megakaryocytic cell line.
- Responsive to megakaryocytic differentiation induction.
- Available from cell line banks.
- Cells grow very well.

References

1 Sato, T. et al. (1989) Brit. J. Haematol. 72, 184–190.
2 Komatsu, N. et al. (1989) Blood 74, 42–48.
3 Drexler, H.G. et al. (1999) DSMZ Catalogue of Cell Lines, 7th edn, Braunschweig, Germany.
4 Toba, K. et al. (1996) Exp. Hematol. 24, 894–901.
5 Sunami, S. et al. (1987) Blood 70, 368–371.
6 Ogawa, S. et al. (1994) Blood 84, 2431–2435.
7 Sugimoto, K. et al. (1992) Blood 79, 2378–2383.

Culture characterization[1]

Culture medium 80% RPMI 1640 + 20% FBS
Doubling time 42 h
Authentication yes (by cytogenetics)
Primary reference Sato et al.[1]
Availability of cell line not known

Clinical characterization[1]

Patient 2-year-old female
Disease diagnosis granulosarcoma → AML M7
Treatment status at relapse
Specimen site peripheral blood
Year of establishment 1991

Immunophenotypic characterization[1]

T-/NK cell marker CD1⁻, CD2⁻, CD3⁻, CD4⁻, CD5⁻, CD7⁻, CD8⁻
B-cell marker CD19⁻
Myelomonocytic marker CD13⁺, CD33⁺
Erythroid-megakaryocytic marker CD41⁺, CD42b⁽⁺⁾, GlyA⁻, PPO⁺
Progenitor/activation/other marker CD34⁺, CD38⁻, HLA-DR⁺
Cytokine receptor MPL⁺

Genetic characterization[1]

Cytogenetic karyotype 46, X, del(X)(q23), der(2)t(1;2)(q22-24;q35), add(5)(q15), der(7)t(7;8)(p22;q21), add(17)(p13), add(19)(q13)

Functional characterization[1]

Colony formation yes
Cytochemistry ACP⁺, ALP⁻, ANAE⁺, MPO⁻, PAS⁺
Inducibility of differentiation TPA → multinucleated cells

Comments

- Expresses predominantly megakaryocytic markers.

Reference

[1] Sato, T. et al. (2000) Leukemia Lymphoma 36, 397–404.

Culture characterization[1]

Culture medium	90% RPMI 1640 + 10% FBS
Doubling time	46 h
Authentication	no
Primary reference	Miura et al.[1]
Availability	original authors

Clinical characterization[1]

Patient	2-year-old male
Disease diagnosis	AML M7 + Down's syndrome
Treatment status	during therapy (refractory)
Specimen site	bone marrow
Year of establishment	1991

Immunophenotypic characterization[1]

T-/NK cell marker	$CD3^-$, $CD4^+$, $CD7^+$, $CD8^-$
B-cell marker	$CD10^-$, $CD19^-$, $CD20^-$
Myelomonocytic marker	$CD13^+$, $CD14^-$, $CD33^+$
Erythroid-megakaryocytic marker	$CD41a^+$, $GlyA^-$, PPO^+
Progenitor/activation/other marker	$CD34^+$

Genetic characterization[1]

Cytogenetic karyotype	del(17p)
Unique gene alteration	*P53* mutation

Functional characterization[1,2]

Colony formation	in methylcellulose
Cytochemistry	ACP^+, ALP^-, $ANBE^+$, MPO^-, PAS^+
Cytokine response	growth stimulation: GM-CSF, IL-3, IL-6, TPO
Inducibility of differentiation	cytokines → meg differentiation
Special features	α-granules$^+$

Comments

- Cell line expressing some megakaryocytic characteristics.
- Cells grow well.

References

[1] Miura, N. et al. (1998) Int. J. Mol. Med. 1, 559–563.
[2] Miura, N. et al. (1998) Brit. J. Haematol. 102 (suppl.), 33.

ELF-153

Culture characterization[1,2]

Culture medium	90% RPMI 1640 + 10% FBS + 2.5 ng/ml GM-CSF
Doubling time	36 h
Authentication	no
Primary reference	Mouthon et al.[1]
Availability of cell line	original authors

Clinical characterization[1,2]

Patient	48-year-old male
Disease diagnosis	acute myelofibrosis → AML M7
Treatment status	at refractory relapse
Specimen site	bone marrow
Year of establishment	1988

Immunophenotypic characterization[1]

T-/NK cell marker	$CD2^-$, $CD3^-$, $CD4^+$
B-cell marker	$CD10^-$, $CD19^-$, $CD20^-$
Myelomonocytic marker	$CD13^+$, $CD14^-$, $CD15^-$, $CD33^+$
Erythroid-megakaryocytic marker	$CD9^+$, $CD31^+$, $CD36^-$, $CD41^+$, $CD42a^-$, $CD42b^-$, $CD61^+$, $CD62^-$, $CD63^+$, $CD107a^+$, $CD107b^+$, $GlyA^-$, PPO^+, $vWF^{(+)}$
Progenitor/activation/other marker	$CD34^+$, $CD38^-$, $HLA\text{-}DR^+$
Adhesion marker	$CD11b^-$, $CD11c^-$, $CD51^+$
Cytokine receptor	$CD116^+$, $CD117^+$, $CD126^+$, $CD130^+$, EPO^-, MPL^+

Genetic characterization[1]

Cytogenetic karyotype	43, XY, −7, −14, −17, del(5)(q13q31), t(12;14)(p11.2;q11.2)

Functional characterization[1–3]

Colony formation	in methylcellulose
Cytochemistry	$Benzidine^-$
Cytokine response	GM-CSF-dependent; growth stimulation: IL-3, IL-4, IL-6, PIXY-321, SCF
Heterotransplantation	into nude mice
Transcription factor	$mRNA^+$: GATA-1, GATA-2
Special features	demarcation $system^+$

Comments

- Megakaryocytic precursor cell line.
- Constitutively growth factor-dependent cell line.
- Cells grow slowly, difficult to culture.

References

[1] Mouthon, M.A. et al. (1994) Blood 84, 1085–1097.
[2] Hassan, H.T. et al. (1995) Int. J. Exp. Pathol. 76, 361–367.
[3] Drexler, H.G. et al. (1997) Leukemia 11, 701–708.

F-36P

Culture characterization[1]

Subclone	**F-36E** (EPO-dependent)
Culture medium	90% RPMI 1640 + 10% FBS + 1 ng/ml GM-CSF or 1 ng/ml IL-3
Authentication	no
Primary reference	Chiba et al.[1]
Availability of cell line	F-36P: RCB 0775; F-36E: RCB 0776

Clinical characterization[1]

Patient	68-year-old male
Disease diagnosis	MDS (RAEB) → AML M6
Treatment status	at diagnosis
Specimen site	pleural effusion
Year of establishment	1989

Immunophenotypic characterization[1]

T-/NK cell marker	CD4$^-$, CD5$^-$, CD8$^-$
B-cell marker	CD10$^-$, CD19$^-$
Myelomonocytic marker	CD13$^+$, CD14$^-$, CD33$^+$
Erythroid-megakaryocytic marker	CD41a$^+$, CD42b$^-$, GlyA$^+$, PPO$^{(+)}$
Progenitor/activation/other marker	CD34$^+$, CD45$^+$
Cytokine receptor	CD25$^-$, CD116$^+$

Genetic characterization[1,2]

Cytogenetic karyotype	43, Y, Xp+, −5, −7, −13, −16, −17, −19, −21, 2q−, 9p+, 10q+
Unique gene alteration	F-36E: *P16INK4A* deletion, *P53* mutation

Functional characterization[1,3]

Cytochemistry	ANAE$^-$, ANBE$^-$, MPO$^-$, PAS$^+$
Cytokine response	GM-CSF/IL-3-dependent; growth stimulation: EPO, IFN-γ, IL-5, PIXY-321
Inducibility of differentiation	EPO → ery differentiation

Comments

- Model for erythroid differentiation.
- Constitutively growth factor-dependent cell line.
- Available from cell line bank.

References

[1] Chiba, S. et al. (1991) Blood 78, 2261–2268.
[2] Ogawa, S. et al. (1994) Blood 84, 2431–2435.
[3] Drexler, H.G. et al. (1997) Leukemia 11, 701–708.

HEL

Culture characterization[1,2]

Culture medium	90% RPMI 1640 + 10% FBS
Doubling time	24–36 h
Viral status	EBV⁻, HBV⁻, HCV⁻, HIV⁻, HTLV-I/II⁻
Authentication	yes (by cytogenetics)
Primary reference	Martin and Papayannopoulou[1]
Availability of cell line	ATCC TIB 180, DSM ACC 11, JCRB 0062

Clinical characterization[1]

Patient	30-year-old male
Disease diagnosis	Hodgkin's disease → AML M6
Treatment status	at relapse (post-BMT)
Specimen site	peripheral blood
Year of establishment	1980

Immunophenotypic characterization[1-5]

T-/NK cell marker	CD1$^-$, CD2$^-$, CD3$^-$, CD4$^+$, CD5$^-$, CD7$^-$, CD8$^-$, CD28$^-$, CD56$^-$, TCRαβ$^-$
B-cell marker	CD10$^-$, CD19$^-$, CD20$^{(+)}$, CD21$^{(+)}$, CD22$^-$, Ig$^-$
Myelomonocytic marker	CD13$^+$, CD14$^-$, CD15$^+$, CD33$^+$
Erythroid-megakaryocytic marker	CD9$^+$, CD36$^+$, CD41a$^+$, CD42a$^-$, CD42b$^{(+)}$, CD61$^+$, CD62$^-$, GlyA$^+$, PPO$^-$, vWF$^{(+)}$
Progenitor/activation/other marker	CD34$^-$, CD38$^-$, CD71$^+$, HLA-DR$^{(+)}$
Adhesion marker	CD11b$^+$, CD44$^+$
Cytokine receptor	CD25$^-$, CD117$^+$, CD122$^+$, CD132$^+$, EPO-R$^+$, MPL$^+$

Genetic characterization[1,2,6]

Cytogenetic karyotype	hypotriploid, 2.3% polyploidy; 63(60–64)<3n>XYY, −2, −9, −10, −10, −11, −14, −16, −16, −17, −19, +20, +21, +2mar, del(2)(q32), t(3;6)(p13;q16), der(5)t(5;17)(q10;q10), der(6)t(1;6)(p13;p21), der(7)add(7)(p14;q32), add(8)(p21), der(9)t(9;?)(?;11)(p24;?) (?;q13), del(11)(q13), add(15)(p11), del(20)(q13), r(20)(p11q11), dup(21)(q11q22.3-qter), psu dic(22;9)t(9;?) (?;22)(p24;?)(?;p11-13); carries masked 5q− and 20q−
Unique gene alteration	*P15INK4B* deletion, *P16INK4A* deletion

Functional characterization[1,7–9]

Colony formation	in methylcellulose, clonable
Cytochemistry	ACP$^+$, ANAE$^+$, ANBE$^+$, Benzidine$^-$, CAE$^-$, MPO$^-$, PAS$^+$, SBB$^-$
Cytokine production	PDGF
Inducibility of differentiation	hemin → erythroid differentiation; TPA → meg differentiation
Transcription factor	mRNA$^+$: GATA-1, GATA-2, SCL
Special features	α-/β-/γ-/ε-/ξ-globin$^+$; δ-aminolevulin synthase mRNA$^+$; Hb Bart's$^+$

Comments

- Widely distributed and used cell line (reference cell line).
- Well-characterized cell line.
- Erythrocytic and megakaryocytic differentiation potential.
- Large cells growing well.
- Available from cell line banks.

References

1 Martin, P. and Papayannopoulou, T. (1982) Science 216, 1233–1235.
2 Drexler, H.G. et al. (1999) DSMZ Catalogue of Cell Lines, 7th edn, Braunschweig, Germany.
3 Micallef, M. et al. (1994) Hematol. Oncol. 12, 163–174.
4 Toba, K. et al. (1996) Exp. Hematol. 24, 894–901.
5 Schumann, R.R. et al. (1996) Blood 87, 2419–2427.
6 Drexler, H.G. (1998) Leukemia 12, 845–859.
7 Hirata, J. et al. (1990) Leukemia 4, 365–372.
8 Endo, K. et al. (1993) Brit. J. Haematol. 85, 653–662.
9 Shimamoto, T. et al. (1995) Blood 86, 3173–3180.

Culture characterization[1-3]

Subclones	**HML-2, HML/SE** (differences in response to cytokines)
Establishment	initially cultured in semi-solid media
Culture medium	90% α-MEM + 10% FBS + 10 ng/ml GM-CSF
Authentication	no
Primary reference	Ma et al.[1]
Availability of cell line	original authors

Clinical characterization[1]

Patient	2-year-old male
Disease diagnosis	AML M7 + Down's syndrome
Treatment status	at diagnosis
Specimen site	peripheral blood

Immunophenotypic characterization[1-3]

T-/NK cell marker	$CD2^-$, $CD3^-$, $CD4^+$, $CD8^-$
B-cell marker	$CD19^-$
Myelomonocytic marker	$CD13^+$, $CD14^+$, $CD15^-$, $CD33^+$
Erythroid-megakaryocytic marker	$CD36^+$, $CD41b^+$, $CD42b^-$, $GlyA^+$, $PPO^{(+)}$, vWF^-
Progenitor/activation/other marker	$CD34^-$, $CD38^-$, $CD45RA^+$, $CD45RO^+$, $CD71^+$, $HLA\text{-}DR^+$
Adhesion marker	$CD11b^{(+)}$
Cytokine receptor	$CD117^+$, $CD126^-$, $CD130^+$

Genetic characterization[1]

Cytogenetic karyotype	47, XY, +21, del(3)(q24q26.1)

Functional characterization[1-3]

Cytochemistry	ACP^+, ALP^-, $ANBE^+$, CAE^-, MPO^-, PAS^+, SBB^-
Cytokine response	GM-CSF-dependent; growth stimulation: SCF; growth inhibition: TGF-β
Inducibility of differentiation	TPA → meg differentiation; SCF + IL-5 + retinoic acid → eosino differentiation; SCF + EPO → ery differentiation (HML/SE)
Special features	eosino POX^-, major basic $protein^{(+)}$, Biebrich $Scarlet^-$, Luxol Fast $Blue^-$, Toluidine $Blue^-$; $PF4^-$, $\beta\text{-}TG^-$

Comments

- Megakaryocytic cell line with erythroid, megakaryocytic and eosinophil differentiation potential.
- Constitutively growth factor-dependent and cytokine-responsive.
- Cells are very difficult to culture.

References

[1] Ma, F. et al. (1998) Brit. J. Haematol. 100, 427–435.
[2] Kamijo, T. et al. (1997) Leukemia Res. 21, 1097–1106.
[3] Sato, T. et al. (1998) J. Biol. Chem. 273, 16921–16926.

HU-3

Culture characterization[1,2]

Subclone	**HU-3/TPO** (TPO-dependent)
Culture medium	90% RPMI 1640 + 10% human serum + 5 ng/ml GM-CSF or 5 ng/ml IL-3 or 100 U/ml TPO
Authentication	no
Primary reference	Morgan et al.[1]
Availability of cell line	original authors

Clinical characterization[1]

Patient	69-year-old female
Disease diagnosis	AML M7
Treatment status	at diagnosis
Specimen site	bone marrow
Year of establishment	1991

Immunophenotypic characterization[1]

T-/NK cell marker	$CD1a^-$, $CD2^-$, $CD3^-$, $CD4^+$, $CD7^-$, $CD8^-$
B-cell marker	$CD10^-$, $CD19^-$, $CD21^+$, $CD23^-$
Myelomonocytic marker	$CD13^+$, $CD14^+$, $CD15^+$, $CD16^-$, $CD32^+$, $CD33^+$, $CD35^+$
Erythroid-megakaryocytic marker	$CD31^+$, $CD36^+$, $CD41a^+$, $CD41b^-$, $CD42a^-$, $CD42b^-$, $CD61^+$, $GlyA^+$
Progenitor/activation/other marker	$CD34^+$, $CD38^+$, $CD45^+$, $CD45RA^+$, $CD71^+$, $HLA\text{-}DR^+$
Adhesion marker	$CD11a^-$, $CD11b^+$, $CD11c^-$, $CD18^-$, $CD44^+$, $CD49b^-$, $CD49d^+$, $CD49e^+$, $CD49f^+$, $CD51^-$, $CD54^+$, $CD58^+$
Cytokine receptor	$CD25^-$, MPL^+

Genetic characterization[1]

Cytogenetic karyotype	73(70–76)XXX, +7, +7, +8, +8, −17, −17, −22, +21mar

Functional characterization[1–4]

Colony formation	clonable
Cytokine production	$RT\text{-}PCR^+$: GM-CSF, IL-1β, IL-6, IL-7, IL-10, IL-13, SCF, TGF-β, TNF-α
Cytokine response	GM-CSF/IL-3-dependent; growth stimulation: EPO, IFN-β, IFN-γ, IL-1α, IL-4, IL-5, IL-6, LIF, NGF, OSM, PIXY-321, SCF, TNF-α, TNF-β, TPO
Inducibility of differentiation	EPO → ery differentiation; TPA → meg differentiation
Transcription factor	$NF\text{-}E2^+$

Comments

- Megakaryocytic cell line with erythroid and megakaryocytic differentiation potential.
- Well characterized regarding immunoprofile and cytokines.
- Constitutively cytokine-dependent cell line responsive to several other cytokines.
- Cells are difficult to culture.

References

[1] Morgan, D.A. et al. (1997) Exp. Hematol. 25, 1378–1385.
[2] Drexler, H.G. et al. (1997) Leukemia 11, 541–551.
[3] Soslau, G. et al. (1997) Cytokine 9, 405–411.
[4] Drexler, H.G. et al. (1997) Leukemia 11, 701–708.

JK-1

Culture characterization[1,2]

Culture medium	80% RPMI 1640 + 20% FBS
Doubling time	48 h
Viral status	EBV⁻, HBV⁻, HCV⁻, HHV-8⁻, HIV⁻, HTLV-I/II⁻
Authentication	no
Primary reference	Okuno et al.[1]
Availability of cell line	DSM ACC 347

Clinical characterization[1]

Patient	62-year-old male
Disease diagnosis	CML
Treatment status	at erythroid blast crisis
Specimen site	tumor
Year of establishment	1987

Immunophenotypic characterization[1-3]

T-/NK cell marker	$CD2^-$, $CD3^-$, $CD4^-$, $CD8^-$
B-cell marker	$CD10^-$, $CD19^-$, $CD20^-$, $CD21^-$
Myelomonocytic marker	$CD13^+$, $CD14^-$
Erythroid-megakaryocytic marker	$CD41^+$, $CD42^+$, $GlyA^+$
Progenitor/activation/other marker	HLA-DR^-
Adhesion marker	$CD11b^-$
Cytokine receptor	$CD117^+$, EPO-R^+, $MPL^{(+)}$

Genetic characterization[1,2,4]

Cytogenetic karyotype	hyperdiploid, no significant polyploidy; 48(45–48)<2n>XY, +8, +22, t(1;7)(q10;q10), der(9)t(9;22)(q34;q11), der(22)t(9;22)(q34;q11)×2; carries two Ph
Unique translocation/fusion gene	Ph^+ t(9;22)(q34;q11) → *BCR-ABL* b2-a2 fusion gene

Functional characterization[1,5]

Colony formation	in methylcellulose (red colonies)
Cytochemistry	ACP^+, $ANBE^-$, $Benzidine^+$, CAE^-, MPO^-, PAS^+
Cytokine response	growth stimulation: EPO
Inducibility of differentiation	spontaneous erythroid differentiation; δ-ALA → ery differentiation
Special features	β-/γ-globin $mRNA^+$; HbF^+

Comments

- Erythroid precursor cell line.
- Potential to differentiate along erythrocytic cell lineage.
- Produces globin and hemoglobin.
- Carries Ph chromosome with *BCR-ABL* b2-a2 fusion gene.
- Available from cell line bank.
- Cells grow well.

References

[1] Okuno, Y. et al. (1990) Cancer 66, 1544–1551.

[2] Drexler, H.G. et al. (1999) DSMZ Catalogue of Cell Lines, 7th edn, Braunschweig, Germany.

[3] Hitomi, K. et al. (1988) Biochem. Biophys. Res. Commun. 154, 902–909.

[4] Drexler, H.G. et al. (1999) Leukemia Res. 23, 207–215.

[5] Hirata, J. et al. (1990) Leukemia 4, 365–372.

JURL-MK1

Culture characterization[1]

Sister cell line	**JURL-MK2** (simultaneous sister cell line – very similar)
Culture medium	80% DMEM + 20% FBS
Doubling time	48 h
Authentication	no
Primary reference	Di Noto et al.[1]
Availability of cell line	original authors

Clinical characterization[1]

Patient	73-year-old male
Disease diagnosis	CML
Treatment status	at blast crisis
Specimen site	peripheral blood
Year of establishment	1993

Immunophenotypic characterization[1]

T-/NK cell marker	CD2⁻, CD7⁻, CD56⁻
B-cell marker	CD19⁻, CD24⁻, CD40⁻
Myelomonocytic marker	CD13⁺, CD33⁺
Erythroid-megakaryocytic marker	CD9⁻, CD36⁺, CD41a⁺, CD42b⁻, CD61⁺, CD62P⁻, CD63⁺, GlyA⁺
Progenitor/activation/other marker	CD30⁻, CD34⁻, CD45RA⁻, CD45RO⁺, CD69⁺, HLA-DR⁺
Adhesion marker	CD11a⁻, CD11b⁻, CD11c⁻, CD43⁺, CD44⁻, CD54⁻, CD62L⁻
Cytokine receptor	CD25⁺, CD117⁺, CD122⁻

Genetic characterization[1,2]

Cytogenetic karyotype	33–43(39), XY, −4, −5, −9, −11, −12, −18, −19, +1mar, der(9)t (9;22)(q34;q11), i(17), der(22)t(9;22)(q34;q11)
Unique translocation/fusion gene	Ph⁺ t(9;22)(q34;q11) → *BCR-ABL* b3-a2 fusion gene

Functional characterization[1]

Cytochemistry	ALP⁻, ANAE⁽⁺⁾, MPO⁻, PAS⁺
Cytokine response	growth stimulation: SCF
Inducibility of differentiation	TPA → meg differentiation

Comments

- Predominantly megakaryocytic cell line.
- Carries Ph chromosome with *BCR-ABL* b3-a2 fusion gene.
- Megakaryocytic differentiation upon induction.
- Cells grow very well.

References

[1] Di Noto, R. et al. (1997) Leukemia 11, 1554–1564.
[2] Drexler, H.G. et al. (1999) Leukemia Res. 23, 207–215.

K-562

Culture characterization[1–3]

Subclones	many drug-resistant subclones
Culture medium	90% Eagle's MEM or RPMI 1640 + 10% FBS
Doubling time	24–30 h
Viral status	EBV$^-$, HBV$^-$, HCV$^-$, HHV-8$^-$, HIV$^-$, HTLV-I/II$^-$
Authentication	yes (by cytogenetics)
Primary reference	Lozzio and Lozzio[1]
Availability of cell line	ATCC CCL 243, DSM ACC 10, JCRB 0019, RCB 0027

Clinical characterization[1,2]

Patient	53-year-old female
Disease diagnosis	CML
Treatment status	at blast crisis
Specimen site	pleural effusion
Year of establishment	1970

Immunophenotypic characterization[1,3–5]

T-/NK cell marker	CD1$^-$, CD2$^-$, CD3$^-$, CD4$^-$, CD5$^-$, CD7$^-$, CD8$^-$, CD28$^-$, CD56$^-$, CD57$^-$, TCR$\alpha\beta^-$, TCRδ^-
B-cell marker	CD10$^-$, CD19$^-$, CD20$^-$, CD21$^-$, CD22$^-$, CD80$^+$, CD86$^-$
Myelomonocytic marker	CD13$^+$, CD14$^-$, CD15$^+$, CD33$^+$
Erythroid-megakaryocytic marker	CD9$^+$, CD36$^-$, CD41$^{(+)}$, CD42b$^-$, CD61$^{(+)}$, GlyA$^+$, PPO$^-$, vWF$^-$
Progenitor/activation/other marker	CD34$^-$, CD38$^-$, CD45$^+$, CD71$^+$, HLA-DR$^-$, TdT$^-$
Adhesion marker	CD11b$^{(+)}$, CD44$^+$, CD54$^+$
Cytokine receptor	CD25$^-$, CD117$^{(+)}$, CD122$^+$, CD132$^+$, EPO-R$^+$

Genetic characterization[1,3,6–11]

Cytogenetic karyotype	61–68<3n>XX, −X, −3, +7, −13, −18, +3mar, del(9)(p11/13), der(14)t(14;?)(p11;?), der(17)t(17;?)(p11/13;?), der(?18)t(15;?18) (q21;?q12), del(X)(p22)
Unique translocation/fusion gene	Ph$^+$ t(9;22)(q34;q11) → *BCR-ABL* b3-a2 fusion gene
Unique gene alteration	*ABL* amplification, *P15INK4B* deletion, *P16INK4A* deletion, *P53* mutation

Functional characterization[1,12–17]

Colony formation	in agar
Cytochemistry	ALP⁻, ANBE⁻, Benzidine⁺, MPO⁻, PAS⁺
Cytokine production	PDGF, TGF-β
Heterotransplantation	into nude or SCID mice
Inducibility of differentiation	hemin → ery differentiation; TPA → meg differentiation
Transcription factor	mRNA⁺: GATA-1, GATA-2, SCL
Special features	mRNA⁺: α-/γ-/ϵ-/ξ-globin, δ-aminolevulin synthase; Hb Bart's⁺, Hb Portland⁺, Hb GowerI⁺, Hb GowerII⁺, HbF⁺

Comments

- First human immortalized myeloid leukemia cell line.
- Widely used and distributed cell line (reference cell line).
- Well-characterized erythroid cell line.
- Erythroid and megakaryocytic differentiation potential.
- Large cells with budding blebs growing very well.
- Available from cell line banks.
- Carries Ph chromosome with *BCR-ABL* b3-a2 fusion gene.
- Production of hemoglobin.
- Highly sensitive target for natural killer assays.

References

1 Lozzio, C.B. and Lozzio, B.B. (1975) Blood 45, 321–334.
2 Lozzio, C.B. and Lozzio, B.B. (1973) J. Natl. Cancer Inst. 50, 535–538.
3 Drexler, H.G. et al. (1999) DSMZ Catalogue of Cell Lines, 7th edn, Braunschweig, Germany.
4 Toba, K. et al. (1996) Exp. Hematol. 24, 894–901.
5 Schumann, R.R. et al. (1996) Blood 87, 2419–2427.
6 Drexler, H.G. et al. (1999) Leukemia Res. 23, 207–215.
7 Shiohara, M. et al. (1996) Leukemia 10, 1897–1900.
8 Nakamaki, T. et al. (1995) Brit. J. Haematol 91, 139–149.
9 Drexler, H.G. (1998) Leukemia 12, 845–859.
10 Ogawa, S. et al (1994) Blood 84, 2431–2435.
11 Bi, S. et al. (1992) Leukemia 6, 839–842.
12 Shimamoto, T. et al. (1995) Blood 86, 3173–3180.
13 Endo, K. et al. (1993) Brit. J. Haematol. 85, 653–662.
14 Hirata, J. et al. (1990) Leukemia 4, 365–372.
15 Nilsson, K. et al. (1977) Int. J. Cancer 19, 337–344.
16 Andersson, L.C. et al. (1979) Int. J. Cancer 23, 143–147.
17 Andersson, L.C. et al. (1979) Nature 278, 364–368.

KH88

Culture characterization[1]

Subclones	**KH88 B4D6, KH88 C2F8** (minor immunological, cytogenetic differences)
Culture medium	90% IMDM + 10% FBS
Doubling time	20–26 h
Authentication	yes (by cytogenetics)
Primary reference	Furukawa et al.[1]
Availability of cell line	original authors

Clinical characterization[1]

Patient	70-year-old male
Disease diagnosis	CML
Treatment status	at erythroid blast crisis
Specimen site	peripheral blood
Year of establishment	1988

Immunophenotypic characterization[1,2]

T-/NK cell marker	CD1$^-$, CD2$^-$, CD3$^-$, CD4$^-$, CD7$^-$, CD8$^-$, CD56$^-$
B-cell marker	CD10$^-$, CD19$^-$, CD20$^-$, CD21$^-$, CD22$^-$
Myelomonocytic marker	CD13$^-$, CD14$^-$, CD15$^-$, CD33$^-$
Erythroid-megakaryocytic marker	CD36$^+$, CD41b$^-$, CD42$^-$, CD61$^-$, GlyA$^-$, PPO$^+$
Progenitor/activation/other marker	CD34$^-$, CD38$^-$, CD71$^+$, HLA-DR$^-$
Adhesion marker	CD11b$^+$
Cytokine receptor	CD25$^{(+)}$, CD122$^-$

Genetic characterization[1,3]

Cytogenetic karyotype	71, XY, del(X)(q22q28), +Y, +3, +5, +6, +8, −9, +10, −12, −14, −14, −15, −15, −17, −18, −20, +21, −22, +3mar, del(2)(p11.1), add(7)(p11.2), del(9)(p13), t(9;22)(q34;q11), i(11)(q10), der(17)t (2;17)(p13;p13)×2, add(18)(p11.2), +der(19)t(7;19)(q10;q10), +add(19)(p13)
Unique translocation/fusion gene	Ph$^+$ t(9;22)(q34;q11) → *BCR-ABL* b3-a2 fusion gene

Functional characterization[1]

Colony formation	in methylcellulose
Cytochemistry	ACP$^+$, ANAE$^+$, Ferritin$^-$, Lactoferrin$^-$, MPO$^-$
Inducibility of differentiation	hemin → ery differentiation; TPA → meg differentiation
Special features	HbA$^+$, HbF$^+$

Comments

- Erythrocytic precursor cell line.
- Carries Ph chromosome with *BCR-ABL* b3-a2 fusion gene.
- Cells produce hemoglobin.
- Cells grow well.

References

[1] Furukawa, T. et al. (1994) Leukemia 8, 171–180.
[2] Toba, K. et al. (1996) Exp. Hematol. 24, 894–901.
[3] Drexler, H.G. et al. (1999) Leukemia Res. 23, 207–215.

KH184

Culture characterization[1]

Culture medium	90% RPMI 1640 + 10% FBS
Doubling time	36–60 h
Authentication	no
Primary reference	Yasunaga et al.[1]
Availability of cell line	original authors

Clinical characterization[1]

Patient	25-year-old male
Disease diagnosis	megakaryocytic sarcoma
Treatment status	at relapse
Specimen site	peripheral blood
Year of establishment	1984

Immunophenotypic characterization[1]

T-/NK cell marker	CD2$^-$, CD4$^-$, CD5$^-$, CD7$^-$
B-cell marker	CD10$^-$, CD19$^-$, CD20$^-$
Myelomonocytic marker	CD13$^{(+)}$, CD14$^-$, CD33$^+$
Erythroid-megakaryocytic marker	CD36$^{(+)}$, CD41a$^+$, CD42b$^+$, GlyA$^-$, PPO$^-$
Progenitor/activation/other marker	CD34$^-$, HLA-DR$^{(+)}$
Adhesion marker	CD11b$^{(+)}$, CD54$^-$

Genetic characterization[1]

Cytogenetic karyotype	44, X, −Y, −1, −2, −2, −6, −16, −17, −18, +19, −20, −20, −22, −22, +der(?Y)t(?Y;22)(p11;q11), t(1;8)(q11;q11), +der(2)t(2;?) (p23;?), +der(6)t(6;?)(p23;?), del(9)(q13q22), t(10;17)(p12;q11), t(11;19)(q23;q13.3), +der(17)t(17;?), +der(20)t(3;20)(q13;p13), +der(22)t(22;?)(p13;?)

Functional characterization[1]

Cytochemistry	ACP$^+$, ANBE$^-$, CAE$^-$, MPO$^-$, PAS$^-$
Special features	mRNA$^+$: GPIb, GPIIb, GPIIIa

Comments

- Megakaryocytic cell line.

Reference

[1] Yasunaga, M. et al. (1994) Ann. Hematol. 68, 145–151.

KMOE-2

Culture characterization[1,2]

Sister cell lines (1) **KMOE-1** (serial sister cell line – from bone marrow – similar features) (2) **KMOE-3N** (after heterotransplantation of KMOE-1)
Culture medium 92% Ham's F12 + 8% FBS/horse serum or 90% RPMI 1640 + 10% FBS
Doubling time 24 h
Viral status EBV⁻, HBV⁻, HCV⁻, HHV-8⁻, HIV⁻, HTLV-I/II⁻
Authentication no
Primary reference Okano et al.[1]
Availability of cell line DSM ACC 33

Clinical characterization[1]

Patient 2-year-old female
Disease diagnosis acute erythemia (erythroblastosis)
Treatment status at diagnosis
Specimen site peripheral blood
Year of establishment 1978

Immunophenotypic characterization[1,2]

T-/NK cell marker $CD3^-$
B-cell marker $CD19^-$, $cyIg^-$, sIg^-
Myelomonocytic marker $CD13^+$
Erythroid-megakaryocytic marker $GlyA^+$
Progenitor/activation/other marker TdT^-
Cytokine receptor $EPO\text{-}R^+$

Genetic characterization[1,2]

Cytogenetic karyotype hypotetraploid, 6% polyploidy; 82(80–88)<4n>XX, −X, −X, −10, −13, −14, −16, −18, −20, −22, +3mar, der(1)t(1;?)(q21;?)×2, del(7)(p15)×2, del(9)(p12), i(17q)

Functional characterization[1,3,4]

Colony formation in agar
Cytochemistry $ANAE^-$, $Benzidine^-$, MPO^-, PAS^+, SBB^-
Heterotransplantation into nude mice
Inducibility of differentiation sodium butyrate → ery differentiation
Special features $mRNA^+$: α-/β-/γ-/δ-globin; HbA^+

Comments

- Erythroid precursor cell line.
- Erythrocytic differentiation potential.
- Cells grow and produce (hemo)globin.
- Available from cell line bank.

References

[1] Okano, H. et al. (1981) J. Cancer Res. Clin. Oncol. 102, 49–55.

[2] Drexler, H.G. et al. (1999) DSMZ Catalogue of Cell Lines, 7th edn, Braunschweig, Germany.

[3] Kaku, M. et al. (1984) Blood 64, 314–317.

[4] Hirata, J. et al. (1990) Leukemia 4, 365–372.

LAMA-84

Culture characterization[1–3]

Subclones	(1) **LAMA-87** (erythroid-eosinophilic phenotypic) (2) **LAMA-88** (eosinophilic-monocytic phenotype)
Culture medium	90% RPMI 1640 + 10% FBS
Doubling time	30 h
Viral status	EBV⁻, HBV⁻, HCV⁻, HHV-8⁻, HIV⁻, HTLV-I/II⁻
Authentication	yes (by cytogenetics)
Primary reference	Seigneurin et al.[1]
Availability of cell line	LAMA-84: DSM ACC 168; LAMA-87: DSM ACC 270

Clinical characterization[1]

Patient	29-year-old female
Disease diagnosis	CML
Treatment status	at blast crisis
Specimen site	peripheral blood
Year of establishment	1984

Immunophenotypic characterization[1]

T-/NK cell marker	$CD2^-$, $CD3^-$, $CD4^-$, $CD5^-$, $CD7^+$, $CD8^-$, $CD56^-$
B-cell marker	$CD10^-$, $CD19^-$, $CD20^+$, $CD21^-$, $CD22^-$, $CD23^-$, $CD24^+$, $CD37^+$, $CD40^+$
Myelomonocytic marker	$CD13^+$, $CD14^-$, $CD15^-$, $CD16^-$, $CD17^+$, $CD32^+$, $CD33^+$, $CD35^+$, $CD65^-$, $CD66a^-$, $CD66b^-$, $CD68^+$
Erythroid-megakaryocytic marker	$CD9^-$, $CD31^+$, $CD41a^+$, $CD41b^+$, $CD42b^-$, $CD61^+$, $CD63^+$, $GlyA^-$, PPO^+, vWF^-
Progenitor/activation/other marker	$CD34^-$, $CD45^+$, $CD69^+$, $CD71^+$, EMA^+, $HLA\text{-}DR^+$, TdT^-
Adhesion marker	$CD11a^+$, $CD11b^+$, $CD11c^-$, $CD43^+$, $CD44^+$, $CD54^-$
Cytokine receptor	$CD25^-$, $CD117^+$

Genetic characterization[1,3–5]

Cytogenetic karyotype	hypertriploid, 3.6% polyploidy; 73/74(69–77)<3n>XX–X, +1, −2, +5, +6, +8, +13, −14, +17, +17, −18, +22, +mar, del(7) (p15), der(9)t(9;22)(q34;q11)×2, i(11q), add(13)(q33), del(17) (p12), der(22)t(9;22)(q34;q11)×4
Unique translocation/fusion gene	4 copies of Ph^+ t(9;22)(q34;q11) → *BCR-ABL* b3-a2 fusion gene
Unique gene alteration	*P15INK4B* deletion, *P16INK4A* deletion, *P53* mutation

Functional characterization[1,2,6]

Colony formation	in agar
Cytochemistry	ACP+, CAE+, Lactoferrin−, Luxol−, MPO−, PAS+, SBB(+), Vimentin+
Heterotransplantation	into nude mice
Inducibility of differentiation	hemin → ery differentiation; DMSO, natrium butyrate, TPA → meg differentiation
Special features	mRNA+: α-/β-/γ-globin

Comments

- Well-characterized erythroid-megakaryocytic precursor cell line (reference cell line).
- Differentiation potential along erythroid, megakaryocytic, eosinophil, basophil and monocytic lineages.
- Carries Ph chromosome with *BCR-ABL* b3-a2 fusion gene.
- Available from cell line bank.
- Cells grow very well.
- Extensive immunological and functional characterization in ref. 6.

References

1 Seigneurin, D. et al. (1987) Exp. Hematol. 15, 822–832.
2 Champelovier, P. et al. (1994) Leukemia Res. 18, 903–918.
3 Drexler, H.G. et al. (1999) DSMZ Catalogue of Cell Lines, 7th edn, Braunschweig, Germany.
4 Aguiar, R.C.T. et al. (1997) Leukemia 11, 233–238.
5 Drexler, H.G. et al. (1999) Leukemia Res. 23, 207–215.
6 Blom, T. et al. (1996) Scand. J. Immunol. 44, 54–61.

M-07e

Culture characterization[1–4]

Subclones	**M-07e** = subclone of **M-07** (cytokine-independent); **M-07e/TPO** (TPO-dependent)
Establishment	M-07: initially on feeder layer (human amniotic fibroblasts); M-07e: initially in semi-solid medium (methylcellulose)
Culture medium	80% IMDM + 20% FBS + 10 ng/ml GM-CSF or 10 ng/ml IL-3
Doubling time	32–50 h
Viral status	EBV⁻, HBV⁻, HCV⁻, HHV-8⁻, HIV⁻, HTLV-I/II⁻
Authentication	no
Primary reference	Avanzi et al.[1]
Availability of cell line	DSM ACC 104

Clinical characterization[1,2]

Patient	6-month-old female
Disease diagnosis	AML M7
Treatment status	at diagnosis
Specimen site	peripheral blood
Year of establishment	1987

Immunophenotypic characterization[3,5]

T-/NK cell marker	$CD3^-$
B-cell marker	$CD19^-$
Myelomonocytic marker	$CD13^+$, $CD14^-$, $CD15^{(+)}$, $CD33^+$
Erythroid-megakaryocytic marker	$CD41a^+$, $CD42b^+$, $GlyA^-$, $PPO^{(+)}$
Progenitor/activation/other marker	$CD34^{(+)}$, $CD45^+$, $CD71^+$, $HLA\text{-}DR^-$
Adhesion marker	$CD44^+$
Cytokine receptor	$CD25^-$, $CD122^+$, $CD132^+$, MPL^+
Comment	extensive immunoprofile on mother cell line M-07 [1]

Genetic characterization[3]

Cytogenetic karyotype	46(45–46)<2n>XX, t(11;21)(p11;p13), add(13)(p13), add(22) (p13)

Functional characterization[1,2,6,7]

Colony formation	in methylcellulose
Cytochemistry	ACP^+, $ANAE^+$, $ANBE^-$, CAE^-, MPO^-, PAS^-
Cytokine receptor	$mRNA^+$: IL-4Rα, IL-9Rα

Cytokine response	GM-CSF/IL-3-dependent; growth stimulation: IFN-α, IFN-β, IFN-γ, IL-2, IL-4, IL-6, IL-9, IL-15, NGF, PIXY-321, SCF, TNF-α, TPO; growth inhibition: TGF-β1
Heterotransplantation	into SCID mice
Special features	spontaneous production of platelet-like particles; demarcation system^{+}

Comments

- Constitutively cytokine-dependent megakaryocytic precursor cell line.
- Widely distributed and used cell line for cytokine studies (reference cell line).
- Proliferatively responsive to a variety of cytokines.
- Recommended cell line for cytokine bioassays[8].
- Available from cell line bank.
- Cells may be difficult to grow.

References

[1] Avanzi, G.C. et al. (1988) Brit. J. Haematol. 69, 359–366.
[2] Avanzi, G.C. et al. (1990) J. Cell. Physiol. 145, 458–464.
[3] Drexler, H.G. et al. (1999) DSMZ Catalogue of Cell Lines, 7th edn, Braunschweig, Germany.
[4] Drexler, H.G. et al. (1997) Leukemia 11, 541–551.
[5] Schumann, R.R. et al. (1996) Blood 87, 2419–2427.
[6] Quentmeier, H. et al. (1996) Leukemia 10, 297–310.
[7] Drexler, H.G. et al. (1997) Leukemia 11, 701–708.
[8] Mire-Sluis, A.R. et al. (1995) J. Immunol. Methods 187, 191–199.

MB-02

Culture characterization[1]

Culture medium	90% RPMI 1640 + 10% human serum + 5 ng/ml GM-CSF
Viral status	EBNA⁻
Authentication	yes (by cytogenetics)
Primary reference	Morgan et al.[1]
Availability of cell line	original authors

Clinical characterization[1]

Patient	70-year-old male
Disease diagnosis	myelofibrosis + myeloid metaplasia → AML M7
Treatment status	at diagnosis
Specimen site	peripheral blood
Year of establishment	1988

Immunophenotypic characterization[1,2]

T-/NK cell marker	CD1a⁻, CD2⁻, CD3⁻, CD4⁻, CD7⁻, CD8⁻
B-cell marker	CD10⁻, CD19⁻, CD20⁻, CD21⁺, CD23⁻
Myelomonocytic marker	CD13⁺, CD14⁻, CD15⁻, CD16⁻, CD32⁺, CD33⁺, CD35⁺
Erythroid-megakaryocytic marker	CD31⁺, CD36⁺, CD41a⁻, CD41b⁻, CD42a⁻, CD42b⁻, CD61⁺, GlyA⁺
Progenitor/activation/other marker	CD34⁺, CD38⁺, CD45⁺, CD45RA⁺, CD71⁺, HLA-DR⁺
Adhesion marker	CD11a⁺, CD11b⁺, CD11c⁻, CD18⁺, CD44⁺, CD49b⁻, CD49d⁺, CD49e⁺, CD49f⁻, CD51⁻, CD54⁺, CD58⁺
Cytokine receptor	CD25⁻, MPL⁻

Genetic characterization[1]

Cytogenetic karyotype	43–50, XY, −3, −4, −10, −11, −12, −14, +15, +16, +der(1)t (1;2)(q42;q23), +del(3)(q21), +der(3)t(3;?)(p25?)

Functional characterization[1–4]

Colony formation	in agar or methylcellulose
Cytochemistry	ACP⁺, ANAE⁻, Benzidine⁻, CAE⁻, MPO⁻, PAS⁺
Cytokine production	RT-PCR⁺: GM-CSF, IL-1β, IL-7, IL-13, TGF-β
Cytokine response	GM-CSF-dependent; growth stimulation: EPO, IL-1β, IL-3, PIXY-321, SCF, TNF-α, TNF-β; growth inhibition: IFN-α, IFN-β, TGF-β1

Inducibility of differentiation	DMSO, EPO → ery differentiation; TPA → meg differentiation
Transcription factor	NF-E2$^+$
Special features	mRNA$^+$: β-/γ-/ε-globin; HbF$^+$

Comments

- Erythroid-megakaryocytic precursor cell line.
- Constitutively growth factor-dependent and responsive to other cytokines.
- Cells are very difficult to culture (only in 24-well plate).
- Differentiation potential along erythroid and megakaryocytic cell lineages.

References

[1] Morgan, D.A. et al. (1991) Blood 78, 2860–2871.
[2] Morgan, D.A. (1994) personal communication.
[3] Drexler, H.G. et al. (1997) Leukemia 11, 701–708.
[4] Soslau, G. et al. (1997) Cytokine 9, 405–411.

Culture characterization[1]

Culture medium	90% RPMI 1640 + 10% FBS
Authentication	yes (identical *BCR* rearrangement)
Primary reference	Okabe et al.[1]
Availability of cell line	original authors

Clinical characterization[1]

Patient	51-year-old female
Disease diagnosis	CML
Treatment status	at megakaryocytic blast crisis
Specimen site	peripheral blood
Year of establishment	1986

Immunophenotypic characterization[1,2]

T-/NK cell marker	CD3$^-$, CD7$^-$
B-cell marker	CD10$^-$, CD19$^+$, CD20$^-$, cyIgM$^-$
Myelomonocytic marker	CD13$^+$, CD14$^-$, CD33$^+$
Erythroid-megakaryocytic marker	CD41a$^+$, CD42b$^+$
Progenitor/activation/other marker	CD34$^+$, CD38$^+$, HLA-DR$^-$
Adhesion marker	CD54$^+$

Genetic characterization[1,3]

Cytogenetic karyotype	55, XX, +6, +8, +11, +15, +19, +2mar, t(1;5)(q25;q15), t(9;22)(q34;q11)
Unique translocation/fusion gene	Ph$^+$ t(9;22)(q34;q11) → *BCR-ABL* b3-a2 fusion gene
Receptor gene rearrangement	*IGH* G, *IGK* G
Unique gene alteration	*P53* rearrangement

Functional characterization[1,2]

Cytochemistry	ACP$^+$, ANBE$^-$, CAE$^-$, MPO$^-$, PAS$^-$, SBB$^-$
Cytokine receptor	RT-PCR$^+$: EPO-R, MPL
Cytokine response	growth stimulation: IL-1β, IL-3; growth inhibition: TGF-β
Inducibility of differentiation	TPA → meg differentiation
Transcription factor	GATA-2 mRNA$^+$

Comments

- Megakaryocytic cell line with differentiation potential.
- Carries Ph chromosome with *BCR-ABL* b3-a2 fusion gene.

References

[1] Okabe, M. et al. (1995) Leukemia Lymphoma 16, 493–503.
[2] Okabe, M. et al. (1995) Leukemia Res. 19, 933–943.
[3] Drexler, H.G. et al. (1999) Leukemia Res. 23, 207–215.

MEG-01

Culture characterization[1–3]

Subclone	**MEG-01s** (slightly different)
Culture medium	90% RPMI 1640 + 10% FBS
Doubling time	36–48 h
Viral status	EBV⁻, HBV⁻, HCV⁻, HHV-8⁻, HIV⁻, HTLV-I/II⁻
Authentication	no
Primary reference	Ogura et al.[1]
Availability of cell line	ATCC CRL 2021, DSM ACC 364, IFO 50151

Clinical characterization[1]

Patient	55-year-old male
Disease diagnosis	CML
Treatment status	at megakaryocytic blast crisis
Specimen site	bone marrow
Year of establishment	1983

Immunophenotypic characterization[1,3–5]

T-/NK cell marker	$CD1^-$, $CD2^-$, $CD3^-$, $CD4^+$, $CD5^-$, $CD7^-$, $CD8^-$, $CD28^-$, $CD56^-$, $CD57^{(+)}$
B-cell marker	$CD10^-$, $CD19^-$, $CD20^-$, $CD21^+$, $CD22^-$
Myelomonocytic marker	$CD13^+$, $CD14^-$, $CD15^+$, $CD16^-$, $CD33^+$, $CD68^-$
Erythroid-megakaryocytic marker	$CD9^+$, $CD36^+$, $CD41a^+$, $CD42a^-$, $CD42b^{(+)}$, $CD61^+$, $CD62P^-$, $GlyA^-$, PPO^+, vWF^+
Progenitor/activation/other marker	$CD34^-$, $CD38^-$, $HLA\text{-}DR^+$, TdT^-
Adhesion marker	$CD11b^+$, $CD44^+$
Cytokine receptor	$CD25^{(+)}$, $CD122^-$

Genetic characterization[1,3,6,7]

Cytogenetic karyotype	hyperdiploid, 12% polyploidy; 54(53–56)<2n>XY, +6, +19, +19, +21, +3-4mar, t(1;15)(p13;p13), ?inv(3)(p25q26), i(4q), add(5)(p15), der(9)t(9;22)(q34;q11)×2, add(10)(p14), dup(13) (q13q33-34), add(14)(p11), der(22)t(9;22)(q34;q11); r/dmin ×1 present at 57%; sideline i(5)(q10)del(5)(q11q13) instead of add(5)
Unique translocation/fusion gene	Ph⁺ t(9;22)(q34;q11) → *BCR-ABL* b2-a2 fusion gene
Unique gene alteration	*P15INK4B* deletion, *P16INK4A* deletion, *P53* mutation

Functional characterization[1,5,8]

Cytochemistry	ACP+, ALP−, ANAE−, ANBE−, CAE−, MPO−, PAS(+)
Cytokine production	β-TG, PDGF, PF-4, TGF-β
Inducibility of differentiation	TPA → meg differentiation
Special features	spontaneous production of platelet-like particles

Comments

- Megakaryocytic cell line (reference cell line).
- Carries Ph chromosome with *BCR-ABL* b2-a2 fusion gene.
- Available from cell line banks.
- Cells grow very well.

References

1 Ogura, M. et al. (1985) Blood 66, 1384–1392.
2 Ogura, M. et al. (1988) Blood 72, 49–60.
3 Drexler, H.G. et al. (1999) DSMZ Catalogue of Cell Lines, 7th edn, Braunschweig, Germany.
4 Toba, K. et al. (1996) Exp. Hematol. 24, 894–901.
5 Micallef, M. et al. (1994) Hematol. Oncol. 12, 163–174.
6 Drexler, H.G. et al. (1999) Leukemia Res. 23, 207–215.
7 Nakamaki, T. et al. (1995) Brit. J. Haematol. 91, 139–149.
8 Takeuchi, K. et al. (1998) Brit. J. Haematol. 100, 436–444.

MEG-A2

Culture characterization[1]

Culture medium	90% IMDM + 10% FBS
Doubling time	26–30 h
Viral status	EBV⁻
Authentication	yes (by cytogenetics)
Primary reference	Abe et al.[1]
Availability of cell line	original authors

Clinical characterization[1]

Patient	24-year-old male
Disease diagnosis	CML
Treatment status	at megakaryocytic blast crisis
Specimen site	peripheral blood
Year of establishment	1991

Immunophenotypic characterization[1]

T-/NK cell marker	$CD1^-$, $CD2^-$, $CD3^-$, $CD4^+$, $CD5^-$, $CD7^+$, $CD8^-$
B-cell marker	$CD10^-$, $CD19^-$, $CD20^-$
Myelomonocytic marker	$CD13^+$, $CD14^-$, $CD33^+$
Erythroid-megakaryocytic marker	$CD41a^+$, $CD42b^-$, $GlyA^-$, PPO^-, vWF^-
Progenitor/activation/other marker	$CD34^+$, $HLA\text{-}DR^+$

Genetic characterization[1,2]

Cytogenetic karyotype	55(51–56), XY, +1, +6, +8, +11, +18, −20, +21, +21, t(1;6) (p13;q22), t(9;22)(q34;q11), del(17)(p11), +dic(19)t(8;19) (p22;q13), +dic(20)t(8;20)(p22;q13), +der(22)t(9;22)(p34;q11)
Unique translocation/fusion gene	Ph^+ t(9;22)(q34;q11) → *BCR-ABL* b3-a2 fusion gene

Functional characterization[1]

Cytochemistry	ACP^+, ALP^-, $ANAE^-$, $ANBE^+$, CAE^-, MPO^-, PAS^+
Cytokine production	RT-PCR^+: G-CSF, GM-CSF, IL-1α, IL-1β, IL-3, IL-4, IL-6, M-CSF, TPO
Cytokine response	growth stimulation: EPO, GM-CSF, IL-3
Inducibility of differentiation	cytokines, TPA → meg differentiation
Special features	demarcation $system^+$

Comments

- Megakaryocytic cell line with several megakaryocytic features.
- Inducible to megakaryocytic differentiation.
- Carries Ph chromosome with *BCR-ABL* b3-a2 fusion gene.

References

[1] Abe, A. et al. (1995) Leukemia 9, 341–349.
[2] Drexler, H.G. et al. (1999) Leukemia Res. 23, 207–215.

Culture characterization[1]

Culture medium 80–90% RPMI 1640 + 10–20% FBS
Doubling time 36 h
Authentication no
Primary reference Higashitani et al.[1]
Availability of cell line not known

Clinical characterization[1]

Patient 3-year-old male
Disease diagnosis AML M0 + Down's syndrome
Treatment status at relapse
Specimen site peripheral blood
Year of establishment 1991

Immunophenotypic characterization[1]

T-/NK cell marker CD1a−, CD2−, CD3−, cyCD3−, CD4+, CD5−, CD7+, CD8−, CD56−, CD57−
B-cell marker CD10−, CD19−, CD20−, CD21−, cyCD22+, CD24+
Myelomonocytic marker CD13+, CD14−, CD15(+), CD16−, CD33+
Erythroid-megakaryocytic marker CD36+, CD41a+, CD42b−, GlyA−, PPO+
Progenitor/activation/other marker CD34+, CD38+, HLA-DR+, TdT+
Adhesion marker CD11b+, CD11c−
Cytokine receptor CD25−, CD117+

Genetic characterization[1]

Cytogenetic karyotype 46, XY, −4, −12, −17, −17, +21, +2mar, +der(17)t(4;17)(4qter→4q21::17p11→17qter), 7p+, 7q+, 13p+

Functional characterization[1]

Cytochemistry ACP+, ANBE−, CAE−, MPO−, PAS+
Cytokine response growth stimulation: GM-CSF
Inducibility of differentiation TPA → meg differentiation

Comments

- Megakaryocytic cell line.
- Responsive to induction of differentiation.

Reference

[1] Higashitani, A. et al. (1995) J. Jap. Pediatr. Soc. 99, 920–928.

MHH 225

Culture characterization[1,2]

Culture medium 90% RPMI 1640 + 10% FBS or serum-free
Doubling time 38–48 h
Authentication yes (by cytogenetics)
Primary reference Hassan et al.[1]
Availability of cell line original authors

Clinical characterization[1,2]

Patient 60-year-old male
Disease diagnosis AML M7
Treatment status at diagnosis
Specimen site bone marrow
Year of establishment 1993

Immunophenotypic characterization[1,2]

T-/NK cell marker CD2⁻, CD3⁻, CD4⁻, CD5⁻, CD7⁻, CD8⁻
B-cell marker CD10⁻, CD19⁻, CD20⁻, CD21⁻
Myelomonocytic marker CD13⁺, CD15⁺, CD33⁺, CD65⁺
Erythroid-megakaryocytic marker CD41⁺, CD42b⁻, CD62⁺, GlyA⁺, PPO⁺
Progenitor/activation/other marker CD34⁺, CD38⁻, HLA-DR⁺, TdT⁻
Adhesion marker CD11a⁻, CD11b⁻, CD11c⁻, CD54⁻
Cytokine receptor CD25⁻

Genetic characterization[1]

Cytogenetic karyotype del(7)(p13), t(9;21)(q10;q10), 11q+, 9p−, r(21), t(9;11)(p?;q23)

Functional characterization[1]

Cytochemistry ACP(+), ANAE⁻, ANBE⁻, PAS⁺, SBB⁻
Cytokine response survival: SCF; growth inhibition: IFN-α, TNF-α
Inducibility of differentiation IL-4 → myeloid differentiation

Comments

- Megakaryocytic cell line.
- Cells grow easily and quickly (also without FBS).
- Myeloid differentiation potential.
- A MHH 225 culture which was received from the original investigator was found to be cross-contaminated (real identity of this culture: immature T-cell line JURKAT)[3]. It is not known whether the original, correct MHH 225 still exists.

References

[1] Hassan, H.T. et al. (1995) Ann. Hematol. 71, 111–117.
[2] Hassan, H.T. et al. (1994) Hematol. Oncol. 12, 61–66.
[3] Drexler, H.G. and Dirks, W.G. (1999) unpublished data.

MKPL-1

Culture characterization[1]

Culture medium	85% RPMI 1640 + 15% FBS
Doubling time	30 h
Viral status	EBNA$^-$
Authentication	yes (by cytogenetics)
Primary reference	Takeuchi et al.[1]
Availability of cell line	original authors

Clinical characterization[1]

Patient	66-year-old male
Disease diagnosis	AML M7
Treatment status	at diagnosis
Specimen site	bone marrow
Year of establishment	1989

Immunophenotypic characterization[1]

T-/NK cell marker	CD3$^-$, CD4$^-$, CD5$^-$, CD8$^-$
B-cell marker	CD10$^-$, CD19$^-$, CD20$^-$
Myelomonocytic marker	CD13$^+$, CD33$^+$
Erythroid-megakaryocytic marker	CD36$^+$, CD41a$^+$, CD42b$^-$, CD61$^+$, GlyA$^-$, PPO$^-$, vWF$^-$
Progenitor/activation/other marker	CD30$^-$
Cytokine receptor	CD25$^-$

Genetic characterization[1]

Cytogenetic karyotype	92, +3mar

Functional characterization[1]

Cytochemistry	ACP$^+$, ALP$^-$, ANAE$^+$, ANBE$^+$, CAE$^-$, MPO$^-$, PAS$^+$
Heterotransplantation	into nude mice
Special features	α-granules$^+$, demarcation system$^+$

Comments

- Megakaryocytic cell line with characteristic features.

Reference

[1] Takeuchi, S. et al. (1992) Leukemia 6, 588–594.

M-MOK

Culture characterization[1,2]

Subclone **M-MOK/TPO** (TPO-dependent)
Establishment initially on feeder cells (embryonic lung fibroblasts)
Culture medium 80% RPMI 1640 + 20% FBS + 10 ng/ml GM-CSF
Viral status EBV$^-$
Authentication no
Primary reference Itano et al.[1]
Availability of cell line original authors

Clinical characterization[1]

Patient 1-year-old female
Disease diagnosis AML M7
Treatment status at relapse
Specimen site bone marrow
Year of establishment 1989

Immunophenotypic characterization[1]

T-/NK cell marker CD1$^-$, CD3$^-$, CD4$^-$, CD7$^-$, CD8$^-$
B-cell marker CD10$^-$, CD19$^-$, CD20$^-$
Myelomonocytic marker CD13$^+$, CD14$^-$, CD33$^+$
Erythroid-megakaryocytic marker CD41b$^+$, CD42b$^+$, GlyA$^-$, PPO$^-$
Progenitor/activation/other marker CD34$^+$, HLA-DR$^-$
Adhesion marker CD11a$^+$, CD11b$^-$, CD11c$^-$, CD18$^+$, CD54$^+$
Cytokine receptor CD117$^+$

Genetic characterization[1]

Cytogenetic karyotype 46, XX, 2q+, 14p+
Receptor gene rearrangement *IGH* G, *TCRB* G

Functional characterization[1,3]

Cytochemistry ANBE$^-$, CAE$^-$, MPO$^-$, PAS$^-$, SBB$^-$
Cytokine response GM-CSF-dependent; growth stimulation: IFN-β, IFN-γ, IL-3, IL-4, IL-9, IL-13, IL-15, PIXY-321, SCF, TNF-α, TNF-β, TPO; growth inhibition: TGF-β1
Inducibility of differentiation TPA → meg differentiation
Transcription factor mRNA$^+$: GATA-2, GATA-3

Comments

- Constitutively cytokine-dependent megakaryocytic cell line.
- Proliferatively responsive to a variety of cytokines.
- Responsive to TPA induction of megakaryocytic differentiation.
- Cells grow well, but may be slow.

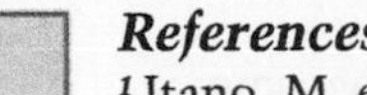

References

[1] Itano, M. et al. (1995) Exp. Hematol. 23, 1301–1309.
[2] Drexler, H.G. et al. (1997) Leukemia 11, 541–551.
[3] Drexler, H.G. et al. (1997) Leukemia 11, 701–708.

MOLM-1

Culture characterization[1]

Culture medium	90% RPMI 1640 + 10% FBS
Doubling time	43 h
Authentication	no
Primary reference	Matsuo et al.[1]
Availability of cell line	original authors

Clinical characterization[1]

Patient	41-year-old male
Disease diagnosis	CML
Treatment status	at blast crisis
Specimen site	bone marrow
Year of establishment	1988

Immunophenotypic characterization[1]

T-/NK cell marker	CD1$^-$, CD2$^-$, CD3$^-$, CD4$^{(+)}$, CD5$^-$, CD7$^+$, CD8$^-$, CD57$^-$
B-cell marker	CD10$^-$, CD19$^-$, CD20$^-$, sIg$^-$
Myelomonocytic marker	CD13$^+$, CD14$^-$, CD15$^-$, CD33$^+$
Erythroid-megakaryocytic marker	CD9$^+$, CD41$^{(+)}$, CD42a$^-$, CD42b$^-$, CD61$^+$, CD62$^{(+)}$, vWF$^-$
Progenitor/activation/other marker	CD34$^+$, CD71$^+$, HLA-DR$^+$, TdT$^-$
Adhesion marker	CD44$^+$
Cytokine receptor	CD25$^+$

Genetic characterization[1–3]

Cytogenetic karyotype	72, X, +1, +2, +2, +2, +4, +4, +6, +8, +8, +10, +11, +12, +13, +13, +14, +15, +16, +19, +19, +20, +21, +21, +22q−, inv(3)(q21q26), +inv(9)t(9;22)(q34;q10), +der(17)t(17;?)(p1?1;?)
Unique translocation/fusion gene	(1) Ph$^+$ t(9;22)(q34;q11) → *BCR-ABL* b2-a2 fusion gene (2) inv(3)(q21q26) → EVI1 overexpression

Functional characterization[1,4]

Cytochemistry	MPO$^-$
Inducibility of differentiation	TPA → meg differentiation

Comments

- Megakaryocytic cell line with differentiation potential.
- Carries Ph chromosome with *BCR-ABL* b2-a2 fusion gene.
- Carries inv(3)(q21q26) leading to *EVI1* overexpression.

References

[1] Matsuo, Y. et al. (1991) Human Cell 4, 261–263.
[2] Ogawa, S. et al. (1996) Oncogene 13, 183–191.
[3] Drexler, H.G. et al. (1999) Leukemia Res. 23, 207–215.
[4] Micallef, M. et al. (1994) Hematol. Oncol. 12, 163–174.

NS-Meg

Culture characterization[1]

Culture medium	90% RPMI 1640 + 10% FBS
Doubling time	72 h
Authentication	yes (by cytogenetics)
Primary reference	Tsuyuoka et al.[1]
Availability of cell line	original authors

Clinical characterization[1]

Patient	44-year-old female
Disease diagnosis	CML
Treatment status	at megakaryocytic blast crisis
Specimen site	peripheral blood
Year of establishment	1992

Immunophenotypic characterization[1]

T-/NK cell marker	CD3$^-$, CD4$^+$, CD7$^+$, CD56$^-$
B-cell marker	CD10$^+$, CD19$^-$
Myelomonocytic marker	CD13$^+$, CD14$^-$, CD33$^+$
Erythroid-megakaryocytic marker	CD36$^+$, CD41a$^+$, CD42a$^-$, CD61$^+$, CD62$^-$, GlyA$^+$, PPO$^+$
Progenitor/activation/other marker	CD34$^+$, HLA-DR$^+$
Adhesion marker	CD11b$^+$, CD18$^+$

Genetic characterization[1,2]

Cytogenetic karyotype	53, XY, +6, +7, +8, +9, +11, +14, +19, +21, t(9;22) (q34;q11), der(17)t(1;17)(p11;p13), +der(22)t(9;22)(q34;q11)
Unique translocation/fusion gene	Ph$^+$ t(9;22)(q34;q11) → *BCR-ABL* b3-a2 fusion gene

Functional characterization[1]

Cytochemistry	ACP$^+$, ANAE$^+$, ANBE$^-$, Benzidine$^-$, MPO$^-$, PAS$^+$, SBB$^-$
Cytokine receptor	EPO-R mRNA$^+$
Cytokine response	growth stimulation: EPO, GM-CSF, IL-3
Inducibility of differentiation	EPO → ery differentiation; cytokines, TPA → meg differentiation
Transcription factor	GATA-1 mRNA$^+$
Special features	mRNA$^+$: α-/γ-globin; spontaneous production of platelet-like particles; α-granules$^+$, demarcation system$^+$; induction of HbF

Comments

- Megakaryocytic cell line with erythroid features.
- Carries Ph chromosome with *BCR-ABL* b3-a2 fusion gene.
- Responsive to erythroid or megakaryocytic differentiation induction.

References

[1] Tsuyuoka, R. et al. (1995) Stem Cells 13, 54–64.
[2] Drexler, H.G. et al. (1999) Leukemia Res. 23, 207–215.

OCIM1

Culture characterization[1]

Establishment	initially culture and passage in semi-solid media (methylcellulose)
Culture medium	90% IMDM + 10% FBS
Authentication	no
Primary reference	Papayannopoulou et al.[1]
Availability of cell line	original authors

Clinical characterization[1]

Patient	62-year-old male
Disease diagnosis	CLL → AML M6
Year of establishment	1984

Immunophenotypic characterization[1,2]

T-/NK cell marker	CD1$^-$, CD2$^-$, CD3$^-$, CD4$^-$, CD5$^-$, CD7$^-$, CD8$^-$
B-cell marker	CD10$^-$, CD19$^-$, sIg$^-$
Myelomonocytic marker	CD13$^+$, CD15$^-$, CD33$^+$
Erythroid-megakaryocytic marker	CD36$^+$, CD41$^{(+)}$, CD42b$^-$, GlyA$^+$, vWF$^-$
Progenitor/activation/other marker	CD34$^+$, CD38$^-$, HLA-DR$^+$, TdT$^-$
Cytokine receptor	CD117$^+$, EPO-R$^+$

Genetic characterization[1]

Receptor gene rearrangement	*TCRB* G

Functional characterization[1-3]

Colony formation	in methylcellulose (red colonies)
Cytochemistry	ANAE$^+$, Benzidine$^+$, CAE$^-$, MPO$^-$, PAS$^+$
Cytokine production	PDGF
Inducibility of differentiation	δ-ALA → ery differentiation; TPA → macro/meg differentiation
Special features	α-/γ-/ε-globin$^+$; induction of Hb Bart's, HbF

Comments

- Erythroid precursor cell line.
- Cells produce (hemo)globin.
- Differentiation along erythroid, macrophage or megakaryocytic pathways.
- Full karyotypic description still required.

References

[1] Papayannopoulou, T. et al. (1988) Blood 72, 1029–1038.
[2] Hirata, J. et al. (1990) Leukemia 4, 365–372.
[3] Tweeddale, M. et al. (1989) Blood 74, 572–578.

OCIM2

Culture characterization[1]

Establishment	initially culture and passage in semisolid media (methylcellulose)
Culture medium	90% IMDM + 10% FBS
Authentication	no
Primary reference	Papayannopoulou et al.[1]
Availability of cell line	original authors

Clinical characterization[1]

Patient	56-year-old male
Disease diagnosis	MDS → AML M6
Year of establishment	1984

Immunophenotypic characterization[1,2]

T-/NK cell marker	CD1⁻, CD2⁻, CD3⁻, CD4⁻, CD5⁻, CD7⁻, CD8⁻
B-cell marker	CD10⁻, CD19⁻, sIg⁻
Myelomonocytic marker	CD13⁺, CD15⁻, CD33⁺
Erythroid-megakaryocytic marker	CD36⁺, CD41⁺, CD42b⁻, GlyA⁺, vWF⁺
Progenitor/activation/other marker	CD34⁺, CD38⁻, HLA-DR⁻, TdT⁻
Adhesion marker	CD18⁺
Cytokine receptor	CD117⁺

Genetic characterization[1,3]

Receptor gene rearrangement	*TCRB* G
Unique gene alteration	*P53* mutation

Functional characterization[1,2,4-6]

Colony formation	in methylcellulose (red colonies)
Cytochemistry	ANAE⁺, Benzidine⁺, CAE⁻, MPO⁻, PAS⁺
Cytokine production	PDGF
Heterotransplantation	into SCID mice
Inducibility of differentiation	δ-ALA → ery differentiation; TPA → macro/meg differentiation
Special features	α-/γ-/δ-/ε-/ζ-globin⁺; induction of Hb Bart's, HbF, Hb Portland

Comments

- Erythroid precursor cell line (similar to OCIM1).
- Cells produce various (hemo)globins.
- Induced differentiation along erythroid, macrophage or megakaryocytic pathways.
- Full karyotypic description still required.

References

[1] Papayannopoulou, T. et al. (1988) Blood 72, 1029–1038.
[2] Hirata, J. et al. (1990) Leukemia 4, 365–372.
[3] Sutcliffe, T. et al. (1998) Blood 92, 2977–2979.
[4] Kamel-Reid, S. et al. (1989) Science 246, 1597–1600.
[5] Papayannopoulou, T. et al. (1987) J. Clin. Invest. 79, 859–866.
[6] Tweeddale, M. et al. (1989) Blood 74, 572–578.

SAM-1

Culture characterization[1]

Culture medium	90% RPMI 1640 + 10% FBS
Authentication	yes (identical *BCR* rearrangement)
Primary reference	Kamesaki et al.[1]
Availability of cell line	original authors

Clinical characterization[1]

Patient	22-year-old male
Disease diagnosis	CML
Treatment status	at blast crisis
Specimen site	peripheral blood
Year of establishment	1987

Immunophenotypic characterization[1,2]

T-/NK cell marker	CD2⁻, CD3⁻, CD4⁻, CD5⁻, CD8⁻, CD57⁻
B-cell marker	CD19⁻, CD20⁻
Myelomonocytic marker	CD13⁻, CD33⁻
Erythroid-megakaryocytic marker	CD41⁺, CD42b⁺, GlyA⁺, PPO⁺
Progenitor/activation/other marker	CD34⁻, CD38⁻, HLA-DR⁻
Adhesion marker	CD11b⁻
Cytokine receptor	CD117⁻, EPO-R⁻

Genetic characterization[1,3]

Cytogenetic karyotype	t(9;22)(q34;q11)
Unique translocation/fusion gene	Ph⁺ t(9;22)(q34;q11) → *BCR-ABL* b3-a2 fusion gene

Functional characterization[1,4]

Cytokine response	growth stimulation: IL-6
Inducibility of differentiation	TPA → meg differentiation
Transcription factor	mRNA⁺: GATA-1, GATA-2
Special features	γ-globin mRNA⁺

Comments

- Cell line expressing erythroid and megakaryocytic markers.
- Carries Ph chromosome with *BCR-ABL* b3-a2 fusion gene.

References

[1] Kamesaki, H. et al. (1996) Blood 87, 999–1005.
[2] Inoue, K. et al. (1997) Blood 89, 1405–1412.
[3] Drexler, H.G. et al. (1999) Leukemia Res. 23, 207–215.
[4] Cheng, T. et al. (1995) Leukemia 9, 1257–1263.

SAS413

Culture characterization[1]

Sister cell line	**SAS527** (serial sister cell line – also at blast crisis – different cytogenetic features)
Culture medium	90% RPMI 1640 + 10% FBS
Viral status	HHV-6⁻, HHV-7⁻
Authentication	yes (by *BCR* Southern blotting)
Primary reference	Yasukawa et al.[1]
Availability of cell line	original authors

Clinical characterization[1]

Patient	46-year-old male
Disease diagnosis	CML
Treatment status	at blast crisis
Specimen site	peripheral blood
Year of establishment	1994

Immunophenotypic characterization[1]

T-/NK cell marker	$CD2^-$, $CD3^-$, $CD4^-$, $CD5^-$, $CD7^-$, $CD8^-$, $CD56^-$
B-cell marker	$CD9^+$, $CD10^-$, $CD19^-$, $CD20^-$, $CD21^-$
Myelomonocytic marker	$CD13^+$, $CD14^+$, $CD16^-$, $CD33^-$
Erythroid-megakaryocytic marker	$CD41a^+$, $GlyA^+$
Progenitor/activation marker	$CD34^-$, $HLA\text{-}DR^-$, TdT^-
Adhesion marker	$CD11b^-$
Cytokine receptor	$CD25^-$

Genetic characterization[1]

Cytogenetic karyotype	60, XY, +1, +6, +6, +8, +8, +10, +12, −18, +19, +19, +20, +20, +21, +21, −22, +2mar, der(9)t(9;22)(q34;q11)×2, der(11)t(3;11)(p21;p13)
Unique translocation/fusion gene	Ph^+ t(9;22)(q34;q11) → *BCR-ABL* fusion gene

Functional characterization[1]

Cytochemistry	$ANBE^{(+)}$, CAE^-, MPO^-
Special features	infectable with HHV-6A and HHV-6B

Comments

- Cell line derived from CML in blast crisis expressing ery-meg immunomarkers.
- Carries Philadelphia chromosome leading to *BCR-ABL* fusion gene.
- Cells are infectable with HHV-6.

Reference

[1] Yasukawa, M. et al. (1999) Blood 93, 991–999.

T-33

Culture characterization[1]

Culture medium	90% RPMI 1640 + 10% FBS or serum-free
Doubling time	24–36 h
Authentication	no
Primary reference	Tange et al.[1]
Availability of cell line	original authors

Clinical characterization[1]

Patient	62-year-old female
Disease diagnosis	CML
Treatment status	at megakaryocytic blast crisis
Specimen site	peripheral blood
Year of establishment	1985

Immunophenotypic characterization[1]

T-/NK cell marker	$CD3^-$, $CD4^-$, $CD5^-$, $CD8^-$, $CD57^-$
B-cell marker	$CD10^-$, $CD19^-$
Myelomonocytic marker	$CD13^+$, $CD14^+$, $CD15^-$, $CD33^+$
Erythroid-megakaryocytic marker	$CD41a^+$, $CD42b^+$, $CD61^+$, $GlyA^-$, PPO^+, vWF^+
Progenitor/activation/other marker	$HLA\text{-}DR^+$

Genetic characterization[1,2]

Cytogenetic karyotype	51, XX, 1p+, +4, +9, +9, +19, +Ph, t(9;22)(q34;q11)
Unique translocation/fusion gene	Ph^+ t(9;22)(q34;q11) → *BCR-ABL* b3-a2 fusion gene

Functional characterization[1]

Cytochemistry	ACP^+, ALP^+, $ANAE^+$, $ANBE^{(+)}$, CAE^-, GLC^+, MPO^-, PAS^+, SBB^+
Cytokine response	growth stimulation: EPO
Inducibility of differentiation	TPA → meg differentiation
Special features	α-granules$^+$, demarcation system$^+$

Comments

- Megakaryocytic cell line presenting various megakaryocytic markers.
- Carries Ph chromosome with *BCR-ABL* b3-a2 fusion gene.
- A T-33 culture which was received from the original investigator was found to be cross-contaminated (real identity of this culture: erythrocytic-megakaryocytic cell line K-562)[3]. It is not known whether the original, correct T-33 still exists.

References

[1] Tange, T. et al. (1988) Cancer Res. 48, 6137–6144.
[2] Drexler, H.G. et al. (1999) Leukemia Res. 23, 207–215.
[3] Drexler, H.G. and Dirks, W.G. (1999) unpublished data.

TF-1

Culture characterization[1–6]

Other name of cell line	**MFD-1**
Subclones	various independent subclones; **TF-1/TPO** (TPO-dependent)
Culture medium	90% RPMI 1640 + 10% FBS + 5 ng/ml GM-CSF
Doubling time	70 h
Viral status	EBV⁻, HBV⁻, HCV⁻, HHV-8⁻, HIV⁻, HTLV-I/II⁻
Authentication	no
Primary reference	Kitamura et al.[1]
Availability of cell line	ATCC CRL 2003, DSM ACC 334

Clinical characterization[1]

Patient	35-year-old male
Disease diagnosis	AML M6
Treatment status	at diagnosis
Specimen site	bone marrow
Year of establishment	1987

Immunophenotypic characterization[1,6,7]

T-/NK cell marker	$CD3^-$, $CD4^-$, $CD5^-$, $CD7^-$, $CD8^-$, $CD57^-$
B-cell marker	$CD10^-$, $CD19^-$
Myelomonocytic marker	$CD13^+$, $CD14^-$, $CD15^-$, $CD33^+$
Erythroid-megakaryocytic marker	$CD41^+$, $CD42a^{(+)}$, $CD42b^+$, $CD61^+$, $CD62^-$, $GlyA^{(+)}$, PPO^+
Progenitor/activation/other marker	$CD34^+$, $CD38^+$, $HLA\text{-}DR^+$
Adhesion marker	$CD11a^-$, $CD11b^+$, $CD11c^+$, $CD18^+$, $CD44^+$, $CD54^+$, $CD56^-$, $CD62L^-$, $CD106^-$
Cytokine receptor	$CD116^+$, $CD117^{(+)}$, $CD123^+$, $CD125^+$

Genetic characterization[1,6,8,9]

Cytogenetic karyotype	highly rearranged hyperdiploid, 12% polyploidy; 52–57<2n>XY/XXY, +3, +5, +6, −8, +12, +15, +19, +19, +20, +20, +3mar, der(1)?dup(1)(p21p31)t(1;8)(p36;q11), t(2;12) (q32;q14), t(3;12)(p13-14;p12-13), add(3)(q21), add(5)(q11-13), der(8)t(1;8)(p36;q11), der(12)t(3;12)(p13-14;p12-13)t(1;12)(q31-32;q24), add(14)(p12), iso(17)(q10)add(17)(q21), add(19)(q13), trp(19)(q12;q13.3), der(21)t(19;21)(q13.1;q22)dup(19)(q13.1q13.3)t(11;19)(q13;q13.3), der(22)t(19;22)(q11;p11)
Unique gene alteration	*P53* mutation

Functional characterization[1,2,10–12]

Colony formation	in agar or methylcellulose
Cytochemistry	ACP^-, $ANBE^-$, CAE^-, Fe^-, MPO^-, PAS^+
Cytokine receptor	$mRNA^+$: IL-1RII, IL-1R antagonist, IL-4Rα, TGF-βRII
Cytokine response	GM-CSF/IL-3-dependent; growth stimulation: EPO, IFN-γ, IL-1, IL-4, IL-5, IL-6, IL-13, LIF, NGF, OSM, PIXY-321, SCF, TNF-α, TNF-β, TPO; growth inhibition: TGF-β
Heterotransplantation	into SCID mice
Inducibility of differentiation	hemin, δ-ALA → ery differentiation; TPA → macro or meg differentiation
Special features	$mRNA^+$: α-/β-/γ-globin; induction of HbF, HbA

Comments

- Multi-cytokine-dependent or -responsive cell line with erythroid features.
- Widely distributed and used cell line for cytokine studies (reference cell line).
- Multi-potential differentiation capacities (erythroid, megakaryocytic, monocytic/macrophage).
- Recommended cell line for cytokine bioassays[13].
- Available from cell line banks.
- Cells grow easily, but undergo quickly apoptosis in the absence of cytokines.

References

1 Kitamura, T. et al. (1989) J. Cell. Physiol. 140, 323–334.
2 Kitamura, T. et al. (1989) Blood 73, 375–380.
3 Gabert, J.A. et al. (1994) Leukemia 8, 1359–1368.
4 Hu, X. et al. (1998) Leukemia Res. 22, 817–826.
5 Drexler, H.G. et al. (1997) Leukemia 11, 541–551.
6 Drexler, H.G. et al. (1999) DSMZ Catalogue of Cell Lines, 7th edn, Braunschweig, Germany.
7 Testa, U. et al. (1998) Leukemia 12, 563–570.
8 Urashima, M. et al. (1998) Blood 92, 959–967.
9 Sugimoto, K. et al. (1992) Blood 79, 2378–2383.
10 Fukuchi, Y. et al. (1998) Leukemia Res. 22, 837–843.
11 Drexler, H.G. et al. (1997) Leukemia 11, 701–708.
12 Murate, T. et al. (1993) Exp. Cell Res. 208, 35–43.
13 Mire-Sluis, A.R. et al. (1995) J. Immunol. Methods 187, 191–199.

TN922

Culture characterization[1]

Culture medium	90% RPMI 1640 + 10% FBS (or serum-free with GM-CSF and transferrin)
Doubling time	36 h
Authentication	yes (by cytogenetics)
Primary reference	Nakayama et al.[1]
Availability	not known

Clinical characterization[1]

Patient	67-year-old male
Disease diagnosis	CML + megakaryocytoma
Treatment status	at megakaryocytic blast crisis
Specimen site	peripheral blood
Year of establishment	1994

Immunophenotypic characterization[1]

T-/NK cell marker	CD2⁻
B-cell marker	CD19⁻
Myelomonocytic marker	CD13⁺, CD14⁻, CD15⁻, CD33⁺, CD68⁺
Erythroid-megakaryocytic marker	CD41⁺, CD61⁽⁺⁾, GlyA⁺
Progenitor/activation marker	CD34⁽⁺⁾, HLA-DR⁺, TdT⁻
Adhesion marker	CD11b⁺, CD11c⁻
Cytokine receptor	CD114⁻, CD116⁺, EPO-R⁻, MPL⁺

Genetic characterization[1]

Cytogenetic karyotype	59, XY, +1, +2, +8, +11, +13, +19, +21, +22, +2mar, +der(6;13)(p10;q10), +add(8)(q24), t(9;22)(q34;q11), add(14)(q32), +del(16)(q2?)
Unique translocation/fusion gene	Ph⁺ t(9;22)(q34;q11) → *BCR-ABL* fusion gene

Functional characterization[1]

Cytochemistry	ALP⁻, ANBE⁻, CAE⁻, Lysozyme⁺, MPO⁻, PAS⁺
Cytokine production	RT-PCR⁺: GM-CSF
Cytokine response	growth stimulation: GM-CSF, IL-3; growth inhibition: IL-6, TGF-β, TNF-α
Inducibility of differentiation	DMSO → ery differentiation; TPA → meg differentiation
Special features	RT-PCR⁺: β-/γ-globin; demarcation system⁺; Hb⁺

Comments

- CML-blast crisis-derived cell line expressing erythroid and megakaryocytic features.
- Carries Philadelphia chromosome leading to *BCR-ABL* fusion gene.

Reference

[1] Nakayama, T. et al. (1999) Anticancer Res. 19, 349–356.

TS9;22

Culture characterization[1]

Culture medium 80% RPMI 1640 + 20% FBS
Authentication yes (by cytogenetics)
Primary reference Ohyashiki et al.[1]
Availability of cell line original authors

Clinical characterization[1]

Patient 56-year-old male
Disease diagnosis CML
Treatment status at blast crisis
Specimen site peripheral blood
Year of establishment 1992

Immunophenotypic characterization[1]

T-/NK cell marker CD1a⁻, CD1b⁻, CD2⁻, CD3⁻, cyCD3⁻, CD4⁻, CD5⁻, CD7⁻, CD8⁻, CD28⁻, CD56⁻, CD57⁻
B-cell marker CD10⁻, CD19⁻, CD20⁻, CD21⁻, CD22⁻, CD24⁺, cyIg⁻, sIg⁻
Myelomonocytic marker CD13⁺, CD14⁻, CD15⁻, CD16⁻, CD33⁺
Erythroid-megakaryocytic marker CD9⁺, CD36⁺, CD41⁺, CD42b⁻, CD61⁽⁺⁾, GlyA⁻
Progenitor/activation/other marker CD34⁺, CD38⁺, CD71⁺, HLA-DR⁺, TdT⁻
Adhesion marker CD11b⁻
Cytokine receptor CD25⁺

Genetic characterization[1,2]

Cytogenetic karyotype 43, XY, −7, −9, −10, −13, −14, −15, +3mar, add(1)(p22), add(4)(p15), t(9;22)(q34;q11), add(17)(p12)
Unique translocation/fusion gene Ph⁺ t(9;22)(q34;q11) → *BCR-ABL* b3-a2 fusion gene

Functional characterization[1,3]

Transcription factor mRNA⁺: GATA-1, GATA-2, SCL

Comments

- Cell line with megakaryocytic markers.
- Carries Ph chromosome with *BCR-ABL* b3-a2 fusion gene.

References

[1] Ohyashiki, K. et al. (1994) Leukemia 8, 2169–2173.
[2] Drexler, H.G. et al. (1999) Leukemia Res. 23, 207–215.
[3] Shimamoto, T. et al. (1995) Blood 86, 3173–3180.

UoC-M1

Culture characterization[1]

Establishment	initially on feeder layer (complete media, human serum, agar) in hypoxic environment
Culture medium	90% McCoy's 5A + 10% FBS
Authentication	yes (by cytogenetics)
Primary reference	Allen et al.[1]
Availability of cell line	not known

Clinical characterization[1]

Patient	68-year-old male
Disease diagnosis	AML M1
Treatment status	at diagnosis
Specimen site	bone marrow
Year of establishment	1992

Immunophenotypic characterization[1]

T-/NK cell marker	$CD1a^-$, $CD2^-$, $CD3^-$, $CD4^-$, $CD5^-$, $CD7^+$, $CD8^-$, $CD56^-$
B-cell marker	$CD10^-$, $CD19^-$, $CD20^-$, $CD24^+$, $sIgL^-$
Myelomonocytic marker	$CD13^+$, $CD14^-$, $CD15^-$, $CD33^+$
Erythroid-megakaryocytic marker	$CD41^-$, $CD42^-$, $CD61^+$, $GlyA^-$
Progenitor/activation/other marker	$CD34^+$, $CD38^+$, $CD45^+$, $HLA\text{-}DR^+$
Cytokine receptor	$CD25^+$

Genetic characterization[1]

Cytogenetic karyotype	42, X, −Y, −7, −9, −16, −19, +21, der(9)t(Y;9)(q1?2;p22)t(9;19)(q1?2;p12orq12), del(5)(q1?2q3?4), der(5)t(5;9)(p1?5;q1?3), dic(11)t(9;11;19), der(14;21)(q10;q10)del(14)(q1?3), der(17)t(7;17)(p14;p12), del(19)(p1?1q1?2), der(21)t(11;21)(q22;q22)dup(21)(q11q22), +der(21)t(16;21)(p11;p11), der(22)t(19;22)(p12orq12;p1?1)
Unique gene alteration	*MLL* amplification (four copies)

Functional characterization[1]

Cytochemistry	ACP^+, $ANAE^+$ (NaF inhibitable), $ANBE^-$, MPO^-

Comments

- Weak expression of megakaryocytic markers.
- Cytogenetically well-characterized cell line.

Reference

[1] Allen, R.J. et al. (1998) Leukemia 12, 1119–1127.

Culture characterization[1-5]

Subclones	**UT-7/EPO, UT-7/GM, UT-7/TPO** (EPO/GM-CSF/TPO-dependent)
Culture medium	90% IMDM + 10% FBS + 1 ng/ml GM-CSF or 1 ng/ml IL-3 or 1 U/ml EPO
Doubling time	36–60 h
Viral status	EBV⁻, HBV⁻, HCV⁻, HIV⁻, HTLV-I/II⁻
Authentication	no
Primary reference	Komatsu et al.[1]
Availability of cell line	DSM ACC 137

Clinical characterization[1]

Patient	64-year-old male
Disease diagnosis	AML M7
Treatment status	at diagnosis
Specimen site	bone marrow
Year of establishment	1988

Immunophenotypic characterization[1]

T-/NK cell marker	$CD2^-$, $CD3^-$, $CD4^-$
B-cell marker	$CD10^-$, $CD19^-$, $CD20^-$
Myelomonocytic marker	$CD13^+$, $CD14^-$, $CD15^+$, $CD16^+$, $CD33^+$, $CD68^+$
Erythroid-megakaryocytic marker	$CD41a^+$, $CD41b^+$, $CD42b^+$, $CD61^+$, $GlyA^+$, PPO^+
Progenitor/activation/other marker	$CD34^+$, $HLA\text{-}DR^+$
Adhesion marker	$CD11b^+$, $CD44^{(+)}$
Cytokine receptor	$CD116^+$, $EPO\text{-}R^+$

Genetic characterization[1,6,7]

Cytogenetic karyotype	flat-moded hypotetraploid; 72–88<4n>XXYY, +1, −8, −9, −9, −9, −11, −11, −12, −13, −13, −14, −18, −21, −22, +6mar, del(1)(q42)×2, t(2;4)(p16-23;q27-31)×1–2, der(2)t(2;5)(p16-23;q12-14)×1–2, add(5)(q11), del(5)(q13), der(6;13)(p10;q10), i(7q)×1–2, add(8)(q24)×1–3, del(9)(q22), add(12)(p12-13), der(13)t(13;13)(p11;q14), add(14)(p11), add(15)(p11), add(16)(p1?), add(17)(p11), add(19)(p13), add(19)(p13), add(19)(q13), i(21q)×1–2
Unique gene alteration	*P16INK4*A deletion, *P53* mutation

Functional characterization[1,8,9]

Cytochemistry	ACP$^+$, ALP$^-$, ANAE$^+$, ANBE$^+$, CAE$^-$, MPO$^-$, PAS$^+$, SBB$^-$
Cytokine response	GM-CSF/IL-3/EPO-dependent; growth stimulation: G-CSF, IFN-β, IFN-γ, IL-4, IL-5, IL-6, NGF, PIXY-321, SCF; growth inhibition: TGF-β1
Inducibility of differentiation	TPA → meg differentiation
Proto-oncogene	MYB mRNA$^+$
Transcription factor	mRNA$^+$: GATA-1, GATA-2
Special features	α-granules$^+$

Comments

- Multi-cytokine-dependent and -responsive cell line with megakaryocytic features.
- Widely distributed and used cell line for cytokine and other studies (reference cell line).
- Pluripotent in its differentiation possibilities.
- Recommended cell line for cytokine bioassays[10,11].
- Available from cell line bank.
- Cells grow slowly.

References

1 Komatsu, N. et al. (1991) Cancer Res. 51, 341–348.
2 Komatsu, N. et al. (1993) Blood 82, 456–464.
3 Komatsu, N. et al. (1996) Blood 87, 4552–4560.
4 Komatsu, N. et al. (1997) Blood 89, 4021–4033.
5 Drexler, H.G. et al. (1997) Leukemia 11, 541–551.
6 Sugimoto, K. et al. (1992) Blood 79, 2378–2383.
7 Ogawa, S. et al. (1994) Blood 84, 2431–2435.
8 Hermine, O. et al. (1992) Blood 80, 3060–3069.
9 Drexler, H.G. et al. (1997) Leukemia 11, 701–708.
10 Mire-Sluis, A.R. et al. (1995) J. Immunol. Methods 187, 191–199.
11 Drexler, H.G. et al. (1999) DSMZ Catalogue of Cell Lines, 7th edn, Braunschweig, Germany.

Y-1K

Culture characterization[1]

Culture medium	90% IMDM + 10% FBS
Doubling time	62 h
Authentication	no
Primary reference	Endo et al.[1]
Availability of cell line	restricted

Clinical characterization[1]

Patient	57-year-old female
Disease diagnosis	CML
Treatment status	at erythroid blast crisis
Specimen site	peripheral blood
Year of establishment	1990

Immunophenotypic characterization[1]

T-/NK cell marker	$CD3^-$, $CD4^+$, $CD8^-$
B-cell marker	$CD19^-$, $CD20^-$
Myelomonocytic marker	$CD13^+$, $CD33^+$
Erythroid-megakaryocytic marker	$CD41^+$, $GlyA^+$
Progenitor/activation/other marker	$CD34^+$, $HLA\text{-}DR^+$

Genetic characterization[1,2]

Cytogenetic karyotype	66(64–67), XX, t(9;22)
Unique translocation/fusion gene	Ph^+ t(9;22)(q34;q11) → *BCR-ABL* b3-a2 fusion gene

Functional characterization[1]

Cytochemistry	ALP^+, $ANBE^+$, $Benzidine^-$, CAE^-, MPO^-, PAS^+
Transcription factor	GATA-1 $mRNA^+$
Special features	$mRNA^+$: γ-globin,δ-aminolevulin synthase

Comments

- Erythroid precursor cell line.
- Carries Ph chromosome with *BCR-ABL* b3-a2 fusion gene.

References

1 Endo, K. et al. (1993) Brit. J. Haematol. 85, 653–662.
2 Drexler, H.G. et al. (1999) Leukemia Res. 23, 207–215.

YN-1

Culture characterization[1]

Establishment	initial culture as colonies in semi-solid media (soft agar)
Culture medium	90% IMDM + 10% FBS
Doubling time	20–24 h
Authentication	no
Primary reference	Endo et al.[1]
Availability of cell line	restricted

Clinical characterization[1]

Patient	35-year-old male
Disease diagnosis	CML
Treatment status	at erythroid blast crisis
Specimen site	peripheral blood
Year of establishment	1980

Immunophenotypic characterization[1]

T-/NK cell marker	CD3⁻, CD4⁻, CD8⁻
B-cell marker	CD19⁻, CD20⁻
Myelomonocytic marker	CD13⁺, CD33⁺
Erythroid-megakaryocytic marker	CD41⁻, GlyA⁺
Progenitor/activation/other marker	CD34⁻, HLA-DR⁻
Cytokine receptor	CD117⁻

Genetic characterization[1,2]

Cytogenetic karyotype	53, XY, +8, +8, +14, +14, +19, +21, +22q−, t(9;22) (q34;q11)
Unique translocation/fusion gene	Ph⁺ t(9;22)(q34;q11) → *BCR-ABL* b3-a2 fusion gene

Functional characterization[1]

Cytochemistry	ALP⁻, ANBE⁻, Benzidine⁺, CAE⁻, MPO⁻, PAS⁺
Inducibility of differentiation	AraC, hemin → ery differentiation
Transcription factor	GATA-1 mRNA⁺
Special features	mRNA⁺: γ-globin, δ-aminolevulin synthase; Hb Bart's⁺, Hb Portland⁺, HbF⁺, HbA⁽⁺⁾

Comments

- Erythroid cell line producing various hemoglobins.
- Carries Ph chromosome with *BCR-ABL* b3-a2 fusion gene.

References

[1] Endo, K. et al. (1993) Brit. J. Haematol. 85, 653–662.
[2] Drexler, H.G. et al. (1999) Leukemia Res. 23, 207–215.

YS9;22

Culture characterization[1]

Culture medium	80% RPMI 1640 + 20% FBS
Authentication	yes (by cytogenetics)
Primary reference	Ohyashiki et al.[1]
Availability of cell line	original authors

Clinical characterization[1]

Patient	23-year-old female
Disease diagnosis	CML
Treatment status	at blast crisis
Specimen site	peripheral blood
Year of establishment	1992

Immunophenotypic characterization[1]

T-/NK cell marker	CD1a⁻, CD1b⁻, CD2⁻, CD3⁻, CD4⁻, CD5⁻, CD7⁻, CD8⁻, CD28⁻, CD56⁺, CD57⁻
B-cell marker	CD10⁻, CD19⁻, CD20⁻, CD21⁻, CD22⁻, CD24⁻, sIg⁻
Myelomonocytic marker	CD13⁺, CD14⁻, CD15⁻, CD16⁻, CD33⁺
Erythroid-megakaryocytic marker	CD9⁺, CD36⁺, CD41a⁺, CD42b⁻, CD61⁺, GlyA⁻
Progenitor/activation/other marker	CD34⁺, CD38⁺, CD71⁺, HLA-DR⁻, TdT⁻
Adhesion marker	CD11b⁻
Cytokine receptor	CD25⁻

Genetic characterization[1,2]

Cytogenetic karyotype	46, X, −X, +19, der(1)t(1;3)(p32;p12), dic(3)(q26;p12), t(9;22)(q34;q11), add(16)(p13)
Unique translocation/fusion gene	Ph⁺ t(9;22)(q34;q11) → *BCR-ABL* b3-a2 fusion gene

Functional characterization[1,3]

Proto-oncogene	EVI1 mRNA⁺
Transcription factor	mRNA⁺: GATA-1, GATA-2, SCL

Comments

- Cell line displaying megakaryocytic markers.
- Carries Ph chromosome with *BCR-ABL* b3-a2 fusion gene.

References

[1] Ohyashiki, K. et al. (1994) Leukemia 8, 2169–2173.
[2] Drexler, H.G. et al. (1999) Leukemia Res. 23, 207–215.
[3] Shimamoto, T. et al. (1995) Blood 86, 3173–3180.

Section IX

Tables

Table I *Additional precursor B-cell lines*

Cell line	Disease (status)	Age, sex, specimen	Core data available on			Comments	Ref.
			Immuno-profile	Karyo-type	Functional aspects		
316	pre B-ALL (2nd R)	37, M, BM	(+)	+	+	partial characterization	*1*
AHLL-1	ALL (R)	18, M, BM	+	(+)		t(1;19)(q23;p13)	2
ALL-3	pre B-ALL (D)	19, M, BM	(+)	+	+	partial characterization	3
ALL-202	pre B-ALL (L2, D)	10, F	+	(+)	+	partial characterization	*4*
ALL-K	pre B-ALL (R)	child, BM	(+)			partial characterization	5
ALL-PO	BCP-ALL (R)	1, BM	(+)	(+)	+	t(4;11)(q21;q23); only in SCID mice	6
BEL-1	BCP-ALL (R)	41, F, PB	+	+	+	t(4;11)(q21;q23)	7
BLIN-3	ALL	BM	(+)			no characterization	8
BLIN-4	ALL	BM	(+)			no characterization	8
C1	pre B-ALL (D)	child, PB	(+)	+	+	= **ALL-C**	5,9
EU-5	pre B-ALL (R)	10, F, BM	(+)		+	partial characterization	*10–13*
EU-6	pre B-ALL (R)	9, F, BM	(+)		+	partial characterization	*10–13*
EU-10	pre B-ALL (D)	14, M, BM	(+)		+	partial characterization	*10,11*
EU-11	pre B-ALL (D)	<1, F, BM	(+)		+	partial characterization	*10,11*
EU-12	pre B-ALL	child	(+)		+	partial characterization	*10–12*
EU-13	pre B-ALL	child, BM	(+)		+	partial characterization	*10–12*
EU-18	pre B-ALL (R)	4, M, BM			+	no characterization	*11*
HB1119	pre B-ALL			(+)		t(11;19)(q23;p13)	*14*
HOON	non-T non-B ALL (R)	9, M, PB/BM	+		+	partial characterization	*15*
HPB-NULL	ALL	47, M, PB	+		+	full immunoprofile	*16*
HYON	non-T non-B ALL (R)	11, M, BM	+		+	partial characterization	*15*
INC	ALL	67, F	+		+	full immunoprofile	*16,17*
JM-1	B-NHL (immunoblastic)	PB	(+)			no characterization	*18*
Karpas 190	ALL (1st R)	5, M, BM			(+)	minimal characterization	*19*
KEN-L-1	ALL		(+)		+	no characterization	*20*
KH-4	non-T non-B ALL (R)	11, M, BM	+	(+)	+	partial characterization	*21*
Kid92	pre B-ALL (R,T)	28, M, PB	+	(+)	+	t(4;11)(q21;q23); full immunoprofile	*16,17*
KIYOTA	cALL		(+)			no characterization	22
K-LL-3	lymphoblastic lymphoma (R)	13, M, BM		(+)	+	partial characterization	23
KM-3	ALL (2nd R)	12, M, PB	+			authentication needed (to exclude line **Reh**)	*24*

Table I ***Continued***

Cell line	Disease (status)	Age, sex, specimen	Core data available on			Comments	Ref.
			Immuno-profile	Karyo-type	Functional aspects		
KOCL-33	ALL (L1)	<1, F	+	(+)	+	t(11;19)(q23;p13); full immunoprofile	***25–27***
KOCL-44	ALL (L1)	<1, F	+	(+)	+	t(11;19)(q23;p13); full immunoprofile	***25–27***
KOCL-45	ALL (L1)	<1, F	+	(+)	+	t(4;11)(q21;q23); full immunoprofile	***25–28***
KOCL-50	ALL (L1)	<1, F	+	(+)	+	t(11;19)(q23;p13); full immunoprofile	***25–27***
KOCL-51	ALL (L1)	<1, M	+	(+)	+	t(11;19)(q23;p13); full immunoprofile	***25–28***
KOCL-58	ALL (L1)	<1, M	+	(+)	+	t(4;11)(q21;q23); full immunoprofile	***25–28***
KOCL-69	ALL (L1)	<1, F	+	(+)	+	t(4;11)(q21;q23)	***26,27***
KOPB-26	ALL (L1)	1, F	+	(+)	+	t(9;11)(p22;q23)	***25,26***
KOPN-1	ALL (L1)	<1, F, PB	+	(+)	+	t(11;19)(q23;p13)	***25,26,29***
KOPN-32	ALL	PB	+	(+)	+	del(17)(p13)	***28,30***
KOPN-35	ALL	PB	(+)		+	partial characterization	***30***
KOPN-39	BCP-ALL				(+)	no characterization	***31***
KOPN-40	ALL	PB	(+)	(+)	+	ins(3;9)	***30***
KOPN-49	ALL	PB	(+)	(+)	+	del(17)(p13)	***30***
KOPN-57bi	cALL	11, M, BM	+	(+)		t(9;22)(q34;q11)	***28,31***
KOPN-60	ALL		+	(+)	(+)	t(1;19)(q23;p13)	***28***
KOPN-61	ALL		+		(+)	partial characterization	***28***
KOPN-62	ALL		+		(+)	partial characterization	***28***
KOPN-63	ALL	PB	(+)	(+)	+	t(1;19)(q23;p13)	***30***
KOPN-66bi	ALL		+	(+)	(+)	t(9;22)(q34;q11)	***28***
KOPN-70	BCP-ALL				(+)	*P16INK4*A mutation	***31***
KOPN-K	ALL	5, M	+		+	full immunoprofile	***16,32***
KOS 20	ALL		(+)		+	partial characterization	***28***
LC1;19	pre B-ALL		(+)	(+)	+	partial characterization	***33***
LC4–1	cALL	13, F, PB	+	+	+	good characterization	***34***
LiLa-1	pre B-ALL	PB	(+)	+	+	t(1;19)(q23;p13)	***35***
LK63	pre B-ALL	PB	(+)	+	+	t(1;19)(q23;p13)	***35***
L-KUM	ALL		(+)			no characterization	***36***
LLM	ALL	0.5, F, BM	+	+	+	t(4;11)(q21;q23)	***37***
L-MEG	ALL		(+)			no characterization	***36***
LR10.6	ALL	child, BM				**cross-contamination = line NALM-6**	***38***
MHH-CALL2	cALL (D)	15, F, PB	+	+	+	cell line bank: DSM ACC 341;	***39,40***

Table I ***Continued***

						(= **MHH-ALL2**)	
MHH-CALL3	pre B-ALL (D)	11, F, BM	+	+	+	cell line bank: DSM ACC 339; (= **MHH-ALL3**)	*39,40*
MHH-CALL4	pre B-ALL (D)	10, M, PB	+	+	+	cell line bank: DSM ACC 337; (= **MHH-ALL4**)	*39,40*
NALL-1	ALL (R)	14, M, PB	+	(+)	+	full immunoprofile	*16,32,41*
NALL-2	ALL				+	no characterization	*42,43*
NALM-17	ALL	9, M, PB	+		+	simultaneous sister line **NALM-18**	*16,32*
OH94	cALL		+			full immunoprofile	*44*
PALL-1	pre B-ALL (2nd R)	24, M, PB	+	+	+	t(9;22)(q34;q11); only in nude mice	*45,46*
PBAD	pre B-ALL	child			+	bone marrow stroma-dependent	*47*
PBEI	pre B-ALL		(+)	+	+	**cross-contamination = line NALM-6**	*48*
PER-145	ALL (L1, 3rd R)	19, M, BM	(+)		+	serial sister line **PER-163** (4th R)	*49,50*
PER-271	BCP-ALL	child	(+)			no characterization	*51*
PER-288	BCP-ALL (D)	child	(+)		+	no characterization	*51,52*
PER-371	BCP-ALL (D)	child			+	no characterization	*52*
RW	cALL (R)	PB	(+)		+	partial immunoprofile	*18*
SCMC-L1	BCP-ALL	child		(+)	+	t(4;11)(q21;q23)	*53,54*
SCMC-L2	ALL	11, M, PB	(+)	(+)		t(9;22)(q34;q11)	*52,55,56*
SCMC-L9	pre B-ALL	child	(+)	(+)	+	t(1;19)(q23;p13)	*22,55*
SCMC-L10	pre B-ALL	child	(+)	(+)	+	t(1;19)(q23;p13)	*22,55*
SCMC-L11	pre B-ALL	6, F	(+)	(+)	+	t(1;19)(q23;p13)	*22,55*
SD-1	BCP-ALL	F, PB	(+)	+		**EBV^{+} B-LCL;** t(9;22)(q34;q11); DSM ACC 366	*57*
SILVANUS	BCP-ALL (L1)	28, F, PB	(+)	+	+	partial characterization	*58,59*
SMS-SB	lymphoblastic lymphoma (R)	16, F, PB	(+)		+	partial immunoprofile	*60*
SN-Ow1	cALL		+			no characterization	*44*
SU-ALL-1	ALL	14, F, PB			(+)	no characterization	*61*
SUP-B16	pre B-ALL (1st R)	10, F, CSF	+	+	+	serial sister line **SUP-B19** (2nd R)	*62,63*
Tahr87	AUL	27, M	+		+	full immunoprofile	*16,32*
THP-3–1	non-T non-B ALL (D)	3, F, PB	+			serial sister line **THP-3-2** (T)	*64*
THP-4	cALL (L1)	2, M, PB	+	(+)	+	t(1;19)(q23;p13)	*65,66*
THP-5	null ALL (D)	13, F, PB	+	+	+	partial immunoprofile	*67*
THP-7	cALL (L1)	13, F, PB/BM	+		+	partial immunoprofile	*65*
THP-8	null ALL (L2)	<1, F, PB	+	(+)	+	t(1;19)(q23;p13)	*55,65*
Udd88	AUL	14, F	+	+		sister line **Udd-Oct**	*32*

Cell line	Disease (status)	Age, sex, specimen	Core data available on			Comments	Ref.
			Immuno-profile	Karyo-type	Functional aspects		
UoC-B7	BCP-ALL (2nd R)	5, M	+	+	+	partial characterization	***62,63***
UoC-B8	BCP-ALL (D)	3, M	+	+	+	partial characterization	***62,63***
UoC-B9	BCP-ALL (D)	9, F	+	+	+	partial characterization	***62,63***
UoC-B11	pre B-ALL (D)	child	+		+	partial characterization	***12,13***
UoC-B12	pre B-ALL (D)	5, M, BM	(+)		+	doubling time of 10–14 days!	***62,68***
UTL-16	cALL		(+)		+	no characterization	***23***
UTL-17	cALL		(+)		+	no caracterization	***23***
UTP-L2	Ph+ ALL			(+)		t(9;22)(q34;q11)	***56***
WH94	BCP-ALL		+	(+)		t(1;19)(q23;p13)	***44***
WK93	cALL		+			full immunoprofile	***44***
YS-1	CML-BC		+			full immunoprofile	***44***

Additional abbreviations: BM, bone marrow; CSF, cerebrospinal fluid; D, at diagnosis; F, female; M, male; PB, peripheral blood; R, at relapse/resistant; T, terminal stage.

References

[1] Kheradpour, A. et al. (1994) Blood 84, 45a.
[2] Sawyer, M.C. et al. (1989) Blood 74, 369a.
[3] Cesano, A. et al. (1991) Blood 77, 2463–2474.
[4] Lange, B. et al. (1987) Blood 70, 192–199.
[5] Brown, G.A. et al. (1995) Cancer Res. 55, 78–82.
[6] Gobbi, A. et al. (1997) Leukemia Res. 21, 1107–1114.
[7] Tang, R.P. et al. (1999) Blood 94, 485a.
[8] Jarvis, L.J. et al. (1997) Blood 90, 1626–1635.
[9] Freedman, M.H. et al. (1993) Blood 81, 3068–3075.
[10] Zhou, M. et al. (1995) Leukemia 9, 1159–1161.
[11] Zhou, M. et al. (1998) Leukemia 12, 1756–1763.
[12] Findley, H.M. et al. (1997) Blood 89, 2986–2993.
[13] Zhou, M. et al. (1995) Blood 85, 1608–1614.
[14] Tkachuk, D.C. et al. (1992) Cell 71, 691–700.
[15] Okamura, J. et al. (1984) Leukemia Res. 8, 97–104.
[16] Matsuo, Y. and Drexler, H.G. (1998) Leukemia Res. 22, 567–579.
[17] Tani, A. et al. (1996) Leukemia 10, 1592–1603.
[18] Park, L.S. et al. (1989) J. Biol. Chem. 264, 5420–5427.
[19] Karpas, A. et al. (1977) Brit. J. Cancer 36, 177–186.
[20] Miyagi, T. et al. (1993) Leukemia 7, 970–977.
[21] Nagasaka-Yabe, M. et al. (1988) Jpn. J. Cancer Res. 79, 59–68.
[22] Ohnishi, H. et al. (1996) Leukemia 10, 1104–1110.
[23] Smith, S.D. and Rosen, D. (1979) Int. J. Cancer 23, 494–503.
[24] Schneider, U. et al. (1977) Int. J. Cancer 19, 621–626.
[25] Iida, S. et al. (1992) Leukemia Res. 16, 1155–1163.
[26] Yamamoto, K. et al. (1994) Blood 83, 2912–2921.
[27] Inukai, T. et al. (1998) Leukemia 12, 382–389.
[28] Kojika, S. et al. (1996) Leukemia 10, 994–999.
[29] Nakazawa, S. et al. (1978) Clin. Haematol. 20, 189.
[30] Wada, H. et al. (1994) Leukemia 8, 53–59.
[31] Saito, M. et al. (1995) Blood 86, 329a.
[32] Minowada, J. and Matsuo, Y. (1999) unpublished data.
[33] Uckun, F.M. et al. (1993) Blood 81, 3052–3062.
[34] Yoshimura, T. et al. (1987) Am. J. Hematol. 26, 47–54.
[35] Salvaris, E. et al. (1992) Leukemia Res. 16, 655–663.

[36] Morita, S. et al. (1996) Leukemia 10, 102–105.
[37] Greil, J. et al. (1999) unpublished data.
[38] Inglés-Esteve, J. et al. (1997) Leukemia 11, 1040–1044.
[39] Tomeczkowski, J. et al. (1995) Brit. J. Haematol. 89, 771–779.
[40] Drexler, H.G. et al. (1999) DSMZ Catalogue of Cell Lines, 7th edn, Braunschweig, Germany.
[41] Miyoshi, I. et al. (1977) Nature 267, 843–844.
[42] Shiohara, M. et al. (1996) Leukemia 10, 1897–1900.
[43] Ikezoe, T. et al. (1998) Leukemia 12, 94–95.
[44] Toba, K. et al. (1996) Exp. Hematol. 24, 894–901.
[45] Miyagi, T. et al. (1989) Int. J. Cancer 43, 1149–1154.
[46] Miyagi, T. et al. (1993) Int. J. Cancer 53, 457–462.
[47] Lairmore, J. et al. (1992) Blood 80, 452a.
[48] Pirruccello, S.J. et al. (1991) Blood 78, 39a.
[49] Kees, U.R. (1987) Cancer Res. 47, 3088–3091.
[50] Kees, U.R. et al. (1989) Cancer Res. 49, 3015–3019.
[51] Kees, U.R. and Ashman, L.K. (1995) Leukemia 9, 1046–1050.
[52] Kees, U.R. et al. (1996) Oncogene 12, 2235–2239.
[53] Ogawa, S. et al.(1994) Blood 84, 2431–2435.
[54] Hayashi, Y. et al. (1993) Int. J. Hematol. 57, 124a.
[55] Kawamura, M. et al. (1995) Blood 85, 2546–2552.
[56] Sato, S. et al. (1994) Leukemia Res. 18, 221–228.
[57] Dhut, S. et al. (1991) Leukemia 5, 49–55.
[58] Renard, N. et al. (1997) Leukemia Res. 21, 1037–1046.
[59] Renard, N. et al. (1994) Blood 84, 2253–2260.
[60] Smith, R.G. et al. (1981) J. Immunol. 126, 596–602.
[61] Epstein, A.L. et al. (1979) Cancer Res. 39, 1748–1759.
[62] Kim, D.H. et al. (1996) Blood 88, 785–794.
[63] Zhang, L.Q. et al. (1993) Leukemia 7, 1865–1874.
[64] Tsuchiya, S et al. (1983) Tohoku J. Exp. Med. 140, 443–444.
[65] Minegishi, M. et al. (1987) Tohoku J. Exp. Med. 151, 283–292.
[66] Yoshinari, M. et al. (1998) Cancer Genet. Cytogenet. 101, 95–102.
[67] Imaizumi, M. et al. (1984) Tohoku J. Exp. Med. 142, 51–58.
[68] Jagasia, A.A. et al. (1996) Leukemia 10, 624–628.

Table II *Additional mature B-cell lines*

Cell line	Disease (status)	Age, sex, specimen	Core data available on: Immuno-profile	Karyo-type	Functional aspects	Comments	Ref.
2B1-TPA	HCL	PB			(+)	EBV⁺; no authentication/verification; sister line **2Q1-TPA** (EBV⁺)	*1*
232 B4-CLL	B-CLL (stage 0)	PB		(+)		EBV⁺; origin/neoplastic nature verified sister line **232 A4-LCL** (**EBV⁺ B-LCL**)	*2*
840	B-NHL (FCL)	65, F, SP			(+)	EBV⁺; origin/neoplastic nature verified	*3*
901	B-NHL (FCL)	67, F, LN			(+)	EBV⁺; origin/neoplastic nature verified	*3*
4972	HCL	30, M, SP	(+)		(+)	EBV⁺; no authentication/verification	*4*
A1	B-CLL	PB				EBV⁺; origin/neoplastic nature verified; sister lines **B1, B2** (EBV⁺)	*5*
AI-60	B-CLL	54, F, PB		+		EBV⁺; origin/neoplastic nature verified; sister lines **AI-AC, AII, AIII** (LCLs)	*6–8*
B104	B-NHL (LP)	child, M, PB	(+)		+	EBV⁻; neoplastic nature not verified; JCRB 0117	*9*
BA-25	B-ALL (L3) (R)	BM	+	(+)	+	EBV⁻; t(2;8)(p12;q24)	*10*
BCBL-2	AIDS-PEL	effusion			(+)	EBV⁺, HHV-8⁺	*11,12*
B-CLL	B-CLL (D)	64, M, PB	+	(+)	+	EBV⁺; origin/neoplastic nature verified	*13*
BHT3	B-NHL (undifferentiated)	AF	(+)	(+)		EBV⁻; t(8;14)(q24;q32); sister lines **BHM** (EBV⁺) + **BHT1** (EBV⁺)	*14,15*
BMB	B-NHL		+			EBV? partial characterization	*16*
Boar88	B-NHL (follicular) (R)		+	+		EBV? t(14;18)(q32;q21)	*17*
BW-90	AML M4 (R)	54, M, PB	+	(+)	+	EBV? probably **EBV⁺ B-LCL**	*18*
CII	B-CLL	47, F, PB		+		EBV⁺; origin/neoplastic nature verified; sister line **CI** (**EBV⁺ B-LCL**)	*6–8*
CJ18	HCL (R)	52, M, PB	(+)	+	+	EBV⁺; no authentication/verification	*19*
CLL 14	B-CLL	PB		+	+	EBV⁺; no authentication/verification	*20*
DB	B-NHL (DLCL)	45, M, AF	(+)		+	EBV⁻; *P53* mutation; ATCC CRL 2289	*21–23*
DLCL-SKI-1	B-NHL (DLCL) (R)		(+)	(+)	(+)	EBV⁻; origin/neoplastic nature verified; t(1;14)(q21;q32)	*24*
E51-10F	B-CLL (D)	PB	(+)	+	(+)	EBV⁺; origin/neoplastic nature verified; sisters **E51-11C, E55-7G**	*25*

Table II **Continued**

EBV-CLL	B-CLL (D)	56, F, PB	(+)	(+)	(+)	EBV^+; no authentication/verification	*26*
EBV-CLL(1)	B-CLL (D)	53, M, PB	(+)	(+)	+	EBV^+; no authentication/verification	*27,28*
EH	HCL	68, F, PB	(+)	(+)	+	EBV^-; no authentication/verification	*29*
EHC4	B-CLL (Rai stage 1)	PB		+	+	EBV^+; origin/neoplastic nature verified; t(14;18)(q32;q21)	*30*
Ei 26	B-CLL	87, F, PB			(+)	EBV^+; origin/neoplastic nature verified	*31*
EW36	B-NHL (undifferentiated)	PE	(+)	(+)		EBV^-; t(8;14)(q24;q32)	*14,15*
Granta 452	B-NHL → B-ALL			(+)	(+)	EBV? t(9;14;18)(p13;q32;q21)	*32,33*
HBL-4	B-NHL (small non-cleaved)	AF			+	EBV^+; origin/neoplastic nature verified	*34*
HC	HCL (D)	69, M, PB	(+)		(+)	EBV? no authentication/verification	*35*
HCF-MLpN	B-NHL (extranodal)	CSF	+		+	EBV^-; no authentication; responds to G-CSF	*36*
HCL 1, 2	HCL (variant type)	PB			(+)	EBV^+; no authentication/verification; probably 2 **EBV^+ B-LCLs**; no characterization	*37,38*
HCL-O	HCL (variant type)	62, M, PB	+	+	+	EBV^+; origin verified, but normal karyotype; secretion of IL-6	*39*
HCLW-3B	HCL (LP)	54, F, PB	+	+	+	EBV^+; origin/neoplastic nature verified	*40*
HCL-Z1	HCL	SP			(+)	EBV^-; no authentication/verification	*41*
HDS	HCL (R)	74, M, PB	+	+	(+)	EBV? no authentication/verification	*42*
HK	HCL	66, M, PB	(+)	+	+	EBV^+; no authentication, verification yes	*29*
HO-85	B-NHL (DLCL)		(+)			EBV? no authentication/verification	*43*
HOB1	B-NHL (immunoblastic) (D)	24, M, TU	(+)	+	+	EBV^-; t(8;14)(q24;q32)	*44,45*
HS-1	ALL		(+)	(+)		EBV? t(8;14)(q24.1;q32.3)	*46*
HT	B-NHL (DLCL)	70, M, AF	(+)		+	EBV^-; *P53* mutation; ATCC CRL 2260	*21–23*
I83-E95	B-CLL	PB	+	(+)		EBV^+; origin/neoplastic nature verified	*2*
IW.2	B-CLL	PB	(+)		(+)	EBV^-; no authentication/verification	*47*
JC-1	HCL		+			EBV? only full immunoprofile	*17*
JD38	B-NHL (undifferentiated)	4, M, PB	(+)	(+)	+	EBV^-; t(8;14)(q24;q32); sister lines **JD39** (EBV^-) + **JD40** (EBV^-)	*48–50*
JHC-2	HCL (Japanese variant)	61, M, PB	+	+	+	EBV^+; origin verified; neoplastic nature not verified	*51*
JLPC119	B-NHL (undifferentiated)	12, M, BM		(+)	+	EBV^-; t(8;14)(q24;q32); sister lines **JLM** (EBV^+) + **JLP** (EBV^+)	*48,49*
JOK-1	HCL	61, M	+	+		EBV? t(8;14)(q24;q32)	*17,52*
JVM-14	B-CLL/PLL (D)	M, PB	(+)	(+)	+	EBV^+; origin/neoplastic nature verified; cell line is lost	53
KHB1	B-CLL	PB	(+)		(+)	not permanent cell lines (**KHB2, KHB3**)	*54*

Table II ***Continued***

Cell line	Disease (status)	Age, sex, specimen	Core data available on			Comments	Ref.
			Immuno-profile	Karyo-type	Functional aspects		
KML-2	B-NHL			(+)		EBV? t(14;18)(q32;q21)	*55*
KS-7	Ph^+ ALL (L2)	37, M, PB		(+)		EBV^+ B-LCL with del(20)(q11.2-q13.1); sister line **KS-10** (**EBV^+ B-LCL**)	*56,57*
KT	B-NHL (diffuse)	39, F, TU				no characterization; culture by serial xenotransplantation in nude mice	*58*
KW1	B-NHL → ALL	F, PB	(+)	+		EBV^+; t(14;18)(q32;q21)	*59*
L 660	Hodgkin → ALL (R)	38, M, PB		+		EBV^+; no authentication, verification yes; sister line **L 660 EBV**; t(8;22)(q24;q12)	*60*
LNPL	B-NHL (DLCL)	18, M, TU	(+)	(+)	+	EBV^-; t(8;14)(q24;q32)	*61*
MANACA-2	B-NHL		+			EBV? no authentication/verification	*17*
MD903	B-ALL (L3)	38, M	(+)	+		EBV^-; t(3;14)(q27;q32)	*62*
MH1B14	MDS	72, F		(+)		EBV^+ B-LCL with del(20)(q11.2-q13.1); sister line **MH1A1** (**EBV^+ B-LCL**)	*56*
MHH-PREB1	B-NHL (lymphoblastic) (D)	5, M, LN	(+)	+	+	EBV^-; t(8;14)(q24;q32); DSM ACC 354	*63*
MINO	B-NHL (MCL)			(+)		EBV? t(11;14)(q13;q32)	*64*
MO1043	B-CLL	71, M, PB	(+)	+	+	EBV? no authentication, verification yes	*65*
MO1094	B-PLL	F, PB	(+)	+		EBV^+; t(11;14)(q13;q32)	*66,67*
MO2058	B-PLL	M, PB	(+)	+		EBV^+; t(11;14)(q13;q32)	*66,67*
NCU-L-1	B-ALL (L3) (D)	71, F, PB	(+)	+	+	EBV^-; t(2;8)(p12;q24)	*68*
NH-AR	B-NHL (nodular, histiocytic)	43, M, LN	(+)	+	+	EBV^+; origin/neoplastic nature verified	*69,70*
NOL-3	B-NHL (DLCL)	36, F, PE	+	(+)	+	EBV? t(11;X)(q23;q13)	*71*
NU-DHL-1	B-NHL (DLCL)	LN	+		+	EBV^-; full immunoprofile, no karyotype	*72,73*
OCI-Ly1	B-NHL (DLCL) (R)	44, M, BM	(+)	(+)	+	EBV^-; t(14;18)(q32;q21)	*74–79*
OCI-Ly2	B-NHL (DLCL) (R)	50, M, BM	(+)		+	EBV^-; partial characterization	*74–78*
OCI-Ly3	B-NHL (DLCL) (R)	52, M, BM	(+)		+	EBV^-; partial characterization	*74–78*
OCI-Ly4	B-NHL (IBL) (D)	21, M, BM	(+)		+	EBV^-; partial characterization	*74–76,78*
OCI-Ly5	B-NHL (D)	LN	(+)			EBV^+; probably **EBV^+ B-LCL**	*74*
OCI-Ly6	B-NHL (DLCL) (D)	LN	(+)		+	EBV^-; partial characterization	*74–76*
OCI-Ly7	B-NHL (DLCL) (R)	48, M, PB	(+)		+	EBV^-; partial characterization	*74–78*
OCI-Ly8	B-NHL (IBL) (R)	58, M, BM	(+)	(+)	+	EBV^-; t(14;18)(q32;q21)	*74–80*
OCI-Ly9	B-NHL (DLCL) (D)	53, M, BM	(+)		+	EBV^-; partial characterization	*74–76,78*
OCI-Ly10	B-NHL (IBL) (D)	66, F, LN	(+)		+	EBV^-; partial characterization	*74–76,78*

OCI-Ly11	B-NHL (DLCL) (D)	LN	(+)		+	EBV^{-}; partial characterization	***74–76***
OCI-Ly14	B-NHL (diffuse mixed) (D)	LN	(+)		+	EBV^{-}; partial characterization	***74,75***
OCI-Ly15	B-NHL (IBL) (D)	BM	(+)			EBV^{+}; probably **EBV^{+} B-LCL**	***74***
OCI-Ly16	B-NHL (DLCL) (D)	LN				EBV^{-}; cell line is lost	***74***
OCI-Ly19	B-NHL (DLCL) (R)	27, F, BM			(+)	EBV^{-}; no characterization	***78***
Oto	B-NHL (FCL)		+			EBV? no authentication/verification	***81***
PD1	B-NHL (lymphoblastic)	PE	(+)	(+)		EBV^{-}; t(8;14)(q24;q32); sister line **PD2** (EBV^{-})	***14,15***
Pfeiffer	B-NHL (DLCL) (LP)		(+)	(+)	(+)	EBV? t(14;18)(q32;q21)	***82***
PLL 1, 2, 3, 4, 5, 6	B-PLL	PB			(+)	EBV^{+}; no authentication/verification; probably 6 **EBV^{+} B-LCLs**; no characterization	***37,38***
PLL1	B-PLL	M, PB	(+)	+	(+)	EBV^{+}; origin/neoplastic nature verified	***83***
PLL2	B-PLL	F, PB	(+)	+	(+)	EBV^{+}; origin/neoplastic nature verified	***83***
PR	HCL	77, M, PB	(+)		(+)	EBV? no authentication/verification	***84,85***
PV-90	MDS (RAEB-T) (D)	54, M, BM	+	+	+	EBV^{+}; no authentication, verification yes	***86***
Rec-1	B-NHL (MCL)			(+)		EBV? t(11;14)(q13;q32)	***87,88***
RI	HCL	50, M, PB	(+)		(+)	EBV? no authentication/verification	***85,89***
RK4	B-CLL (stage 0)	PB		+	+	EBV^{+}; origin/neoplastic nature verified	***30***
RS11846	B-NHL			(+)		EBV? t(14;18)(q32;q21)	***90***
SA4	B-CLL (stage 4)	PB		+	+	EBV^{+}; origin/neoplastic nature verified; sister line **SA4-2** (EBV^{+})	***30***
SDK	ALL L2 (D)	6, M, PB	(+)	(+)	+	EBV^{-}; t(8;14)(q24;q32)	***91***
SeD	B-CLL	62, F, SP			(+)	EBV^{+}; origin/neoplastic nature verified	***31***
SK-DHL2A	B-NHL (diffuse histiocytic) (R)	39, M, AF		(+)	+	EBV? t(8;14)(q24;q32); subclone **SK-DHL2B** (tetraploid)	***92***
SU-DHL-3	B-NHL (diffuse histiocytic) (R)	35, M, AF	(+)	+	+	EBV^{-}; no immunoprofile	***93–97***
SU-DHL-11	B-NHL (diffuse histiocytic)	F	(+)	+		EBV? t(11;14)(q23;q32)	***93***
SU-DHL-12	B-NHL (diffuse histiocytic)	F	(+)	+		EBV? t(14;18)(q32;q21)	***93***
SU-DHL-13	B-NHL (diffuse histiocytic)	F	(+)	+		EBV? t(14;18)(q32;q21)	***93***
SU-DHL-14	B-NHL (diffuse histiocytic)	F	(+)	+		EBV? 14q32 alteration	***93***
SU-DHL-15	B-NHL (diffuse histiocytic)	F	(+)	+		EBV? no authentication/verification	***93***
SU-DUL5	B-NHL (lymphoblastic)	30, M		+	+	EBV? t(8;18;14)(q24;q21;q32)	***33,98–100***
SUP-B8	B-ALL (D)	13, F, BM			+	EBV^{-}; ATCC CRL 12073	***101***
TANAKA	B-NHL		+			EBV? no authentication/verification	***17***
TEM110	B-NHL (undifferentiated)	7, F, BM		(+)	+	EBV^{-}; 14q+; sister line **TEMPC**	***48***
THP-9	B-NHL (D)	7, M, AF	+		+	EBV^{-}; no authentication/verification	***102,103***

Table II ***Continued***

Cell line	Disease (status)	Age, sex, specimen	Core data available on			Comments	Ref.
			Immuno-profile	Karyo-type	Functional aspects		
Toledo	B-NHL (DLCL)	PB	(+)		(+)	EBV? no authentication/verification	***82***
U-715 M	lymphoblastic lymphosarcoma	44, F, LN	(+)	(+)	+	EBV⁻; t(14;18)(q32;q21)	***104,105***
ÚHKT-7	HCL (D)	41, M, PB	(+)		(+)	EBV⁺; no authentication/verification	***106***
UTMB-460	normal	BM				**cross-contamination! = line CCRF-CEM**	***107***
VAL	B-ALL	50, F, BM		+		EBV? t(3;4)(q27;p11), t(8;14;18)(q24;q32;q21)	***108–110***
VL51	B-NHL (SLVL) (D)	70, F, SP	+	+	+	EBV⁺; authentication yes, normal karyotype	***111***
Wa C3,CD5+	B-CLL (stage 1) (D)	PB	(+)	(+)		EBV⁺; origin/neoplastic nature verified	***2***
WIEN 133	B-ALL (L3)	12, M, BM	(+)	+		EBV⁻; t(8;14;12)(q24;q32;q24); *BCL7A* rearrangement	***33,112,113***
WL2	B-NHL (follicular)	PB	+	(+)	+	EBV? t(14;18)(q32;q21)	***114***
WM1	Waldenström's	F, PB	(+)	+	(+)	EBV⁺; no authentication/verification	***83***
ZK-H	HCL (D)	69, M, PB		(+)	(+)	EBV⁺; no authentication/verification; sister line **ZK-N (EBV⁺ B-LCL)**	***115***

Authentication refers to conclusive proof of derivation from one particular patient; verification refers to unequivocal proof of the neoplastic origin of the cell line.

Additional abbreviations: AF, ascitic fluid; BM, bone marrow; EBV?, EBV status not known; F, female; LN, lymph node; LP, leukemic phase; M, male; PB, peripheral blood; PE, pleural effusion; R, at relapse or refractory; SLVL, splenic lymphoma with circulating villous lymphocytes; SP, spleen; TU, tumor.

References

1 Sairenji, T. et al. (1983) Am. J. Hematol. 15, 361–374.
2 Wendel-Hansen, V. et al. (1994) Leukemia 8, 476–484.
3 Smith, L.J. et al. (1987) Cancer Res. 47, 2062–2066.
4 Lemon, S.M. et al. (1979) Ann. Int. Med. 90, 54–55.
5 Lewin, N. et al. (1988) Int. J. Cancer 41, 892–895.
6 Najfeld, V. et al. (1980) Int. J. Cancer 26, 543–549.
7 Karande, A. et al. (1980) Int. J. Cancer 26, 551–556.
8 Fialkow, P.J. et al. (1978) Lancet ii, 444–446. (case report)
9 Kim, K.M. et al. (1991) J. Immunol. 146, 819–825.
10 Yan, Y. et al. (1997) Blood 90, 62b.
11 Renne, R. et al. (1996) Nature Medicine 2, 342–346.
12 Drexler, H.G. et al. (1998) Leukemia 12, 1507–1517.
13 Caligaris-Cappio, F. et al. (1987) Leukemia Res. 11, 579–588.
14 Magrath, I.T. et al. (1980) J. Natl. Cancer Inst. 64, 465–476.
15 Magrath, I.T. et al. (1980) J. Natl. Cancer Inst. 64, 477–483.
16 Park, L.S. et al. (1989) J. Biol. Chem. 264, 5420–5427.
17 Minowada, J. and Matsuo, Y. (1999) unpublished data.
18 Zinzar, S. et al. (1998) Leukemia Res. 22, 677–685.
19 Skinnider, L.F. et al. (1985) Scand. J. Haematol. 35, 430–436.
20 Rickinson, A.B. et al. (1982) Clin. Exp. Immunol. 50, 347–354.
21 Beckwith, M. et al. (1990) J. Natl. Cancer Inst. 82, 501–509.
22 Li, C.C. et al. (1995) Leukemia 9, 650–655.
23 ATCC Website: <www.atcc.org>

24 Goy, A. et al. (1998) Blood 92, 183b.
25 Levitt, M.L. et al. (1983) Cancer Res. 43, 1195–1203.
26 Creszenzi, M. et al. (1988) Blood 71, 9–12.
27 Lee, C.L.Y. et al. (1986) Cancer Res. 46, 2497–2501.
28 Ghose, T. et al. (1988) Am. J. Hematol. 28, 146–154.
29 Faguet, G.B. et al. (1988) Blood 71, 422–429.
30 Zheng, C.Y. et al. (1996) Brit. J. Haematol. 93, 681–683.
31 Hurley, J.N. et al. (1978) Proc. Natl. Acad. Sci. USA 75, 5706–5710.
32 Stranks, G. et al. (1995) Blood 85, 893–901.
33 Willis, T.G. et al. (1997) Blood 90, 2456–2464.
34 Gaidano, G. et al. (1996) Leukemia 10, 1237–1240.
35 Gransar, A. et al. (1983) Am. J. Clin. Pathol. 79, 733–737.
36 Morikawa, K. et al. (1996) Brit. J. Haematol. 94, 250–257.
37 Walls, E.V. et al. (1989) Int. J. Cancer 44, 846–853.
38 Walls, E.V. et al. (1990) Leukemia Res. 14, 389–391.
39 Tokioka, T. et al. (1994) Acta Haematol. 92, 8–13.
40 Gribbin, T.E. et al. (1989) Leukemia 3, 643–647.
41 Lang, A.B. et al. (1985) Exp. Cell Biol. 53, 61–68.
42 Vedanthan, S. et al. (1991) Leukemia Lymphoma 5, 407–413.
43 Gingrich, R.D. et al. (1990) Blood 75, 2375–2387.
44 Ho, Y.S. et al. (1990) Brit. J. Cancer 61, 655–658.
45 Ming, P.L. et al. (1987) Am. J. Hum. Genet. 41, A33.
46 Taniwaki, M. et al. (1995) Blood 85, 3223–3228.
47 Yasuda, Y. et al. (1988) Acta Haematol. 80, 145–152.
48 Benjamin, D. et al. (1983) Blood 61, 1017–1019.
49 Benjamin, D. et al. (1982) J. Immunol. 129, 1336–1342.
50 Benjamin, D. et al. (1984) Proc. Natl. Acad. Sci. USA 81, 3547–3551.
51 Shibayama, H. et al. (1997) Leukemia Lymphoma 25, 373–380.
52 Andersson, L. et al. (1981) in Leukemia Markers (Knapp, W. et al. eds), Academic Press, New York, pp. 847–851.
53 Melo, J.V. et al. (1988) Clin. Exp. Immunol. 73, 23–28.
54 Kawamura, N. et al. (1986) J. Clin. Invest. 78, 1331–1338.
55 Ikezoe, T. et al. (1998) Leukemia 12, 94–95.
56 Asimakopoulos, F.A. et al. (1994) Blood 84, 3086–3094.
57 Morris, C.M. et al. (1989) Blood 74, 1768–1773.
58 Tsutsumi, Y. et al. (1982) Blood 59, 1220–1224.
59 Willem, P. (1992) Blood 80, 445a.
60 Fonatsch, C. et al. (1982) Int. J. Cancer 30, 321–327.
61 Dillman, R.O. et al. (1982) Cancer Res. 42, 1368–1373.
62 Nakamura, Y. et al. (1997) Leukemia 11, 1993–1994.
63 Drexler, H.G. et al. (1999) DSMZ Catalogue of Cell Lines, 7th edn, Braunschweig, Germany.
64 Ford, R.J. et al. (1998) Blood 92, 314a.
65 Kawata, A. et al. (1993) Leukemia Res. 17, 883–894.
66 Meeker, T.C. et al. (1991) Leukemia 5, 733–737.
67 Withers, D.A. et al. (1991) Mol. Cell. Biol. 11, 4846–4852.
68 Nitta, M. et al. (1990) Leukemia Lymphoma 3, 67–71.
69 Watanabe, S. et al. (1980) Cancer 46, 2438–2445.
70 Watanabe, S. et al. (1980) Cancer Res. 40, 2588–2595.
71 Iida, S. et al. (1992) Leukemia Res. 16, 1155–1163.
72 Winter, J.N. et al. (1984) Blood 63, 140–146.
73 Li, C.C. et al. (1995) Leukemia 9, 650–655.
74 Tweeddale, M.E. et al. (1987) Blood 69, 1307–1314.
75 Tweeddale, M.E. et al. (1989) Blood 74, 572–578.
76 Yee, C. et al. (1989) Blood 74, 798–804.
77 Farrugia, M.M. et al. (1994) Blood 83, 191–198.
78 Chang, H. et al. (1995) Leukemia Lymphoma 19, 165–171.
79 Cattoretti, G. et al. (1995) Blood 86, 45–53.
80 Schmidt-Wolf, I.G.H. et al. (1991) J. Exp. Med. 174, 139–149.
81 Toba, K. et al. (1996) Exp. Hematol. 24, 894–901.
82 Gabay, C. et al. (2000) Leukemia Lymphoma (in press).
83 Finerty, S. et al. (1982) Int. J. Cancer 30, 1–7.
84 Golde, D.W. et al. (1976) New Engl. J. Med., 296, 92–93.
85 Saxon, A. et al. (1978) J. Immunol. 120, 777–782.
86 Bergamaschi, G. et al. (1991) Brit. J. Haematol. 78, 167–172.
87 Raynaud, S. et al. (1993) Genes Chromosomes Cancer 8, 80–87.
88 Rimokh, R. et al. (1994) Blood 83, 1871–1875.
89 Golde, D.W. et al. (1977) Brit. J. Haematol. 35, 359–365.
90 Tanaka, S. et al. (1992) Blood 79, 229–237.
91 Kottaridis, S. et al. (1985) Leukemia Res. 9, 113–122.
92 Nishikori, M. et al. (1984) Cancer Genet. Cyogenet. 12, 39–50.
93 Kaiser-McGaw Hecht, B. et al. (1985) Cancer Genet. Cytogenet. 14, 205–218.
94 Epstein, A.L. et al. (1974) Cancer 34, 1851–1972.
95 Epstein, A.L. et al. (1976) Cancer 37, 2158–2176.
96 Epstein, A.L. et al. (1978) Cancer 42, 2379–2391.

[97] Epstein, A.L. et al. (1979) Cancer Res. 39, 1748–1759.
[98] Kiem, H.P. et al. (1990) Oncogene 5, 1815–1819.
[99] Lambrechts, A.C. et al. (1994) Leukemia 8, 1164–1171.
[100] van Ooteghem, R.B.C. et al. (1994) Cancer Genet. Cytogenet. 74, 87–94.
[101] Brophy, N.A. et al. (1994) Leukemia 8, 327–335.
[102] Minegishi, M. et al. (1987) Tohoku J. Exp. Med. 151, 283–292.
[103] Morita, S. et al. (1996) Leukemia 10, 102–105.
[104] Nilsson, K. and Sundström, C. (1974) Int. J. Cancer 13, 808–823.
[105] Sambade, C. et al. (1995) Int. J. Cancer 63, 710–715.
[106] Stöckbauer, P. et al. (1981) Neoplasma 28, 385–395.
[107] Juneja, H.S. et al. (1986) Leukemia Res. 10, 1209–1219.
[108] Kerckaert, J.P. et al. (1993) Nature Genet. 5, 66–70.
[109] Dallery, E. et al. (1995) Oncogene 10, 2171–2178.
[110] Dyer, M.J.S. et al. (1996) Leukemia 10, 1198–1208.
[111] Inokuchi, K. et al. (1995) Leukemia Res. 19, 817–822.
[112] Nacheva, E. et al. (1987) Cancer Genet. Cytogenet. 28, 145–153.
[113] Zani, V.J. et al. (1996) Blood 87, 3124–3134.
[114] Warburton, P. et al. (1990) Blood 76, 378a.
[115] Miyoshi, I. et al. (1981) Cancer 47, 60–65.

Table III *Panel of selected Burkitt's lymphoma cell lines*

Cell line	Disease	Age, sex, specimen	Year	EBV status	Availability	Comments	Ref.
BJA-B	African BL	5, F, TU	1973	EBV⁻	DSM ACC 72	t(2;8)(p12;q24); *P53* mutation	***1–3***
CA46	American BL	AF		EBV⁻	ATCC CRL 1648, DSM ACC 73	t(8;14)(q24;q32)	***3–7***
Daudi	African BL	16, M, TU	1967	EBV⁺	ATCC CCL 213, DSM ACC 78, JCRB 9071	t(8;14)(q24;q32); *P53* mutation	***3,4,8–10***
D.G.-75	American BL	10, M, PE	1975	EBNA⁻	DSM ACC 83	t(8;14)(q24;q32)	***3,11***
EB1	African BL	9, F, LN	1963	EBV⁺	ATCC HTB 60, DSM ACC 80	t(8;14)(q24;q32)	***3,4,12–14***
EB2	African BL	7, F, TU	1963	EBV⁺	ATCC HTB 61		***4,15–17***
EB3	African BL	3, M, TU	1965	EBV⁺	ATCC CCL 85, IFO 50028, JCRB 5028	*P53* mutation	***4,10,18***
Jijoye	African BL	7, M, TU		EBV⁺	ATCC CCL 87	alternative spelling: **Jiyoye**	***4,19***
Namalwa	African BL	child, TU	1967	EBV⁺	ATCC CRL 1432, DSM ACC 24, IFO 50040, JCRB 5040	t(8;14)(q24;q32)	***3,4***
Raji	African BL	12, M, TU	1963	EBV⁺	ATCC CCL 86, DSM ACC 319, IFO 50046, JCRB 9012	t(8;14)(q24;q32); *P53* mutation; first hematopoietic cell line	***2–4,20***
Ramos	American BL	3, M, TU	1972	EBV⁻	ATCC CRL 1596, JCRB 9119	*P53* mutation	***2,4,21,22***
SU-AmB-2	American BL	16, M, LN		EBNA⁺		rare EBV⁺ American Burkitt	***23***

Additional abbreviations: AF, ascitic fluid; BL, Burkitt's lymphoma; F, female; LN, lymph node; M, male; PE, pleural effusion; TU, tumor tissue.

References

[1] Menezes, J. et al. (1975) Biomedicine 22, 276–284.
[2] Farrell, P.J. et al. (1991) EMBO J. 10, 2879–2887.
[3] Drexler, H.G. et al. (1999) DSMZ Catalogue of Cell Lines, 7th edn, Braunschweig, Germany.
[4] ATCC Website: <www.atcc.org>
[5] Magrath, I.T. et al. (1980) J. Natl. Cancer Inst. 64, 465–476.
[6] Magrath, I.T. et al. (1980) J. Natl. Cancer Inst. 64, 477–483.
[7] Benjamin, D. et al. (1982) J. Immunol. 129, 1336–1342.
[8] Nadkarni, J.S. et al. (1969) Cancer 23, 64–79.
[9] Klein, E. et al. (1968) Cancer Res. 28, 1300–1310.
[10] Gaidano, G. et al. (1991) Proc. Natl. Acad. Sci. USA 88, 5413–5417.
[11] Ben-Bassat, H. et al. (1977) Int. J. Cancer 19, 27–33.
[12] Epstein, M.A. and Barr, Y.M. (1964) Lancet i, 252–253.
[13] Epstein, M.A. and Barr, Y.M. (1965) J. Natl. Cancer Inst. 34, 231–240.
[14] Epstein, M.A. and Achong, B.G. (1965) J. Natl. Cancer Inst. 34, 241–253.
[15] Epstein, M.A. et al. (1964) Path. Biol. Semaine Hop. 12, 1233–1234.
[16] Epstein, M.A. et al. (1965) J. Exp. Med. 121, 761–777.
[17] Epstein, M.A. et al. (1965) Brit. J. Cancer 19, 108–115.
[18] Epstein, M.A. et al. (1965) Wistar Inst. Symp. Monogr. 4, 59–70.
[19] Kohn, G. et al. (1967) J. Natl. Cancer Inst. 38, 209–222.
[20] Pulvertaft, R.J.V. (1964) Lancet i, 238–240.
[21] Klein, G. et al. (1975) Intervirology 5, 319–334.
[22] Klein, G. et al. (1974) Proc. Natl. Acad. Sci. USA 71, 3283–3287.
[23] Epstein, A.L. et al. (1976) Proc. Natl. Acad. Sci. USA 73, 228–232.

Table IV *Additional myeloma and plasma cell leukemia cell lines*

Cell line	Disease (status)	Age, sex, specimen	Core data available on: Immuno-profile	Karyo-type	Functional aspects	Comments	Ref.
ARH-77	plasma cell leukemia	33, F, PB	+		+	widely used line, but **EBV⁺ B-LCL**; ATCC CRL 1621	*1–4*
ARH-DR	multiple myeloma		(+)		+	EBV?	*5*
ARK	myeloma			(+)		EBV?	*6*
ARP-1	multiple myeloma	BM	(+)		+	EBV? partial immunoprofile	*5,7*
ARP-2	multiple myeloma		(+)		+	EBV? partial immunoprofile	*5*
DP-6	multiple myeloma (R, T)	78, F, PB	+		+	EBV^-; IL-6-dependent	*8,9*
GM1312	myeloma	BM	(+)		+	**EBV^+ B-LCL**	*10*
GM1500	myeloma	BM	(+)		+	**EBV^+ B-LCL**	*10*
HL407	multiple myeloma (R)	62, F, BM	+		+	EBV^-; secretes IgAλ and IL-6	*11*
Hs	multiple myeloma		+		+	EBV?	*12,13*
HS-Sultan	plasmacytoma					**cross-contamination! = line Jijoye**; ATCC CRL 1484	*14*
ILKM2	multiple myeloma (TH)	75, F, BM	(+)		+	EBV^-; IL-6-dependent	*15*
ILKM3	multiple myeloma (TH)	43, M, BM	(+)		+	EBV^-; IL-6-dependent	*15*
IM-9	multiple myeloma	F, BM	+	+	+	widely used line, but **EBV^+ B-LCL**; ATCC CCL 159, DSM ACC 117	*3,4,16,17*
INA-6	plasma cell leukemia	80, M, PE	+	+	+	EBV^-; IL-6-dependent; t(11;14)(q13;q32)	*18*
JIM-3	myeloma	PE	(+)	(+)	(+)	simultaneous sister lines **JIM-1, -2, -4**; *IGH-FGFR3/IGH-MMSET* fusion genes	*19–22*
JK-6	myeloma (leukemic phase)	75, M, PB	+		+	EBV^-; IL-6-dependent	*18*
Karpas 25	plasma cell leukemia			(+)		t(2;11)(p11;q23-24)	*23*
Karpas 1272	myeloma				(+)	EBV?	*24*
KAS-6/1	multiple myeloma (R, T)	56, F, PB	+		+	EBV^-; IL-6-dependent; serial sister line **KAS-6/2**	*8,9*
KHM-1A	multiple myeloma (R)	53, M, PE	(+)		+	EBV^-; simultaneous sister line **KHM-1B** produces amylase	*25,26*
KHM-4	multiple myeloma (R)	51, F, PE	(+)		+	EBV? produces ammonia	*27*
KHM-11	multiple myeloma (R)	52, M, PE	+		+	EBV^-; secretes IL-6; t(4;14)(p16;q32); adherent subclone **KHM-11ad**	*28,29*

KMM-56	multiple myeloma (R)	62, M, PE			+	EBV?	***30***
KMS-5	multiple myeloma	80, M, PE	+		+	EBV⁻	***31***
KP-6	multiple myeloma (R, T)	54, M, PE	+		+	EBV⁻; IL-6-responsive	***8,9***
KPC-32	multiple myeloma	M, BM	+	(+)	+	EBV⁻	***32***
KPMM1	myeloma				(+)	EBV? *P16INK4*A deletion	***33***
LA-49	multiple myeloma	59, F, PE			+	cell line lost	***34,35***
LB842	multiple myeloma				+	EBV⁻; no characterization	***36***
LB843	multiple myeloma		+		+	EBV⁺; probably **EBV⁺ B-LCL**	***36,37***
LB844	multiple myeloma				+	EBV⁻; no characterization	***36***
LB845	multiple myeloma		+		+	EBV⁺; probably **EBV⁺ B-LCL**	***36,37***
LB851	multiple myeloma		+		+	EBV⁻	***36,37***
LES	multiple myeloma	BM	(+)		+	EBV? partial immunoprofile	7
LOPRA-2	plasma cell leukemia	PB	+			EBV? partial immunoprofile	***38***
MC/CAR	plasma cell leukemia (D)	81, M, PB	(+)	+	+	EBV⁺; probably **EBV⁺ B-LCL**; ATCC CRL 8083	***3,39***
ME-1	multiple myeloma				+	HHV-8⁺	***40***
MER	multiple myeloma		(+)		+	EBV? partial immunoprofile	***5***
MIT	multiple myeloma	BM	(+)		+	EBV? partial immunoprofile	***7***
MM5.1	multiple myeloma (R, T)	55, M, BM	+		+	EBV⁻; stroma-dependent; stroma-independent subclone **MM5.2**; t(4;14)(p16.3;q32.3)	***20,41***
MM-A1	multiple myeloma (T)	39, M, PE	(+)		+	EBV⁻; IL-6-dependent; no cytogenetics	***42***
MM-C1	multiple myeloma (T)	50, M, BM	(+)		+	EBV⁻; IL-6-dependent; no cytogenetics	***42***
MM-M1	multiple myeloma			(+)		EBV? t(11;14)(q13;q32)	***6***
MM-S1	multiple myeloma (T)	56, F, PE	(+)		+	EBV⁻; IL-6-dependent; no cytogenetics	***42,43***
MM-Y1	multiple myeloma (T)	61, F, BM	(+)		+	EBV⁻; IL-6-dependent; no cytogenetics	***42***
MOLP-2	multiple myeloma (R, T)	55, M, PB	+		+	EBV? sister line **MOLP-3**	***44***
Nak-Pc2	multiple myeloma	68, M, PB	+			EBV? full immunoprofile	***12***
NOP-1	multiple myeloma	M, BM	(+)	(+)	+	EBV⁻; secretion of IgAκ	***45***
OCI-My1	multiple myeloma (T)				+	EBV⁻; no characterization	***9,46***
OCI-My2	multiple myeloma (T)				+	EBV⁻; secretion of IL-6	***46***
OCI-My3	multiple myeloma (T)				+	EBV⁻; secretion of IL-6	***46***
OCI-My4	multiple myeloma (T)				+	EBV⁻; no characterization	***9,46***
OCI-My5	multiple myeloma (T)				+	EBV⁻; t(14;16)(q32;q23)	***9,46,47***
OCI-My6	multiple myeloma (T)				+	EBV⁻; no characterization	***46***
OCI-My7	multiple myeloma (T)				+	EBV⁻; no characterization	***46***
OH-2	multiple myeloma (R, T)	52, F, PE	(+)		+	EBV⁻; IL-6-dependent; secretion of IgGκ	***48–50***

Table IV *Continued*

Cell line	Disease (status)	Age, sex, specimen	Core data available on			Comments	Ref.
			Immuno-profile	Karyo-type	Functional aspects		
OPM-3	plasma cell leukemia	PB	(+)	+	+	**EBV⁺ B-LCL**	*51*
PA 3	myeloma				+	EBV? no characterization	*52*
Sa5	plasma cell leukemia	62, F, BM	+	+	+	EBV? secretion of Igκ	*53*
SIK	multiple myeloma	BM	(+)		+	EBV? partial immunoprofile	*5*
SMITH	multiple myeloma		(+)		+	EBV? partial immunoprofile	*5*
TRAMA	multiple myeloma		(+)		+	EBV? partial immunoprofile	*5*
TU-1	multiple myeloma (R)	59, M, BM	(+)		+	EBV? secretion of cytokines	*54*
U-1957	plasma cell leukemia (R, T)	60, M, PE	+	(+)	+	EBV^-; IL-6-dependent; simultaneous sister line **U-1958**	*55*
U-1996	myeloma (D)	70, F, ascites	+	(+)	+	EBV^-	*55*
U-2030	myeloma (R)	PE	+		+	EBV^-; sister line **U-2031 (EBV^+ B-LCL)**; cell line is lost	*56*
UCLA #1	plasma cell leukemia	PB	(+)		+	EBV? no characterization	*57*
UM-1	myeloma (T)	BM		(+)		EBV? t(8;14)(q24;q32)	*22*
UM-2	myeloma (T)	Pericard		(+)		EBV? t(7;14)	*22*
UM-3	myeloma (D)	BM		(+)		EBV? t(14;20)	*22*
UM-6	myeloma (D)	BM		(+)		EBV? del(14)(q32.3)	*22*
UMJF-2	multiple myeloma (D)	74, M, BM	+	+	(+)	EBV^+; probably **EBV^+ B-LCL**	*58*
VC5	plasma cell leukemia	84, F, PB	+	+	+	EBV^-; well-characterized line	*59*
XG-8	secondary plasma cell leukemia	F, PB	+	(+)	+	EBV? IL-6-dependent; t(11;14)(q13;q32)	*60*
XG-9	secondary plasma cell leukemia	M, BM	+		+	EBV? IL-6-dependent	*60*

Additional abbreviations: BM, bone marrow; D, at diagnosis; EBV?, EBV status not known; F, female; M, male; PB, peripheral blood; PE, pleural effusion; R, at relapse or refractory; T, terminal; TH, during therapy.

References

[1] Drewinko, B. et al. (1984) Cancer 54, 1883–1892.
[2] Drewinko, B. et al. (1984) Cancer 54, 1893–1903.
[3] Pellat-Deceunynk, C. et al. (1995) Blood 86, 4001–4002.
[4] Drexler, H.G. et al. (1999) Leukemia 13, 1601–1607.
[5] Flick, J.T. et al. (1993) Blood 82, 259a.
[6] Chesi, M. et al. (1996) Blood 88, 674–681.
[7] Hardin, J. et al. (1994) Blood 84, 3063–3070.
[8] Westendorf, J.J. et al. (1996) Leukemia 10, 866–876.
[9] Westendorf, J.J. et al. (1995) Blood 85, 3566–3576.
[10] Goldstein, M. et al. (1985) Blood 66, 444–446.
[11] Scibienski, R.J. et al. (1992) Leukemia 6, 940–947.

[12] Minowada, J. and Matsuo, Y. (1999) unpublished data.
[13] Shima, Y. et al. (1995) Blood 85, 757–764.
[14] Harris, N.S. (1974) Nature 250, 507–509.
[15] Shimizu, S. et al. (1989) J. Exp. Med. 169, 339–344.
[16] van Boxel, J.A. and Buell, D.N. (1974) Nature 251, 443–444.
[17] Drexler, H.G. et al. (1999) DSMZ Catalogue of Cell Lines, 7th edn, Braunschweig, Germany.
[18] Gramatzki, M. et al. (1994) Blood 84, 173a.
[19] Hamilton, M.S. et al. (1991) Leukemia 5, 768–771.
[20] Chesi, M. et al. (1997) Nature Genet. 16, 260–265.
[21] Chesi, M. et al. (1998) Blood 92, 3025–3034.
[22] Kuipers, J. et al. (1999) Cancer Genet. Cytogenet. 109, 99–107.
[23] Kearney, L. et al. (1992) Blood 80, 194a.
[24] Stranks, G. et al. (1995) Blood 85, 893–901.
[25] Matsuzaki, H. et al. (1998) Blood 72, 978–982.
[26] Hata, H. et al. (1990) Acta Haematol. 83, 133–136.
[27] Matsuzaki, H. et al. (1992) Int. Medicine 31, 339–343.
[28] Hata, H. et al. (1992) Leukemia 8, 1768–1773.
[29] Kuribayashi, N. et al. (1999) Acta Haematol. 101, 113–118.
[30] Niho, Y. et al. (1984) Int. J. Cell Cloning 2, 161–172.
[31] Namba, M. et al. (1989) In Vitro Cell. Dev. Biol. 25, 723–729.
[32] Goto, H. et al. (1994) Blood 84, 1922–1930.
[33] Ogawa, S. et al. (1995) Blood 86, 1548–1556.
[34] Jobin, M.E. et al. (1974) J. Exp. Med. 140, 494–507.
[35] Ishii, K. et al. (1992) Am. J. Hematol. 41, 218–224.
[36] Bataille, R. et al. (1989) Cancer 63, 877–880.
[37] Duperray, C. et al. (1989) Blood 73, 566–572.
[38] Lohmeyer, J. et al. (1992) Blood 80, 124a.
[39] Ritts, R.E. Jr. et al. (1983) Int. J. Cancer 31, 133–141.
[40] Hyjek, E. et al. (1998) Blood 92, 96a.
[41] Van Riet, I. et al. (1997) Leukemia 11, 284–293.
[42] Okuno, Y. et al. (1991) Leukemia 5, 585–591.
[43] Okuno, Y. et al. (1992) Exp. Hematol. 20, 395–400.
[44] Matsuo, Y. et al. (1993) Human Cell 6, 310–312.
[45] Ogura, M. et al. (1989) Blood 74, 81a.
[46] Hitzler, J.K. et al. (1991) Blood 78, 1996–2004.
[47] Chesi, M. et al. (1998) Blood 91, 4457–4463.
[48] Börset, M. et al. (1994) Eur. J. Haematol. 53, 31–37.
[49] Börset, M. et al. (1996) Cytokine 8, 430–438.
[50] Hjorth-Hansen, H. et al. (1999) Brit. J. Haematol. 106, 28–34.
[51] Yoshida, H. et al. (1998) Brit. J. Haematol. 103, 804–812.
[52] Siebert, R. et al. (1995) Brit. J. Haematol. 91, 350–354.
[53] Kimura, Y. et al. (1990) Blood 76, 356a.
[54] Koskela, K. et al. (1998) unpublished cell line.
[55] Jernberg, H. et al. (1987) Blood 69, 1605–1612.
[56] Jernberg, H. et al. (1987) Int. J. Cancer 39, 745–751.
[57] Xu, F.H. et al. (1998) Blood 92, 241–251.
[58] Farnen, J.F. et al. (1991) Leukemia 5, 574–584.
[59] Fixe, P. et al. (1997) unpublished cell line.
[60] Zhang, X.G. et al. (1994) Blood 83, 3654–3663.

Table V *Additional immature T-cell lines*

Cell line	Disease (status)	Age, sex, specimen	Core data available on: Immuno-profile	Karyo-type	Functional aspects	Comments	Ref.
1301	ALL			+		t(8;14)(q24;q32)	*1,2*
ALL 320	Ataxia telangiectasia → T-ALL	18, F		+		t(7;14)(q35;q32)	*3,4*
Amsalem	T-ALL (D)	12, F, PB	(+)		+	no karyotype	*5*
Andia	T-ALL	12, M, BM/PB	(+)		+	not permanent cell line	*6*
EU-7	T-ALL (R)	16, F, BM	+	+	+	**cross-contamination! = line CCRF-CEM**	*7,8*
Gem	T-ALL	16, M, PB	(+)	+	+	not permanent cell line	*6*
HATL	T-ALL (R)	9, F, PB	(+)	(+)	+	*P53* mutation; sister line **HABL (EBV⁺ B-LCL)**	*9*
HPB-MLT	T-lymphoid malignancy (D)	62, F, PB	+		(+)	*P53* mutation; no karyotype	*10,11*
IARC 301	T-NHL (high grade)	LN	(+)		+	secretion of IL-2	*12*
K2-MDS	MDS (RAEB) → AML M4eo (R)	63, M, PB	+	+	+	questionable authentication	*13*
K3P	T-ALL			(+)		t(10;14)(q24;q11) *HOX11-TCRD*	*14*
Karpas 241	NHL → AML M5 (R)	35, M, PB	(+)	(+)	+	tetraploidy, no authentication	*15*
KH-1	T-cell lymphoma (R)	9, M, PB	(+)	+	+	normal karyotype 46, XY	*16*
KOPT-4			+		+	no detailed characterization; *P53* mutation	*17,18*
KOPT-5			+		(+)	no detailed characterization	*17*
KOPT-11			+		(+)	no detailed characterization	*17*
LALW-2	T-ALL (R,T)	12, M, PB	(+)	+	+	only serial *in vivo* passage in nude mice; t(11;14)(p13;q12)	*19*
L-MAT	ALL				+	no detailed characterization; *P53* mutation	*18,20*
L-SAK	ALL				+	no detailed characterization; *P53* mutation	*18,20*
L-SMY	T-ALL				+	no detailed characterization; *P15INK4B/P16INK4A* rearranged	*18,21*
MDS	MDS (CMML)	73, M, PE	+		+	**cross-contamination! = line JURKAT**	*22,23*
MHH-TALL1	T-ALL	11, M, PB	(+)	(+)	+	t(1;9;22)(q32;q34;q11)	*24,25*
MHH-TALL2	T-ALL	PB	(+)		+	responsive to SCF	*24*
MKB-1	AML	17, F, PB	+	+	+	**cross-contamination! = line CCRF-CEM**	*26*
MOLT 10	ALL	4, F, PB	+			no karyotype	*11*
MOLT 11	ALL	11, M, PB	+			cell line is lost	*11*

MOLT 15	AML M5a	33, M, PB	+		+	**cross-contamination! = line CTV-1**	***27,28***
PER-315	T-ALL (L1/L2) (D)	5, M, BM	+		+	not permanent cell line; sister **PER-423**	***29,30***
PER-420	T-ALL				+	no characterization; IL-2-dependent	***30***
PER-423	T-ALL (D)	5, M, BM	+		+	IL-2-dependent	***30–32***
PER-427	T-ALL (D)				+	IL-2-dependent	***30,32***
PER-487	ALL	child	+		+	IL-7/SCF-dependent	***33***
REX	ALL		+			no detailed characterization	***34,35***
Rym	T-ALL	4, M, PB	(+)	+	+	not permanent cell line	***6***
SPI-801	T-ALL (L2) (R)	10, M, PB	+	+	+	**cross-contamination! = line K-562**; DSM ACC 86	***28,36,37***
SPI-802	T-ALL (L2) (R)	10, M, PB	+	+	+	**cross-contamination! = line K-562**; DSM ACC 92	***28,36,37***
SUP-T5	T-lymphoblastic lymphoma (R)	12, M, PE	+	+	+	cell line is lost	***38,39***
THP-6	ALL (L2)	6, F, BM	+		+	no karyotype; *NRAS, P53* mutations	***18,40,41***
Timpani	T-ALL	8, M, BM	(+)	(+)	+	not permanent cell line	***6***
ÚHKT-42	AUL (R)	13, M, PB	+		+	good immunological characterization	***42***
UTP-5	T-ALL				+	no detailed characterization	***18***
UTP-6	T-ALL				+	no detailed characterization	***18***
Vepa	T-ALL	15, F, BM	(+)	+	+	not permanent cell line	***6***

Additional abbreviations: BM, bone marrow; D, at diagnosis; F, female; LN, lymph node; M, male; PB, peripheral blood; PE, pleural effusion; R, at relapse or refractory; T, terminal.

References

[1] Brattsand, G. et al. (1993) Leukemia 7, 569–579.
[2] van Ooteghem, R.B.C. et al. (1994) Cancer Genet. Cytogenet. 74, 87–94.
[3] Vitolo, V. et al. (1984) Haematologica 69, 696–700.
[4] Russo, G. et al. (1988) Cell 53, 137–144.
[5] Ben-Bassat, H. et al. (1985) Int. J. Cancer 35, 27–33.
[6] Gjerset, R. et al. (1990) Cancer Res. 50, 10–14.
[7] Zhou, M. et al. (1995) Blood 85, 1608–1614.
[8] Zhou, M. et al. (1995) Leukemia 9, 1159–1161.
[9] Yeargin, J. et al. (1993) J. Clin. Invest. 91, 2111–2117.
[10] Morikawa, S. et al. (1978) Int. J. Cancer 21, 166–170.
[11] Minowada, J. and Matsuo, Y. (1999) unpublished data.
[12] Duprez, V. et al. (1985) Proc. Natl. Acad. Sci. USA 82, 6932–6936.
[13] Matsuda, M. et al. (2000) Leukemia Res. 24, 103–108.
[14] Dubé, I. et al. (1991) Blood 78, 2996–3003.
[15] Karpas, A. et al. (1980) Brit. J. Haematol. 44, 415–424.
[16] Nagasaka, M. et al. (1982) Int. J. Cancer 30, 173–180.
[17] Kojika, S. et al. (1996) Leukemia 10, 994–999.
[18] Kawamura, M. et al. (1999) Leukemia Res. 23, 115–126.
[19] White, L. et al. (1984) J. Natl. Cancer Inst. 72, 1029–1038.
[20] Morita, S. et al. (1996) Leukemia 10, 102–105.
[21] Xu, F. et al. (1999) Brit. J. Haematol. 105, 155–162.
[22] Banerjee, R. et al. (1992) Proc. Natl. Acad. Sci. USA 89, 9996–10000.
[23] Drexler, H.G. and Dirks, W.G. (1999) unpublished data.
[24] Tomeczkowski, J. et al. (1998) Leukemia 12, 1221–1229.
[25] Drexler, H.G. et al. (1999) Leukemia Res. 23, 207–215.

Table V Continued

[26] Matsuo, Y. et al. (1989) Human Cell 2, 423–429.
[27] Drexler, H.G. and Minowada, J. (1989) Hematol. Oncol. 7, 115–125.
[28] Drexler, H.G. et al. (1999) Leukemia 13, 1601–1607.
[29] Kees, U.R. et al. (1990) Leukemia 4, 292–296.
[30] Kees, U.R. et al. (1993) Leukemia Res. 17, 51–59.
[31] Kees, U.R. et al. (1995) Leukemia 9, 1046–1050.
[32] Kees, U.R. et al. (1996) Oncogene 12, 2235–2239.
[33] Kees, U.R. and Ford, J. (1999) Immunology 96, 202–206.
[34] Acuto, O. et al. (1983) Cell 34, 717–726.
[35] Acuto, O. et al. (1984) Proc. Natl. Acad. Sci. USA 81, 3851–3855.
[36] Komiyama, A. et al. (1982) Blood 60, 1429–1436.
[37] Drexler, H.G. et al. (1999) DSMZ Catalogue of Cell Lines, 7th edn, Braunschweig, Germany.
[38] Smith, S.D. et al. (1988) Blood 71, 395–402.
[39] Smith, S.D. et al. (1989) Blood 73, 2182–2187.
[40] Minegishi, M. et al. (1987) Tohoku J. Exp. Med. 151, 283–292.
[41] Minegishi, M. et al. (1988) Leukemia Res. 12, 227–232.
[42] Stöckbauer, P. et al. (1994) Human Cell 7, 40–46.

Table VI *Additional mature T-cell lines*

Cell line	Disease (status)	Age, sex, specimen	Core data available on: Immuno-profile	Karyo-type	Functional aspects	Comments	Ref.
CTCL-2	Sézary syndrome (D)	60, F, PB	+	+	+	normal karyotype; immortalization not confirmed	***1,2***
HH	CTCL (R,T)	61, M, PB	(+)		+	no karyotype; HTLV-I$^-$	***3***
KT-3	Lennert's lymphoma, leukemic phase (T)	53, M, PB	+		+	no karyotype; IL-6-dependent	***4–6***
MO (= **Mo-T**)	HCL	33, M, SP	(+)		+	production of GM-CSF, HTLV-II (ATCC CRL 8066); sister line **Mo-B** (HTLV-II$^+$, **EBV$^+$ B-LCL** from PB, ATCC CCL 245)	***7–9***
OCI-Ly12	T-NHL (small non-cleaved) (D)	37, M, LN	(+)		+	no karyotype; production of GM-CSF, IL-6	***10–13***
OCI-Ly13.1	T-NHL (DLCL) (D)	28, F, BM	(+)		+	no karyotype; responsive to cytokines serial sister line **OCI-Ly13.2** (at R)	***10–14***
OCI-Ly17	T-NHL (DLCL) (D) with hypereosinophilia	72, M, PB	+		+	no karyotype; production of IL-5, IL-6; *P53* mutation	***12–17***
SZ-4 to -33	Sézary syndrome	PB	(+)			series of 9 poorly characterized HTLV-I$^+$ lines; immortalization not confirmed	***18–20***

Additional abbreviations: BM, bone marrow; D, at diagnosis; F, female; LN, lymph node; M, male; PB, peripheral blood; R, at relapse or refractory; SP, spleen; T, terminal.

References

[1] Poiesz, B.J. et al. (1980) Proc. Natl. Acad. Sci. USA 77, 6815–6819.
[2] Minowada, J. and Matsuo, Y. (1999) unpublished data.
[3] Starkebaum, G. et al. (1991) Int. J. Cancer 49, 246–253.
[4] Shimizu, S. et al. (1988) Blood 71, 196–203.
[5] Shimizu, S. et al. (1988) Blood 72, 1826–1828.
[6] Inoue, K. et al. (1997) Blood 89, 1405–1412.
[7] Saxon, A, et al. (1978) Ann. Intern. Med. 88, 323–326.
[8] Saxon, A. et al. (1978) J. Immunol. 120, 777–782.
[9] Chen, I.S.Y. et al. (1983) Proc. Natl. Acad. Sci. USA 80, 7006–7009.
[10] Tweeddale, M.E. et al. (1987) Blood 69, 1307–1314.
[11] Tweeddale, M.E. et al. (1989) Blood 74, 572–578.
[12] Yee, C. et al. (1989) Blood 74, 798–804.
[13] Chang, H. et al. (1995) Leukemia Lymphoma 19, 165–171.
[14] Chang, H. et al. (1994) Blood 83, 452–459.
[15] Chang, H. et al. (1992) Leukemia Lymphoma 8, 97–107.
[16] Chang, H. et al. (1992) Leukemia Lymphoma 8, 129–136.
[17] Farrugia, M.M. et al. (1994) Blood 83, 191–198.
[18] Abrams, J.T. et al. (1991) J. Immunol. 146, 1455–1462.
[19] Abrams, J.T. et al. (1991) J. Invest. Dermatol. 96, 31–37.
[20] Ghosh, S. et al. (1994) Blood 84, 2663–2671.

Table VII *Panel of selected adult T-cell leukemia-lymphoma cell lines*

Cell line	Disease	Age, sex, specimen	Year	HTLV-I status	Comments	Ref.
ATL-1K	ATLL	82, M, PB		+	IL-2-independent; *P53* mutation	*1*
ATL-2M	ATLL	56, M, PB		+	IL-2-dependent	*1*
ATL-3I	ATLL	71, M, PB		+	IL-2-independent	*1*
ATL-4K	ATLL	62, M, PB		+	IL-2-dependent	*1*
ATL-5S	ATLL	34, M, PB		+	IL-2-independent	*1*
ATL-6A	ATLL	33, F, PB		+	IL-2-dependent	*1*
ATL-7S	ATLL	68, M, LN		+	IL-2-dependent	*1*
ATL-8K	ATLL	63, F, PB		+	IL-2-dependent	*1*
ATL-9Y	ATLL	59, M, PB		+	IL-2-dependent	*1*
ATL-10Y	ATLL	47, M, PB		+	IL-2-dependent	*1*
KH-2	ATLL (T)	48, M, PB	1979	+	$EBNA^-$	*2*
ME	ATLL (D)	39, M, PB	1981	+	EBV^+; subclones **ME-2** (precursor B-cell) and **ME-11A** (B-cell) also $EBV^+/HTLV\text{-}I^+$	*3*
MT-1	ATLL	69, M, PB	1978	+	$EBNA^-$	*4*
MT-2	normal	infant, M, UCB	1980	+	established by cocultivation with CM of MT-1; normal karyotype; $EBNA^-$	*5,6*
MU	ATLL (R)	37, M, PB	1986	+	$EBNA^-$	*7*

Additional abbreviations: D, at diagnosis; F, female; LN, lymph node; M, male; PB, peripheral blood; R, refractory/resistant; UCB, umbilical cord blood; T, terminal.

References

[1] Hoshino, H. et al. (1983) Proc. Natl. Acad. Sci. USA 80, 6061–6065.
[2] Nagasaka, M. et al. (1982) Int. J. Cancer 30, 173–180.
[3] Koeffler, H.P. et al. (1984) Blood 64, 482–490.
[4] Miyoshi, I. et al. (1980) Gann 71, 155–156.
[5] Miyoshi, I. et al. (1981) Nature 294, 770–771.
[6] Miyoshi, I. et al. (1981) Gann 72, 978–981.
[7] Koizumi, S. et al. (1992) J. Natl. Cancer Inst. 84, 690–693.

Table VIII *Additional natural killer cell leukemia-lymphoma cell lines*

Cell line	Disease (status)	Age, sex, specimen	Core data available on			Comments	Ref.
			Immuno-profile	Karyo-type	Functional aspects		
EBT-8	LGL leukemia (D)	36, M, PB	+	+	+	IL-2-dependent; not NK cell line, but cytotoxic T-cell line ($CD8^+$)	*1*
EP	NK cell lymphoma (R) (leukemic phase)		+		+	IL-2-dependent	*2*
KAI3	severe chronic active EBV-infection	1, M, PB	(+)	+	+	EBV^+; neoplastic karyotype of cell line, but not of primary cells	*3*
KH-143	NK/myeloid leukemia		+	+	+	rather myelocytic line	*4*
NK3.3	normal donor	PB	+		+	not neoplastic, normal cells	*5–7*
SPI-801	T-ALL (R)	10, F, PB	+	+	+	**cross-contamination! = line K-562**	*8,9*
SPI-802	T-ALL (R)	10, F, PB	+	+	+	**cross-contamination! = line K-562**	*8,9*
TALL-103/2	T-NHL lymphoblastic (R)	6, M, BM	+	+	+	IL-2-dependent; not NK cell line, but cytotoxic T-cell line ($CD8^+$)	*10–12*
TALL-104	T-ALL (R)	2, M, PB	+	(+)	+	IL-2-dependent; not NK cell line, but cytotoxic T-cell line ($CD8^+$)	*12,13*
TALL-107	T-ALL (D)	8, M, BM	+	(+)	+	IL-2-dependent; not NK cell line, but cytotoxic T-cell line ($CD8^+$)	*14*
TKS-1	NK cell LGL leukemia	21, M, PB	+	+	+	cells stopped to grow; IL-2-dependent	*15,16*
VP	NK cell lymphoma (R)		+		+		*2*
[Unnamed line]	NK cell leukemia (D)	71, M, PB	(+)	(+)	+	probably not continuous cell line	*17*
[Unnamed lines]	undefined LGL expansions					probably not continuous cell lines	*18*

Additional abbreviations: BM, bone marrow; D, at diagnosis; F, female; M, male; PB, peripheral blood; PE, pleural effusion; R, at relapse or refractory.

References

[1] Asada, H. et al. (1994) Leukemia 8, 1415–1423.
[2] Hsi, E.D. et al. (1997) Blood 90, 258b.
[3] Tsuge, I. et al. (1999) Clin. Exp. Immunol. 115, 385–392.
[4] Takiguchi, T. et al. (1998) Brit. J. Haematol. 102, 24.
[5] Kornbluth, J. et al. (1982) J. Immunol. 129, 2831–2837.
[6] Leiden, J. et al. (1988) Immunogenetics 27, 231–238.
[7] Zhou, J. et al. (1997) Cell. Immunol. 178, 108–116.
[8] Komiyama, A. et al. (1982) Blood 60, 1429–1436.
[9] Gignac, S.M. et al. (1993) Leukemia Lymphoma 10, 359–368.
[10] Santoli, D. et al. (1990) J. Immunol. 144, 4703–4711.
[11] O'Connor, R. et al. (1990) J. Immunol. 145, 3779–3787.

Table VIII ***Continued***

[12] O'Connor, R. et al. (1991) Blood 77, 1534–1545.
[13] Cesano, A. et al. (1991) Blood 77, 2463–2474.
[14] Cesano, A. et al. (1993) Blood 81, 2714–2722.
[15] Kojima, H. et al. (1994) Leukemia 8, 1999–2004.
[16] Kojima, H. (1999) personal communication.
[17] Fernandez, L.A. et al. (1986) Blood 67, 925–930.
[18] Pistoia, V. et al. (1986) J. Clin. Immunol. 6, 457–466.

Table IX *Additional Hodgkin's disease cell lines*

Cell line	Disease (status)	Age, sex, specimen	Core data available on			Comments	Ref.
			Immuno-profile	Karyo-type	Functional aspects		
AG-F	neuroblastoma (stage IV), later Hodgkin's disease	adult, F, BM	+	+	+	**cross-contamination! = line CCRF-CEM**	*1,2*
AICHI-4	Hodgkin's disease (LP)	50, M, LN	+			no verification of malignant nature of cell line; probably **EBV⁺ B-LCL**	*3,4*
Co	Hodgkin's disease (NS, IV)	F, LN	+	(+)	+	**cross-contamination! = line CCRF-CEM; (= Cole)**	*2,5*
FQ	Hodgkin's disease					**cross-contamination! = monkey line**	*6,7*
HD-Mar	Hodgkin's disease (MC, R)	20, M, PE	+	+	+	$CD15^-$, $CD30^-$; rather derived from T-cell lymphoblastic lymphoma	*4,8*
Ho	Hodgkin's disease (NS, II)	F, LN	+		+	incomplete characterization; (= **Holden**)	*9–11*
HuT11	Hodgkin's disease (MC, IIA)	6, F, LN		(+)	+	incomplete characterization	*10,12*
L 439	Hodgkin's disease (NS, IVB)	36, M, PE				cell line is lost	*13,14*
L 591	Hodgkin's disease (NS, IVB)	31, F, PE	+	(+)	+	probably **EBV⁺ B-LCL**	*10,11,15*
RB	Hodgkin's disease					**cross-contamination! = monkey line**	*6,7*
RSp	Hodgkin's disease (MC, IV)	14, M, SP				**EBV⁺ B-LCL**; sister line **RN** from LN	*16*
RY	Hodgkin's disease					**cross-contamination! = unknown line**	*6,7*
SpR	Hodgkin's disease					**cross-contamination! = monkey line**	*6,7*
ZO	Hodgkin's disease (NS, II, R)	26, F, PE	+	+	+	immortalization not confirmed	*10,11,17*

Additional abbreviations: BM, bone marrow; F, female; LN, lymph node; LP, lymphocyte predominant; M, male; MC, mixed cellularity; NS, nodular sclerosis; PE, pericardial or pleural effusion; R, at relapse or refractory; SP, spleen.

References

[1] Gazitt, Y. et al. (1993) Leukemia 7, 2034–2044.
[2] Drexler, H.G. et al. (1999) Leukemia 13, 1601–1607.
[3] Ito, Y. et al. (1968) J. Natl. Cancer Inst. 41, 1367–1375.
[4] Minowada, J. and Matsuo, Y. (1999) unpublished data.
[5] Jones, D. et al. (1985) Hematol. Oncol. 3, 133–145.
[6] Long, J.C. et al. (1977) J. Exp. Med. 145, 1484–1500.
[7] Harris, N.L. et al. (1981) Nature 289, 228–230.
[8] Ben-Bassat, H. et al. (1980) Int. J. Cancer 25, 583–590.
[9] Jones, D. B. et al. (1989) Recent Results Cancer Res. 117, 62–66.
[10] Drexler, H.G. (1993) Leukemia Lymphoma 9, 1–25.
[11] Diehl, V. et al. (1990) Semin. Oncol. 17: 660–672.
[12] Roberts, A.N. et al. (1978) Cancer Res. 38, 3033–3043.
[13] Schaadt, M. et al. (1980) Int. J. Cancer 26, 723–731.
[14] Schaadt, M. et al. (1989) Recent Results Cancer Res. 117, 53–61.
[15] Diehl, V. et al. (1982) Cancer Treat. Rep. 66, 615–632.
[16] Friend, C. et al. (1978) Cancer Res. 38, 2581–2591.
[17] Poppema, S. et al. (1989) Recent Results Cancer Res. 117, 67–74.

Table X *Additional anaplastic large cell lymphoma cell lines*

Cell line	Disease (status)	Age, sex, specimen	Core data available on			Comments	Ref.
			Immuno-profile	Karyo-type	Functional aspects		
AMS3	Ki-1 lymphoma	23, M, TU	(+)	+	+	only in SCID mice; t(2;5)(p23;q35)	***1,2***
HSC-M1	Ki-1 lymphoma			(+)	+	t(2;5)(p23;q35)	***3,4***
HX	malignant histiocytosis		+		+	ALCL or true histiocytic?	***5***
JK	ALCL	58, M, TU	(+)		+	resistant to TGF-β	***6***
L82	ALCL (R)	24, F, PE	(+)			t(2;5)(p23;q35); T-cell type	***7***
MH-1	malignant histiocytosis	61, M, BM	(+)		+	ALCL or true histiocytic?	***8,9***
SU-DHL-2	histiocytic lymphoma	73, F, PE	+	+	+	probably ALCL (unconfirmed)	***10–13***
UCONN-L2	ALCL (R)	LN	+	(+)		t(2;5)(p23;q35)	***14,15***
WSU-ALCL	ALCL (R)	18, PE	+	(+)	+	t(2;5)(p23;q35); T-cell type	***15,16***

Additional abbreviations: BM, bone marrow; F, female; LN, lymph node; M, male; PE, pleural effusion; R, at relapse or refractory; TU, tumor.

References

[1] Shiota, M. et al. (1994) Oncogene 9, 1567–1574.
[2] Itoh, T. et al. (1993) Cancer 72, 2686–2694.
[3] Al Hashmi, I. et al. (1993) Blood 82, 132a.
[4] Gruss, H.J. et al. (1994) Blood 83, 2045–2056.
[5] Su, I.J. et al. (1988) Am. J. Pathol. 132, 192–198.
[6] Schiemann, W.P. et al. (1999) Blood 94, 2854–2861.
[7] Orscheschek, K. et al. (1995) Lancet 345, 87–90.
[8] Kim, H. et al. (1982) Am. J. Pathol. 106, 204–223.
[9] Kadin, M.E. et al. (1984) Lab. Invest. 50, 29a.
[10] Epstein, A.L. et al. (1974) Cancer 34, 1851–1872.
[11] Epstein, A.L. et al. (1978) Cancer 42, 2379–2391.
[12] Winter, J.N. et al. (1984) Blood 63, 140–146.
[13] Kaiser-McCaw Hecht, B. et al. (1985) Cancer Genet. Cytogenet. 14, 205–218.
[14] Morris, S.M. et al. (1994) Science 263, 1281–1284.
[15] Dirks, W.G. et al. (1996) Leukemia 10, 142–149.
[16] Al-Katib, A. et al. (1994) Blood 84, 638a.

Table XI *Additional myelocytic cell lines*

Cell line	Disease (status)	Age, sex, specimen	Core data available on: Immuno-profile	Karyo-type	Functional aspects	Comments	Ref.
2L1	t-AML M5	6, M, BM	+				*1*
8261	AML	PB			+		*2*
AML 5q-	AML M2				+		*3,4*
AML-CL	AML M1 (D)	27, F, PB	(+)	+		only growing in SCID mice	*5,6*
AML-PS	AML M1 (3rd R)	61, M, BM	(+)	+		only growing in SCID mice	*5,6*
AP-1060	AML M3 (R)	45, M, BM		+	+	t(15;17)(q22;q11)	*7*
CC-AML	AML M1 (R)	PB	+		+		*8*
Ei501	AML M3 (D)	17, F, BM	+	+	+	t(15;17)(q22;q12); probably lost due to yeast infection	*9*
EU-4	AML (R)	child	(+)		+		*10,11*
HSM-911	acute mixed leukemia (R)	65, F, BM	+			cytokine-dependent	*12*
IRTA17	AML M2 (R)	25, F, BM	+	+	+	t(16;21)(p11;q22); sister line **IRTA21** cytokine-dependent; lost to earthquake	*13*
KF-19	AML M2 (R)	35, M, PE	+		+		*14*
KH-143	NK-myeloid leukemia		+	+	+		*15*
KM-MDS	MDS	BM	+	+	+		*15*
KOPM-30	AML	PB		(+)	+	t(9;22)(q34;q11)	*16*
KOPM-55	AML		+	(+)		t(9;22)(q34;q11)	*17*
M20	AML M2 (D)	10, F, PB	(+)		+		*18,19*
M24	AML M4	PB			+		*19*
M26	AML M4	PB			+		*19*
MDS-KZ	MDS (RAEBT) → AML	73, F, BM	+	(+)	+	Vit. K induces apoptosis	*20*
MHH 203	AML	adult, PB			+		*21*
MML-1	AML M1	BM	+	+			*22,23*
MO-91	AML M0				+		*3,4*
MOLM-6	CML (BC)	44, M, PB	+			sister lines **MOLM-7, -8, -9, -10, -11, -12**	*24*
MPD	non-CML MPD		+		+		*25*
MZ93	CML (BC)		+				*26*
OCI/AML4	Hodgkin → AML M4 (D)	35, F, PB	+		+		*27*
OCI/AML6	MDS → AML M4 (D)	68, F, PB			+	cytokine-dependent	*21,28*

Table XI ***Continued***

Cell line	Disease (status)	Age, sex, specimen	Core data available on: Immuno-profile	Karyo-type	Functional aspects	Comments	Ref.
OU-AML-1	AML M4 (D)	26, F, PB	+		(+)	**cross-contamination! = line OCI/AML2**	*29*
OU-AML-2	AML M2 (D)	52, F, PB	+		(+)	**cross-contamination! = line OCI/AML2**	*29*
OU-AML-3	AML M4 (D)	48, M, PB	+		(+)	**cross-contamination! = line OCI/AML2**	*29*
OU-AML-4	AML M2 (1st R)	39, M, PB	+		(+)	**cross-contamination! = line OCI/AML2**	*29*
OU-AML-5	AML M5 (2nd R)	70, M, PB	+		(+)	**cross-contamination! = line OCI/AML2**	*29*
OU-AML-6	AML M1 (1st R)	47, F, PB	+		(+)	**cross-contamination! = line OCI/AML2**	*29*
OU-AML-7	AML M4 (1st R)	63, F, PB	+		(+)	**cross-contamination! = line OCI/AML2**	*29*
OU-AML-8	AML M4 (D)	63, F, PB	+		(+)	**cross-contamination! = line OCI/AML2**	*29*
PB-1049	CML (BC)	50, M, PB	(+)	+		**EBV$^+$ B-LCL**; t(9;22) Ph$^+$	*30*
PhB1	CML (BC)	67, F, BM	+	+		**EBV$^+$ B-LCL**; t(9;22) Ph$^+$	*31*
RDFD-2	AML M1	56, M, PB			+		*32*
RED-3	T-ALL → AML (2nd R)	24, M, PB	+		+	**cross-contamination with HL-60?**	*33*
SR-91	T-ALL (R)	22, M, PB	+	+	+	**cross-contamination! = line AML-193**	*34*
TI-1	AML M2 (D)	42, M, PB	+	+	+	**cross-contamination! = line K-562**	*35*
TMM	CMML (BC)	62, M, PB	+	+	+	**EBV$^+$ B-LCL** (not leukemic; Ph$^-$)	*36*
UTP-L12	AML			(+)		t(16;21)(p11;q22) *TLS/FUS-ERG*	*37*
YJ	CMML (BC)	69, M, PB	(+)	+	+	**cross-contamination! = line HL-60**	*38*

Additional abbreviations: BM, bone marrow; BC, at blast crisis; D, at diagnosis; F, female; M, male; PB, peripheral blood; PE, pericard effusion; R, at relapse/resistant/refractory.

References

[1] Felix, C.A. et al. (1998) Blood 91, 4451–4456.
[2] Barak, Y. et al. (1974) Brit. J. Haematol. 27, 543–549.
[3] Okabe, M. et al. (1992) Leukemia Lymphoma 8, 57–63.
[4] Okabe, M. et al. (1995) Leukemia Res. 19, 933–943.
[5] Giavazzi, R. et al. (1995) Int. J. Cancer 61, 280–285.
[6] Rambaldi, A. et al. (1993) Blood 82, 545a.
[7] Kim, S.H. et al. (1998) Blood 92, 609a.
[8] Srikanth, S. et al. (1993) Blood 82, 23a.
[9] Weidmann, E. et al. (1997) Leukemia 11, 709–713.
[10] Zhou, M. et al. (1995) Blood 85, 1608–1614.
[11] Zhou, M. et al. (1995) Leukemia 9, 1159–1161.
[12] Takahashi, M. et al. (1997) Leukemia Res. 21, 1115–1123.
[13] Hiyoshi, M. et al. (1995) Brit. J. Haematol. 90, 417–424.
[14] Fukuda, T. et al. (1996) Leukemia Res. 20, 931–939.
[15] Takiguchi, T. et al. (1998) Brit. J. Haematol. 102, 24.
[16] Wada, H. et al. (1994) Leukemia 8, 53–59.
[17] Kojika, S. et al. (1996) Leukemia 10, 994–999.
[18] Treves, A.J. et al. (1985) Exp. Hematol. 13, 281–288.
[19] Treves, A.J. et al. (1986) Immunol. Letters 12, 225–230.
[20] Nishimaki, J. et al. (1999) Leukemia 13, 1399–1405.
[21] Drexler, H.G. et al. (1997) Leukemia 11, 701–708.

[22] Komada, Y. et al. (1995) Blood 86, 3848–3860.
[23] Komada, Y. and Sakurai, M. (1997) Leukemia Lymphoma 25, 9–21.
[24] Tsuji-Takayama, K. et al. (1994) Human Cell 7, 167–171.
[25] Paul, C.C. et al. (1998) Blood 92, 147b.
[26] Toba, K. et al. (1996) Exp. Hematol. 24, 894–901.
[27] Koistinen, P. et al. (1991) Leukemia 5, 704–711.
[28] Tohda, S. et al. (1993) J. Cell. Physiol. 154, 410–418.
[29] Zheng, A. et al. (1999) Brit. J. Cancer 79, 407–415.
[30] Yamada, T. et al. (1985) Jpn. J. Cancer Res. 76, 365–373.
[31] Knuutila, S. et al. (1993) Acta Haematol. 90, 190–194.
[32] Lucas, D.L. et al. (1983) Blood 62, 152a.
[33] Mallet, M.K. et al. (1989) Leukemia 3, 511–515.
[34] Klingemann, H.G. et al. (1994) Leukemia Lymphoma 12, 463–470.
[35] Taoka, T. et al. (1992) Blood 80, 46–52.
[36] McCarty, T.M. et al. (1987) Blood 70, 1665–1672.
[37] Ichikawa, H. et al. (1994) Cancer Res. 54, 2865–2868.
[38] Yamaguchi, Y. et al. (1998) Leukemia 12, 1430–1439.

Table XII *Additional monocytic cell lines*

Cell line	Disease (status)	Age, sex, specimen	Core data available on			Comments	Ref.
			Immuno-profile	Karyo-type	Functional aspects		
2MAC	normal donor	PB				apparently normal, non-neoplastic cells	*1*
AMoL I	AML M5 (D)	child, BM	(+)				*2*
AMoL II	AML M5 (D)	child, BM	(+)				*2*
DD	malignant histiocytosis	PB	+		+	at leukemic phase	*3*
HBM-MI-1	diffuse cutaneous mastocytosis	1, M, BM				SV40-immortalized, non-leukemic lines; sister line **HBM-MI-2**	*4,5*
HL-92	AML M4	M, PB				cell line lost	*6*
J6-1	AML M5		(+)		+		*7*
J-111	AML M5	25, F, PB				**cross-contamination! = line HeLa**	*8*
JOSK-I	AML M4	72, M, PB				**cross-contamination! = line U-937**	*9*
JOSK-K	AML M5	54, M, PB				**cross-contamination! = line U-937**	*9*
JOSK-M	CML (BC)	37, M, PB				**cross-contamination! = line U-937**	*9*
JOSK-S	AML M5	66, F, PB				**cross-contamination! = line U-937**	*9*
K1m	normal donor	PB				apparently normal, non-neoplastic cells	*10*
Karpas 230	AML M4 (D)	35, M, PB	(+)				*11*
KMT-2	normal donor	CB				apparently normal, non-neoplastic cells	*12*
KOCL-48	ALL L2 → AML M4	1, F	+			t(4;11)(q21;q23)	*13–15*
Na	AML M2 → AML M4 (R)	69, F, BM		(+)			*16*
OCI/AML3	AML M4 (D)	57, M, PB	+				*17,18*
OTC-4	AML M4	PB			+		*19*
PLB-985	AML M4	38, F, PB	+	+	+	**cross-contamination! = line HL-60**	*20*
YAP	psoriasis vulgaris	42, M, PB	+	+		patient without hematological disease, but multiple chromosomal alterations	*21*

Additional abbreviations: BM, bone marrow; BC, at blast crisis; CB, cord blood; D, at diagnosis; F, female; M, male; PB, peripheral blood; R, at relapse/resistant.

References

[1] Dialynas, D.P. et al. (1997) Cell. Immunol. 177, 182–193.
[2] Giller, R.H. et al. (1984) J. Clin. Immunol. 4, 429–438.
[3] Kávai, M. et al. (1992) Cell. Immunol. 139, 531–540.

[4] Townsend, M. et al. (1993) Brit. J. Haematol. 85, 452–461.
[5] Krilis, S.A. et al. (1991) Blood 78, 290–303.
[6] Salahuddin, S.Z. et al. (1982) Leukemia Res. 6, 729–741.
[7] Wu, K.F. et al. (1994) Leukemia Res. 18, 843–849.
[8] Osgood, E.E. and Brooke, J.H. (1955) Blood 10, 1010–1022.
[9] Ohta, M. et al. (1986) Cancer Res. 46, 3067–3074.
[10] Dialynas, D.P. et al. (1996) Blood 88, 158a.
[11] Karpas, A. et al. (1978) Brit. J. Cancer 37, 308–315.
[12] Tamura, S. et al. (1990) Blood 76, 501–507.
[13] Iida, S. et al. (1992) Leukemia Res. 16, 1155–1163.
[14] Yamamoto, K. et al. (1994) Blood 83, 2912–2921.
[15] Kojika, S. et al. (1996) Leukemia 10, 994–999.
[16] Kimura, Y. et al. (1990) Blood 76, 289a.
[17] Wang, C. et al. (1989) Leukemia 3, 264–269.
[18] Wang, C. et al. (1991) Leukemia 5, 493–499.
[19] Ono, Y. et al. (1994) Blood 84, 574a.
[20] Tucker, K.A. et al. (1987) Blood 70, 372–378.
[21] Hamamoto, Y. et al. (1996) Arch. Dermatol. Res. 288, 225–229.

Table XIII *Additional erythrocytic-megakaryocytic cell lines*

Cell line	Disease (status)	Age, sex, specimen	Core data available on: Immuno-profile	Karyo-type	Functional aspects	Comments	Ref.
AML-HJ	AML M7	BM	+	+			*1,2*
B-403	AML	PB				not neoplastic line (**EBV+ B-LCL**)	*3*
Dami	AML M7	57, M, PB				**cross-contamination! = line HEL**	*4,5*
EST-IU	secondary AML	BM				not permanent cell line	*6,7*
GRW	AML M7 + Down´s (R)	4, F, PB	+	+	+	cytokine-dependent	*8*
HIMeg-1	CML (CP)	25, F, PB	+		+		*9,10*
KG-91	AML	PB	(+)		+		*11*
KOPMK-53	CML (BC)		(+)				*12*
Meg-J	CML (BC)	24, M, PB	(+)		+		*13,14*
MOLM-7	CML (BC)	44, M, PB	+			sister lines **MOLM-6, -8, -9, -10, -11, -12**	*15*
M-TAT	MDS (RAEBT) (R)	3, M, PB			+	cytokine-dependent	*16,17*
MTT-95	AM M7 (D)	52, M, BM	+	+	+	probably overgrown by **EBV+ B-LCL**	*18*
P-320	AML M7	PB				not neoplastic cell line (**EBV+ B-LCL**)	*3*
RM10	CML (BC)	44, F, BM				**cross-contamination! = line K-562**	*19,20*
RS-1	AML M7 (R)	2, F, PB	(+)	(+)	+	**cross-contamination! = line K-562**	*21*
S-1214	ALL	PB				not neoplastic cell line (**EBV+ B-LCL**)	*3*
SET-2	essential thrombocythemia	71, F, PB	+	+	+	spontaneous platelet-production	*22*
SKH1	CML (BC)			(+)		t(3;21)(q26;q22)	*23*
TK91	CML (BC)		+				*24*
TW14-8	AML M6 + Rothmund-Thomson syndrome	25, M, PB	(+)				*25*

Additional abbreviations: BC, blast crisis; BM, bone marrow; CP, chronic phase; D, at diagnosis; F, female; M, male; PB, peripheral blood; R, at relapse/resistant.

References

[1] Scheid, C. et al. (1995) Blood 86, 778a.
[2] Fuchs, M. et al. (1996) Brit. J. Haematol. 93, Suppl. 2, 342.
[3] Morgan, D.A. and Brodsky, I. (1985) J. Cell Biol. 100, 565–573.
[4] Greenberg, S.M. et al. (1988) Blood 72, 1968–1977.
[5] MacLeod, R.A.F. et al. (1997) Leukemia 11, 2032–2038.
[6] Sledge Jr., G.W. et al. (1986) Cancer Res. 46, 2155–2159.
[7] Roth, B.J. et al. (1988) Blood 72, 202–207.

[8] Zipursky, A. et al. (1993) Blood 82, 122a.
[9] Song, L.N. and Chen, T. (1992) Biochem. Pharmacol. 43, 2292–2295.
[10] Cheng, T. et al. (1995) Leukemia 9, 1257–1263.
[11] Kollia, P. et al. (1997) Blood 90, 156b.
[12] Kojika, S. et al. (1996) Leukemia 10, 994–999.
[13] Kobayashi, S. et al. (1993) Blood 81, 889–893.
[14] Takanashi, M. et al. (1997) Ann. Hematol. 75, 149–153.
[15] Tsuji-Takayama, K. et al. (1994) Human Cell 7, 167–171.
[16] Minegishi, N. et al. (1994) J. Biol. Chem. 269, 27700–27704.
[17] Morita, S. et al. (1996) Leukemia 10, 102–105.
[18] Mizobuchi, N. et al. (1999) Acta Medica Okayama 53, 95–98.
[19] Hirata, J. et al. (1990) Leukemia 4, 365–372.
[20] Drexler, H.G. et al. (1999) Leukemia 13, 1601–1607.
[21] Skinnider, L.F. et al. (1997) Acta Haematol. 98, 26–31.
[22] Uozumi, K. et al. (2000) Leukemia 14, 142–152.
[23] Mitani, K. et al. (1994) EMBO J. 13, 504–510.
[24] Toba, K. et al. (1996) Exp. Hematol. 24, 894–901.
[25] Winkler, U. et al. (1997) Blood 90, 236b.

Table XIV *Growth factor-dependent cell lines*

Cell line	Cell type	Absolute cytokine dependency on	Growth stimulation or inhibition by cytokines
AML-193	monocytic	GM-CSF, IL-3	stimulation: G-CSF, IGF-I, PIXY-321; inhibition: TGF-β1
ANBL-6	plasma cell	IL-6	inhibition: IFN-α, IFN-γ, TGF-β, TNF-α
AS-E2	erythrocytic	EPO	not responsive to other cytokines
DP-6	plasma cell	IL-6	stimulation: IGF-I, IL-1α, IL-1β, IL-10, TNF-α; inhibition: IFN-α, TGF-β
DS-1	mature B-cell	IL-6	
ELF-153	megakaryocytic	GM-CSF	stimulation: IL-3, IL-4, IL-6, PIXY-321, SCF
EP	NK cell	IL-2	stimulation: IL-15
F-36P	erythrocytic	GM-CSF, IL-3	stimulation: EPO, IFN-γ, IL-5, PIXY-321
FKH-1	myelocytic	G-CSF, GM-CSF	stimulation: IL-3, SCF
FLAM-76	plasma cell	IL-6	inhibition: IFN-α
GF-D8	myelocytic	GM-CSF, IL-3	stimulation: IFN-γ, PIXY-321, SCF; inhibition: TGF-β1
GM/SO	myelocytic	GM-CSF	stimulation: IFN-γ, IL-1α, IL-4, IL-13, PIXY-321, SCF; inhibition: TGF-β1
GRW	megakaryocytic	SCF	
HANK1	NK cell	IL-2	
HML	megakaryocytic	GM-CSF	stimulation: SCF; inhibition: TGF-β
HSM-2	plasma cell	IL-6	
HU-3	megakaryocytic	GM-CSF, IL-3	stimulation: EPO, IFN-β, IFN-γ, IL-1α, IL-4, IL-5, IL-6, LIF, NGF, OSM, PIXY-321, SCF, TNF-α, TNF-β, TPO
ILKM2	plasma cell	IL-6	
ILKM3	plasma cell	IL-6	
INA-6	plasma cell	IL-6	inhibition: IFN-γ
JJN-2	plasma cell	IL-6	
JK-6	plasma cell	IL-6	stimulation: IFN-α, IL-7; inhibition: IFN-γ
KAS-6/1	plasma cell	IL-6	stimulation: IFN-α, IFN-γ, IGF-I, IL-3, IL-10, LIF, OSM, TNF-α
Kit 225	mature T-cell	IL-2	
KPMM2	plasma cell	IL-6	inhibition: IFN-α, IFN-γ
KT-3	mature T-cell	IL-6	stimulation: IL-2, IL-4
M-07e	megakaryocytic	GM-CSF, IL-3	stimulation: IFN-α, IFN-β, IFN-γ, IL-2, IL-4, IL-6, IL-9, IL-15, NGF, PIXY-321, SCF, TNF-α, TPO; inhibition: TGF-β1
MB-02	erythrocytic	GM-CSF	stimulation: EPO, IL-1β, IL-3, PIXY-321, SCF, TNF-α, TNF-β; inhibition: IFN-α, IFN-β, TGF-β1

MDS92	myelocytic	IL-3	stimulation: GM-CSF, SCF, TPO
MM-A1	plasma cell	IL-6	
MM-C1	plasma cell	IL-6	
MM-S1	plasma cell	IL-6	
MM-Y1	plasma cell	IL-6	
M-MOK	megakaryocytic	GM-CSF	stimulation: IFN-β, IFN-γ, IL-3, IL-4, IL-9, IL-13, IL-15, PIXY-321, SCF, TNF-α, TNF-β, TPO; inhibition: TGF-β1
M-TAT	erythrocytic	GM-CSF	stimulation: EPO, IL-3, SCF
MUTZ-2	myelocytic	SCF	stimulation: bFGF, FL, G-CSF, IFN-β, IFN-γ, IGF-I, IL-6, M-CSF, TNF-α
MUTZ-3	monocytic	GM-CSF, IL-3	stimulation: G-CSF, IFN-β, IFN-γ, IGF-I, IL-4, IL-6, IL-7, M-CSF, MIP-1α, PIXY-321, SCF, TPO
My-La	mature T-cell	IL-2	
NK-92	NK cell	IL-2	
NKL	NK cell	IL-2	
NK-YS	NK cell	IL-2	
OCI/AML1	myelocytic	G-CSF	stimulation: GM-CSF, IFN-β, IGF-1, IL-3, IL-4, IL-6, M-CSF, PIXY-321, SCF; inhibition: TGF-β1, TNF-α, TNF-β
OCI/AML5	myelocytic	GM-CSF, IL-3	stimulation: FL, G-CSF, IFN-β, IFN-γ, IL-6, M-CSF, PIXY-321; inhibition: TGF-β1, TNF-α
OCI/AML6	myelocytic	GM-CSF, IL-3	stimulation: FL, G-CSF, IFN-γ, IL-4, IL-13, M-CSF, PIXY-321, SCF
OH-2	plasma cell	IL-6	stimulation: IGF-I, IL-2, IL-10, IL-15, TNF-α, TNF-β; inhibition: TGF-β
OHN-GM	myelocytic	GM-CSF	
OIH-1	myelocytic	G-CSF, GM-CSF	stimulation: IL-3, SCF
PCM6	plasma cell	IL-6	inhibition: IFN-α
PER-420	immature T-cell	IL-2	
PER-423	immature T-cell	IL-2	
PER-427	immature T-cell	IL-2	
PER-487	immature T-cell	IL-7 + SCF	
Se-Ax	mature T-cell	IL-2	stimulation: IL-7, IL-15; inhibition: TNF-α
SKNO-1	myelocytic	GM-CSF	stimulation: G-CSF, IFN-α, IFN-β, IFN-γ, IL-3, IL-5, IL-6, IL-13, M-CSF, PIXY-321, SCF; inhibition: TNF-α
TALL-101	immature T-cell?	GM-CSF	stimulation: IL-3, IL-5
TALL-103/2	immature T-cell	IL-2	inhibition: IFN-γ, IL-4
TALL-104	immature T-cell	IL-2	
TALL-107	immature T-cell	IL-2	inhibition: IL-4
TF-1	erythrocytic	GM-CSF, IL-3	stimulation: EPO, IFN-γ, IL-1, IL-4, IL-5, IL-6, IL-13, LIF, NGF, OSM, PIXY-321, SCF, TNF-α, TNF-β, TPO; inhibition: TGF-β

Table XIV *Continued*

Cell line	Cell type	Absolute cytokine dependency on	Growth stimulation or inhibition by cytokines
TKS-1	NK cell	IL-2	
TMD2	mature B-cell	IL-3	
TSU1621MT	myelocytic	G-CSF	stimulation: GM-CSF, IL-3, IL-4, SCF; inhibition: IL-5
U-1957	plasma cell	IL-6	stimulation: IGF-I
UCSD/AML1	myelocytic	GM-CSF	stimulation: CNTF, IFN-β, IL-3, IL-4, IL-6, IL-12, IL-13, IL-15, LIF, M-CSF, OSM, PIXY-321, SCF; inhibition: TGF-β1
UG3	monocytic	GM-CSF, IL-3	stimulation: G-CSF, M-CSF
UT-7	megakaryocytic	GM-CSF, IL-3, EPO	stimulation: G-CSF, IFN-β, IFN-γ, IL-4, IL-5, IL-6, NGF, PIXY-321, SCF; inhibition: TGF-β1
XG-1	plasma cell	IL-6	stimulation: GM-CSF, IFN-α, IL-3, TNF-α; inhibition: IFN-γ
XG-2	plasma cell	IL-6	stimulation: IFN-α; inhibition: IFN-γ
XG-3P	plasma cell	IL-6	stimulation: GM-CSF, IL-3; inhibition: IFN-γ
XG-4	plasma cell	IL-6	stimulation: CNTF, IFN-α, IL-11, LIF, OSM; inhibition: IFN-γ
XG-5	plasma cell	IL-6	stimulation: GM-CSF, IFN-α, IL-3; inhibition: IFN-γ
XG-6	plasma cell	IL-6	stimulation: IL-11, LIF, OSM
XG-7	plasma cell	IL-6	
XG-8	plasma cell	IL-6	stimulation: IL-3
XG-9	plasma cell	IL-6	stimulation: GM-CSF, IL-3
YNH-1	myelocytic	G-CSF, GM-CSF, IL-3	stimulation: EPO, IL-4, SCF, TPO

For a detailed description of these cell lines and the literature references, see the respective chapters in this book.

Further Reading

Drexler, H.G. et al. (1997) Leukemia 11, 701–708.
Hassan, H.T. and Drexler, H.G. (1995) Leukemia Lymphoma 20, 1–15.
Ihle, J.N. and Askew, D. (1989) Int. J. Cell Cloning 7, 68–91.
Mire-Sluis, A.R. et al. (1995) J. Immunol. Methods 187, 191–199.
Oval, J. and Taetle, R. (1990) Blood Reviews 4, 270–279.

Table XV *Cell lines with unique translocations and fusion genes*

Chromosomal abnormality	Fusion gene or altered genes	Fusion gene type	Cell type	Cell lines
del(1)(p32)	*SIL-SCL*		immature T-cell	**CCRF-CEM, H-SB2, MOLT 16, RPMI 8402**
t(1;7)(p34;q34)	*LCK-TCRB*		immature T-cell	**EU-9, H-SB2, SUP-T12**
t(1;12)(q25;p13)	*ETV6-ARG*		myelocytic	**HT93**
t(1;14)(p32;q11)	*SCL-TCRD*		immature T-cell	**DU.528**
t(1;14)(q21;q32)	*BCL9-IGH*		precursor B-cell	**CEMO-1**
			mature B-cell	**DLCL-SKI-1**
t(1;19)(q23;p13)	*E2A-PBX1*		precursor B-cell	**697, AHLL-1, ALL-2B, KMO-90, KOPN-60, KOPN-63, LiLa-1, LK63, MHH-CALL-3, PER-278, Pre-Alp, RCH-ACV, SCMC-L9, SCMC-L10, SCMC-L11, SUP-B27, THP-4, THP-8, UoC-B3, WH94**
t(2;5)(p23;q35)	*NPM-ALK*		ALCL	**AMS3, DEL, HSC-M1, JB6, Karpas 299, Ki-JK, SR-786, SU-DHL-1, SUP-M2, UCONN-L2, WSU-ALCL**
t(2;8)(p12;q24)	*IGK-MYC*		mature B-cell	**BA-25, BJA-B, Ci-1, NCU-L-1**
t(2;8)(q34;q24)	*TCL4-MYC*		mature T-cell	**HUT 78**
t(2;18)(p11;q21)	*IGK-BCL2*		mature B-cell	**YM**
inv(3)(q21q26)	*EVI1*		myelocytic	**Kasumi-4**
			monocytic	**MUTZ-3**
			ery-megakaryocytic	**MOLM-1**
t(3;3)(q21;q26)	*EVI1*		myelocytic	**HNT-34, UCSD/AML1**
t(3;4)(q27;p11)	*BCL6-TTF*		mature B-cell	**VAL**
t(3;7)(q27;q22)	*EVI1*		myelocytic	**Kasumi-3**
t(3;11)(q27;q23)	*BCL6-BOB1*		mature B-cell	**Karpas 231**
t(3;14)(q27;q32)	*BCL6-IGH*		mature B-cell	**MD903**
t(3;21)(q26;q22)	*EVI1-AML1*		ery-megakaryocytic	**SKH1**
t(4;11)(q21;q23)	*MLL-AF4*		precursor B-cell	**ALL-2B, ALL-PO, B1, BEL-1, EU-8, Kid92, KOCL-45, KOCL-58, KOCL-69, LLM, RS4;11, SCMC-L1, SEM**
			monocytic	**KOCL-48, MV4-11**
t(4;14)(p16;q32)	*FGFR3-IGH-MMSET*		plasma cell	**JIM-3, KHM-11, KMS-11, KMS-18, LP-1, MM5.1, NCI-H929, OPM-2, UTMC-2**
t(6;9)(p23;q34)	*DEK-CAN*		myelocytic	**FKH-1**
t(6;11)(q27;q23)	*MLL-AF6*		myelocytic	**CTS**
			monocytic	**ML-2**

Table XV ***Continued***

Chromosomal abnormality	Fusion gene or altered genes	Fusion gene type	Cell type	Cell lines
t(6;12)(q23;p13)	*STL-ETV6/TEL*		precursor B-cell	**SUP-B2**
t(6;14)(p25;q32)	*MUM1/IRF4-IGH*		plasma cell	**FR4, SK-MM-1, XG-7**
t(7;9)(q34;q34)	*TCRB-TAL2*		immature T-cell	**SUP-T1, SUP-T3**
t(7;10)(q35;q24)	*TCRB-HOX11*		immature T-cell	**PER-255**
t(7;14)(q35;q32)	*TCRB-TCL1*		immature T-cell	**ALL 320**
t(7;19)(q34;p13)	*TCRB-LYL1*		immature T-cell	**SUP-T7**
t(8;14)(q24;q11)	*MYC-TCRA*		immature T-cell	**JK-1, KE-37, MOLT 16, TALL-101, TALL-103/2, TALL-105, TALL-106**
			mature T-cell	**SKW-3**
t(8;14)(q24;q32)	*MYC-IGH*		precursor B-cell	**380, NALM-33**
			mature B-cell	**BAL-KHs, BALL-1, BALL-2, BALM-7, BALM-13, BALM-18, BHT3, CA46, Daudi, D.G.-75, DoHH2, DS, EB1, EW36, FMC-Hu-1-B, HBL-1, HBL-2, HOB1, HS-1, JLPC119, JOK-1, KHM-2B, LNPL, MC116, MHH-PREB-1, MN-60, Namalwa, OCI-Ly18, PD1, Raji, ROS-50, SDK, SK-DHL2A, SU-DUL5, Tanoue, TEM110, Tree92, U 2904, VAL, WIEN 133, WSU-WM, Z-138**
			plasma cell	**Karpas 620, UM-1**
t(8;21)(q22;q22)	*AML1-ETO*		myelocytic	**Kasumi-1, SKNO-1**
t(8;22)(q24;q11)	*MYC-IGL*		mature B-cell	**BALM-16, DS-1, HBL-3, KAL-1, KHM-10B, L 660**
t(9;11)(p22;q23)	*MLL-AF9*		precursor B-cell	**KOPB-26**
			monocytic	**IMS-M1, Mono Mac 6, THP-1, UG3**
t(9;14)(p13;q32)	*PAX5-IGH*		mature B-cell	**KIS-1**
t(9;22)(q34;q11)	*BCR-ABL*	b2-a2	precursor B-cell	**BV173**
			immature T-cell	**CML-T1**
			myelocytic	**Kasumi-4, KBM-7, KCL-22, KT-1, KYO-1, YOS-M**
			monocytic	**RWLeu-4**
			ery-megakaryocytic	**JK-1, MEG-01, MOLM-1**
		b3-a2	precursor B-cell	**ALL-1, NALM-1, NALM-27**
			immature T-cell	**TK-6**
			myelocytic	**EM-2, GM/SO, KOPM-28, KU812**
			monocytic	**KBM-5**
			ery-megakaryocytic	**AP-217, JURL-MK1, K-562, KH88, LAMA-84, MC3, MEG-A2, NS-Meg, SAM-1, SKH1, T-33, TS9;22, Y-1K, YN-1, YS9;22**

		M-bcr	myelocytic	**HNT-34**
			ery-megakaryocytic	**RM10**
		e1-a2	precursor B-cell	**ALL/MIK, KOPN-30bi, KOPN-57bi, MR-87, NALM-20, NALM-29, OM9;22, PALL-1, PALL-2, SUP-B13, TOM-1, Z-33, Z-119, Z-181**
		e1-a3	precursor B-cell	**MY**
		m-bcr	precursor B-cell	**LEF1, SCMC-L2, TMD5, UTP-L2**
			myelocytic	**KOPM-30**
		e19-a2	myelocytic	**AR230**
		not known	precursor B-cell	**DUNATIS, KOPN-66bi, NALM-24**
			immature T-cell	**MHH-TALL1**
			myelocytic	**CML-C-1, KOPM-55, KPB-M15**
			ery-megakaryocytic	**TN922**
t(10;11)(p13;q14)	*CALM-AF10*		monocytic	**KP-MO-TS, P31/Fujioka, U-937**
t(10;14)(q24;q11)	*HOX11-TCRD*		immature T-cell	**ALL-SIL, K3P, SUP-T4**
ins(11;9)(q23;p22p23)	*MLL-AF9*		monocytic	**MOLM-13**
t(11;14)(p13;q11)	*TTG2-TCRD*		immature T-cell	**KOPT-K1, LALW-2, TALL-104**
t(11;14)(p15;q11)	*TTG1-TCRD*		immature T-cell	**RPMI 8402**
t(11;14)(q13;q32)	*BCL1-IGH*		mature B-cell	**Granta 519, HBL-2, JeKo-1, JVM-2 (MTC), MINO, MO1094 (MTC), MO2058 (MTC), NCEB-1, RC-K8, Rec-1, SP-53, SU-DHL-11**
			plasma cell	**DOBIL-6, FLAM-76, INA-6, KMS-12–PE, MEF-1, MM-M1, MOLP-5, NOP-2, SK-MM-2, XG-1, XG-2, XG-5, XG-6, XG-8**
t(11;19)(q23;p13)	*MLL-ENL*		precursor B-cell	**BS, HB1119, KOCL-33, KOCL-44, KOCL-50, KOCL-51, KOPN-1, KOPN-8**
			immature T-cell	**SUP-T13**
t(12;21)(p13;q22)	*TEL-AML1*		precursor B-cell	**Reh, SUP-B26, UoC-B4, UoC-B12**
t(12;22)(p13;q12)	*MN1-TEL*		myelocytic	**UCSD/AML1**
inv(14)(q11q32)	*TCRA-IGH*		immature T-cell	**HT-1, SUP-T1**
t(14;18)(q32;q21)	*IGH-BCL2*	MBR	mature B-cell	**BEVA, DoHH2, DS, FL-218, FL-318, Granta 452, HF-1, HF-4a, Karpas 231, Karpas 353, Karpas 422, KW1, OCI-Ly1, OCI-Ly8, RL, SU-DUL5, TK, VAL, WSU-FSCCL, WSU-NHL**
		mcr	mature B-cell	**ROS-50, U 2904**
		not known	precursor B-cell	**380, OZ**
			mature B-cell	**BALM-3, Bay91, Boar88, EHC4, FL-18, KHM-2B, KML-2, ONHL-1, Pfeiffer, RS11846, SU-DHL-4, SU-DHL-6, SU-DHL-12, SU-DHL-13, U-715 M, WL2, Z-138**
t(15;17)(q22;q11)	*PML-RARA*		myelocytic	**AP-1060, Ei501, HT93, NB4, UF-1**

Table XV ***Continued***

Chromosomal abnormality	Fusion gene or altered genes	Fusion gene type	Cell type	Cell lines
inv(16)(p13q22)	*CBFB-MYH11*		monocytic	**ME-1**
t(16;21)(p11;q22)	*TLS/FUS-ERG*		myelocytic	**IRTA17, TSU1621MT, UTP-L12, YNH-1**
t(17;19)(q21;p13)	*E2A-HLF*		precursor B-cell	**HAL-01, UoC-B1**
t(X;11)(q13;q23)	*MLL-AFX*		immature T-cell	**Karpas 45**

Involvement of the genes indicated has not been always shown for each cell line, but can be assumed to have occurred in most cases given the specific chromosomal breakpoints.

For a detailed description of these cell lines and the literature references, see the respective chapters in this book.

Further Reading

Barr, F.G. (1996) Nature Genetics 12, 113–114.
Cline, M.J. (1994) New Engl. J. Med. 330, 328–336.
Drexler, H.G. (1994) Leukemia Res. 18, 919–927.
Drexler, H.G. et al. (1995) Leukemia 9, 480–500.
Drexler, H.G. et al. (1995) Leukemia Lymphoma 19, 359–380.
Drexler, H.G. et al. (1999) Leukemia Res. 23, 207–215.
Faderl, S. et al. (1998) Blood 91, 3995–4019.
Hilden, J.M. and Kersey, J.H. (1994) Leukemia Lymphoma 14, 189–195.
Look, A.T. (1997) Science 278, 1059–1064.
Rabbitts, T.H. (1994) Nature 372, 143–149.
Ridge, S.A. and Wiedemann, L.M. (1994) Leukemia Lymphoma 14, 11–17.
Sawyer, C.L. (1998) Leukemia Res. 22, 1113–1122.
Zeleznik-Le, N.J. et al. (1995) Semin. Hematol. 32, 201–219.

Table XVI *False and misinterpreted cell lines*

False cell line	Purported malignancy	Real identity	Actual malignancy or actual cell type	DNA fingerprinting at DSMZ: Cells from originators	Identical fingerprint	Identical karyotype	Comment [a]
207	BCP-ALL	**Reh**	ALL	+	+		A
AG-F	Hodgkin	**CCRF-CEM**	T-ALL			+	B
AICHI-4	Hodgkin	**AICHI-4**	EBV+ B-LCL				
ARH-77	myeloma	**ARH-77**	EBV+ B-LCL				
B-403	AML	**B-403**	EBV+ B-LCL				
Co	Hodgkin	**CCRF-CEM**	T-ALL		+	+	B
Dami	AML M7	**HEL**	AML M6		+	+	B
EU-7	T-ALL	**CCRF-CEM**	T-ALL	+	+	+	B
FQ	Hodgkin	**unknown**	monkey cell line				
GM1312	myeloma	**GM1312**	EBV+ B-LCL				
GM1500	myeloma	**GM1500**	EBV+ B-LCL				
HS-Sultan	myeloma	**Jijoye**	Burkitt				
IM-9	myeloma	**IM-9**	EBV+ B-LCL				
J-111	AML M5	**HELA**	cervix carcinoma				
JOSK-I	AML M4	**U-937**	histiocytic lymphoma	+	+	+	B
JOSK-K	AML M5	**U-937**	histiocytic lymphoma	+	+	+	B
JOSK-M	CML-BC	**U-937**	histiocytic lymphoma	+	+	+	B
JOSK-S	AML M5	**U-937**	histiocytic lymphoma	+	+	+	B
KE-37	T-ALL	**CCRF-CEM**	T-ALL		+		A
KM-3	BCP-ALL	**Reh**	BCP-ALL		+		A
L 591	Hodgkin	**L 591**	EBV+ B-LCL				
LR10.6	BCP-ALL	**NALM-6**	BCP-ALL	+	+	+	B
MC/CAR	myeloma	**MC/CAR**	EBV+ B-LCL				
MDS	CMML	**JURKAT**	T-ALL	+	+		
MHH 225	AML M7	**JURKAT**	T-ALL	+	+		A
MKB-1	AML	**CCRF-CEM**	T-ALL	+	+	+	B
MOBS-1	AML M5	**U-937**	histiocytic lymphoma		+		
MOLT 15	T-ALL	**CTV-1**	AML M5	+	+		
NOI-90	NK-NHL	**Reh**	BCP-ALL	+	+		
OU-AML-1	AML M4	**OCI/AML2**	AML M4	+	+		

Table XVI *Continued*

False cell line	Purported malignancy	Real identity	Actual malignancy or actual cell type	DNA fingerprinting at DSMZ:		Identical karyotype	Comment [a]
				Cells from originators	Identical fingerprint		
OU-AML-2	AML M2	**OCI/AML2**	AML M4	+	+		
OU-AML-3	AML M4	**OCI/AML2**	AML M4	+	+		
OU-AML-4	AML M2	**OCI/AML2**	AML M4	+	+		
OU-AML-5	AML M5	**OCI/AML2**	AML M4	+	+		
OU-AML-6	AML M1	**OCI/AML2**	AML M4	+	+		
OU-AML-7	AML M4	**OCI/AML2**	AML M4	+	+		
OU-AML-8	AML M4	**OCI/AML2**	AML M4	+	+		
P-320	AML M7	**P-320**	EBV^+ B-LCL				
PBEI	BCP-ALL	**NALM-6**	BCP-ALL	+	+		
PLB-985	AML M4	**HL-60**	AML M2	+	+	+	B
RB	Hodgkin	**unknown**	monkey cell line				
RC-2A	AML M4	**CCRF-CEM**	T-ALL		+		A
RM10	CML-BC	**K-562**	CML-BC			+	
RS-1	AML M7	**K-562**	CML-BC	+	+		
RSp	Hodgkin	**RSp**	EBV^+ B-LCL				
RY	Hodgkin	**unknown**	unknown cell line				
S-1214	AML	**S-1214**	EBV^+ B-LCL				
SPI-801	T-ALL	**K-562**	CML-BC	+	+	+	B
SPI-802	T-ALL	**K-562**	CML-BC	+	+	+	B
SpR	Hodgkin	**unknown**	monkey cell line				
SR-91	T-ALL	**AML-193**	AML M5	+	+		
T-33	CML-BC	**K-562**	CML-BC	+	+		A
TI-1	AML M2	**K-562**	CML-BC	+	+		
TMM	CML-BC	**TMM**	EBV^+ B-LCL				
UMJF-2	myeloma	**UMJF-2**	EBV^+ B-LCL				
UTMB-460	normal	**CCRF-CEM**	T-ALL			+	B
YJ	CMML	**HL-60**	AML M2	+	+	+	B

For a detailed description of these cell lines and for the literature references, see the respective chapters in this book.

[a] Comments:

A: The correct cell line may still exist.

B: The false cell line shows the karyotype of the real cell line in the original publication.

Further Reading

Drexler, H.G. et al. (1993) Leukemia 7, 2077–2078. (cell line J-111)
Drexler, H.G. et al. (1999) DSMZ Catalogue of Cell Lines, 7th edn, Braunschweig, Germany. (JOSK-cell lines, SPI-cell lines, cell line TMM)
Drexler, H.G. et al. (1999) Leukemia 13, 1601–1607. (overview on false hematopoietic cell lines)
DSMZ Website: www.dsmz.de/mutz/mutzhome.htm
Gignac, S.M. et al. (1993) Leukemia Lymphoma 10, 359–368. (SPI-cell lines)
Häne, B. et al (1992) Leukemia 6, 1129–1133. (various cell lines)
Harris, N.L. et al. (1981) Nature 289, 228–230. (various Hodgkin cell lines)
MacLeod, R.A.F. et al. (1997) Leukemia 11, 2032–2038. (cell line Dami)
MacLeod, R.A.F. et al. (1997) Blood 90, 2850–2851. (cell line Dami)
MacLeod, R.A.F. et al. (1999) Int. J. Cancer 83, 555–563.
Nelson-Rees, W.A. and Flandermeyer, R.R. (1976) Science 191, 96–98. (cell line J-111)
Pellat-Deceunynck, C. et al. (1995) Blood 86, 4001–4002. (various myeloma cell lines)

Table XVII *Recommended reference cell lines*

Cell line	Available at cell bank	Cell growth	Recommended for
Precursor B-cell lines:			
697	DSMZ	+	• RT-PCR control: t(1;19)(q23;p13) *E2A-PBX1*
BV173	DSMZ	++	• prototypical precursor B-cell line • RT-PCR control: t(9;22)(q34;q11) *BCR-ABL* b2-a2
NALM-1	DSMZ, JCRB	++	• prototypical precursor B-cell line
NALM-6	DSMZ	++	• prototypical precursor B-cell line
Reh	ATCC, DSMZ	++	• prototypical precursor B-cell line • RT-PCR control: t(12;21)(p13;q22) *TEL-AML1*
SUP-B15	ATCC, DSMZ	+	• RT-PCR control: t(9;22)(q34;q11) *BCR-ABL* e1-a2
Mature B-cell lines:			
CRO-AP/2	DSMZ	+++	• prototypical PEL cell line: EBV^+, $HHV\text{-}8^+$
Daudi	ATCC, DSMZ, JCRB	+++	• prototypical Burkitt´s lymphoma cell line • typical t(8;14)(q24;q32) *MYC-IGH*
DoHH2	DSMZ	+++	• RT-PCR control: t(14;18)(q32;q21) *IGH-BCL2* (MBR)
Granta 519	DSMZ	++	• RT-PCR control: t(11;14)(q13;q32) *BCL1-IGH*
HC1	DSMZ	++	• prototypical HCL cell line
JVM-2	DSMZ	++	• prototypical B-PLL cell line
Karpas 422	DSMZ	+	• prototypical DLCL cell line
Myeloma cell lines:			
L 363	DSMZ	+++	• prototypical plasma cell line (IgG-type)
NCI-H929	ATCC, DSMZ	(+)	• prototypical myeloma cell line (IgAκ-type)
OPM-2	DSMZ	++	• prototypical myeloma cell line (Igλ-type)
RPMI 8226	ATCC, DSMZ, IFO, JCRB	++	• prototypical myeloma cell line (Igλ-type)
U-266	ATCC, DSMZ	++	• prototypical myeloma cell line (IgEλ-type)
Immature T-cell lines:			
CCRF-CEM	ATCC, DSMZ, IFO, JCRB	+++	• prototypical immature T-cell line (TCRαβ)
JURKAT	DSMZ, RCB	+++	• prototypical immature T-cell line (TCRαβ)
MOLT 3	ATCC, DSMZ, JCRB, RCB	+++	• prototypical immature T-cell line
Peer	DSMZ, JCRB	++	• prototypical immature T-cell line (TCRγδ)

RPMI 8402	ATCC, DSMZ	+++	• carries t(11;14)(p15;q11) and del(1)(p32)
SUP-T1	ATCC, DSMZ	+++	• carries t(7;9)(q34;q34) and inv(14)(q11q32)
Mature T-cell lines:			
HUT 102	ATCC	+	• secretion of HTLV-I
HUT 78	ATCC	+	• permissive for HIV replication (subclone H9)
SKW-3	DSMZ, RCB	++	• prototypical mature T-cell line
Natural killer cell leukemia-lymphoma cell lines:			
NK-92	not available at cell bank	+	• prototypical NK cell line (IL-2-dependent)
YT	DSMZ	++	• prototypical NK cell line (NK, ADCC activity)
Hodgkin's disease cell lines:			
HDLM-2	DSMZ	++	• prototypical HD cell line (T-cell type)
KM-H2	DSMZ	+++	• prototypical HD cell line (B-cell type)
L 428	DSMZ	+++	• prototypical HD cell line (B-cell type)
Anaplastic large cell lymphoma cell lines:			
DEL	DSMZ	+++	• prototypical ALCL cell line (null-cell type) • RT-PCR control: t(2;5)(p23;q35) *NPM-ALK*
Karpas 299	DSMZ	+++	• prototypical ALCL cell line (T-cell type) • RT-PCR control: t(2;5)(p23;q35) *NPM-ALK*
SU-DHL-1	DSMZ	+++	• prototypical ALCL cell line (null-type) • RT-PCR control: t(2;5)(p23;q35) *NPM-ALK*
Myelocytic cell lines:			
EM-2	DSMZ	+++	• prototypical myelocytic cell line • RT-PCR control: t(9;22)(q34;q11) *BCR-ABL* b3-a2
EoL-1	DSMZ, RCB	++	• prototypical eosinophilic cell line
HL-60	ATCC, DSMZ, IFO, JCRB, RCB	+++	• prototypical myelocytic cell line (promyelocytic model) • pluripotent differentiation into various cell lineages
HMC-1	DSMZ	+	• prototypical mast cell line
Kasumi-1	DSMZ	+	• RT-PCR control: t(8;21)(q22;q22) *AML1-ETO*
KG-1	ATCC, DSMZ, JCRB, RCB	+++	• prototypical myelocytic cell line • model for induced differentiation
KU812	DSMZ, IFO, JCRB, RCB	++	• prototypical basophilic cell line

Table XVII *Continued*

Cell line	Available at cell bank	Cell growth	Recommended for
NB4	DSMZ	+++	• pluripotent differentiation into various cell lineages • RT-PCR control: t(15;17)(q22;q11) *PML-RARA* • model for retinoic acid-induced differentiation
Monocytic cell lines:			
ME-1	not available at cell bank	(+)	• RT-PCR control: inv(16)(p13q22) *CBFB-MYH11*
ML-2	DSMZ	++	• prototypical monocytic cell line • RT-PCR control: t(6;11)(q27;q23) *MLL-AF6*
Mono Mac 6	DSMZ	++	• prototypical monocytic cell line • RT-PCR control: t(9;11)(p21;q23) *MLL-AF9*
MV4-11	ATCC, DSMZ	++	• RT-PCR control: t(4;11)(q21;q23) *MLL-AF4*
THP-1	ATCC, DSMZ, JCRB, RCB	+++	• prototypical monocytic cell line • RT-PCR control: t(9;11)(p21;q23) *MLL-AF9*
U-937	ATCC, DSMZ, IFO, JCRB, RCB	+++	• prototypical monocytic cell line • model for induction of macrophage differentiation
Erythrocytic-megakaryocytic cell lines:			
HEL	ATCC, DSMZ, JCRB	+++	• prototypical cell line for bipotent (ery-meg) differentiation • constitutive/inducible production of various (hemo)globins
K-562	ATCC, DSMZ, JCRB, RCB	+++	• constitutive/inducible production of various (hemo)globins • RT-PCR control: t(9;22)(q34;q11) *BCR-ABL* b3-a2 • target for natural killer assays
LAMA-84	DSMZ	+++	• pluripotent differentiation into various cell lineages
M-07e	DSMZ	+	• multi-cytokine-dependent/-responsive • indicator cell line for specific cytokine bioassays
MEG-01	ATCC, DSMZ, IFO	+++	• prototypical megakaryocytic cell line • RT-PCR control: t(9;22)(q34;q11) *BCR-ABL* b2-a2
TF-1	ATCC, DSMZ	+	• multi-cytokine-responsive (cytokine indicator cell line) • pluripotent differentiation capacities
UT-7	DSMZ	+	• multi-cytokine-responsive (cytokine indicator cell line) • pluripotent differentiation capacities

Cell growth: +++ = uncomplicated, rapid cell proliferation; ++ = uncomplicated, but slow cell proliferation; + = difficult cell line, needs attention; (+) = very difficult cell line and/or very slow proliferation.

For a detailed description of these cell lines and the literature references, see the respective chapters in this book.

Websites/E-Mails of Cell Line Banks
ATCC: www.atcc.org; help@atcc.org
DSMZ: www.dsmz.de; mutz@dsmz.de
JCRB + IFO = HSRRB: www.nihs.go.jp; cellbank@nihs.go.jp; hsrrb@nihs.go.jp
RIKEN: www.rtc.riken.go.jp; cellbank@rtc.riken.go.jp

Index

Bold entries indicate well-characterized cell lines that are described in detail in sections II-VIII.

Printed and bound by CPI Group (UK) Ltd, Croydon, CR0 4YY
03/10/2024
01040413-0020